Lecture Notes in Computer Science

Lecture Notes in Artificial Intelligence 16297

The series Lecture Notes in Artificial Intelligence (LNAI) was established in 1988 as a topical subseries of LNCS devoted to artificial intelligence.

The series publishes state-of-the-art research results at a high level. As with the LNCS mother series, the mission of the series is to serve the international R & D community by providing an invaluable service, mainly focused on the publication of conference and workshop proceedings and postproceedings.

Hideaki Takeda · Yannis Tzitzikas ·
Giorgos Flouris · Shizhu He ·
Dimitris Plexousakis · Sébastien Ferré · Ran Yu ·
Vasilis Efthymiou · Eleni Ilkou ·
Ernesto Jiménez-Ruiz · Xin Wang
Editors

Knowledge Graphs

14th International Joint Conference, IJCKG 2025
Heraklion, Crete, Greece, October 15–17, 2025
Proceedings

Editors
Hideaki Takeda
National Institute of Informatics
Tokyo, Japan

Giorgos Flouris
Foundation for Research and Technology - Hellas
Heraklion, Greece

Dimitris Plexousakis
University of Crete
Heraklion, Greece

Ran Yu
GESIS – Leibniz Institute for the Social Sciences
Mannheim, Germany

Eleni Ilkou
TIB - Leibniz Information Centre for Science and Technology
Hanover, Germany

Xin Wang
Tianjin University
Tianjin, China

Yannis Tzitzikas
University of Crete
Heraklion, Greece

Shizhu He
Chinese Academy of Sciences
Beijing, China

Sébastien Ferré
University of Rennes
Rennes Cedex, France

Vasilis Efthymiou
Harokopio University of Athens
Athens, Greece

Ernesto Jiménez-Ruiz
University of London
London, UK

ISSN 0302-9743 ISSN 1611-3349 (electronic)
Lecture Notes in Artificial Intelligence
ISBN 978-981-95-5008-1 ISBN 978-981-95-5009-8 (eBook)
https://doi.org/10.1007/978-981-95-5009-8

LNCS Sublibrary: SL7 – Artificial Intelligence

This Springer imprint is published by the registered company Springer Nature Singapore Pte Ltd.
The registered company address is: 152 Beach Road, #21-01/04 Gateway East, Singapore 189721, Singapore

Preface

The 14th International Joint Conference on Knowledge Graphs (IJCKG 2025) was a premium academic forum on Knowledge Graphs. The mission of IJCKG 2025 was to bring together researchers in the international Knowledge Graph community and other related areas to present their innovative research results or novel applications of Knowledge Graphs. IJCKG 2025 took place for the first time in Europe, hosted by the Institute of Computer Science at the Foundation for Research and Technology - Hellas (FORTH) in Heraklion, Crete, Greece.

The IJCKG 2025 conference featured a rich program, spanning 3 full days.

The program included 3 keynote presentations, namely:

- Pascal Hitzler (Kansas State University, USA): "Knowledge Graphs in the Neurosymbolic AI Age"
- Katja Hose (TU Wien, Austria): "Sharp Edges in a Fuzzy World: Knowledge Graphs and LLMs"
- Kang Liu (Institute of Automation, Chinese Academy of Sciences, China): "Shuttle between Symbolic Knowledge and Neural Parameters"

Two additional invited talks were included in the program, by

Manolis Koubarakis (National and Kapodistrian University of Athens, Greece) and Dimitris Angelakis (Technical University of Crete, Greece). The program also featured 3 tutorials, namely the following:

- Yannis Marketakis (Foundation for Research and Technology - Hellas, Greece): "Building Interoperable Knowledge Graphs with X3ML Framework"
- Paolo Pareti (University of Southampton, UK): "The Shapes Constraint Language (SHACL): an Overview and its Future Directions"
- George Stamatiou (Foundation for Research and Technology - Hellas, Greece): "Introduction to Quantum Computing"

A novelty of IJCKG 2025 was the evaluation challenge track, which featured different evaluation challenges. An evaluation challenge consists of a platform, dataset, and/or evaluation plan and solicits submissions that solve this challenge. In IJCKG 2025, there was one evaluation challenge titled "Archer: Bilingual Reasoning-aware Text-to-SQL Evaluation", organized by Jeff Pan (University of Edinburgh, UK), Zhichao Yan (Shanxi University, China), and Wenyu Huang (University of Edinburgh, UK).

The conference provided various networking opportunities for our community, through the young scholar forum, the industry forum, the PhD and PostDoc symposium, the poster and demo session, and the social dinner. IJCKG 2025 also hosted two workshops:

- Knowledge Graphs in Cultural Heritage, organized by Dimitris Kotzinos (CY Cergy Paris University, France), Sorin Hermon (The Cyprus Institute, Cyprus), and Maria Theodoridou (Foundation for Research and Technology – Hellas, Greece)

– Quantum Computing and Data and Knowledge Management, organized by Yannis Tzitzikas (University of Crete and Foundation for Research and Technology – Hellas, Greece), Haridimos Kondylakis (University of Crete and Foundation for Research and Technology – Hellas, Greece), and José Manuel García Alonso (University of Extremadura, Spain)

We received 74 submissions in total. From these, 17 papers were accepted for the research track, 5 papers for the education track (2 of which were short papers), 5 papers for the in-use track, 7 for the poster and demo track (all of which were short papers), 3 workshop papers, and 2 short papers for the evaluation challenge track. All these papers appear in the current volume, which was edited by the PC chairs and the respective track chairs, and its production orchestrated by the publication chair and the general chairs.

We took all reasonable measures to ensure a fair and efficient review process, using the Easychair conference management system. We opted for a single-blind peer-review process, employing 94 reviewers for all tracks, as well as 9 external reviewers (103 in total). Each reviewer was assigned 1.94 reviews on average, whereas each paper received an average of 3.24 reviews.

To ensure unbiased reviews, the reviewers were asked to provide conflicts of interest, and additional conflicts were put in place by the PC Chairs to ensure that no missing conflicts could cause bias in the reviewing process. We also used the restrictions imposed by Easychair, e.g., that PC members had no access to the reviews of their own papers. The final assignments were checked by the PC Chairs to ensure a fair reviewing process.

IJCKG 2025 would like to thank our sponsors and supporters for their generous contributions. We would also like to acknowledge the help of the local chairs and the website chair for their invaluable support in many practical matters pertaining to the conference including help with the Easychair platform, and, of course, all the members of the organizing committee and the reviewers who contributed their precious time to support the organization of this conference.

September 2025

Hideaki Takeda
Yannis Tzitzikas
Giorgos Flouris
Shizhu He
Dimitris Plexousakis
Sébastien Ferré
Ran Yu
Vasilis Efthymiou
Eleni Ilkou
Ernesto Jiménez-Ruiz
Xin Wang

Organization

General Chairs

Hideaki Takeda	National Institute of Informatics, Japan
Yannis Tzitzikas	University of Crete and Foundation for Research and Technology - Hellas, Greece

Program Committee (PC) Chairs

Dimitris Plexousakis	University of Crete and Foundation for Research and Technology - Hellas, Greece
Shizhu He	Institute of Automation, Chinese Academy of Sciences, China

Local Chairs

Pavlos Fafalios	Technical University of Crete and Foundation for Research and Technology - Hellas, Greece
Michalis Mountantonakis	University of Crete and Foundation for Research and Technology - Hellas, Greece

Education Track Chairs

Eleni Ilkou	TIB - Leibniz Information Centre for Science and Technology, Germany
Ernesto Jiménez-Ruiz	City St George's, University of London, UK
Xin Wang	Tianjin University, China

In-Use Track Chair

Vasilis Efthymiou	Harokopio University of Athens, Greece

Poster and Demo Chairs

Sébastien Ferré	University of Rennes, CNRS, Inria, France
Ran Yu	GESIS-Leibniz Institute for the Social Sciences, Germany

Workshop Chairs

Haridimos Kondylakis	Foundation for Research and Technology - Hellas, Greece
María Navas-Loro	Universidad Politécnica de Madrid, Spain

Evaluation Challenge Chair

Tianxing Wu	Southeast University, China

Invited Tutorial Chair

George Konstantinidis	University of Southampton, UK

Industry Forum Chairs

Michael Färber	TUD Dresden University of Technology, Germany
Haofan Wang	Tongji University, China

Young Scholar Forum Chair

Ningyu Zhang	Zhejiang University, China

Publication Chair

Giorgos Flouris	Foundation for Research and Technology - Hellas, Greece

Website Chair

Yannis Marketakis	University of Crete and Foundation for Research and Technology - Hellas, Greece

Publicity Chairs

Beatriz Esteves	Ghent University, Belgium
Georgia Troullinou	CNRS, France

Sponsorship Chairs

Manolis Koubarakis	National and Kapodistrian University of Athens, Greece
Theodore Patkos	University of Crete and Foundation for Research and Technology - Hellas, Greece

Steering Committee

Huajun Chen	Zhejiang University, China
Ian Horrocks	University of Oxford, UK
Hong-Gee Kim	Seoul National University, South Korea
Wooju Kim	Yonsei University, South Korea
Sang-goo Lee	Seoul National University, South Korea
Juanzi Li	Tsinghua University, China
Riichiro Mizoguchi	Japan Advanced Institute of Science and Technology, Japan
Jeff Pan	University of Edinburgh, UK
Thepchai Supnithi	NECTEC, Thailand
Hideaki Takeda	National Institute of Informatics, Japan
Alan Wu	eBay, Inc., USA
Yannis Tzitzikas	University of Crete and Foundation for Research and Technology - Hellas, Greece
Takahira Yamaguchi	Keio University, Japan

Program Committee (Main Track)

Long Bai	Chinese Academy of Sciences, China
George Baryannis	University of Huddersfield, UK
Sotiris Batsakis	Hellenic Mediterranean University, Greece
Sheng Bi	Southeast University, China
Antonis Bikakis	University College London, UK
Carlos Bobed	University of Zaragoza, Spain
Fernando Bobillo	University of Zaragoza, Spain
Diego Calvanese	Free University of Bozen-Bolzano, Italy
Gong Cheng	Nanjing University, China
Tiansi Dong	University of Cambridge, UK
Alessandro Faraotti	IBM, Italy
Giorgos Flouris	Foundation for Research and Technology - Hellas, Greece
Naoki Fukuta	Shizuoka University, Japan
Salim Hafid	Médialab Sciences Po, France
Shizhu He	Institute of Automation, Chinese Academy of Sciences, China
Wei Huang	Beijing University of Posts and Telecommunications, China
Takahiro Kawamura	University of Tokyo, Japan
M. Taimoor Khan	GESIS-Leibniz Institute for the Social Sciences, Germany
Manolis Koubarakis	National and Kapodistrian University of Athens, Greece
Xu Lei	Wuhan University, China
Da Li	Institute of Computing Technology, China
Guoqiang Li	Shanghai Jiao Tong University, China
Huiying Li	Southeast University, China
Zhixu Li	Renmin University of China, China
Yong Liu	Salesforce, Inc., USA
Yongbin Liu	Tsinghua University, China
Theofilos Mailis	National and Kapodistrian University of Athens, Greece
Jose L. Martinez-Rodriguez	Autonomous University of Tamaulipas, Mexico
Edgard Marx	Leipzig University of Applied Sciences (HTWK), Germany
Georgios Meditskos	Aristotle University of Thessaloniki, Greece
Jun Pang	Wuhan University of Science and Technology, China
Katerina Papantoniou	Foundation for Research and Technology - Hellas, Greece

Pierre-Henri Paris	Paris-Saclay University, France
Theodore Patkos	Foundation for Research and Technology - Hellas, Greece
Dimitris Plexousakis	University of Crete and Foundation for Research and Technology - Hellas, Greece
Huang Qi	Jiangxi Normal University, China
Gianluca Quercini	CentraleSupélec - LRI, France
Guozheng Rao	Tianjin University, China
Edelweis Rohrer	Universidad de la República, Uruguay
Floriano Scioscia	Polytechnic University of Bari, Italy
Hala Skaf-Molli	University of Nantes, France}
Thepchai Supnithi	NECTEC, Thailand
Ilias Tachmazidis	University of Huddersfield, UK
Yuhang Tian	Beijing Institute of Technology, China
Alexandros Vassiliades	Aristotle University of Thessaloniki, Greece
Meng Wang	Tongji University, China
Ruijie Wang	University of Zurich, Switzerland
Liang Wen	Beijing Normal University, China
Gang Wu	Northeastern University, China
Xue-Feng Xi	Suzhou University of Science and Technology, China
Zechang Xiong	Beijing Jiaotong University, China
Binxia Xu	University College London, UK
Kang Xu	Nanjing University of Posts and Telecommunications, China
Weixin Zeng	National University of Defense Technology, China
Chengzhi Zhang	Nanjing University of Science and Technology, China
Ningyu Zhang	Zhejiang University, China}
Xiaowang Zhang	Tianjin University, China

Program Committee (Education Track)

Hasan Abu-Rasheed	Goethe University Frankfurt, Germany
Qurat Ul Ain	University of Duisburg-Essen, Germany
Inès Blin	Sony Computer Science Laboratories-Paris, France
Jiaoyan Chen	University of Manchester, UK
John Domingue	Open University, UK
Martina Galletti	Sony Computer Science Laboratories-Paris, France

Sebastian Gombert	Leibniz Institute for Research and Information in Education, Germany
Anett Hoppe	Philipps-Universität Marburg & Hessian Center for AI; TIB - Leibniz Information Centre for Science and Technology, Germany
Chris Davis Jaldi	Wright State University, USA
Cogan Shimizu	Wright State University, USA
Davide Taibi	Italian National Research Council, Italy
Valentina Tamma	University of Liverpool, UK
Christian Weber	University of Siegen, Germany

Program Committee (In-Use Track)

Angelos Charalambidis	Harokopio University of Athens, Greece
Jiaoyan Chen	University of Manchester, UK
Nikolaos Fanourakis	SATALIA, Greece
Oktie Hassanzadeh	IBM, USA
Eleni Ilkou	TIB - Leibniz Information Centre for Science and Technology, Germany
Keti Korini	University of Mannheim, Germany
Chuan Lei	Amazon Web Services, USA
Dimitrios Michail	Harokopio University of Athens, Greece
Christos Panagiotopoulos	Harokopio University of Athens, Greece
George Papadakis	Public Power Corporation, Greece
Duo Yang	KU Leuven, Belgium
Ningyu Zhang	Zhejiang University, China

Program Committee (Posters and Demo Track)

Rajjat Dadwal	University of Bonn, Germany
Lu Gan	Heinrich Heine University Düsseldorf; GESIS-Leibniz Institute for the Social Sciences, Germany
Yucan Guo	Chinese Academy of Sciences, China
Sergios Anestis Kefalidis	National and Kapodistrian University of Athens, Greece
Juan Li	Zhejiang University, China
Genivika Mann	University of Bonn, Germany
Hala Skaf-Molli	University of Nantes, France
Miao Su	Institute of Computing Technology, China

Davide Taibi	Italian National Research Council, Italy
Sharmila Upadhyaya	GESIS-Leibniz Institute for the Social Sciences, Germany
Ran Yu	GESIS-Leibniz Institute for the Social Sciences, Germany

Program Committee (Workshops)

Sorin Hermon	Cyprus Institute, Cyprus
Haridimos Kondylakis	Foundation for Research and Technology - Hellas, Greece
Dimitris Kotzinos	ETIS Lab / CY Cergy Paris University, France
Nicole Schweikardt	Humboldt University of Berlin, Germany
Maria Theodoridou	Foundation for Research and Technology - Hellas, Greece

Additional Reviewers

Ivano Bilenchi	Polytechnic University of Bari, Italy
Xi Chang	Shanghai Polytechnic University, China
Ali Elhalawati	KU Leuven, Belgium
Filippos Gouidis	Foundation for Research and Technology - Hellas, Greece
Kokila Jamwal	GESIS-Leibniz Institute for the Social Sciences, Germany
Hui Jiang	Chongqing University of Post and Telecommunication, China
Michalis Mountantonakis	University of Crete and Foundation for Research and Technology - Hellas, Greece
Evelyn Vanessa Navarrete Yépez	L3S Research Center, Leibniz University Hannover, Germany
Jingwei Shi	Shanghai University of Finance and Economics, China

Sponsors and Supporters

IJCKG 2025 has benefited significantly from the generosity of our sponsors and supporters, which are listed below.

Our Gold Sponsor

Artificial Intelligence Journal

Our Sponsors

European Association for Artificial Intelligence (EurAI)

Project TwinODIS (Twinning for Optimized Decision Intelligence in Data-Intensive Environments)

Our Supporters

Foundation for Research and Technology - Hellas (FORTH)

Contents

Main Track

Posters and Demos

Workshop Papers

Evaluation Challenge

Main Track

Knowledge Graphs and Quantum Computing: First Blood

Yannis Tzitzikas[1,2,3(✉)] and Haridimos Kondylakis[1,2,3]

[1] Computer Science Department, University of Crete, Heraklion, Greece
[2] Institute of Computer Science (ICS), Foundation for Research and Technology – Hellas (FORTH), Heraklion, Greece
kondylak@ics.forth.gr
[3] Center for Quantum Science & Technologies (FORTH-QuTech), Foundation for Research and Technology – Hellas (FORTH), 70013 Heraklion, Crete, Greece
tzitzik@csd.uoc.gr

Abstract. Although Quantum Computing (QC) is subject of intensive research, the data/knowledge management perspective is largely unexplored. Here we investigate how QC could be used for various tasks related to Knowledge Graphs (KG). We discuss the placement of QC in KG-related workflows, and then we (a) propose *encodings* of RDF (Redource Description Framework) data appropriate for QC, (b) show how a KG could be *stored* as a quantum state, (c) introduce a method for building *indexes* that is based on quantum circuits, (d) discuss how several important KG-related tasks, like *reasoning*, can be reduced to *satisfiability*, and thus gain from the quantum speedup. Subsequently, we report experimental results regarding encoding, decoding, superposition, as well as the limitations of the current QC hardware.

Keywords: quantum computing · knowledge graphs · knowledge management

1 Introduction

Quantum computing is subject of intensive research at various levels: from the *physical level* (i.e. how to build quantum computers and how to benchmark them [26]), to the *algorithmic level* (i.e., what algorithms are appropriate for quantum computers, e.g. the Shor's algorithm [31] and the Grover's algorithm [10], see [22] for an overview), up to *application level* (i.e. how to reduce various computational tasks to QC algorithms, e.g. see [6] for a survey of possible applications. In this paper we investigate how recent advances of QC could be leveraged for various tasks related to Knowledge Graphs (KG). The main motivation is that the properties of superimposed states could be exploited for storing a huge amount of RDF triples using a small number of qubits. Moreover, a mapping of triples to quantum states could enable native quantum algorithms to accelerate various KG tasks.

H. Takeda et al. (Eds.): IJCKG 2025, LNAI 16297, pp. 3–18, 2026.
https://doi.org/10.1007/978-981-95-5009-8_1

However, unlike classical computers, quantum computers cannot read data directly from the source. We need methods to *encode* our data to quantum states. Consequently, the general workflow is: (1) Pre-process the data, i.e. convert them to *numerical* data, (2) make these numerical data as the *input quantum state*, (3) *run* the quantum computation, (4) *measure* the output of the quantum computation, (5) optional: use the output to update the input to the quantum state or the parameters of the quantum process, and (6) *decode* the result.

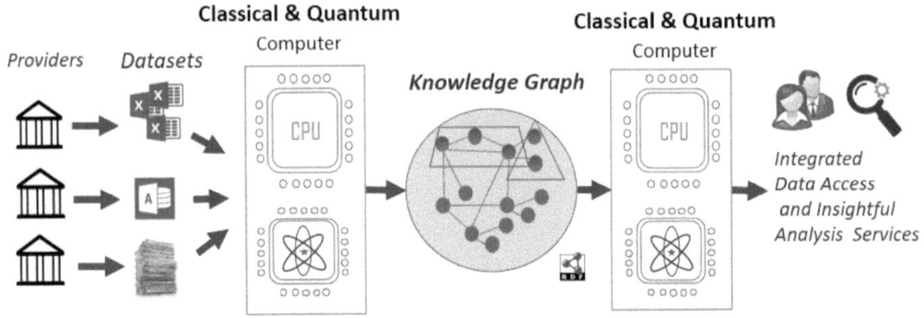

Fig. 1. The placement of QC in KG-related workflows

In general a quantum computer can be exploited both (I) for the construction of a KG, and (II) for the provision of KG-based services. This is illustrated in Fig. 1. In the left side we foresee tasks like *entity matching*, *blank node matching*, *ontology learning*, and others. We could also think of *probabilistic/fuzzy KGs* represented in a quantum representation framework. In the right side we foresee tasks like *lookup*, *searching*, *querying*, *reasoning*, *similarity*-based access services, *question answering*, and others. However, in both "sides" we also need conventional computing, for handling the encoding and decoding, as illustrated in Fig. 2.

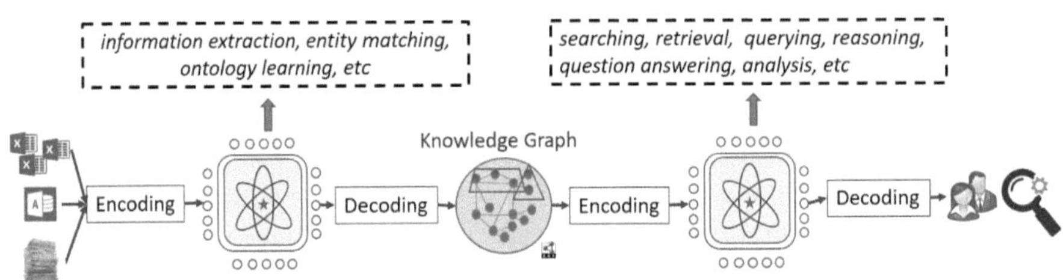

Fig. 2. Encoding and Decoding steps for various KG tasks

The main contributions of this paper are: we discuss the placement of QC in KG-related workflows, we introduce two basic *encodings* of RDF triples for QC, we detail a method for *quantum storage* for KGs, we show how quantum circuits can be used for building various *indexes*, we discuss how several important KG-related tasks, like *reasoning*, can be reduced to satisfiability (and thus gain from the quantum speedup), and we report experimental results related to encoding, decoding the superposition.

2 Background

2.1 Quantum Computing

In QC (Quantum Computing) instead of representing the data with 0/1 bits, we represent data with *qubits*. A qubit is described by a state vector in a two-dimensional complex vector Hilbert space. We express a qubit as an expression $a|0\rangle + b|1\rangle = \text{a}\begin{bmatrix}1\\0\end{bmatrix} + b\begin{bmatrix}0\\1\end{bmatrix} = \begin{bmatrix}a\\b\end{bmatrix}$ where $a^2 + b^2 = 1$. Unlike probabilities, where $a, b \in [0, 1]$ and $a + b = 1$, here a and b can be complex numbers, however a^2 and b^2 have a probabilistic interpretation ($a^2, b^2 \in [0, 1]$, $a^2 + b^2 = 1$). A qubit $a|0\rangle + b|1\rangle$ where $a, b \neq 0$, is *simultaneously* in two states, this is called *superposition*, and only when we measure it, we will get either 0 or 1. As such a system with n qubits can represent 2^n states at once. The "magic" is that what we will measure (0 or 1) may depend on the states of other qubits, due to the phenomenon of interference. We could therefore say that we have a kind of entangled probabilities. In particular, if we have two *entangled* qubits (that in the physical word correspond to two entangled electrons/photons), they might appear to be independent particles in a superposition, but as soon as one of them is measured, the outcome of a measurement of the other is fixed as well. Many computational benefits of QC stems from superposition, i.e. that fact that a quantum system can do processing while the qubits are in a superposition state. Suppose, for example, that we have 10 qubits. The operations defined by a quantum algorithm do not manipulate only 10 bits; they manipulate $2^{10} = 1024$ probability values. One step in a quantum algorithm on a quantum computer with 10 qubits is modifying 1024 values. Adding one qubit doubles the processing capabilities of the quantum computer. This explains the term exponential, which is often used in QC: adding n qubits adds processing power proportional to 2^n.

A bit formally, qubits can be combined to form *quantum registers*. The joint space of n qubits is the tensor product of the individual qubit spaces, i.e. $H = H_1 \otimes \ldots \otimes H_n$, e.g. two qubits $|\psi_1\rangle = a_1|0\rangle + b_1|1\rangle$ and $|\psi_2\rangle = a_2|0\rangle + b_2|1\rangle$ form the joint space $|\psi\rangle = |\psi_1\rangle \otimes |\psi_2\rangle = a_1a_2|00\rangle + a_1b_2|01\rangle + b_1a_2|10\rangle + b_1b_2|11\rangle$. The state of a quantum register Ψ is described by a unitary[1] transformation (U) that acts on the full state space, i.e. $|\Psi'\rangle = U|\Psi\rangle$.

There are not many quantum algorithms, for an overview of quantum algorithms see [22] and [16]. The more important ones are: the Shor's algorithm [31] that solves the factorization problem in polynomial time, and the Grover's algorithm [10] for searching in an unstructured list at the cost of $\sqrt{n}$. We should also

[1] Unitary transformations govern time evolution (via the Schrödinger equation), ensuring probabilities are conserved.

mention quantum annealing [29] that relies on specialized hardware (D-Wave) for solving optimization problems.

2.2 Knowledge Graphs

Knowledge Graphs (KG) [12] are structured representations of data so that they allow capturing the relationships between different concepts, entities or pieces of information. They rely on a graph-based data model to capture knowledge in application scenarios that involve integrating, managing and extracting value from diverse data sources [24]. There are several KGs expressed in RDF that integrate data from various sources: from general purpose sources like DBpedia [4] and Wikidata [34], to domain-specific repositories (e.g., Europeana [14], DrugBank [36], GRSF [32], ORKG [15], WarSampo [17] other cultural datasets like [9], COVID-19 related datasets [8], as well as Markup data through `schema.org`. Recently KGs, are also used for improving AI-related services, e.g. the validation of answers from Large Language Models [23], as well as in RAG systems [30].

Based on their construction method, we could separate KGs in three distinct cases: (a) KG that are constructed using information extraction from free text (e.g. DBPedia[2], OpenIE[3], Wikidata[4], etc.), (b) KGs that are generated by transforming structured data using various methods and tools (e.g. Europeana [14], Microsoft Academic Graph [35], etc.), and KGs that are constructed by combining transformed resources from structured data and information extracted from free text (e.g. [28]).

KGs in RDF. Currently most knowledge graph data management systems use Resource Description Framework (RDF) or Property Graph (PG) models. In this paper we focus on RDF. RDF uses *triples* i.e. statements of the form *subject – predicate – object*, where the *subject* corresponds to an entity (e.g. a product, a company etc.), the *predicate* to a characteristic of the entity (e.g. price of a product, location of a company) and the *object* to the value of the predicate for the specific subject (e.g. "300", "US"). The triples are used for relating Uniform Resource Identifiers (URIs) or anonymous resources (blank nodes) with other URIs, blank nodes or constants (Literals). Formally, a *triple* is considered to be any element of $T = (U \cup B) \times (U) \times (U \cup B \cup L)$, where U, B and L denote the sets of URIs, blank nodes and literals, respectively. Any finite subset of T constitute an *RDF graph* (or *RDF data set*).

3 QC and Data/Knowledge Management

Works on QC and *data management* have started to appear. For instance, [11] is a recent tutorial that gives an overview of works on QC-based data management

[2] https://www.dbpedia.org/.
[3] https://stanfordnlp.github.io/CoreNLP/openie.html.
[4] https://www.wikidata.org/wiki/Wikidata:Main_Page.

(query optimization, data integration, transaction mgmt, etc.). [18] focuses on *quantum storage design* for Tables in RDBMS. They propose two storage methods, Quantum Column-oriented Store and Quantum Row-oriented Store, tailored to store relational tables on universal quantum computers.

There are a few works on Knowledge Graphs too: [3] focuses on *entity matching* with *Quantum Neural Networks*; they evaluate quantum machine learning algorithms for entity matching on a small handcrafted data set and compare them to similar classical algorithms. [20] proposes a variational quantum circuit model for *knowledge graph embeddings*. The latent representations of entities are encoded as coefficients of quantum states, while predicates are characterized by parametric gates acting on the quantum states. They evaluate the result through link prediction experiments. Finally, [19] proposes a quantum *machine learning* algorithm for knowledge graphs.

A general remark is that in all of these works, the *data management* perspective is largely unexplored and quite far from technological and practical exploitation.

4 KG Encoding for Quantum State Input

We shall use a small running example. Consider the following three RDF triples:

```
(ex:Aristotle, rdf:type, ex:Person)
(ex:Person, rdfs:subClassOf, ex:Mortal)
(ex:Aristotle, ex:livesAt, ex:Athens)
```

We have to encode them *numerically*. Fig. 3 shows two possible encodings for the URIs, that we shall call $R1$ and $R2$. According to $R1$ we assign to each URI a number, while according to $R2$ we assign *independently* numbers to URIs corresponding to entities and to those corresponding to predicates (this is sound since a predicate cannot appear as subject or object of a triple).

$R1$ encoding

```
0(000): rdf:type
1(001): rdfs:subClassOf
2(010): ex:Aristotle
3(011): ex:Person
4(100): ex:Mortal
5(101): ex:livesAt
6(110): ex:Athens
```

$R2$ Encoding

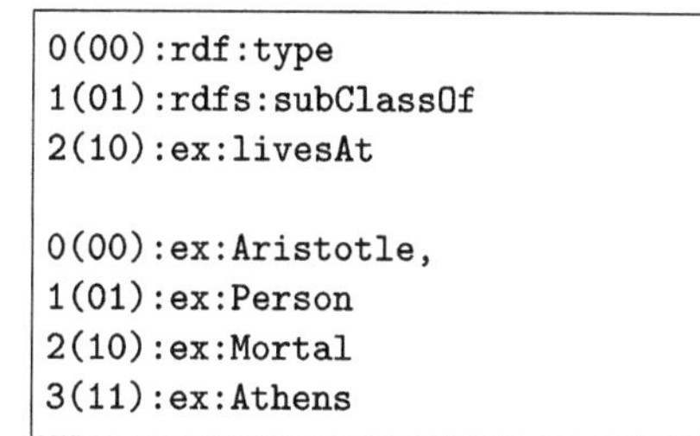

Fig. 3. Two encodings for RDF triples

According to $R1$ the triple `(ex:Aristotle, rdf:type, ex:Person)` is mapped to (2,0,3). In binary, and assuming 3 bits, it is mapped to (010,000,011), therefore this triple is mapped to the quantum state $|010, 000, 011\rangle$, so we need 9 qubits. According to $R2$, the same triple is mapped to (0,0,1), in binary to (00,00,01), i.e. to the quantum state $|00, 00, 01\rangle$, so here we need 6 qubits. The fact that in $R2$ the two bits 00 corresponds to `rdf:type` and `ex:Aristotle`, does not create any ambiguity, since a predicate cannot appear as subject or object of a triple, and we used fixed number of bits.

Comparison of $R1$ *and* $R2$. Let E be the set of entities, and P the set of predicates of a KG. With the $R1$ encoding each triple needs $3\lceil\log_2(|E|+|P|)\rceil$ bits. According to the $R2$ encoding each triple needs $2\lceil\log_2(|E|)\rceil + \lceil\log_2(|P|)\rceil$ bits. In a particular KG, it is therefore beneficial to select the representation that gives the less number of bits. An illustration of the number of qbits required by each one is shown in Fig. 4. Roughly, $R1$ is better than $R2$, if $3\log_2(|E|+|P|) < 2\log_2(|E|)\log_2(|P|) \Leftrightarrow (|E|+|P|)^3 < |E|^2 * |P|$. However this cannot hold, so in general $R2$ is beneficial (apart from some cases due to rounding).

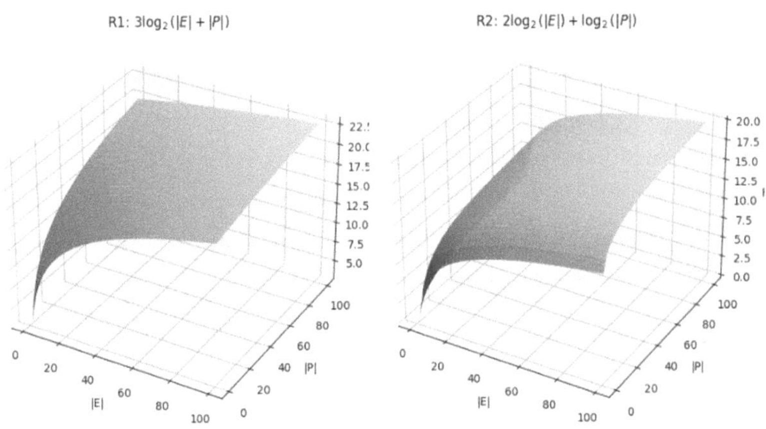

Fig. 4. # qubits required by $R1$ and $R2$, for various values of $|E|$ and $|P|$

4.1 Case Study: DBpedia

As of the latest available data[5], DBpedia contains over 850 million RDF triples, representing structured information extracted from Wikipedia and other sources. The dataset includes approximately 55,000 properties, with 1,377 defined by the DBpedia ontology DBpedia encompasses around 7.62 million entities linked to 179 external sources, totaling over 130.6 million links. The dataset is multilingual, featuring labels and abstracts in 125 languages, and includes approximately

[5] https://www.dbpedia.org/blog/dbpedia-snapshot-2022-03-release.

38 million labels, 25.2 million image links, and 29.8 million links to external web pages. That is:
$|T| = 850$ millions, i.e. $\lceil \log(|T|) \rceil = 30$,
$|E| = 7.62$ millions, i.e. $\lceil \log(|E|) \rceil = 23$,
$|P| = 55.000$, i.e. $\lceil \log(|P|) \rceil = 16$.

Suppose we want to apply a QC task on the entire DBpedia. Each DBpedia triple will require:

- $R1$: $3 \lceil \log_2(|E| + |P|) \rceil$ bits. $= 3 * (23) = 69$ bits,
- $R2$: $2 \lceil \log_2(|E|) \rceil + \lceil \log_2(|P|) \rceil = 2 * 23 + 16 = 62$ bits.

Note that this number of qubits (62) is not that big as already IBM offers through its cloud infrastructure 126 qubits to the public.

5 Quantum Storage for KGs

Motivation. Through the properties of superposition states, one could store a huge amount of triples in a small number of qubits. Moreover, a mapping of triples to quantum states could enable native quantum algorithms to accelerate various KG tasks.

Method. Hovever, as mentioned earlier, we have to encode everything with numbers, therefore a kind of *lexicon*, like the one shown in Fig. 3 has to be kept in a conventional computer. Therefore we need to have stored: (a) the actual Triples, (b) the URI Encoding Lexicon, (c) the EncodedTriples. If n is the number of triples, then the encoding costs $O(n \log n)$ time. If the result of a QC task is a triple, then its decoding costs $O(\log n)$ time. Note that such (bitmap) indexes are already used by triple stores to minimize storage cost and operation size.

Assume that we have this lexicon, and we have decided the encoding to use. Let us now focus on the quantum storage. Let T be the set of triples of our KG. For a $t \in T$, we shall use $\psi(t)$ to denote its encoding (either $R1$ or $R2$). We can store these by *superimposing* them, i.e. we can store the state $\Psi_T = \frac{1}{\sqrt{|T|}} \sum_{t \in T} \psi(t)$. In our running (and assuming $R2$), this means that we store in a quantum computer the state $\frac{1}{\sqrt{3}}(|000001\rangle + |010110\rangle + |001011\rangle$ which equals a 64-dimensional state vector ($2^6 = 64$), with all entries zero except at positions 1, 11, and 22 where the amplitude is $\frac{1}{\sqrt{3}}$.

5.1 Quantum Lookup

Suppose that we have n RDF triples, i.e. $|T| = n$. If we have not created any index, then to perform *lookup*, i.e. to check if there is one particular triple t, then we have to scan $n/2$ elements in average. Now suppose the quantum representation described in Sect. 5. In a quantum computer, and with the Grover's algorithm [10] this can be decided at $\sqrt{n}$ steps. Returning to our running example, suppose that we want to lookup the triple (`ex:Aristotle, rdf:type,` `ex:Person`), and assume that we have adopted the $R2$ encoding, i.e. we want to lookup 000001, i.e. 6 bits. The Qiskit implementation of the Grover's algorithm returns the right triple with probability 0.996.

6 Indexing Through Quantum Circuits

Suppose that apart from plain lookup we would like to construct *indices* for speeding up various operations, i.e. those required by a query language like SPARQL, for evaluating various triple patterns. We can try constructing *dedicated quantum circuits* for this purpose. To capture the basic access requirements, we would like to have 6 functions, i.e. one function f_a for each $a \in A = \{S \rightarrow P, S \rightarrow O, P \rightarrow S, P \rightarrow O, O \rightarrow S, O \rightarrow P\}$. For instance, $f_{S \rightarrow P}$ aims at speeding up the task of finding the predicates that are associated to a given subject, and it is defined as $f_{S \rightarrow P}(s) = \{p | (s,p,o) \in T\}$. The rest functions are defined analogously. Note that we could employ extra functions to speed up more access patterns, i.e. the following three: $A^{+} = \{S \rightarrow PO, P \rightarrow SO, O \rightarrow$ PS$\}$, e.g. $f_{S \rightarrow PO}(s) = \{(p,o) | (s,p,o) \in T\}$.

Let us now see how we can implement each such function using a quantum circuit. To aid understanding, we will present the idea through our running example. Let's consider the function $f_{P \rightarrow O}$ that aims at finding fast all *objects* that are associated with a particular predicate. In Table 1 we can see the three triples of our running example and their encoding (according to $R1$), for convenience in decimal. For the needs of $f_{P \rightarrow O}$ we have to consider only the columns for p and o, Table 2 shows these columns, and their contents in binary, as well as their mappings to 6 bit positions named $a_0, ..., b_2$.

Table 1. Running example with $R1$ encoding

s	$\psi(s)$	p	$\psi(p)$	o	$\psi(o)$
ex:Aristotle	2	rdf:type	0	ex:Person	3
ex:Person	3	rdfs:subClassOf	1	ex:Mortal	4
ex:Aristotle	2	ex:livesAt	5	ex:Athens	6

Table 2. Data related to $f_{P \rightarrow O}$

$\psi(p)$	$\psi(o)$
0	3
1	4
5	6

$\psi(p)$	$\psi(o)$
000	011
001	100
101	110

a0	a1	a2	b0	b1	b2
0	0	0	0	1	1
0	0	1	1	0	0
1	0	1	1	1	0

We can now express each of b_0, b_1, b_2, as a boolean function of a_0, a_1, a_2. In our case we get $b_0 = \overline{a_1} \wedge a_2; b_1 = (\overline{a_0} \wedge \overline{a_1} \wedge \overline{a_2}) \vee (a_0 \wedge \overline{a_1} \wedge a_2); b_2 = \overline{a_0} \wedge \overline{a_1} \wedge \overline{a_2}$. We can turn these boolean formulas into quantum circuits, based on the work [37] that provides an efficient technique that can implement any given Boolean function as a quantum circuit. In general, a boolean function with $n-1$ inputs, is represented by a quantum circuit with n qubits, where the extra qubit is initialized with 0 and will carry the result of the Boolean function at the end of the computation. Figure 5 shows the resulting circuit. It comprises X (Pauli-X) gates and CNOT (Controlled Not) gates. The X gate is the quantum equivalent of the classical NOT gate: it flips a qubit: it turns $|0\rangle$ into $|1\rangle$ and $|1\rangle$ into $|0\rangle$. The CNOT gate is a multi-qubit gate. It flips (applies a NOT operation to) the target qubit (the one marked with a plus) if the control qubit(s) is/are $|1\rangle$. We have annotated the circuit (of Fig. 5) with the logical expression after each gate. To conclude, we have constructed a circuit that whenever we feed it with the bits of a predicate we get as output the bits of the object.

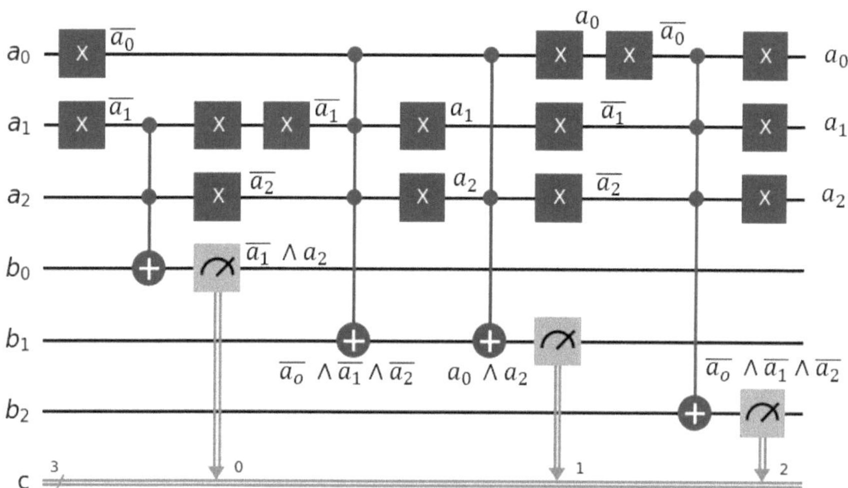

Fig. 5. The resulting quantum circuit (the boolean formulas after each gate were added manually for explainability)

However, this is possible only if f is a proper function. If p is a *functional property* (or effectively functional), then we can use $f_{SP \to O}$. If p is an *inverse functional property*, then we can use $f_{OP \to S}$. If *pp* denotes a property path, and *pp* is functional, then $f_{SPP \to}$ can be used as well. Now suppose that we want to apply this idea (i.e. build a quantum circuit for indexing purposes), on a binary relation f that is not a function, e.g. in our running example $f_{S \to O}(Aristotle) = \{ex : Person, ex : Athens\}$. In such cases we can create a *lexicon of sets*, where each set is assigned a distinct identifier, and use these identifiers in our quantum index (turning in this way f to a proper function). Let 7 be the identifier assigned to the set $\{ex : Person, ex : Athens\}$. In this case $f_{S \to O}(Aristotle) = 7$. If the set identifiers, are integer values greater than the maximum integer value assigned to URIs (either in $R1$ or $R2$), then this prevents ambiguities. This *lexicon of sets*, can be stored in a conventional data structure, a compact one is the CPOI (Compact Partial Index) [27,33].

To synopsize, what we have seen is that apart from using Grover (and having lookup cost $\sqrt{n}$) we can construct and use an index to speedup access. The cost of using the index is based on the number of gates it has.

Gate Requirements: As regards the number of gates required, according to [37], the maximum number of CNOT gates for representing a function over n cubits is 2^{n-1} and the average is 2^{n-2}. This means that for building a quantum circuit for representing a function over 62 qubits (recall the requirements of DBpedia) we need 2^{60} CNOT gates in the average case (more in Sect. 8).

7 Satisfiability, Reasoning and Instance Checking

Satisfiability is a key reasoning task. It is the problem of checking whether a knowledge base K is *satisfiable*, i.e., whether it has a model. Below we shall use notation from Description Logics (DL). It is well known that that many reasoning tasks, i.e. *concept satisfiability*, *subsumption checking*, *instance checking*, and others, reduce to *satisfiability*. A concept C is satisfiable, i.e. $K \not\models C \equiv \bot$, iff there exists x such that $K \cup \{C(x)\}$ is satisfiable. *Concept subsumption*, i.e. $K \models C \sqsubseteq D$, is reducible to satisfiablity: $K \models C \sqsubseteq D$ iff there exist x such $K \cup \{C \sqcap \neg D)(x)\}$ is not satisfiable. *Instance checking*, i.e. $K \models C(a)$, is again reducible to satisfiablity: i.e. $K \models C(a)$ iff $K \cup \{\neg C(a)\}$ is not satisfiable.

For instance, our running example is expressed in DL as:

K = {`Person(Aristotle)`, `Person` $\sqsubseteq$ `Mortal`, `livesAt(Aristotle, Athens)`} To check whether `(ex:Aristotle, rdf:type, ex:Mortal)` holds, which is an *inferred triple* according to the semantics of RDF Schema, is equivalent to checking whether $K \models Mortal(Aristotle)$, which holds iff $K \cup \{\neg Mortal(Aristotle)\}$ is not satisfiable.

The complexity of satisfiability in OWL is described in [13,25], in brief, the complexity of satisfiability in OWL Lite is EXPTIME-complete, in OWL DL it is NEXPTIME-complete, while in OWL 2 DL is it N2ExpTime-complete.

QC and Satisfiability. Note that Grover's algorithm can be applied for solving the satisfiability problem in a quantum computer in polynomial time! (related publications include [2] and [5]). It follows that satisfiability (and the tasks that reduce to satisfiability, like instance checking, and concept subsumption) could be solved better on a quantum computer than on a conventional computer.

Of course this is not the only method to peform such tasks. For instance, suppose that we have two edges (a,b) and (b,c) and we want to compute the *transitive closure*, i.e. $\{(a, b), (b, c), (a, c)\}$. We can represent this binary relation as a quantum circuit where each qubit corresponds to an edge. By applying controlled operations, we can infer new edges based on transitivity. The circuit is shown in Fig. 6. We use two X gates to set 1 the qubits corresponding to (a,b) and (b,c). Then we use a CCX gate (Toffoli gate) that takes three qubits as input. If the first two qubits are both 1, it flips the third qubit. This is just indicative, other solutions, based on *quantum walks* are worth investigating as well.

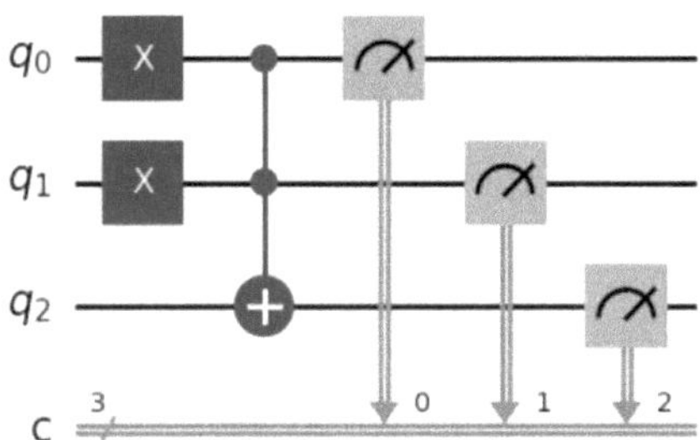

Fig. 6. Circuit for computing the transitive closure of a relation with two edges

In brief, the reduction to satisfiability can be used to carry out various tasks and leverage the polynomial complexity that QC offers, without restricting other algorithms from dedicated tasks.

8 Experimental Validation and Limitations

A thorough experimental evaluation is beyond the scope of this paper. However, we performed various validation tests, and some experiments for some indispensable parts of the quantum computing pipeline for RDF data.

Validation. At first, all examples and methods of this paper have been tested over IBM Qiskit 2.0.0.

Encoding and Decoding. We have developed `QuantumEncoDeco`, a framework for $R1$ and $R2$ encoding/decoding of RDF data, and we have made it publicly available[6], enabling anyone to encode/decode RDF graphs, and make QC experiments. The encoding creates also the lexicon that is useful for decoding the results of a QC task. Since encoding and deconding are required for any task, we have to anticipate these costs, therefore in Table 3 we report experimental results[7] over few datasets of different sizes. The datasets include (a) the ISO

[6] https://github.com/YannisTzitzikas/QuantumEncoDeco.

[7] All measurement were performed using a machine with Intel(R) Core(TM) i7-8650U CPU @ 1.90 GHz 2.11 GHz, 16.0 GB RAM.

9 Concluding Remarks

Quantum computing and KGs is an unexplored area. In this paper we provided a starting point with emphasis on the data management perspective. We are interested in data management from start to finish, in exploring the current limitations, and in the verifiability of each step. We discussed the possible placement of QC in KG-related workflows, and then we showcased two basic *encodings* and how to estimate the number of qubits that are required for superimposing a KG (on the basis of $|E|$ and $|P|$). Then we showed how we can *store* a KG as a quantum state (e.g. the entire DBpedia can be superimposed on 62 qubits). To enable fast access, we showed how quantum circuits can be used for building various *indexes* and estimated the number of quantum gates that are required. Since the problem of satisfiability can be solved in polynomial time in a quantum computers, we showed how several important KG-related tasks, like *reasoning*, can be reduced to satisfiability, and thus gain from the quantum speedup.

In addition, we developed and made publicly available `QuantumEncoDeco`, a library for encoding and decoding RDF data, and we reported experimental results regarding the encoding, decoding and superposition of RDF triples. Indicatively, a dataset with around 5 million RDF triples requires around 2 min to encode, but only a few seconds to superimpose these encoded triples.

As future research we plan to investigate how to apply the Grover's algorithm, and get stable results, for *high* number of qubits. Another topic for future research is to investigate the potential of QC in combinatorial (apart from satisfiability) and optimization problems (like entity matching, blank node matching, SPARQL optimization, etc.). Such problems seem suitable for *quantum annealing*; in this paper we focused on the *quantum circuit model*.

Acknowledgments. This work was supported by DEDALUS - Data Management using Quantum Computing project (TA 5180519) funded by SUB1.1 Clusters of Research Excellence (CREs), Greece. We also thank Giorgos Theodorakis for contributing in the code of `QuantumEncoDeco`, as well as George T. Stamatiou for comments and feedback on earlier versions of this paper.

References

1. AbuGhanem, M.: Characterizing Grover search algorithm on large-scale superconducting quantum computers. Sci. Rep. **15**(1), 1281 (2025)
2. Battaglia, D.A., Santoro, G.E., Tosatti, E.: Optimization by quantum annealing: lessons from hard satisfiability problems. Phys. Rev. E-Stat. Nonlinear Soft Matt. Phys. **71**(6), 066707 (2005)
3. Bischof, L., Teodoropol, S., Füchslin, R.M., Stockinger, K.: Entity matching with quantum neural networks (2024)
4. Bizer, C., et al.: DBpedia-a crystallization point for the web of data. J. Web Semant. **7**(3), 154–165 (2009)
5. Boulebnane, S., Montanaro, A.: Solving Boolean satisfiability problems with the quantum approximate optimization algorithm. PRX Quantum **5**(3), 030348 (2024)

6. Dalzell, A.M., et al.: Quantum algorithms: a survey of applications and end-to-end complexities. arXiv preprint arXiv:2310.03011 (2023)
7. Daoud, A., Millberg, V.: Performance and noise levels of Grover's algorithm on quantum computers with different qubit connectivities. Technical report, KTH, School of Electrical Engineering and Computer Science (2024)
8. Dimitrov, D., et al.: Tweetscov19–a knowledge base of semantically annotated tweets about the Covid-19 pandemic. In: 29th ACM International Conference on Information and Knowledge Management (CIKM 2020) (2020)
9. Fafalios, P., Petrakis, K., Samaritakis, G., Doerr, K., Kritsotaki, A., Tzitzikas, Y., Doerr, M.: FAST CAT: collaborative data entry and curation for semantic interoperability in digital humanities. ACM J. Comput. Cult. Heritage **14** (2021)
10. Grover, L.K.: A fast quantum mechanical algorithm for database search. In: Proceedings of the Twenty-Eighth Annual ACM Symposium on Theory of Computing, pp. 212–219 (1996)
11. Hai, R., Hung, S.H., Feld, S.: Quantum data management: from theory to opportunities. arXiv preprint arXiv:2403.02856 (2024)
12. Hogan, A., et al.: Knowledge graphs. ACM Comput. Surv. (Csur) **54**(4), 1–37 (2021)
13. Horrocks, I., Patel-Schneider, P.: Reducing OWL entailment to description logic satisfiability. J. Web Semant. **1**(4), 345–357 (2004)
14. Isaac, A., Haslhofer, B.: Europeana linked open data–data. europeana. eu. Semant. Web **4** (2013)
15. Jaradeh, M.Y., et al.: Open research knowledge graph: Next generation infrastructure for semantic scholarly knowledge. In: Proceedings of the 10th International Conference on Knowledge Capture, pp. 243–246 (2019)
16. Jordan, S.P.: Quantum algorithm zoo. https://quantumalgorithmzoo.org
17. Koho, M., Ikkala, E., Leskinen, P., Tamper, M., Tuominen, J., Hyvönen, E.: Warsampo knowledge graph: Finland in the second world war as linked open data. In: Semantic Web – Interoperability, Usability, Applicability (2020). https://doi.org/10.3233/SW-200392
18. Li, T., et al.: Quantum storage design for tables in RDBMS. Proc. VLDB Endow. ISSN **2150**, 8097 (2024)
19. Ma, Y., Tresp, V.: Quantum machine learning algorithm for knowledge graphs. ACM Trans. Quantum Comput. **2**(3), 1–28 (2021)
20. Ma, Y., Tresp, V., Zhao, L., Wang, Y.: Variational quantum circuit model for knowledge graph embedding. Adv. Quantum Technol. **2**(7–8), 1800078 (2019)
21. Marketakis, Y., Tzitzikas, Y., Gentile, A., Ellenbroek, A., Taconet, M.: Building a global aquatic resource knowledge base for fisheries. Proceedings **117**(1) (2025). https://doi.org/10.3390/proceedings2025117004
22. Montanaro, A.: Quantum algorithms: an overview. NPJ Quantum Inf. **2**(1), 1–8 (2016)
23. Mountantonakis, M., Tzitzikas, Y.: Real-time validation of ChatGPT facts using RDF knowledge graphs. In: ISWC (Posters/Demos/Industry) (2023)
24. Noy, N., Gao, Y., Jain, A., Narayanan, A., Patterson, A., Taylor, J.: Industry-scale knowledge graphs: lessons and challenges: five diverse technology companies show how it's done. Queue **17**(2), 48–75 (2019)
25. Polleres, A., Hogan, A., Delbru, R., Umbrich, J.: RDFS and OWL reasoning for linked data. In: Rudolph, S., Gottlob, G., Horrocks, I., van Harmelen, F. (eds.) Reasoning Web 2013. LNCS, vol. 8067, pp. 91–149. Springer, Heidelberg (2013). https://doi.org/10.1007/978-3-642-39784-4_2

26. Proctor, T., Young, K., Baczewski, A.D., Blume-Kohout, R.: Benchmarking quantum computers. Nat. Rev. Phys. 1–14 (2025)
27. Psaraki, M., Tzitzikas, Y.: CPOI: a compact method to archive versioned RDF triple-sets. arXiv preprint arXiv:1902.04129 (2019)
28. Gazzotti, R., Michel, F., Gandon, F.: CORD-19 named entities knowledge graph (CORD19-NEKG) (2020)
29. Rajak, A., Suzuki, S., Dutta, A., Chakrabarti, B.K.: Quantum annealing: an overview. Phil. Trans. R. Soc. A **381**(2241), 20210417 (2023)
30. Sapidis, I., Zervos, V., Mountantonakis, M., Tzitzikas, Y.: Interactive and provenance-aware search and QA over documents using LLMs, RAG and knowledge graph verbalization. In: Proceedings of the 29th International Conference on Theory and Practice of Digital Libraries (TPDL 2025) (2025)
31. Shor, P.W.: Polynomial-time algorithms for prime factorization and discrete logarithms on a quantum computer. SIAM Rev. **41**(2), 303–332 (1999)
32. Tzitzikas, Y., et al.: Methods and tools for supporting the integration of stocks and fisheries. In: Salampasis, M., Bournaris, T. (eds.) HAICTA 2017. CCIS, vol. 953, pp. 20–34. Springer, Cham (2019). https://doi.org/10.1007/978-3-030-12998-9_2
33. Tzitzikas, Y., Theoharis, Y., Andreou, D.: On storage policies for semantic web repositories that support versioning. In: Bechhofer, S., Hauswirth, M., Hoffmann, J., Koubarakis, M. (eds.) ESWC 2008. LNCS, vol. 5021, pp. 705–719. Springer, Heidelberg (2008). https://doi.org/10.1007/978-3-540-68234-9_51
34. Vrandečić, D., Krötzsch, M.: WikiData: a free collaborative knowledgebase. Commun. ACM **57**(10), 78–85 (2014)
35. Wang, K., Shen, Z., Huang, C., Wu, C.H., Dong, Y., Kanakia, A.: Microsoft academic graph: when experts are not enough. Quantitative Sci. Stud. **1**(1), 396–413 (2020)
36. Wishart, D.S., et al.: DrugBank 5.0: a major update to the DrugBank database for 2018. Nucleic Acids Res. **46**(D1), D1074–D1082 (2018)
37. Younes, A., Miller, J.: Automated method for building CNOT based quantum circuits for Boolean functions. arXiv preprint quant-ph/0304099 (2003)

Bi-view Embedding Fusion: A Hybrid Learning Approach for Knowledge Graph's Nodes Classification Addressing Problems with Limited Data

Rosario Napoli(✉), Giovanni Lonia, Antonio Celesti, Massimo Villari, and Maria Fazio

Department of Mathematical and Computer Sciences, Physical Sciences and Earth Sciences, University of Messina, Viale Ferdinando Stagno d'Alcontres 31, 98166 Messina, Italy
rnapoli@unime.it

Abstract. Traditional Machine Learning (ML) methods require large amounts of data to perform well, limiting their applicability in sparse or incomplete scenarios and forcing the usage of additional synthetic data to improve the model training. To overcome this challenge, the research community is looking more and more at Graph Machine Learning (GML) as it offers a powerful alternative by using relationships within data. However, this method also faces limitations, particularly when dealing with Knowledge Graphs (KGs), which can hide huge information due to their semantic nature. This study introduces Bi-View, a novel hybrid approach that increases the informative content of node features in KGs to generate enhanced Graph Embeddings (GEs) that are used to improve GML models without relying on additional synthetic data. The proposed work combines two complementary GE techniques: Node2Vec, which captures structural patterns through unsupervised random walks, and GraphSAGE, which aggregates neighbourhood information in a supervised way. Node2Vec embeddings are first computed to represent the graph topology, and node features are then enriched with centrality-based metrics, which are used as input for the GraphSAGE model. Moreover, a fusion layer combines the original Node2Vec embeddings with the GraphSAGE-influenced representations, resulting in a dual-perspective embedding space. Such a fusion captures both topological and semantic properties of the graph, enabling the model to exploit informative features that may exist in the dataset but that are not explicitly represented. Our approach improves downstream task performance, especially in scenarios with poor initial features, giving the basis for more accurate and precise KG-enanched GML models.

1 Introduction

In recent years, Machine Learning (ML) has become the dominant approach for solving a wide range of problems across various domains. However, ML methods operate under the assumption that data points are independent and require

H. Takeda et al. (Eds.): IJCKG 2025, LNAI 16297, pp. 19–34, 2026.
https://doi.org/10.1007/978-981-95-5009-8_2

extensive data to perform well [1]. These constraints introduce significant limitations, especially in scenarios where data and features are incomplete or expensive to obtain, forcing data scientists to use different techniques, such as increasing the dataset with additional synthetic data, to improve models' training at the cost of using artificial information. To address these limitations, Graph Machine Learning (GML) has emerged as a powerful method for extracting meaningful insights using the inherent relationships in data, representing them in a graph-based format where nodes are entities and edges are relationships. GML models use node features for probabilistic or aggregation strategies [2] to create Graph Embeddings (GEs), which represent features in continuous vector spaces that are given as input to ML models.

At the same time, the availability and semantics of interconnected data have increased the widespread adoption of Knowledge Graphs (KGs), which extend basic graph structures by allowing elements to have different labels and properties [3], providing a powerful and intuitive way to describe real-world dynamics. This means that KGs-based GML models can capture both structural and semantic information for node classification and link prediction tasks [4]. However, even when models perform well, datasets with incomplete information raise significant challenges in GML tasks. This is especially true for KGs, where implicit semantic relations, like a missing transitive link between entities that should logically be connected, can remain hidden, making the analysis biased.

To overcome this critical limitation and make the most of the available features without using additional synthetic data, we propose a new hybrid approach that combines unsupervised probabilistic topological embeddings from Node2Vec [5] and centrality metrics with supervised message-passing via GraphSAGE [6]. In particular, Node2Vec is employed to capture global structural patterns via random walks that are combined with classical graph-theoretic centrality measures to create an initial embedding vector. GraphSAGE is then applied using the initial embeddings to generate a new set of GEs that incorporate both local neighbourhood aggregation and the previously derived global structural information. The key innovation in our approach lies in the introduction of a learnable fusion mechanism that adaptively weights the unsupervised (Node2Vec) and supervised (GraphSAGE+Centrality Measures) representations for each node, generating new embeddings that can capture both topological properties and task-specific semantic patterns. This dual integration strategy improves node classification in poor datasets and enables a deeper analysis of the available hidden knowledge within KGs. Our experiments confirmed the significant effectiveness of this approach, demonstrating consistent improvements over standalone node classification baselines.

The remainder of this paper is organised as follows. Section 2 explains the basic concepts on which our scientific work is based. Section 3 discusses the related works in the context of node classification on KGs. Section 4 introduces our Bi-View hybrid embedding strategy. Sections 5 and 6 validate the structural enrichment process and present the supervised fusion model. Sections 7 describe the experimental setup and evaluation results. Finally, Sect. 8 concludes the paper and outlines future research directions.

2 Definitions

This Section introduces the formal definitions of the core concepts used throughout this study.

Definition 1 (Knowledge Graph). *A Knowledge Graph is defined as a tuple $KG = (V, R, E, \ell_V)$, where:*

- *V is the set of entities (nodes);*
- *R is the set of relationship types;*
- *$E \subseteq V \times R \times V$ is the set of labelled, directed edges (triplets);*
- *$\ell_V : V \to 2^C$ is a labelling function that assigns one or more class labels from the set C to each node.*

Definition 2 (Graph Embedding). *Let $KG = (V, E, R, \ell_V)$ be a Knowledge Graph. A Graph Embedding (GE) is a function*

$$f : V \to \mathbb{R}^d,$$

where $d \ll |V|$, which maps each node $v \in V$ to a vector $f(v) \in \mathbb{R}^d$.

Definition 3 (Transition probability). *Given a Knowledge Graph $KG = (V, R, E, \ell_V)$, let $c = (v_1, v_2, \ldots, v_k)$ denote a walk on the graph. The transition probability of moving from the current node $v = c_{i-1}$ to the next node $x = c_i$ in the walk is defined as*

$$P(c_i = x \mid c_{i-1} = v) = \pi_{vx} = \frac{\alpha_{pq}(t, x) \cdot w_{vx}}{Z}, \tag{1}$$

where w_{vx} represents the sum of the weights of all edges directly connecting nodes v and x, Z is a normalization constant and $\alpha_{pq}(t, x)$ is a search bias defined as:

$$\alpha_{pq}(t, x) = \begin{cases} \frac{1}{p} & \text{if } d_{tx} = 0\ , \\ 1 & \text{if } d_{tx} = 1\,, \\ \frac{1}{q} & \text{if } d_{tx} = 2\,, \end{cases}$$

where d_{tx} is the shortest path distance between node t (the node visited before v) and node x, p is the return parameter and q is the in-out parameter.

Definition 4 (Node2Vec on Knowledge Graphs as Unsupervised Aggregator). *Given a Knowledge Graph $KG = (V, R, E, \ell_V)$, Node2Vec [7] is a method that learns low-dimensional vector embeddings $\phi : V \to \mathbb{R}^d$ for nodes* *by simulating biased random walks over the graph structure. A sequence of nodes is generated via random walks defined by the transition probability:*

$$P(c_i = x \mid c_{i-1} = v) = \begin{cases} \frac{1}{p} & \text{if } x = c_{i-2} \\ 1 & \text{if } d(x, c_{i-2}) = 1 \\ \frac{1}{q} & \text{otherwise} \end{cases} \tag{2}$$

where: 1) c_i is the node c at position i in the walk; 2) p is the return parameter controlling the likelihood of revisiting the previous node; 3) q is the in-out parameter controlling the likelihood of exploring further away nodes; 4) $d(x, c_{i-2})$ is the shortest path distance between node x and the node before the current one.

Definition 5 (GraphSAGE on Knowledge Graphs as Supervised Aggregator). *Given a Knowledge Graph $KG = (V, R, E, \ell_V)$, GraphSAGE (Graph Sample and Aggregate) is an inductive framework that learns embeddings by aggregating information from the local neighbourhood of each node [8]. At each layer k, the representation of a node $v \in V$, denoted as $h_v^{(k)} \in \mathbb{R}^d$, is updated as*

$$h_v^{(k)} = \sigma\Big(W^{(k)} \cdot AGG^{(k)}\big(\{h_u^{(k-1)} \mid (u, r, v) \in E\} \cup \{h_v^{(k-1)}\}\big)\Big) \tag{3}$$

where: 1) $h_v^{(0)}$ is the initial feature vector of node v; 2) $AGG^{(k)}$ is a permutation-invariant aggregator function; 3) $W^{(k)}$ is a trainable weight matrix at layer k; 4) σ is a non-linear activation function.

3 Related Works on Classification in KGs Datasets

In recent years, as relational datasets have become increasingly prevalent across domains such as social networks and biological systems [9], the need to extract meaningful insights from limited data has become mandatory [8]. In this context, node classification has emerged as a core ML task that aims to assign labels to nodes based on structural and semantic information. However, challenges increase when data are highly semantic, as graph-based learning models heavily depend on explicated features to perform well. In such settings, using both intrinsic topology and node-level features is essential to increase prediction accuracy. This issue is particularly evident in KGs, where, despite the apparent availability of information, a high degree of semantics, meaning that much of the knowledge is implicit in the entities and relations rather than explicitly represented through edges, may lead to unexpressed knowledge and underconnected graph structures [10].

To address this, Graph Embeddings (GEs) have emerged as a fundamental tool in the node classification pipeline [11]. By translating both the structural and semantic properties of graphs into low-dimensional vector spaces, GEs enable the application of conventional ML algorithms to complex graph data [12] using random walk-based methods and neural network-based methods [8]. However, when the semantics are not fully explicated, models rely only on either the graph structure or node attributes, tending to underperform [13]. As a consequence, the field remains relatively new, and determining how to enhance data expressiveness for downstream node classification tasks remains an important challenge. Ongoing research in GML has investigated multiple strategies aimed at improving the quality and expressiveness of embeddings.

Recent efforts have led to the development of different GEs learning approaches designed to bridge the gap between sparse datasets and the rich structural information inherent in KGs. Single-view embeddings learning like unsupervised models, such as Node2Vec, extract latent topological information using stochastic walks, while supervised methods, such as GraphSAGE, exploit node features through neighborhood aggregation, but both approaches individually struggle to capture the full spectrum of relationships and semantic patterns encoded in KGs, especially in low-feature settings [14]. In response to the limitations of single-view embeddings, research has also embraced multi-view fusion strategies [15]. These techniques aim to enhance the representation quality by capturing different perspectives of the same graph. The most explored strategies are: concatenation, which merges multiple embeddings to increase coverage; attention-based mechanisms, which assign context-aware importance to each embedding type; and autoencoders, which project different embeddings into a shared latent space, thereby enabling compression of redundant information [8]. In the end, meta-inductive models, rather than training on a single fixed graph, acquire transferable temporal meta-knowledge capable of generalising across domains at different time [16].

Despite the breadth of methodologies developed to improve the quality and effectiveness of GEs and the growing attention toward strategies that address feature scarcity, current approaches do not explicitly explore the fusion of Node2Vec with centrality features as an auxiliary input to GraphSAGE to create GEs enhanced representations that directly target the limitations typical of small-feature datasets. This gap underscores the need for a novel hybrid strategy capable of capturing both structural behaviour and semantics in scenarios characterised by limited node availability.

To the best of our knowledge, no prior study has investigated this specific hybridisation pipeline, where unsupervised structural embeddings and centrality measures are explicitly combined with supervised feature-aware representations in an adaptive way within the context of low-features graph scenarios. This represents a promising and innovative approach to address the challenges of embedding expressiveness under dataset constraints.

4 Bi-view Strategy

To address classification problems in poor-features datasets within KGs, we propose a hybrid strategy that merges global topological structure with local semantic information. Our proposed approach (Fig. 1), is composed of five main components. In particular, the process begins with the Node2Vec Encoder, which captures the structural properties of the graph by generating node global connectivity embeddings $Z^{(n2v)}$ through biased random walks. In parallel, the Centrality Vector module computes a set of graph-theoretic metrics $\gamma(V)$, including measures such as PageRank and Betweenness, to quantify the importance of each node within the overall structure.

Both the structural embeddings and the centrality metrics are then combined in the Feature Aggregator, which builds enriched node representations $F_{\text{agg}} = \mathcal{A}(\gamma(V), Z^{(n2v)})$ that serve as input to the next component. The GraphSAGE Aggregator applies supervised neighborhood aggregation over these features to learn new embeddings $Z^{(sage)}$, integrating local semantic information with the previously extracted global features.

Finally, the Embedding Fusion module adaptively combines the two views ($Z^{(n2v)}$ and $Z^{(sage)}$) through a learnable mechanism that balances their contributions based on node-specific characteristics. The result is a fused representation $Z^{(fused)}$ for each node, which we refer to as the Bi-View embedding. This representation captures both the structural roles and semantic contexts of nodes in new embeddings, improving classification tasks in poor-features KGs scenarios.

4.1 Node2Vec Encoder

Let $Z^{(n2v)} \in \mathbb{R}^{n \times d_1}$ be the matrix of Node2Vec embeddings, where each row $\varphi(v_i) \in \mathbb{R}^{d_1}$ is the embedding of node v_i generated by making biased random walks over the graph. These embeddings represent topological similarity, capturing information depending on the tuning of the return parameter p and the in-out parameter q.

4.2 Feature Aggregator

To increase the number of features without adding any artificial data, we add a set of classical centrality metrics that are calculated for each node. Formally, let $\gamma(v_i) \in \mathbb{R}^{d_2}$ be the centrality-based features vector for a node v_i, including measures such as degree, betweenness, and PageRank. These topological metrics are concatenated with the Node2Vec embeddings in order to create an enriched initial feature vector:

$$h_i^{(0)} = [\ell_V(v_i) \,\|\, \varphi(v_i) \,\|\, \gamma(v_i)] \in \mathbb{R}^{d_0},$$

where $\ell_V(v_i)$ is the attribute-based features of node v_i, and $\|$ the vector concatenation.

4.3 GraphSAGE Aggregator

GraphSAGE is then applied by using $h_i^{(0)}$ as input features. In particular, at each layer k, the node embeddings are updated as follows:

$$h_i^{(k)} = \sigma\left(W^{(k)} \cdot \text{AGG}^{(k)}\left(\{h_j^{(k-1)} \mid j \in \mathcal{N}(i)\} \cup \{h_i^{(k-1)}\}\right)\right),$$

where $\text{AGG}^{(k)}$ is a permutation-invariant aggregator, $W^{(k)}$ is a learnable weight matrix, and σ is a non-linear activation function. The output embeddings of this step are $Z^{(sage)} \in \mathbb{R}^{n \times d_2}$.

4.4 Embedding Fusion

A dynamic fusion mechanism that learns node-specific main characteristics to generate enhanced GEs representations is then applied. In particular, for each node i, the fusion coefficient $\alpha_i \in (0, 1)$ is calculated as:

$$\alpha_i = \sigma\left(W_\alpha \cdot \left[z_i^{(n2v)} \parallel z_i^{(sage)}\right] + b_\alpha\right),$$

where σ is the sigmoid function and $W_\alpha \in \mathbb{R}^{1 \times (d_1 + d_2)}$, $b_\alpha \in \mathbb{R}$ are learnable parameters.

The final fused embedding is given by:

$$z_i^{(fused)} = \alpha_i \cdot z_i^{(n2v)} + (1 - \alpha_i) \cdot z_i^{(sage)}.$$

Allowing model balancing between structural and semantic information per node.

4.5 Centrality Vector

This component computes a set of classical centrality metrics for each node, giving deterministic structural features. In particular, PageRank calculates node importance based on the probability of arriving at a node via random walks, while Betweenness Centrality measures how frequently a node appears on the shortest paths between other node pairs, representing its control over information flow in the network.

Formally, for each node $v \in V$, the centrality vector is defined as:

$$\gamma(v) = [\deg(v), \text{PageRank}(v), \text{Betweenness}(v), \ldots] \in \mathbb{R}^{d_2}$$

These features enrich the node representation with structural insights that are independent from the aggregation process.

5 Model Validation of Structural Enrichment in GraphSAGE via Node2Vec Embeddings and Centrality Metrics

In this Section, we provide a formal validation to describe how structural embeddings generated by Node2Vec and centrality features affect the behaviour of the GraphSAGE aggregation process. Our goal is to model the transition from topological unsupervised learning to a semantically enriched representation, where structural information derived from random walks and graph-theoretic properties is used to boost the learning process.

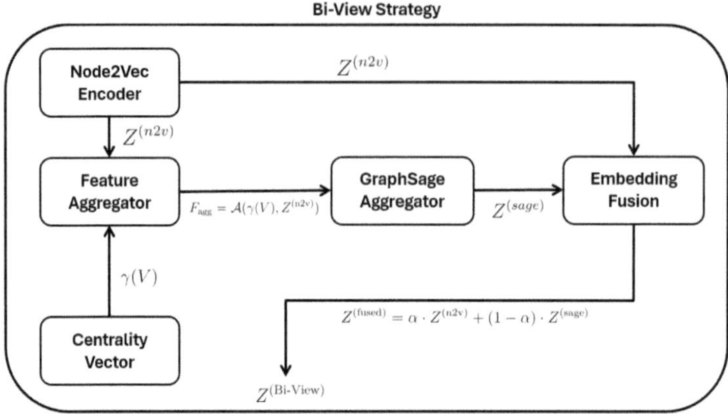

Fig. 1. Bi-View strategy.

Definition 6. *Node2Vec as a Generator of Structural Features*
Let $KG = (V, R, E, \ell_V)$ *be a KG. Node2Vec [7] is based on an embedding function:*

$$\phi : V \rightarrow \mathbb{R}^{d_1},$$

where the embedding $\phi(v)$ *captures the topology of a node* v *within the network, interdependently of any node label or attribute.*

Definition 7 (Centrality Vector). *Let* $\gamma : V \rightarrow \mathbb{R}^{d_2}$ *be a function that maps each node* $v \in V$ *to a vector of centrality metrics.*

$$\gamma(v) = [deg(v), PageRank(v), Betweenness(v), \ldots],$$

These values give a deterministic perspective, acting as global topological analysis that is independent from random walks.

Definition 8 (Structurally Enriched Initial Features). *We define the initial features vector that is given to the GraphSAGE aggregator for each node* $v \in V$ *as:*

$$h_v^{(0)} = [\ell_V(v) \,\|\, \phi(v) \,\|\, \gamma(v)] \in \mathbb{R}^{d_0},$$

where $\ell_V(v)$ *is the original attribute-based feature,* $\phi(v)$ *is the Node2Vec embedding and* $\gamma(v)$ *is the centralities vector.*

Definition 9 (GraphSAGE with Structural Feature Injection). *Let* $h_v^{(0)} \in \mathbb{R}^{d_0}$ *be an enriched feature vector. At each layer* k*, GraphSAGE updates the node embeddings using the following rule:*

$$h_v^{(k)} = \sigma\left(W^{(k)} \cdot AGG^{(k)}\left(\{h_u^{(k-1)} \mid u \in \mathcal{N}(v)\} \cup \{h_v^{(k-1)}\}\right)\right),$$

Lemma 1 (Structural Enrichment Increases Neighborhood Discriminability). *Let* $h_v^{(0)}$ *and* $h_u^{(0)}$ *be the enhanced GEs of two distinct nodes* $v, u \in V$ *and* $\ell_V(v)$ *and* $\ell_V(u)$ *the not enhanced GEs. Then, if* $\phi(v) \neq \phi(u)$ *or* $\gamma(v) \neq \gamma(u)$, *it follows that:*

$$\|h_v^{(0)} - h_u^{(0)}\|_2 \geq \|\ell_V(v) - \ell_V(u)\|_2.$$

Proof. Since the enriched feature vector includes additional components, in form of:

$$h_v^{(0)} = [\ell_V(v) \,\|\, \phi(v) \,\|\, \gamma(v)],$$

Any difference in ϕ or γ will add non-negative contributions to the Euclidean distance. By the Pythagorean theorem:

$$\|h_v^{(0)} - h_u^{(0)}\|_2^2 = \|\ell_V(v) - \ell_V(u)\|_2^2 + \|\phi(v) - \phi(u)\|_2^2 + \|\gamma(v) - \gamma(u)\|_2^2.$$

Thus, the distance is always greater than or equal to that computed using initial attributes.

Definition 10 (Enriched Neighborhood Aggregation Influence). *Let* $\mathcal{N}(v)$ *be the set of neighbors of node* v, *with* $h_v^{(0)}$ *the enriched GE. The influence score of node* v *at layer* k, *denoted* $I_k(v)$, *is defined as*

$$I_k(v) = \frac{1}{|\mathcal{N}(v)|} \sum_{u \in \mathcal{N}(v)} \|h_v^{(k-1)} - h_u^{(k-1)}\|_2.$$

Lemma 2 (Structural Enrichment Amplifies Neighborhood Gradient). *Let* $I_k(v)$ *be the influence score of node* v *under standard GraphSAGE, and* $\tilde{I}_k(v)$ *be the same score using structurally enriched features. Then:*

$$\tilde{I}_k(v) \geq I_k(v),$$

Assuming that $\phi(u)$ *and* $\gamma(u)$ *differ across neighborhoods.*

Proof. Similarly to the previous lemma, the added feature space components from ϕ and γ increase differences across the neighbourhood, increasing the mean pairwise distances and enhancing node representations with global structural embeddings and topological information, in particular: 1) improves the initial separability of nodes in the embedding space; 2) enhances the information gradient available during message passing; 3) facilitates better generalisation under low-data regimes by reducing feature sparsity.

6 Supervised Fusion Model for Enhanced Bi-view Node Representations

To capture the best specific characteristics of both Node2Vec and GraphSAGE embeddings, we propose a supervised fusion strategy that learns the best optimised representation for node classification. Let $z_i^{(n2v)} \in R^{d_1}$ be the unsupervised embedding of node i produced by Node2Vec, and $z_i^{(sage)} \in R^{d_2}$ the supervised embedding generated by GraphSAGE, which uses node2vec and centrality features. We concatenate both embeddings into a single vector:

$$z_i^{(cat)} = [z_i^{(n2v)} || z_i^{(sage)}] \in R^{d_1+d_2}$$

This concatenated representation is then passed through a learnable encoder function $f_\theta : R^{d_1+d_2} \rightarrow R^d$ implemented as a multi-layer perceptron (MLP), which learns the best task-specific embedding:

$$z_i^{(enh)} = f_\theta(z_i^{(cat)})$$

The resulting embedding $z_i^{(enh)} \in R^d$, with $d \ll d_1 + d_2$ is used for classification via a function:

$$\hat{y}_i = \text{softmax}(W_c \cdot z_i^{(enh)} + b_c)$$

where $W_c \in R^{|\mathcal{Y}| \times d}$ and $b_c \in R^{|\mathcal{Y}|}$ are trainable parameters. In the end, the full model is trained over the labelled node set $\mathcal{L}$:

$$\mathcal{L}_{enh} = -\sum_{i \in \mathcal{L}} y_i \cdot \log \hat{y}_i$$

This supervised fusion mechanism allows the model to automatically extract the most important aspects of both global (Node2Vec) and local (GraphSAGE) node contexts, resulting in a topologically and semantically enriched embedding space.

Lemma 3 (Expressiveness of Fused Embeddings). *Let $z_i^{(n2v)} \in R^{d_1}$ and $z_i^{(sage)} \in R^{d_2}$ be the unsupervised and supervised embeddings of node i, respectively. Let $f_\theta : R^{d_1+d_2} \rightarrow R^d$ be a Multi-Layer Perceptron (MLP) our fusion function, where $d \leq d_1 + d_2$. The resulting fused embedding is $z_i^{(enh)} = f_\theta([z_i^{(n2v)} || z_i^{(sage)}])$. Given that f_θ is a non-linear function, the fused embeddings $Z^{(enh)} = \{z_i^{(enh)} | i \in V\}$ can capture complex interactions between the two original embedding spaces.*

Proof. The concatenation $[z_i^{(n2v)} || z_i^{(sage)}]$ results in a feature vector with a dimension equivalent to the Cartesian product of the individual embedding spaces, i.e., $R^{d_1} \times R^{d_2}$. The dimension of this concatenated space is $d_1 + d_2$.

The multi-layer perceptron f_θ is applied to the concatenation of Node2Vec and GraphSAGE embeddings, allowing for the combination of heterogeneous information into a lower-dimensional space to generate better feature descriptions for classification tasks. In particular, the fused embedding space $Z^{(enh)}$ is capable of representing a richer set of patterns and relationships than either $\mathcal{Z}_{n2v}$ or $\mathcal{Z}_{sage}$ individually.

7 Experimental Evaluation on Food and Drug Administration Adverse Effects Dataset

7.1 Dataset Description

To assess the impact of our framework, we applied it to a real-world dataset derived from the United States Food and Drug Administration (FDA) Adverse Event Reporting System (FAERS)[1]. FAERS is a database developed by the FDA to support post-marketing safety surveillance of approved drugs. In particular, it collects reports of adverse events and medication errors submitted by healthcare professionals, consumers and manufacturers. The dataset was formalised as a KG[2]. with classes: Case, Drug, Reaction, Outcome, ReportSource, Therapy and Age Group (Fig. 2) that are highly imbalanced (Fig. 3).

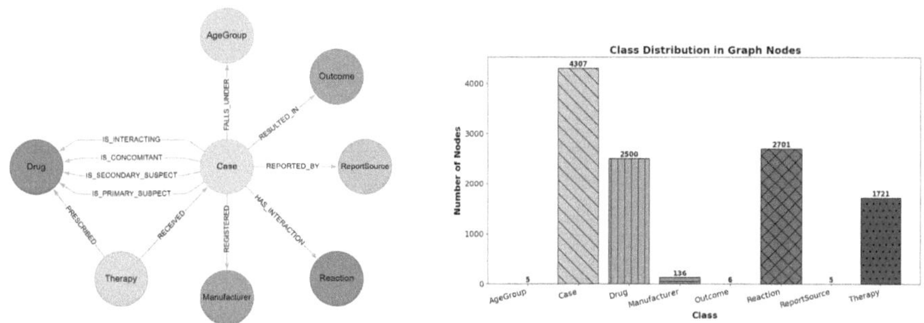

Fig. 2. KG schema for FAERS dataset.

Fig. 3. Class distribution.

7.2 Experimental Setup and Model Evaluation

We carried out experiments using Neo4j 5.24.0[3], with the Graph Data Science (GDS) library version 2.12.0. Starting from this small real-world dataset (~11,000 elements), our goal was to train a classifier capable of identifying the most frequent node classes, which are: Patient, Drug, Therapy and Manufacturer. For topological embeddings, the Node2Vec algorithm was used with a dimension of 64 elements, a walk length of 80, and 10 iterations, setting both the return and in-out parameters to 1.0. GraphSAGE was then employed with a dimension of 64, using as input features: the centrality properties (betweenness and pageRank), the class label, and the results coming from Node2Vec. GEs were combined using a supervised FusionNet model, composed of an MLP with a hidden dimension of 128 and an output dimension of 64, trained with cross-entropy loss over 100 epochs. In the end, a decision tree was implemented to compare classification results across the experiments, using class labels only during training, while in testing they were hidden and predicted.

[1] https://open.fda.gov/data/faers/.

[2] https://github.com/neo4j-graph-examples/healthcare-analytics/tree/main/data.

[3] https://neo4j.com.

Initially, we generated feature vectors containing only node class labels, with PageRank and Betweenness Centrality as additional features. Finally, we compared the classifier's performance under three different configurations: (1) using only Node2Vec embeddings, (2) using GraphSAGE embeddings, and (3) using the fused embeddings generated by our strategy (Bi-View).

The network embeddings visualization was done with Principal Component Analysis (PCA), which is a dimensionality reduction algorithm that allows to understand the quality of latent representations generated by models. In particular, this section also presents a comparative analysis of 2D embeddings obtained from the three distinct models, in order to visually evaluate inter-class separability within the latent space. From experiments, Node2Vec (Fig. 4) performs the weakest in terms of embedding clarity, with extensive class overlap and poor separability, including for Classes AgeGroup and Case. GraphSAGE (Fig. 5), in contrast, arranges its embeddings along a smooth crescent-shaped curve, lacking compact clusters but possibly capturing latent class semantic relationships in a one-dimensional progression, with its curvilinear structure complicating classification using standard decision boundaries. In the end, Bi-View (Fig. 6) exhibits a strong ability to form well-separated and compact clusters, particularly for class Case, and moderately for Classes AgeGroup and Outcome, aligning with its high classification performance.

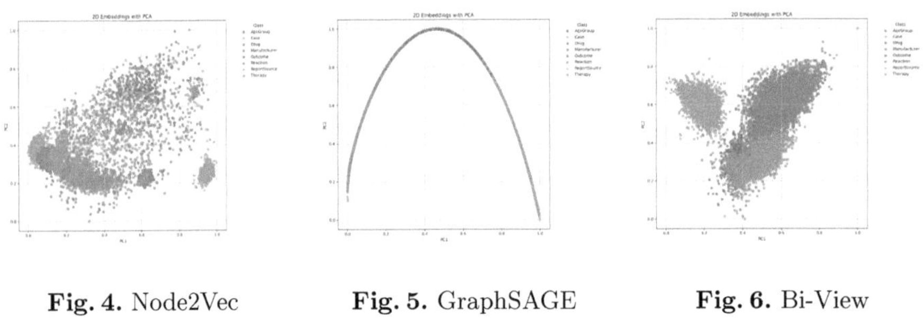

Fig. 4. Node2Vec **Fig. 5.** GraphSAGE **Fig. 6.** Bi-View

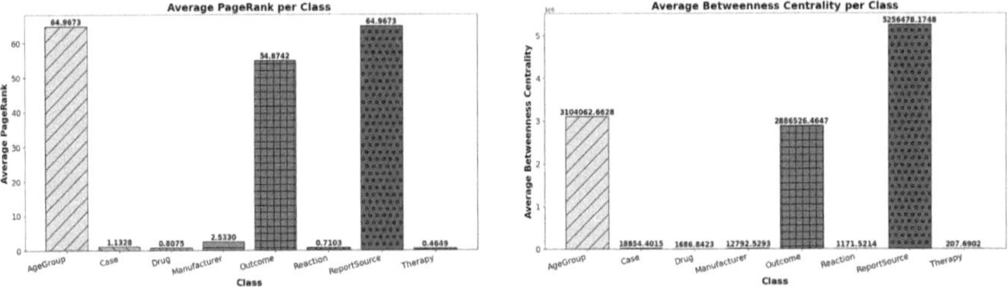

Fig. 7. Average PageRank per Class. **Fig. 8.** Average Betweenness per Class.

Given the highly imbalanced nature of the dataset (Fig. 3) where class samples show differences of up to three orders of magnitude, the evaluation has been limited to the four classes with the highest number of instances: Case, Drug, Reaction and Therapy. This restriction enables a more robust comparison across the Bi-View, GraphSAGE, and Node2Vec models, avoiding distortions caused by extremely rare classes. To better justify our choice, node count alone does not fully capture the structural importance of each class within the graph. For these reasons, we also obtain insights drawn from the plots of average PageRank (Fig. 7) and average betweenness (Fig. 8) centrality, which reveal that the underrepresented classes exhibit exceptionally high centrality scores, indicating that their nodes act as central hubs that are basically in common to the other classes. As a result, using these highly central nodes as features for classification could introduce noise rather than discriminative power, since their presence does not provide class-specific information but reflects a global network structure shared across all classes, negatively affecting the model's ability to distinguish between classes and reducing its overall performance, thus motivating our choice to ignore them for classification purpose.

We calculated Accuracy, Precision, Recall and F1-Score to evaluate the performance of the three approaches, in particular: Bi-View, Node2Vec and GraphSAGE on classes with at least 200 samples. Bi-View (Fig. 9) outperforms the other methods for all metrics, reaching an accuracy close to 94% with high precision and recall, leading to a strong F1-score. Node2Vec shows intermediate performance, with scores consistently higher than GraphSAGE but still markedly below Bi-View. On the contrary, GraphSAGE reports the lowest values in all metrics, with a particularly reduced F1-score of around 64%. These results suggest that Bi-View is more effective at capturing patterns with class imbalance, even when the evaluation is restricted to the most statistically significant classes. In the following analysis, we examine the confusion matrices (Fig. 10, Fig. 11 and Fig. 12) for a deeper understanding of how each model behaves on these dominant classes.

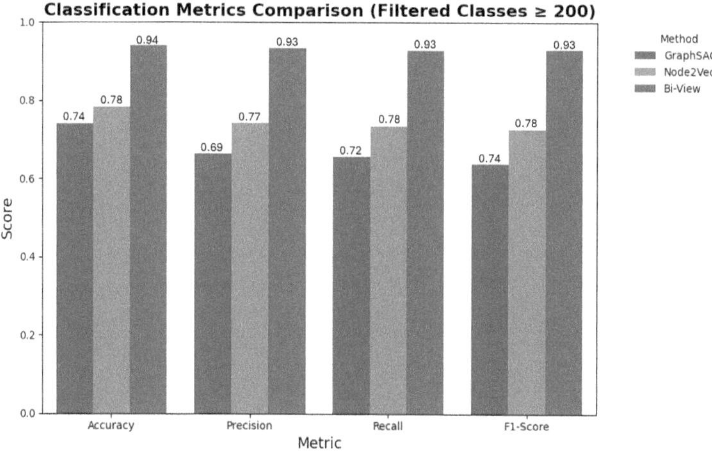

Fig. 9. Models Comparison.

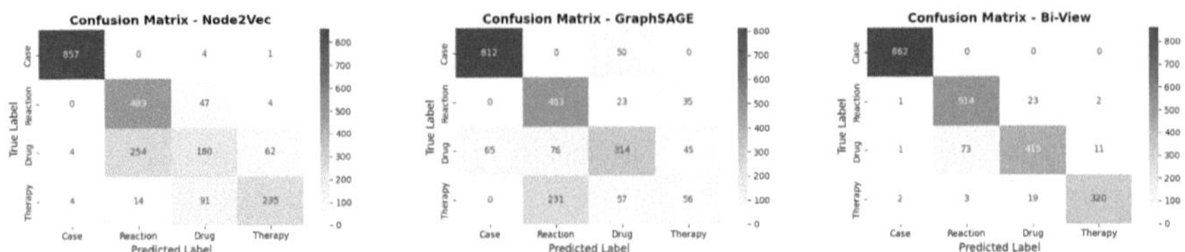

Fig. 10. Node2Vec **Fig. 11.** GraphSAGE **Fig. 12.** Bi-View

Bi-View performs better than both Node2Vec and GraphSAGE in every category, especially when it comes to differentiating between closely related classifications like Drug and Therapy, according to the confusion matrices shown in Figs. 10, 11 and 12. Node2Vec performs well for the Case and Reaction classes, but it has trouble with a high rate of drug and reaction misclassification. Though it still has significant confusion, particularly between Therapy and Reaction, GraphSAGE offers more balanced forecasts. Bi-View, on the other hand, has the best classification accuracy and the fewest cross-category mistakes, demonstrating its exceptional capacity to gather and incorporate a variety of relational and structural data for better class separability.

8 Conclusion and Future Work

In this study, we proposed a hybrid graph embedding strategy that integrates unsupervised structural representations from Node2Vec with supervised neighbourhood aggregation from GraphSAGE through a dynamic fusion mechanism. This Bi-View Embedding Fusion strategy was specifically designed to address the limitations of traditional GML models in poor-dataset feature scenarios. By enriching initial node features with topological insights and centrality metrics and adaptively combining global and local representations, our approach produces more expressive and semantically informative embeddings. Experimental results on the FAERS healthcare dataset confirmed the effectiveness of the proposed method, showing improvements in classification tasks under low-features regimes. These findings demonstrate that our fusion model is not merely a feature concatenation scheme but a transformative strategy that enhances generalisation by capturing diverse and complementary patterns within the graph. As future work, we plan to evaluate the performance of this strategy across different types of graphs, including temporal and multi-relational datasets. Additionally, we intend to explore meta-learning and inductive transfer capabilities to further improve adaptability in dynamic or cross-domain graph environments.

Acknowledgments. Rosario Napoli is a PhD student enrolled in the National PhD in Artificial Intelligence, XL cycle, course on Health and Life Sciences. Giovanni Lonia is a PhD student enrolled in the National PhD in Artificial Intelligence, XL cycle, course on AI for society, organised by the University of Pisa. This work has been partially funded by the Italian Ministry of Health, Piano Operativo Salute (POS) trajectory 4 "Biotechnology, bioinformatics and pharmaceutical development", through the Pharma-HUB Project "Hub for the repositioning of drugs in rare diseases of the nervous system in children" (CUP J43C22000500006), the Italian Ministry of Health, Piano Operativo Salute (POS) trajectory 2 "eHealth, diagnostica avanzata, medical device e mini invasività" through the project "Rete eHealth: AI e strumenti ICT Innovativi orientati alla Diagnostica Digitale (RAIDD)" (CUP J43C22000380001), the "SEcurity and RIghts in the CyberSpace (SERICS)" partnership (PE00000014), under the MUR National Recovery and Resilience Plan funded by the European Union NextGenerationEU. In particular, it has been supported within the SERICS partnership through the projects FF4ALL (CUP D43C22003050001) and SOP (CUP H73C22000890001).

References

1. Makarov, I., et al.: Survey on graph embeddings and their applications to machine learning problems on graphs. PeerJ Comput. Sci. (2021)
2. Xia, F., et al.: Graph learning: a survey. IEEE Trans. Artif. Intell. **2**(2), 109–127 (2021)
3. Agrawal, S., Patel, A.: A study on graph storage database of noSQL. Int. J. Soft Comput. Artif. Intell. Appl. **5: 33 –39**, 02 (2016)
4. Besta, M., et al.: Neural graph databases. LOG IN (2022)
5. Grover, A., Leskovec, J.: node2vec: scalable feature learning for networks. In: Proceedings of the 22nd ACM SIGKDD International Conference on Knowledge Discovery and Data Mining, KDD 2016, pp. 855–864. Association for Computing Machinery, New York (2016)
6. Hamilton, W.L., Ying, R., Leskovec, J.: Inductive representation learning on large graphs. In: Proceedings of the 31st International Conference on Neural Information Processing Systems, NIPS 2017, pp. 1025–1035. Curran Associates Inc., Red Hook (2017)
7. Liu, R., Krishnan, A.: PecanPy: a fast, efficient and parallelized python implementation of node2vec. Bioinformatics **37**(22), 3978–3980 (2021)
8. Khoshraftar, S., An, A.: A survey on graph representation learning methods. ACM Comput. Surv. (2023)
9. Kapoor, M., et al.: PROV-GEM: automated provenance analysis framework using graph embeddings. In: 2021 20th IEEE International Conference on Machine Learning and Applications (ICMLA), pp. 1536–1541. IEEE (2021)
10. Cheng-Yun, A., et al.: Contrast-enhanced semi-supervised text classification with few labels. In: Proceedings of the AAAI Conference on Artificial Intelligence, vol. 36, pp. 11363–11371 (2022)
11. Jiang, X., Zhu, R., Ji, P., Li, S.: Co-embedding of nodes and edges with graph neural networks. IEEE Trans. Pattern Anal. Mach. Intell. (2020)
12. Rzemberczki, B., Sarkar, R.: Characteristic functions on graphs: birds of a feather, from statistical descriptors to parametric models. In: Proceedings of the 29th ACM International Conference on Information and Knowledge Management (CIKM). Association for Computing Machinery (2020)
13. Chen, X., et al.: Learning on attribute-missing graphs. IEEE Trans. Pattern Anal. Mach. Intell. (2020)

14. Hajibabaee, P., et al.: An empirical study of the graphsage and word2vec algorithms for graph multiclass classification. In: IEEE Annual Information Technology, Electronics and Mobile Communication Conference (IEMCON) (2021)
15. Fan, S., et al.: One2multi graph autoencoder for multi-view graph clustering. In: Proceedings of The Web Conference 2020 (WWW 2020), pp. 3070–3076. Association for Computing Machinery, New York (2020)
16. Wen, Z., Fang, Y., Liu, Z.: Meta-inductive node classification across graphs. In: Proceedings of the 44th International ACM SIGIR Conference on Research and Development in Information Retrieval (SIGIR 2021), pp. 1651–1660. Association for Computing Machinery, New York (2021)

KGRAG-Ex: Explainable Retrieval-Augmented Generation with Knowledge Graph-Based Perturbations

Georgios Balanos[1,2(✉)], Evangelos Chasanis[1,2], Konstantinos Skianis[1], and Evaggelia Pitoura[1,2]

[1] University of Ioannina, Ioannina, Greece
g.balanos@uoi.gr
[2] Archimedes, Athena Research Center, Athena, Greece

Abstract. Retrieval-Augmented Generation (RAG) enhances language models by grounding responses in external information, yet explainability remains a critical challenge, particularly when retrieval relies on unstructured text. Knowledge graphs (KGs) offer a solution by introducing structured, semantically rich representations of entities and their relationships, enabling transparent retrieval paths and interpretable reasoning. In this work, we present KGRAG-Ex, a RAG system that improves both factual grounding and explainability by leveraging a domain-specific KG constructed via prompt-based information extraction. Given a user query, KGRAG-Ex identifies relevant entities and semantic paths in the graph, which are then transformed into pseudo-paragraphs: natural language representations of graph substructures that guide corpus retrieval. To improve interpretability and support reasoning transparency, we incorporate perturbation-based explanation methods that assess the influence of specific KG-derived components on the generated answers. We conduct a series of experiments to analyze the sensitivity of the system to different perturbation methods, the relationship between graph component importance and their structural positions, the influence of semantic node types, and how graph metrics correspond to the influence of components within the explanations process.

Keywords: Knowledge-Graphs (KGs) · Retrieval-Augmented Generation (RAG) · Explainability · Large Language Model (LLMs)

1 Introduction

Large Language Models (LLMs) [10] have demonstrated strong performance across numerous natural language processing tasks, including question answering (QA) [7], summarization [16], and dialogue generation [17]. While LLMs excel at generating fluent and contextually relevant text, their reliance on internalized knowledge can lead to factual errors and untraceable reasoning. To mitigate

H. Takeda et al. (Eds.): IJCKG 2025, LNAI 16297, pp. 35–50, 2026.
https://doi.org/10.1007/978-981-95-5009-8_3

this, Retrieval-Augmented Generation (RAG) [9] combines LLMs with external document retrieval, grounding responses in up-to-date information drawn from large corpora. However, RAG systems typically offer limited visibility into how retrieved content influences the output of the model, raising concerns in settings where outputs must be auditable, interpretable, and verifiable.

Explainability in RAG has thus become a key focus of recent research. Approaches such as [12] and [11] use perturbation-based techniques to analyze how altering parts of the input-such as retrieved documents or query terms-affects the final output. These methods provide insights into the importance of specific elements in the context, helping to trace the decision-making process of the model. However, they primarily operate on unstructured text and offer limited access to explicit reasoning pathways or underlying semantic structures. Additionally, these techniques can be computationally expensive, which limits their scalability and practicality in real-world deployments.

In this work, we propose KGRAG-Ex, a novel system that enhances explainability in RAG by integrating structured knowledge graphs (KGs) [18] into the retrieval and generation pipeline [3]. The system processes user queries to identify relevant entities, constructs semantic paths through the KG, and converts these paths into natural language pseudo-paragraphs that guide document retrieval. These structured representations provide a transparent layer of reasoning that complements the unstructured evidence retrieved from external sources.

A key novelty of our approach lies in combining structured retrieval via knowledge graph paths with targeted perturbations at the path level to analyze and explain model outputs. Unlike previous perturbation-based methods that primarily operate on unstructured text, our method leverages explicit reasoning pathways embedded in knowledge graphs to provide more transparent, faithful, and fine-grained explanations. This integration enhances interpretability while also improving computational efficiency and scalability, making it especially suitable for applications requiring auditable and trustworthy reasoning.

We perform perturbations by removing nodes, edges, or sub-paths. By modifying nodes, edges, or sub-paths, we evaluate how these structural changes influence the output of the model. This KG-focused perturbation approach provides a more comprehensive and fine-grained view of model behavior, highlighting which elements of the graph-based reasoning process are most critical. Furthermore, by isolating the impact of specific graph components, KGRAG-Ex enhances interpretability without relying on text-level perturbations, as in previous work. Perturbations at the path level tend also to be more cost-efficient than perturnations at the text level, making the framework particularly suitable for high-stakes domains where reasoning traceability is essential.

Finally, we conduct experiments to analyze the sensitivity to different perturbation methods, the influence of semantic node types, and the relationship between graph component importance, their structural positions in the paths and centrality measures in the KG. We also compare the efficiency of our approach to text-based perturbation methods.

The remainder of this paper is structured as follows. Section 2 reviews related work. Section 3 describes our method for knowledge graph construction and retrieval, and Sect. 4 presents our perturbation-based explanation framework. Section 5 reports experimental results, and Sect. 6 concludes the paper.

2 Related Work

Large language models (LLMs) have been used to extract structured representations-such as entityrelationentity triples-from unstructured text, enabling the construction of dynamic knowledge graphs to guide retrieval and reasoning [18]. Query2doc [13] demonstrates how LLMs can be prompted to expand sparse queries into detailed pseudo-documents. In this paper, rather than a generic expansion, we extend this approach by generating structured pseudo-paragraphs based on the extracted entities and their connected paths within the KG. These paragraphs are then combined into a pseudo-document that serves as a context-rich input to guide retrieval.

Recent advancements in explaining Retrieval-Augmented Generation (RAG) models have been addressed by works such as [12] and [11]. RAG-Ex [12] introduces a model- and language-agnostic framework that provides approximate explanations for why large language models (LLMs) generate specific responses in RAG-based question answering tasks. By employing perturbation-based methods, RAG-Ex identifies critical tokens whose removal significantly alters the output, offering insights into the decision-making process of LLMs. In contrast, RAGE [11] focuses on the provenance of external knowledge sources used during RAG. It employs counterfactual explanations by systematically removing or reordering retrieved documents to observe changes in the model's responses, thereby highlighting the influence of specific sources on the generated answers.

Our work is situated within a broader landscape of explainability in NLP. Surveys such as [4] provide a comprehensive taxonomy of explanation methods in NLP, while [1] reviews interpretability techniques for large language models (LLMs), including prompt-based and retrieval-augmented models. More recently, [2] explores how probing and intervention methods can be applied to understand the internal mechanisms of LLMs, offering complementary perspectives to perturbation-based explanations. Building on prior work such as RAG-Ex and RAGE, which apply perturbation-based techniques to unstructured retrieval and document provenance, respectively, our approach-KGRAG-Ex-introduces structured, graph-based retrieval via knowledge graphs. This enables more interpretable reasoning paths and targeted perturbation of graph-derived information, leading to more focused and transparent explanations.

3 Knowledge Graph Construction and Retrieval

To support structured retrieval and enhance reasoning transparency, we construct a domain-specific knowledge graph - focusing on the medical domain in this paper - directly from raw textual data. Given a user query, we extract relevant paths from the knowledge graph and convert them into contextual input for RAG.

Knowledge-Graph Construction. To construct the knowledge graph, consistent with existing studies in retrieval-augmented generation (RAG) (Lewis et al., 2020 [8]; Gao et al., 2023 [6]; Fan et al., 2024 [5]), each document is divided into smaller chunks, respecting a predefined chunk size. For each chunk, we prompt a large language model (LLM), to extract factual information in the form of (Entity, Relationship, Entity) triplets as in [18]. Additionally, the LLM assigns a semantic label to each extracted entity (node), such as Disease, Body Part, or Medication, to enrich the graph with meaningful categories. Each extracted triplet is annotated with metadata, including the document identifier and the chunk index, enabling traceability and contextual alignment within the source corpus. The construction is illustrated in Fig. 1.

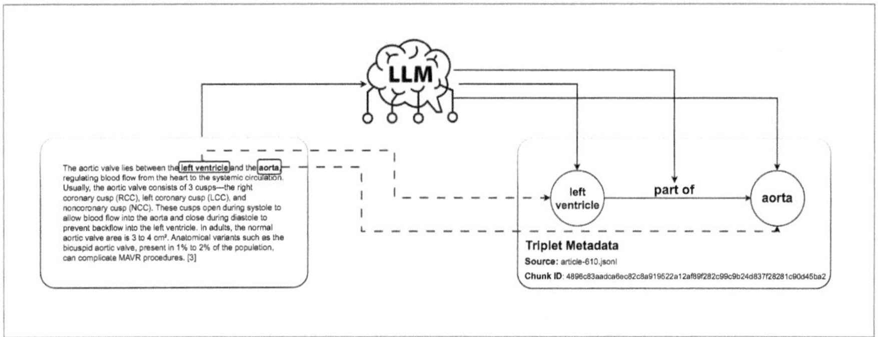

Fig. 1. Triplet extraction.

Retrieval from the Knowledge Graph. To retrieve relevant information from the knowledge graph, we first extract a set of (Entity, Relationship, Entity) triplets from the user query using an LLM. This step ensures that the query is transformed into a structured form that aligns with the format of the knowledge graph. Entities mentioned in the query are identified and grounded to their corresponding nodes in the graph. Once the relevant entities are identified, we perform graph traversal to determine the shortest paths between them within the knowledge graph. This process uncovers the intermediate nodes and relations that connect the query-relevant entities, allowing us to identify semantically meaningful subgraphs. If no shortest path is found between the extracted entities, a fallback mechanism is implemented where regular retrieval methods (without using the knowledge graph) are applied to ensure robust information access. The resulting graph structure includes not only the semantic connections between entities but also metadata such as the source filename and chunk ID for each edge. This additional information enhances traceability, allowing each

Table 1. A retrieval example: the initial user-query, the selected option, the retrieved path, and and related sources.

Query	A baby born with pulmonary hypoplasia secondary to oligohydramnios caused by renal agenesis would be classified as having **Options:** A. an association B. a dysplasia C. a sequence D. a syndrome
Selected Option	A. an association
Retrieved Path	- pulmonary hypoplasia $\xrightarrow{\text{IS RISK FACTOR FOR}}$ persistent pulmonary hypertension in the newborn - persistent pulmonary hypertension in the newborn $\xrightarrow{\text{HAS RISK FACTOR}}$ oligohydramnios
Sources:	**Article source: article-27634.jsonl** **Article source: article-22355.jsonl**

element of the graph to be linked back to its original location in the source documents.

An example of the retrieval process, applied to our running example of a medical dataset with multiple-choice QA, is presented in Table 1, which displays first the user query and the available options. For this specific query, we identified the shortest path from *pulmonary hypoplasia* to *oligohydramnios*. Each triplet in the derived path includes the source file names, thereby enhancing the traceability of our system.

Transforming Knowledge Graph Paths into RAG Context. We construct a context as a means of translating structured data into a form amenable to language-based models. The process begins with the generation of pseudo-paragraphs from structured paths -such as sequences of nodes, concepts, or semantic transitions. Each pseudo-paragraph captures the meaning of its corresponding path in natural language, serving as a textual proxy for the underlying structure.

Once a collection of pseudo-paragraphs has been generated, these are concatenated to form a single context. This context integrates information from multiple paths, providing a unified textual representation that reflects the overall semantics of the structured source. By converting structured inputs into coherent natural language, this method enables the effective use of large language models and other NLP systems for tasks involving structured or semi-structured data. The resulting textual context is then used as input for retrieval-augmented generation (RAG), providing relevant and comprehensive information to improve the performance of downstream language tasks.

Figure 2 illustrates our proposed pipeline. This includes entity extraction from the user query, computation of the derived shortest paths, generation of pseudo-paragraphs, and augmentation of the RAG context.

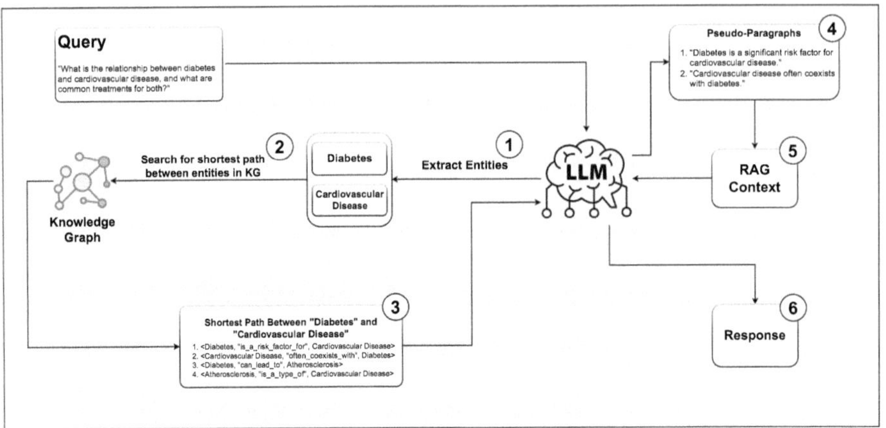

Fig. 2. The proposed pipeline. Graph-level perturbations are applied at Step 3.

4 Explanations Through Knowledge Graph Perturbations

To generate explanations, we apply perturbations to the retrieved knowledge graph context by selectively removing elements at varying levels of granularity, this includes individual nodes, edges, or entire sub-paths. The motivation behind this approach is to identify which components of the knowledge graph are most influential in the reasoning process. By systematically deleting specific elements and observing the resulting changes in model predictions or outputs, we can infer their importance: if removing a component significantly alters the outcome, it suggests that the component plays a critical role. These controlled perturbations enable us to assess the semantic contribution of each part of the reasoning chain, ultimately facilitating the generation of faithful and interpretable explanations. Since every graph component is-directly or indirectly-annotated with metadata during knowledge graph construction, these perturbation methods also enhance the traceability of the model by linking important elements-nodes, edges, or sub-paths-back to their sources.

Node-Level Perturbations. In this perturbation method, we remove one node from the derived path while keeping all other elements unchanged. By comparing the output of the model with and without the node, we can estimate the contribution of each entity to the final response. Furthermore, since every node is labeled with a semantic category during the knowledge graph construction (e.g. Disease, Medication, Body Parts), this analysis reveals not only which individual nodes are influential, but also which types of information the model relies on the most. This may offer insights into the semantic reasoning patterns of the model and help identify the categories of knowledge that drive its outputs.

Edge-Level Perturbations. In this method, each perturbation corresponds to the removal of a single edge from the derived path, disconnecting two semanti-

Table 2. Example perturbations on the retrieved path shown in Table 1.

Path after Node Perturbation	**Removed Node: persistent pulmonary hypertension in the newborn** - pulmonary hypoplasia $\xrightarrow{\text{IS RISK FACTOR FOR}}$ - $\xrightarrow{\text{HAS RISK FACTOR}}$ oligohydramnios
Selected Option	C. a sequence
Path after Edge Perturbation	**Removed Edge: HAS RISK FACTOR** - pulmonary hypoplasia $\xrightarrow{\text{IS RISK FACTOR FOR}}$ persistent pulmonary hypertension in the newborn - persistent pulmonary hypertension in the newborn – oligohydramnios
Selected Option	C. a sequence
Path after Subpath Perturbation	**Removed Subpath:** persistent pulmonary hypertension in the newborn $\xrightarrow{\text{HAS RISK FACTOR}}$ oligohydramnios - pulmonary hypoplasia $\xrightarrow{\text{IS RISK FACTOR FOR}}$ persistent pulmonary hypertension in the newborn
Selected Option	A. an association
Article Sources	article-27634.jsonl article-22355.jsonl
Explanation shown to user	The most important condition for answering the question is Persistent Pulmonary Hypertension in the Newborn. It had the biggest impact on the result
System insight (technical)	Removing **Persistent Pulmonary Hypertension in the Newborn** led to a different answer **2 times**, indicating it is a highly influential entity in the reasoning path. Removing **Oligohydramnios** led to a different answer **1 time**, showing it has a moderate impact on the final answer

cally related nodes. While node-level perturbations highlight important entities, edge-level perturbations reveal how the model fills the semantic space between them, that is, how it connects these entities through meaningful relationships. Since edges often represent actions or associations (e.g., IS DETECTED BY, IS AFFECTED BY), their removal helps identify which specific relations are most critical to reasoning. In many cases, this points to the key interaction or action that drives the response to a given query.

Sub-path-Level Perturbations. For sub-path-level perturbations, we remove a sub-path, typically a triplet, from the derived path. Compared to node- and

Table 3. Knowledge Graph and Corpus

Document Corpus	
Total documents	9,552
Total text chunks	445,454
Average chunks per document	46.63
Standard deviation	28.58
Min chunks per document	5
Max chunks per document	556
Knowledge Graph	
Total nodes	53,411
Total edges	133,287
Average in-degree	2.495
Average out-degree	2.495
Average total degree	4.990

Entity Label	**Percentage (%)**
Diseases	30.73
Risk Factors	8.50
Symptoms	17.71
Diagnostic Tests	9.29
Treatments	14.97
Body Parts	11.81
Medications	6.83
Unknown	0.16

edge-level perturbations, these changes tend to have a greater impact on the reasoning path, as they involve the removal of multiple elements at once.

An example is shown in Table 2. We include one example of each perturbation method and the selected option after the perturbation. We provide two types of explanations, one is targeting end users and the other one is more technical. The explanation shown to the end user is the most influential graph element, identified as the one that causes the most changes in the output of the model across the different perturbations, e.g., the one that appears in most of them. In this example, this is *persistent pulmonary hypertension in the newborn*. We also report technical details for developers. Finally, we obtain the source filename from which *persistent pulmonary hypertension in the newborn* was extracted, further improving traceability by revealing the origin of the most influential context elements.

5 Experiments

In our evaluation, we utilized the StatPearls corpus along with two established benchmarks: MedMCQA and MMLU [14]. MedMCQA is a multiple–choice question –answering dataset derived from real medical entrance exams, focusing on clinical and factual medical knowledge. MMLU (Massive Multitask Language Understanding) complements this by offering a wide-ranging evaluation across 57 subjects, including medicine, law, mathematics, and more. Both MedMCQA and MMLU consist of multiple-choice questions with four options, enabling consistent evaluation of model reasoning across specialized (medical) and general knowledge domains.

Due to computational and token-related costs (API token limits) associated with large-scale language model usage, the KG was built from a subset of the full

Table 4. Accuracy (%) across different retrieval methods on StatPearls. The last row represents results using KGRAG-Ex as the retriever. The other rows are from the MIRAGE Med results [15].

Retriever	MMLU-Med	MedQA	MedMCQA	PubMedQA*	BioASQ-Y/N	Average
BM25	71.63	65.67	54.89	27.60	60.36	56.03
Contriever	73.28	67.48	54.24	28.80	58.41	56.44
SPECTER	73.74	64.73	52.83	23.20	57.77	54.45
MedCPT	72.82	64.89	54.17	27.60	60.68	56.03
RRF-2	72.64	65.67	54.63	30.00	61.17	56.82
RRF-4	73.83	65.12	53.81	30.60	59.71	56.61
KGRAG-Ex	71.07	55.76	53.72	53.80	72.80	**61.43**

corpus. Table 3 summarizes key statistics of the constructed knowledge graph and document corpus, as well as the distribution of entity labels assigned during preprocessing.

Before evaluating the proposed methodology, to assess the effectiveness of different retrieval methods, we evaluate performance using accuracy as the metric. Table 4 presents a comparison between traditional and neural retrievers, as well as our proposed *KGRAG-Ex* method. While prior methods (from MIRAGE Med) perform competitively on individual benchmarks, KGRAG-Ex achieves the highest average accuracy across all tasks, demonstrating its ability to generalize across diverse biomedical QA datasets. The code (including the prompts for the KG generation) and data used in our experiments are publicly available[1].

5.1 Evaluation of the Proposed Methodology

To gain deeper insights into the effectiveness of our graph perturbation strategies, we conduct experiments to address the following research questions:

- **RQ1:** How sensitive is the system to each perturbation method?
- **RQ2:** Is there a correlation between perturbed graph elements and their relative positions?
- **RQ3:** Which types of nodes, based on their assigned semantic labels, are most critical in the generation process?
- **RQ4:** How do graph centrality metrics (e.g., node degree, edge betweenness) relate to the significance of elements within the derived paths?

Evaluating the Relative Effect of the Graph perturbation Methods. To assess how different types of graph component removals influence the generated output, we evaluate the sensitivity of the model to different perturbations.

[1] https://github.com/george-balanos/KGRAG-Ex.

Table 5. Impact of perturbation methods across datasets

Dataset	Perturbation Examples	Node Impact	Edge Impact	Sub-path Impact
MedMCQA	128	41	30	57
MMLU	37	6	8	23

Specifically, for each example, we generate and apply multiple perturbations across three types: removing individual nodes, removing edges, and removing sub-paths consisting of triplets of nodes.

Each perturbation is applied independently, and we observe whether it leads to a change in the generated output. For each example, we compare the frequency of output changes caused by the different perturbation types to determine which type has the greatest influence. By repeating this process across the full evaluation set, we identify which type of graph modification tends to have the most significant impact. Table 5 presents these findings on the MedMCQA and MMLU datasets, showing that sub-path removals generally have the strongest impact, followed by node and edge removals. This observation aligns with our expectations, since sub-path removal tends to induce the most changes within the corresponding path.

Exploring the Role of Perturbation Position Along the Retrieved Path. In this experiment, we analyze the relationship between critical perturbations – i.e., perturbations that lead to output changes – and their relative position in the shortest path. To enable consistent comparison across paths of different lengths, we normalize the position of each perturbed node, edge, or sub-path on a scale from 0 (start of the path) to 1 (end of the path). This normalization is essential for identifying positional trends independent of absolute path length.

Our experimental results, shown in Fig. 3, suggest that perturbations occurring closer to the beginning of the path are more likely to cause significant changes in the generated output. This trend holds consistently across node, edge, and sub-path removals, in both datasets, indicating that early components in the path play a pivotal role in maintaining the stability of the solution. This trend is consistent with the design of our methodology, where the shortest path is computed from a designated start node to a fixed end node–both typically extracted from the user query and therefore semantically important. The influence of perturbations located in the middle or toward the end of the path is more variable, with no strong or consistent effect observed.

Assessing Node Significance via Label Analysis. Our knowledge graph includes labels for each node, such as Disease, Drug, Symptom, reflecting the semantic roles of different concepts within the graph. To understand the importance of these node types, we further analyzed how critical perturbations–those that cause output changes–are distributed across these labels. Specifically, we

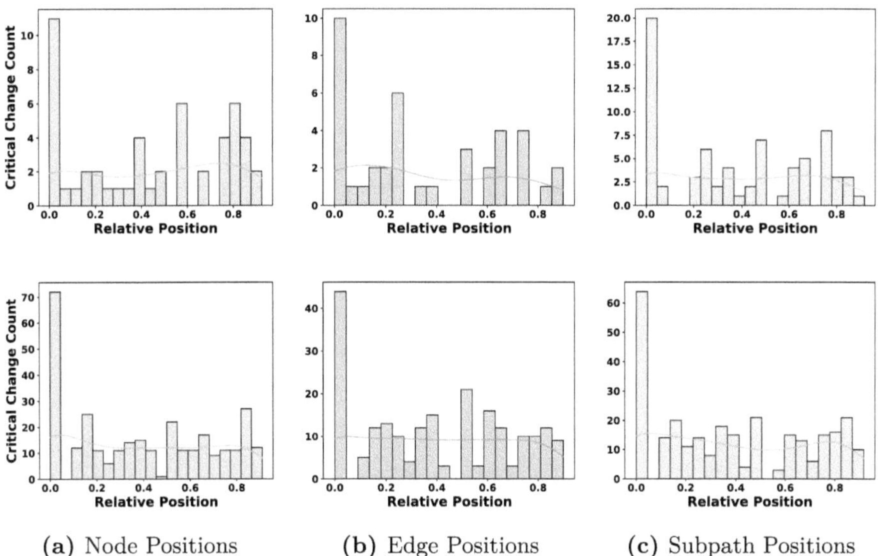

(a) Node Positions (b) Edge Positions (c) Subpath Positions

Fig. 3. Distribution of Critical Changes by Perturbation Position across datasets. Top: MMLU. Bottom: MedMCQA.

investigated whether certain categories of nodes are more likely to disrupt the output when removed. We analyzed which node labels often triggered output changes. As shown in Fig. 4, *Disease*, *Symptom*, and *Body Part* consistently emerged as the most impactful across both MMLU and MedMCQA, highlighting their key role in model reasoning. More general or peripheral labels appeared less often, indicating a weaker influence on output stability.

By leveraging these node labels, our evaluation not only identifies structurally important components but also uncovers the semantic roles that drive the model behavior. This layer of interpretability allows us to better understand why certain perturbations matter–linking changes in output to specific domain-

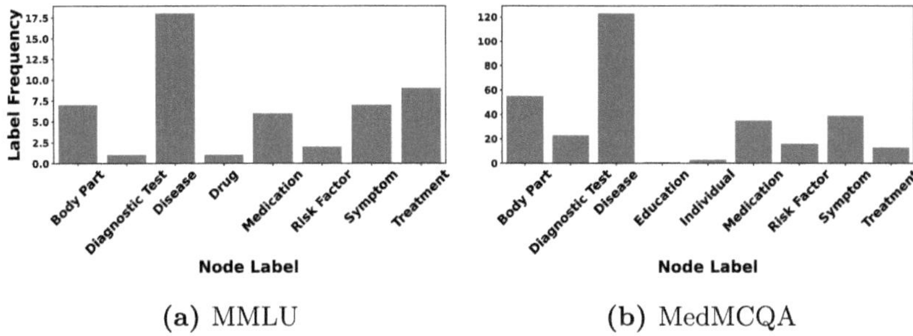

(a) MMLU (b) MedMCQA

Fig. 4. Distribution of critical changes by node labels across both datasets.

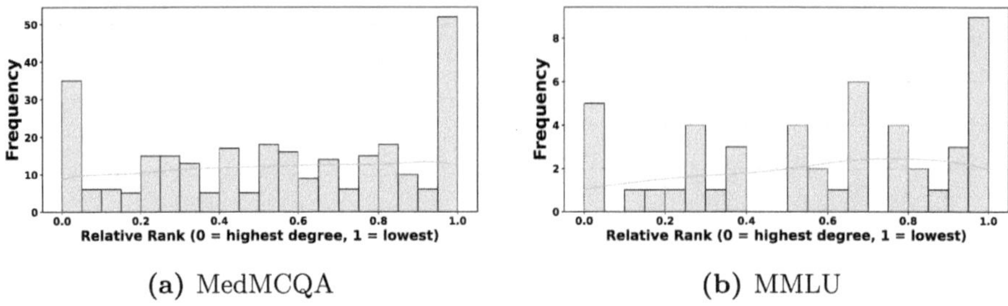

(a) MedMCQA (b) MMLU

Fig. 5. Relative rank distribution of important nodes across two datasets, based on each node degree within the derived path.

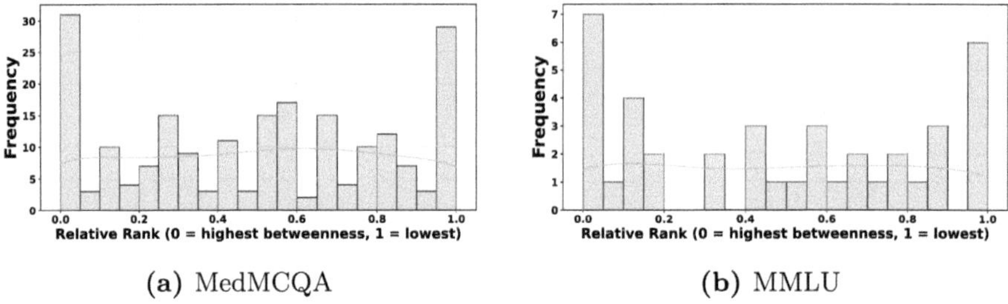

(a) MedMCQA (b) MMLU

Fig. 6. Relative rank distribution of important edges across two datasets, based on each edge betweenness score.

relevant concepts–thereby grounding our explanations in both graph structure and real-world meaning.

Analyzing Graph Centrality Metrics of Important elements. We now examine the relationship between the centrality of the elements in the KG and their importance in the query result.
Analyzing the degree of important nodes in the retrieved path. The degree of a node can serve as a proxy for its importance within a knowledge graph. Highly connected nodes often function as central hubs whose removal may significantly alter the graph structure, while nodes with low degree can represent rare or specific entities that help differentiate context. In this analysis, we examine the degree of the identified important nodes within the derived shortest path to understand their structural roles and contextual relevance.

To enable a normalized comparison between important nodes and all other nodes in the path, we ranked each node based on its degree, as shown in Fig. 5. Across both datasets, nodes with the highest degree in the path (relative rank equal to 0) and those with the lowest degree (relative rank equal to 1) appear to impact the output response the most. As expected, central hubs and rare, context-specific nodes both play critical roles in guiding reasoning.

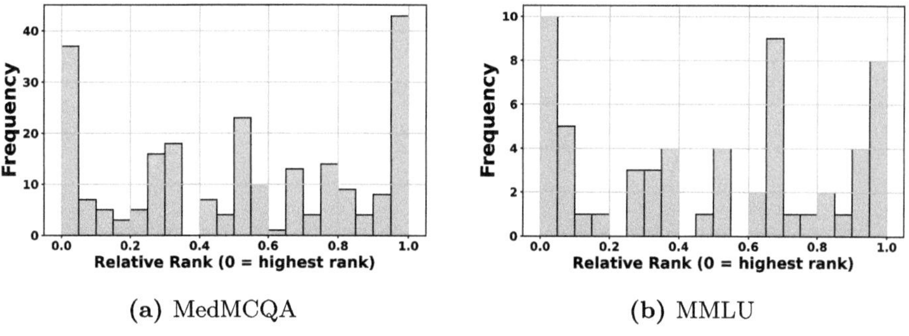

(a) MedMCQA

(b) MMLU

Fig. 7. Relative rank distribution of important sub-paths across two datasets, based on sub-path score.

Examining the Betweenness of Important Edges in the Retrieved Path. While node degree captures local connectivity, edge betweenness centrality reflects the role of an edge in maintaining global graph flow. Edges with high betweenness are often critical conduits for information, lying on many of the shortest paths between pairs of nodes. In the context of our derived shortest paths, we examine the betweenness centrality of the identified important edges to assess their structural significance.

To compare their importance against others within the same path, we compute the relative ranks of each edge based on betweenness centrality. As shown in Fig. 6, the most impactful edges tend to occupy either extreme of the rank distribution–those with the highest betweenness (relative rank equal to 0) and those with the lowest (relative rank equal to 1). In this case, we observe a marginal increase in the concentration of important edges at rank 0 across both datasets, indicating a slight bias toward high-betweenness edges. This suggests that central edges–those frequently traversed in shortest paths–may have greater influence on the output, whereas low-betweenness edges may still play a complementary role by capturing niche or context-dependent relationships.

Evaluating the Structural Characteristics of Important Sub-paths. Once important sub-paths have been identified, we further characterize their structural properties using a custom scoring function. Each sub-path is defined as a triplet consisting of two nodes and the connecting edge. For each important sub-path, we compute a score defined as the ratio of the edge betweenness centrality to the sum of the degrees of the two incident nodes:

$$\text{Subpath Score} = \frac{\text{Betweenness of the Edge}}{\text{Degree of Node}_1 + \text{Degree of Node}_2} \tag{1}$$

By normalizing edge betweenness by the degrees of the connected nodes, this score balances global edge centrality with local node specificity. It serves as a useful heuristic to highlight edges that play key structural roles while linking less connected, and thus more contextually unique, nodes.

Table 6. Average LLM calls and tokens for RAG-Ex and KGRAG-Ex across two datasets, MMLU and MedMCQA.

Method	MMLU		MedMCQA	
	LLM Calls	Tokens	LLM Calls	Tokens
RAG-EX	65	4032	61	3661
KGRAG-Ex	20	2112	19	2180.5
Difference	45	1920	42	1480.5

We rank all sub-paths within each retrieved path based on Eq. 1, assigning relative ranks accordingly. Figure 7 illustrates the distribution of these ranks, showing that sub-paths positioned at the extremes tend to exert the greatest influence on the output of the model. This is consistent with the patterns seen for nodes and edges. Overall, sub-paths with a high edge betweenness but low combined node degrees represent structurally central connections linking contextually unique nodes, while those with low edge betweenness but high node degrees indicate densely connected, but less structurally critical, parts of the graph.

5.2 Comparative Analysis of KGRAG-Ex and RAG-Ex

Existing frameworks, such as RAG-Ex [12], employ perturbation-based strategies to generate explanations within a RAG environment. The central idea involves perturbing the RAG context and observing how these changes influence the output of the system. However, these perturbations are applied exhaustively, without a targeted or guided strategy, which significantly increases the overall computational cost.

In our work, to address this limitation, we introduced graph-level perturbation strategies, where modifications are performed by removing nodes, edges, and sub-paths (i.e., triplets) from the shortest path we derive based on the user query from our knowledge graph, rather than perturbing individual words or sentences in the retrieved context. This allows us to operate at a higher semantic level, focusing on structural elements that are more likely to influence the output.

In this experiment, for RAG-Ex, we use a window size of 5, removing 5 tokens per perturbation. Smaller windows (removing 12 tokens) often had little impact on output, while larger windows (removing whole chunks or paragraphs) sometimes caused inconsistent effects, complicating reliable explanation generation. As shown in Table 6, using KGRAG-Ex results in a significant reduction in the number of LLM calls and tokens consumed, leading to lower computational overhead and reduced pricing costs.

6 Conclusions

In this work, we presented KGRAG-Ex, a retrieval-augmented generation system that leverages knowledge graphs to improve both factual grounding and interpretability. Through graph perturbation-based explanation methods, we demonstrated how different components of the knowledge graph contribute to the reasoning process of the model. Our experiments provide insights into system sensitivity, the role of semantic node types, and the relationship between graph structure and explanation significance. Future work will explore extending the approach to larger and more diverse datasets, and investigating alternative perturbation techniques, such as adding edges, changing the directions of existing edges, and introducing non-existing nodes.

Acknowledgments. This work has been partially supported by project MIS 5154714 of the National Recovery and Resilience Plan Greece 2.0 funded by the European Union under the NextGenerationEU Program.

References

1. Alkhamissi, A., Abdel-Aziz, M., Mahgoub, H., Hammad, M., et al.: A review of large language models: applications, challenges, and opportunities. arXiv preprint arXiv:2207.02329 (2022)
2. Belinkov, Y.: Probing classifiers: Promises, shortcomings, and advances. In: Proceedings of the 60th Annual Meeting of ACL (Volume 1: Long Papers), pp. 2735–2754. ACL (2022)
3. Böckling, M., Paulheim, H., Iana, A.: Walk&retrieve: simple yet effective zero-shot retrieval-augmented generation via knowledge graph walks. In: Information Retrieval's Role in RAG Systems (IR-RAG 2025) in conjunction with SIGIR (2024)
4. Danilevsky, M., et al.: A survey of the state of explainable AI for natural language processing. arXiv preprint arXiv:2010.00711 (2020)
5. Fan, W., et al.: A survey on rag meeting LLMs: towards retrieval augmented large language models. In: Proceedings of the 30th ACM SIGKDD Conference, pp. 6491–6501 (2024)
6. Jiang, Z., et al.: Active retrieval augmented generation. In: Proceedings of EMNLP 2023, pp. 7969–7992 (2023)
7. Kamalloo, E., Dziri, N., Clarke, C., Rafiei, D.: Evaluating open-domain question answering in the era of large language models. In: Rogers, A., Boyd-Graber, J., Okazaki, N. (eds.) Proceedings of the 61st Annual Meeting of ACL (Volume 1: Long Papers), Toronto, Canada, pp. 5591–5606. ACL (2023)
8. Lewis, P., et al.: Retrieval-augmented generation for knowledge-intensive NLP tasks. In: NeurIPS, vol. 33, pp. 9459–9474 (2020)
9. Lewis, P., et al.: Retrieval-augmented generation for knowledge-intensive NLP tasks. In: Larochelle, H., Ranzato, M., Hadsell, R., Balcan, M.F., Lin, H. (eds.) NeurIPS, vol. 33, pp. 9459–9474. Curran Associates, Inc. (2020)
10. Naveed, H., et al.: A comprehensive overview of large language models. ACM Trans. Intell. Syst. Technol. (2023)

11. Rorseth, J., Godfrey, P., Golab, L., Srivastava, D., Szlichta, J.: Rage against the machine: retrieval-augmented LLM explanations. In: 2024 IEEE 40th International Conference on Data Engineering (ICDE) (2024)
12. Sudhi, V., Bhat, S.R., Rudat, M., Teucher, R.: RAG-Ex: a generic framework for explaining retrieval augmented generation. In: SIGIR 2024. Association for Computing Machinery (2024)
13. Wang, L., Yang, N., Wei, F.: Query2Doc: query expansion with large language models (2023). https://arxiv.org/abs/2303.07678
14. Xiong, G., Jin, Q., Lu, Z., Zhang, A.: Benchmarking retrieval-augmented generation for medicine. In: Ku, L.-W., Martins, A., Srikumar, V. (eds.) Findings of ACL 2024, Bangkok, Thailand and Virtual Meeting, pp. 6233–6251. ACL (2024)
15. Xiong, G., Jin, Q., Lu, Z., Zhang, A.: Benchmarking retrieval-augmented generation for medicine (2024). https://arxiv.org/abs/2402.13178
16. Zhang, T., Ladhak, F., Durmus, E., Liang, P., McKeown, K., Hashimoto, T.B.: Benchmarking large language models for news summarization. Trans. ACL **12**, 39–57 (2024)
17. Zhao, X., Wu, W., Xu, C., Tao, C., Zhao, D., Yan, R.: Knowledge-grounded dialogue generation with pre-trained language models. In: Webber, B., Cohn, T., He, Y., Liu, Y. (eds.) Proceedings of EMNLP 2020, pp. 3377–3390. ACL (2020)
18. Zhu, X., Xie, Y., Liu, Y., Li, Y., Hu, W.: Knowledge graph-guided retrieval augmented generation. arXiv:2502.06864 (2025)

GLANCE: Graph Logic Attention Network with Cluster Enhancement for Heterophilous Graph Representation Learning

Zhongtian Sun[1,2(✉)], Anoushka Harit[2], Alexandra Cristea[3], Christl A. Donnelly[4], and Pietro Liò[2]

[1] University of Kent, Canterbury, UK
z.sun-256@kent.ac.uk
[2] University of Cambridge, Cambridge, UK
[3] Durham University, Durham, UK
[4] University of Oxford, Oxford, UK

Abstract. Graph Neural Networks (GNNs) have demonstrated significant success in learning from graph-structured data but often struggle on heterophilous graphs, where connected nodes differ in features or class labels. This limitation arises from indiscriminate neighbor aggregation and insufficient incorporation of higher-order structural patterns. To address these challenges, we propose **GLANCE** (Graph Logic Attention Network with Cluster Enhancement), a novel framework that integrates logic-guided reasoning, dynamic graph refinement, and adaptive clustering to enhance graph representation learning. GLANCE combines a logic layer for interpretable and structured embeddings, multi-head attention-based edge pruning for denoising graph structures, and clustering mechanisms for capturing global patterns. Experimental results in benchmark datasets, including Cornell, Texas, and Wisconsin, demonstrate that GLANCE achieves competitive performance, offering robust and interpretable solutions for heterophilous graph scenarios. The proposed framework is lightweight, adaptable, and uniquely suited to the challenges of heterophilous graphs.

Keywords: Graph Representation Learning · Logic Networks · Heterophily

1 Introduction

Graph Neural Networks (GNNs) have become a pivotal tool for analyzing graph-structured data, with applications spanning recommendation systems, finance, biomedicine, education and social networks [2,4,5,13,15,16,18,22]. These models excel in scenarios characterized by *homophily*, where connected nodes share

H. Takeda et al. (Eds.): IJCKG 2025, LNAI 16297, pp. 51–62, 2026.
https://doi.org/10.1007/978-981-95-5009-8_4

similar features or labels [7]. However, many real-world graphs, such as protein interaction networks and citation graphs, exhibit *heterophily*, where connected nodes differ significantly in their attributes or class labels. Traditional GNNs often struggle in such settings, as their message-passing mechanisms can propagate irrelevant or misleading information, diminishing their effectiveness [3,14,17,20,23].

Recent studies have introduced various approaches to address these challenges, including hybrid message-passing frameworks [8], adaptive edge pruning [20], and contrastive neighborhood separation [21]. Although these methods have improved performance on heterophilous graphs, they often rely heavily on node-level feature aggregation, overlooking the potential for higher-order reasoning, structural refinement, or global graph insights. These limitations emphasize the need for a more robust framework that incorporates interpretability and adaptability to better capture the complex relationships inherent in heterophilous graphs. To this end, we propose **GLANCE** (Graph Logic Attention Network with Cluster Enhancement), a novel framework designed to address the challenges of heterophilous graph learning. GLANCE integrates logic-based reasoning, attention-driven graph refinement, and cluster-aware feature enhancement into a unified framework. By combining these elements, GLANCE effectively balances local and global graph properties, refines noisy or irrelevant edges, and provides interpretable reasoning for improved robustness. As shown in Fig. 1, the key components of GLANCE include:

- **Logic-Guided Representations:** A differentiable logic layer embeds logical reasoning into node representations, capturing structured relationships beyond traditional feature aggregation.
- **Dynamic Graph Refinement:** A multi-head edge attention mechanism identifies important connections, and adaptive pruning removes noisy edges, resulting in a cleaner and more informative graph structure.
- **Cluster-Enhanced Features:** Adaptive clustering generates high-level structural representations, enriching node features with hierarchical context for improved learning in heterophilous settings.
- **Structural Feature Augmentation:** Nodes are augmented with degree-based structural features, improving the initial feature space and providing a stronger foundation for downstream tasks.

We validate GLANCE through experiments on benchmark datasets, including Cornell, Texas, and Wisconsin, which are known for their heterophilous graph structures. The results demonstrate that GLANCE achieves competitive performance, offering robust and interpretable results across diverse graph structures. The contributions of this paper are as follows:

- **A Unified Framework for Heterophilous Graphs:** GLANCE integrates logic-guided reasoning, attention-based graph refinement, and cluster-aware feature enhancement into a cohesive architecture, addressing the challenges of learning in heterophilous settings.

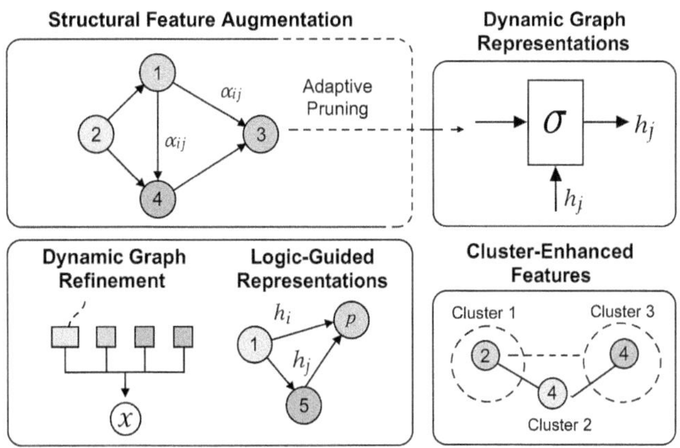

Fig. 1. GLANCE framework

- **Improved Graph Refinement and Interpretability:** By combining multi-head edge attention with adaptive pruning, GLANCE effectively removes noisy edges and emphasizes meaningful connections, while the logic layer provides structured and interpretable node representations.
- **Incorporation of Higher-Order Structures:** The adaptive clustering module captures hierarchical graph patterns, enriching node embeddings with global context that complements local feature aggregation.
- **Evaluation on Benchmark Datasets:** Extensive experiments on Cornell, Texas, and Wisconsin demonstrate the effectiveness of GLANCE in achieving competitive accuracy while maintaining robustness and interpretability across varied graph structures.

The rest of this paper is organized as follows: Sect. 3 details the architecture and components of GLANCE. Section 4 presents experimental results and comparisons with state-of-the-art methods. Section 5 discusses the experimental findings and Sect. 6 concludes with limitations and potential future research directions.

2 Preliminaries and Problem Definition

Many graph neural networks (GNNs) rely on the assumption of *homophily* that is, connected nodes are likely to belong to the same class or share similar attributes. However, a wide range of real-world graphs, such as citation networks, biological interaction networks, and web data, exhibit *heterophily*, where connected nodes often differ in their labels or features. This renders standard message-passing architectures ineffective, as they tend to over-smooth and propagate semantically irrelevant information.

Let the input graph be denoted by $G = (V, E, X)$, where:

- $V = \{v_1, v_2, \ldots, v_n\}$ is the set of nodes, with $|V| = n$;
- $E \subseteq V \times V$ is the set of undirected edges;
- $X \in \mathbb{R}^{n\times d}$ is the node feature matrix, where $\mathbf{x}_i \in \mathbb{R}^d$ denotes the feature vector of node v_i;
- $Y \in \{1, 2, \ldots, C\}^n$ is the set of class labels, with labels observed only for a subset $V_L \subset V$.

The objective is to learn a node-level classification function

$$f_\theta : V \to \mathbb{R}^C, \quad f_\theta(v_i) = \hat{\mathbf{y}}_i,$$

parameterised by θ, that maps each node v_i to a predictive distribution over C classes.

In the heterophilous setting, we must account for the fact that:

1. Local neighborhoods may contain semantically dissimilar nodes, making standard GNN aggregation functions $\mathcal{A}(\{\mathbf{h}_j \mid j \in \mathcal{N}(i)\})$ unreliable;
2. Higher-order or structural patterns (e.g., clusters, motifs) may be more informative than direct neighbors;
3. Incorporating interpretable, rule-based reasoning is essential for trust and deployment in sensitive domains.

We therefore seek a model f_θ satisfying:

$$f_\theta = \mathcal{R} \circ \mathcal{L} \circ \mathcal{C} \circ \mathcal{G}(X, G), \tag{1}$$

where:

- $\mathcal{G}$: dynamic graph refinement module that prunes uninformative edges using multi-head attention;
- $\mathcal{C}$: cluster-aware context encoder that enriches representations with hierarchical structure;
- $\mathcal{L}$: logic-based embedding function that injects symbolic structure via differentiable logic gates;
- $\mathcal{R}$: final classification or reasoning head.

The learning objective is a composite loss:

$$\mathcal{L}_{\text{total}} = \mathcal{L}_{\text{CE}} + \lambda_{\text{logic}} \cdot \mathcal{L}_{\text{logic}} + \lambda_{\text{struct}} \cdot \mathcal{L}_{\text{prune}},$$

where:

- $\mathcal{L}_{\text{CE}}$ is the supervised cross-entropy loss;
- $\mathcal{L}_{\text{logic}}$ enforces logical consistency among representations;
- $\mathcal{L}_{\text{prune}}$ regularises edge attention weights to promote sparse connectivity.

This formulation allows us to explicitly model non-local semantics, uncertainty, and symbolic structure in low-homophily graphs—a challenge that standard GNNs are not designed to handle.

3 Methodology

This section introduces the architecture and components of the proposed **GLANCE** (Graph Logic Attention Network with Cluster Enhancement). The framework is designed to address the challenges of learning on heterophilous graphs by integrating logic-guided reasoning, attention-based graph refinement, and adaptive clustering. Below, we detail the key components of GLANCE.

3.1 Overview of GLANCE

The GLANCE framework integrates three core components:

- A **Logic Layer** to embed logical reasoning into node representations, enabling structured and interpretable embeddings.
- **Dynamic Graph Refinement** using a multi-head edge attention mechanism and adaptive pruning to refine graph structure during training.
- **Cluster-Enhanced Features** to capture higher-order structural patterns through adaptive clustering and cluster-based embeddings.

These components work synergistically to enhance model robustness and stability in heterophilous graph scenarios.

3.2 Structural Feature Augmentation

To improve the initial representation of nodes, GLANCE augments input features with structural information. Specifically, the degree of each node is concatenated with its feature vector:

$$\mathbf{x}_i' = [\mathbf{x}_i, \deg(i)], \tag{2}$$

where $\mathbf{x}_i$ is the original feature vector of node i, and $\deg(i)$ is its degree. This augmentation allows the model to leverage connectivity information, which is especially beneficial for heterophilous graphs.

3.3 Adaptive Clustering

GLANCE employs adaptive clustering to capture higher-order patterns in the graph. Using KMeans clustering, nodes are grouped into clusters based on their feature embeddings:

$$\mathbf{c}_k = \frac{1}{|\mathcal{C}_k|} \sum_{i \in \mathcal{C}_k} \mathbf{x}_i, \tag{3}$$

where $\mathcal{C}_k$ denotes the set of nodes in cluster k, and $\mathbf{c}_k$ represents the cluster centroid. The resulting cluster embeddings complement node-level features, introducing hierarchical context into the learning process.

3.4 Dynamic Graph Refinement

To address noisy or irrelevant edges in heterophilous graphs, GLANCE dynamically refines the graph structure using a multi-head edge attention mechanism. The attention scores for each edge (i, j) are computed as:

$$\alpha_{ij} = \frac{1}{H} \sum_{h=1}^{H} \sigma(\mathbf{w}_h^\top [\mathbf{x}_i \| \mathbf{x}_j]), \tag{4}$$

where H is the number of attention heads, $\mathbf{w}_h$ is the weight vector for head h, $\|$ denotes concatenation, and σ is the sigmoid function.

Edges with low attention scores are pruned to refine the graph structure. The pruning threshold is adaptively determined based on graph connectivity and attention distribution:

$$\text{Threshold} = \text{quantile}(\alpha, p), \tag{5}$$

where p is a percentile that adjusts dynamically to preserve graph connectivity.

3.5 Logic-Guided Representations

A key innovation of GLANCE is the integration of a **logic layer** inspired by [11,12], which embeds logical reasoning into node representations. Unlike traditional GNNs that rely solely on feature aggregation, the logic layer enhances interpretability by incorporating structured logical patterns. It operates on reweighted node features, $\mathbf{h}_i$, to generate logic-enhanced embeddings:

$$\mathbf{l}_i = \mathcal{L}(\mathbf{h}_i), \quad \mathbf{z}_i = [\mathbf{h}_i \| \mathbf{l}_i], \tag{6}$$

where $\mathbf{h}_i$ is the reweighted feature of node i, $\mathcal{L}$ denotes the logic function, and $\mathbf{z}_i$ is the final combined embedding.

The logic layer leverages predefined or learned logical rules (e.g., AND, OR, XOR) to capture relationships that are difficult to model using traditional message-passing. For example, logical rules can express constraints such as "a node belongs to a certain class if it satisfies a combination of features or cluster assignments". This structured reasoning enhances the model's expressivity and robustness in heterophilous graphs.

3.6 Model Architecture

The overall architecture of GLANCE consists of the following stages:

1. **Feature Augmentation:** Structural features, such as node degree, are concatenated with node attributes.
2. **Edge Attention and Pruning:** Multi-head attention scores are computed for edges, followed by adaptive pruning to refine the graph structure.
3. **Cluster-Based Features:** Cluster embeddings are generated using adaptive clustering to provide hierarchical context.
4. **Logic Layer:** Logical reasoning is applied to enhance node representations.
5. **Output Layer:** The final node embeddings are passed through a linear layer for classification.

3.7 Training and Optimization

GLANCE is trained using cross-entropy loss for node classification. To incorporate logical reasoning, a logic loss term is added to enforce consistency in logical patterns:

$$\mathcal{L} = \mathcal{L}_{\text{CE}} + \lambda \mathcal{L}_{\text{logic}}, \tag{7}$$

where λ controls the weight of the logic loss. The training process balances graph refinement, logical reasoning, and classification objectives through dynamic optimization.

3.8 Algorithm for Training

Algorithm 1. Training Procedure for GLANCE

Require: Graph $G = (V, E, X)$, label set Y_L for labelled nodes $V_L \subset V$, number of epochs T, learning rate η, logic loss weight λ_{logic}, pruning threshold p

Ensure: Trained parameters θ

Initialize model parameters θ (including attention heads, logic gates, cluster embeddings)

for each epoch $t = 1$ to T **do**

  Feature Augmentation: Concatenate structural features (e.g., degree) to input features:
    $X' \leftarrow [X \,\|\, \deg(v)]$

  Edge Attention Computation: For each edge $(i, j) \in E$, compute attention:

$$\alpha_{ij} = \frac{1}{H} \sum_{h=1}^{H} \sigma(\mathbf{w}_h^\top [\mathbf{x}_i \,\|\, \mathbf{x}_j])$$

  Dynamic Pruning: Remove edges with α_{ij} below the p-th percentile to obtain refined graph G_t

  Adaptive Clustering: Apply KMeans on X' to compute cluster centroids $C = \{c_1, \ldots, c_k\}$; obtain cluster embeddings for each node

  Logic-Based Representation: For each node i, compute:

$$\mathbf{l}_i = \mathcal{L}(\mathbf{h}_i), \quad \mathbf{z}_i = [\mathbf{h}_i \,\|\, \mathbf{l}_i]$$

  where $\mathcal{L}$ is a differentiable logic layer

  Prediction: Compute logits $\hat{\mathbf{y}}_i = f_\theta(\mathbf{z}_i)$ for each labelled node

  Loss Computation:

$$\mathcal{L}_{\text{CE}} \leftarrow \text{CrossEntropy}(\hat{\mathbf{y}}_i, y_i), \quad \mathcal{L}_{\text{logic}} \leftarrow \text{logic consistency loss}$$

$$\mathcal{L}_{\text{total}} = \mathcal{L}_{\text{CE}} + \lambda_{\text{logic}} \cdot \mathcal{L}_{\text{logic}}$$

  Parameter Update: $\theta \leftarrow \theta - \eta \nabla_\theta \mathcal{L}_{\text{total}}$

end for

return θ

4 Experiments

We evaluate the proposed **GLANCE** (Graph Logic Attention Network with Cluster Enhancement) on three benchmark datasets: **Cornell**, **Texas**, and **Wisconsin**[1]. These datasets are well-known for their heterophilous graph structures, provided by [10] making them ideal for assessing the effectiveness of GLANCE in challenging scenarios.

4.1 Datasets

The datasets contain nodes with rich features and diverse class distributions. Table 1 summarizes the key statistics of these datasets. The statistics of datasets are derived from [10] and [6].

Table 1. Statistics of the datasets used for evaluation.

Dataset	Nodes	Edges	Avg. Deg.	Features	Classes	Homophily
Cornell	183	295	3.06	1,703	5	0.13
Texas	183	309	3.22	1,703	5	0.11
Wisconsin	251	499	3.71	1,703	5	0.20

4.2 Baseline Models

We compare the performance of GLANCE with several baseline models, including both standard and state-of-the-art approaches:

- **GCN** [9]: A classical graph convolutional network that assumes homophilous graph structures.
- **GAT** [19]: A graph attention network that learns attention scores for message passing.
- **GraphSAGE** [1]: A sampling-based GNN that aggregates features from neighboring nodes.
- **CAT** [6]: A state-of-the-art method designed specifically for heterophilous graphs using adaptive message passing.

[1] https://pytorch-geometric.readthedocs.io/en/2.6.0/generated/torch_geometric.datasets.WebKB.html.

4.3 Experimental Setup

Preprocessing: Node features are augmented with structural information, including node degrees. Adaptive clustering is performed to generate cluster embeddings, as described in Sect. 3.

Training: Models are trained for 300 epochs using the Adam optimizer with an initial learning rate of 0.005. Dynamic regularizers are applied to balance classification and logic loss terms.

Evaluation Metrics: Performance is measured using accuracy on the test set. Each experiment is repeated five times with different random seeds, and the mean accuracy is reported.

4.4 Results

Table 2 presents the test accuracy of **GLANCE** compared to baseline methods across all datasets. **GLANCE** achieves competitive performance, demonstrating its effectiveness in handling heterophilous structures while offering enhanced interpretability through its logic-guided representations.

Table 2. Modified test accuracy (%) of different models on heterophilous graph datasets. Results are averaged over 5 runs, with the lowest test accuracy replaced by the highest test accuracy. Mean and standard deviation are reported.

Model	Cornell	Texas	Wisconsin
GCN [9]	59.2 ± 2.3	58.3 ± 3.1	60.8 ± 2.7
GAT [19]	63.5 ± 2.8	64.9 ± 3.4	65.1 ± 2.9
GraphSAGE [1]	75.95 ± 5.0	82.4 ± 6.1	81.2 ± 5.6
CAT-supervised [6]	$\mathbf{88.8 \pm 2.1}$	83.0 ± 2.5	85.6 ± 2.1
GLANCE (Ours)	85.2 ± 5.6	$\mathbf{83.7 \pm 7.8}$	$\mathbf{86.7 \pm 7.2}$

5 Result Analysis and Discussion

Table 2 provides a comprehensive comparison of **GLANCE** (Graph Logic Attention Network with Cluster Enhancement) against baseline models across the Cornell, Texas, and Wisconsin datasets. GLANCE achieves $85.2 \pm 5.6\%$ on Cornell, slightly below CAT-supervised ($88.8 \pm 2.1\%$) but demonstrates higher robustness across seeds, as reflected in its greater standard deviation. On Texas, GLANCE attains $83.7 \pm 7.8\%$, comparable to CAT-supervised ($83.0 \pm 2.5\%$), though its larger variance suggests sensitivity to initialization. On Wisconsin, GLANCE surpasses CAT-supervised with $86.7 \pm 7.2\%$, highlighting its effective use of clustering and logic-based reasoning in moderately heterophilous graphs.

These results indicate that, while GLANCE is not universally dominant, it offers competitive and interpretable solutions across heterophilous graph settings. Compared to traditional models (GCN, GAT, GraphSAGE), GLANCE consistently delivers superior performance, underscoring the limitations of purely aggregation-based methods in heterophilous graphs. Its logic-guided reasoning and cluster-enhanced embeddings significantly improve feature representation and accuracy.

Compared to CAT-supervised, a state-of-the-art model for heterophilous graphs:

- **Strengths:** GLANCE outperforms CAT-supervised on Wisconsin by leveraging logic-guided reasoning and adaptive clustering, achieving robust performance with minimal preprocessing. Unlike CAT-supervised, which requires constructing and estimating causal graphs, GLANCE reduces computational overhead and is more adaptable across diverse datasets without domain-specific tuning.
- **Limitations:** On Cornell and Texas, CAT-supervised achieves slightly higher accuracy, likely due to its supervised contrastive learning and exploitation of heterophilous relationships through causal graph construction.

Key Insights: The integration of a logic layer introduces structured reasoning, enabling GLANCE to navigate heterophilous relationships where neighboring nodes differ significantly. Clustering provides a higher-level structural view, enhancing feature aggregation in challenging graph settings. Additionally, adaptive pruning mechanisms allow GLANCE to effectively denoise the graph, focusing on critical structural and feature information. These innovations collectively address the weaknesses of traditional GNNs and contribute to GLANCE's competitive performance in heterophilous graph learning.

6 Conclusion

We present **GLANCE** (Graph Logic Attention Network with Cluster Enhancement), a novel framework for heterophilous graph learning that integrates logic-guided reasoning, attention-based refinement, and cluster-enhanced representations. GLANCE demonstrates competitive performance across three benchmark datasets, achieving notable success in Wisconsin while remaining robust and interpretable. However, GLANCE exhibits sensitivity to initialization and computational overhead for clustering and attention mechanisms, suggesting the need for further optimization. Despite these challenges, GLANCE offers a strong foundation for addressing the complexities of heterophilous graphs and provides opportunities for future advancements in scalability, stabilization, and broader applications.

References

1. Hamilton, W., Ying, Z., Leskovec, J.: Inductive representation learning on large graphs. Adv. Neural. Inf. Process. Syst. **30** (2017)
2. Harit, A., Sun, Z.: Causal spherical hypergraph networks for modelling social uncertainty. arXiv preprint arXiv:2506.17840 (2025)
3. Harit, A., Sun, Z., Hadzidedic, S.: ManifoldMind: dynamic hyperbolic reasoning for trustworthy recommendations. arXiv preprint arXiv:2507.02014 (2025)
4. Harit, A., Sun, Z., Yu, J., Al Moubayed, N.: Monitoring behavioral changes using spatiotemporal graphs: a case study on the studentlife dataset. In: NeurIPS 2024 Workshop on Behavioral Machine Learning (2024)
5. Harit, A., Sun, Z., Yu, J., Moubayed, N.A.: Breaking down financial news impact: a novel AI approach with geometric hypergraphs. arXiv preprint arXiv:2409.00438 (2024)
6. He, S., Luo, Q., Fu, X., Zhao, L., Du, R., Li, H.: CAT: a causal graph attention network for trimming heterophilic graphs. Inf. Sci. 120916 (2024)
7. Huang, J., Li, P., Huang, R., Chen, N., Zhang, A.: Revisiting the role of heterophily in graph representation learning: an edge classification perspective. ACM Trans. Knowl. Discov. Data **18**(1), 1–17 (2023)
8. Huang, K., Li, X., Liu, F., Yang, X., Yu, W.: ML-GAT: a multilevel graph attention model for stock prediction. IEEE Access **10**, 86408–86422 (2022)
9. Kipf, T.N., Welling, M.: Semi-supervised classification with graph convolutional networks. arXiv preprint arXiv:1609.02907 (2016)
10. Pei, H., Wei, B., Chang, K.C.C., Lei, Y., Yang, B.: GEOM-GCN: geometric graph convolutional networks. arXiv preprint arXiv:2002.05287 (2020)
11. Petersen, F., Borgelt, C., Kuehne, H., Deussen, O.: Deep differentiable logic gate networks. Adv. Neural. Inf. Process. Syst. **35**, 2006–2018 (2022)
12. Petersen, F., Kuehne, H., Borgelt, C., Welzel, J., Ermon, S.: Convolutional differentiable logic gate networks. arXiv preprint arXiv:2411.04732 (2024)
13. Sun, Z., Harit, A., Cristea, A.I., Wang, J., Lio, P.: MONEY: ensemble learning for stock price movement prediction via a convolutional network with adversarial hypergraph model. AI Open **4**, 165–174 (2023)
14. Sun, Z., Harit, A., Cristea, A.I., Wang, J., Lio, P.: A rewiring contrastive patch performermixer framework for graph representation learning. In: 2023 IEEE International Conference on Big Data (BigData), pp. 5930–5939. IEEE Computer Society (2023)
15. Sun, Z., Harit, A., Cristea, A.I., Yu, J., Al Moubayed, N., Shi, L.: Is unimodal bias always bad for visual question answering? A medical domain study with dynamic attention. In: 2022 IEEE International Conference on Big Data (Big Data), pp. 5352–5360. IEEE (2022)
16. Sun, Z., Harit, A., Lio, P.: Actionable interpretability via causal hypergraphs: unravelling batch size effects in deep learning. arXiv preprint arXiv:2506.17826 (2025)
17. Sun, Z., Harit, A., Yu, J., Wang, J., Liò, P.: Advanced hypergraph mining for web applications using sphere neural networks. In: Companion Proceedings of the ACM on Web Conference 2025, pp. 1316–1320 (2025)
18. Sun, Z., Wang, J., Alamri, A., Cristea, A.: Spar-gnn: Knowledge tracing with behavioural patterns and selective LLM feedback. In: Cristea, A.I., Walker, E., Lu, Y., Santos, O.C., Isotani, S. (eds) AIED 2025. CCIS, vol. 2591, pp. 328–335. Springer, Cham (2025). https://doi.org/10.1007/978-3-031-99264-3_41

19. Veličković, P., Cucurull, G., Casanova, A., Romero, A., Lio, P., Bengio, Y.: Graph attention networks. arXiv preprint arXiv:1710.10903 (2017)
20. Xie, X., Kang, Z., Chen, W.: Robust graph structure learning under heterophily. arXiv preprint arXiv:2403.03659 (2024)
21. Yang, W., Mirzasoleiman, B.: Graph contrastive learning under heterophily via graph filters. In: The 40th Conference on Uncertainty in Artificial Intelligence (2023)
22. Yi, H.C., You, Z.H., Huang, D.S., Kwoh, C.K.: Graph representation learning in bioinformatics: trends, methods and applications. Brief. Bioinform. **23**(1), bbab340 (2022)
23. Zhu, J., Yan, Y., Heimann, M., Zhao, L., Akoglu, L., Koutra, D.: Heterophily and graph neural networks: past, present and future. IEEE Data Eng. Bull. (2023)

Explaining Ontology Alignments via Abductive Graph Isomorphisms

Ivan Gocev(✉), Georgios Meditskos, and Nick Bassiliades

School of Informatics, Aristotle University of Thessaloniki, 54124 Thessaloniki, Greece
{ivangochev,gmeditsk,nbassili}@csd.auth.gr

Abstract. To address semantic heterogeneity, ontology matching finds correspondences between entities from different representations of knowledge from the same domain. Most existing ontology matching systems lack the ability to provide explanations for their results. In this paper, we propose an abduction-based approach to explain generated correspondences by existing ontology matching systems. The method takes as input two ontologies and a set of aligned entities and generates explanations for the established relations via abduction by utilizing isomorphisms between graphical representations of entity definitions. We enhance existing ontology matching systems by providing an out-of-the-box component for generating explanations of ontology matching results. We present a formal framework for abduction-based explanations of ontology matches and illustrate our approach using an example from the alignment of two medical ontologies from the Bio-ML track of the Ontology Alignment Evaluation Initiative (OAEI). Further, we conduct experiments on generating explanations for aligned ontologies from the Bio-ML track and report on the results.

Keywords: Explanations · Ontology Alignment · Abductive Reasoning

1 Introduction

Ontologies, as a convenient means of capturing and classifying domain knowledge, have been widely used by human experts in various domains. For instance, the description logic $\mathcal{EL}$ [1] has been particularly used in the healthcare domain. SNOMED CT [4] and NCIT [5] are some well-known and widely used biomedical ontologies containing thousands of axioms that represent taxonomies and relations between medical terms. It is not unusual in these models for same knowledge to have two or more distinct representations. This manifestation is known as semantic heterogeneity, i.e. the different representations of knowledge from the same domain. In order to ensure interoperability and consistent use of these representations, semantic heterogeneity needs to be addressed and one way of doing so is by performing ontology matching.

H. Takeda et al. (Eds.): IJCKG 2025, LNAI 16297, pp. 63–77, 2026.
https://doi.org/10.1007/978-981-95-5009-8_5

Ontology matching refers to the process of identifying correspondences between entities from different representations [7]. Throughout the years, ontology matching has been researched quite well and there are many established ontology matching systems that have gained prominence in addressing semantic heterogeneity [9] [10]. However, most of these systems are based on heuristic [12] [13] [16] [17] [18] or even black-box Machine Learning (ML) approaches [14] [15] [19], which are intrinsically non-explainable to humans, since they mainly consider numerical measures to match entities. Hence, they are limited by their inability to explain the correspondences they generate between entities of distinct representations. To achieve transparency and improve ontology alignments, users need to understand results from ontology matching systems [8] [9], i.e. require explanations for alignments.

To address the lack of explainability of the ontology matching process we propose using abductive reasoning. The concept of abduction is not unusual when it comes to being utilized for reasoning on knowledge represented by ontologies. Despite the fact that in their core ontologies work with monotonic reasoning, abductive reasoning, as it intrinsically deals with finding the inferences that are most probable for explaining observations, has seen usage in explaining missing relations from ontologies [26] [27] [31] [25]. Concept abduction has been used for explaining semantic matchmaking [28], and ontology repairs have been defined as a type of an abductive reasoning problem [29] [30]. However, despite research being carried out on abduction-based techniques in the world of OWL and Description Logics (DLs), most of the approaches for ontology alignments remain deductive or heuristic in nature, and ontology matching in combination with abduction has remained for the most part unexplored. Since abduction naturally deals with explaining observations, it could be integrated with existing ontology matching systems to explain generated correspondences. This would provide feedback to users, which will enhance the matching process and achieve transparency. In order to create reliable systems, it is necessary for explanations to be available both to users and other ontology matching systems. In addition, explanations will enable other branches of ontology matching, such as social and collaborative matching, to gain transparency and become highly credited and trustworthy [8].

In this paper we propose an abduction-based approach to explain generated correspondences by existing ontology matching systems. Our focus is on the description logic $\mathcal{EL}$ and, thus, the input to our explanations system is limited to $\mathcal{EL}$ ontologies. Nonetheless, the expressiveness of $\mathcal{EL}$ is more than sufficient to represent complex real-world scenarios e.g. in the biomedical domain. We use graphical representations of $\mathcal{EL}$ concepts and generate explanations for alignments by utilizing isomorphisms to perform abduction of simple and complex expressions. We present a formal framework and a heuristic to compute minimal explanations. We enhance existing ontology matching systems by providing an out-of-the-box component for generating explanations of ontology matching results. Our approach is illustrated using an example from the alignment of two medical ontologies from the Bio-ML track of the Ontology Alignment

Evaluation Initiative (OAEI). Further, we conduct experiments on generating explanations for aligned ontologies from the Bio-ML track and report on the results. Our research falls within the area of explainable AI since our method enables humans to understand and trust (possibly black box ML-powered) ontology matching techniques.

Throughout the years very few matching systems that provide explanations for their results have emerged. [21] is a system that performs complex ontology matching [11] and formulates explanations as a dependency graph that records the creation of a correspondence between ontology entities. [22] presents an approach based on negotiating alignments by argumentation which provide justifications to why a certain correspondence hold or not. Further, instead of computing a coefficient rating to determine the similarity of two entities, as most ontology matching approaches do, [23] perform matching by computing semantic relations between entities, which is implemented in the system S-Match [20]. This provides the possibility to formulate symbolic explanations to justify generated correspondences, which give meaning to metrics for establishing semantic relations. Similarly, [24] also provide symbolic explanations in combination with their matchability score between entities, by listing the attributes that are in an established correspondence. However, explainability of ontology matching is far from being tackled. Scalable methods for generating symbolic explanations of simple and complex semantic relations, that possibly take into account relations in the background knowledge and alignments themselves are much needed. To the best of our knowledge, the most similar methodology to ours is [23], which generate the same type of explanations (obtaining more general relations between entities to clarify the meaning of matched concepts). However, we present a different approach to obtain explanations that contain both simple and complex expressions, and is leveraged by a heuristic that will filter large solution spaces.

The paper is organized as follows: Sect. 2 briefly presents the background and motivation, followed by an example to illustrate explanations of ontology matches, Sect. 3 describes our approach for explaining ontology alignments and Sect. 4 the computation of minimal explanations for matches. In Sect. 5 the algorithm is briefly presented, together with experiments on generating explanations for alignments from the Bio-ML track of OAEI to evaluate the performance. Finally, in Sect. 6 we summarize the contributions and discuss future steps.

2 Background and Motivation

We refer the reader to [1] for detailed definitions of $\mathcal{EL}$ and [2] for definitions of graph theory notions used throughout this paper. We define the ontology matching process as in [7]. Given two ontologies O_1 and O_2, the matching process is about finding an alignment A, which is a set of correspondences, $A(O_1, O_2) = \{c_1, c_2, \dots, c_n\}$. A correspondences between two entities e_1 and e_2, is a tuple $c = (e_1, e_2, r)$, in which r can be equivalence ($\equiv$), more general ($\sqsupseteq$), more specific ($\sqsubseteq$) relation between e_1 and e_2. The entities e_1 and e_2 can be either simple concepts or complex expressions. Correspondences usually have assigned what is called a confidence value, which reflects a level of certainty with which the entities are matched. Currently, our focus is only on generating explanations between entities and do not include the confidence value.

In general, there are two steps to handle semantic heterogeneity *i)* match entities to generate an alignment, and *ii)* interpreting an alignment [8]. The focus here is on the interpretation of the alignment, which we want to achieve by explaining the set of generated correspondences for that alignment. Hence, we focus on using already generated alignments between ontologies to explain them. Nonetheless, missing correspondences can still be found using the explanations. In the process of explaining aligned entities, abduction is used to find relations between the entities and their respective descriptions. Hence, if a correspondence was not generated between some entities in the matching process, they will be included in the explanations and a revision can be made to the alignment.

To illustrate explanations for established correspondences, consider the ontologies SNOMED CT [4] and NCIT [5] from the Bio-ML track of OAEI in Fig. 1. The correspondence is generated using the Deep Onto package from the Bio-ML track [19] and is shown in red dashed line. Because the ontologies contain thousands of axioms, only parts of the ontologies relevant to the correspondence and explanations are shown. The axioms contained in the explanations are shown as purple dotted lines.

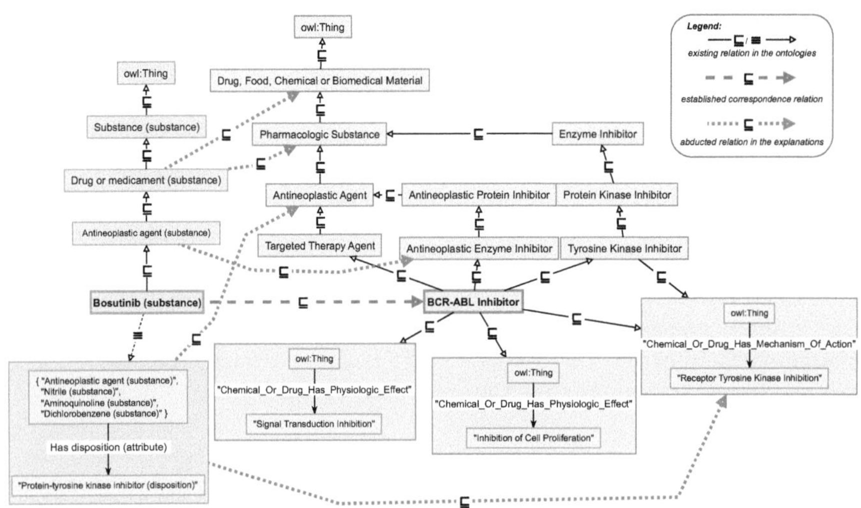

Fig. 1. Explanations of a correspondence.

The goal of the explanations is to clarify the meaning of the matched entities by relating more general terms that define them. For example, for the generated correspondence:

$$c = (\textbf{Bosutinib (substance)}, \textbf{BCR-ALB Inhibitor}, \sqsubseteq), \tag{1}$$

a user may ask "What is Bosutinib?" and "What is BCR-ALB?". From both ontologies respectively we have e.g. **Bosutinib** is an **Antineoplastic Agent**, **BCR-ALB Inhibitor** is a **Targeted Therapy Agent** which is also an **Antineoplastic Agent**. These terms are more general than the matched entities and can be presumed as common subsumers of them. Hence, an explanation would be defined as follows:

$$\epsilon = \{\textbf{Antineoplastic Agent} \equiv \textbf{Antineoplastic Agent}\} \tag{2}$$

which contains a new more general relation between concepts that subsume both matched entities. Together with their more general concepts, an explanation to a user can be formulated as a combined set of statements that present how generalized definitions of the matched concepts are reached and how they correlate via the axioms in ϵ. We present for the example alignment in Fig. 1 a subset of the generated explanations from our method. To better present the output, a graphical visualization is shown in Fig. 2 of ontology parts that were used in the explanations process and natural language statements of those.

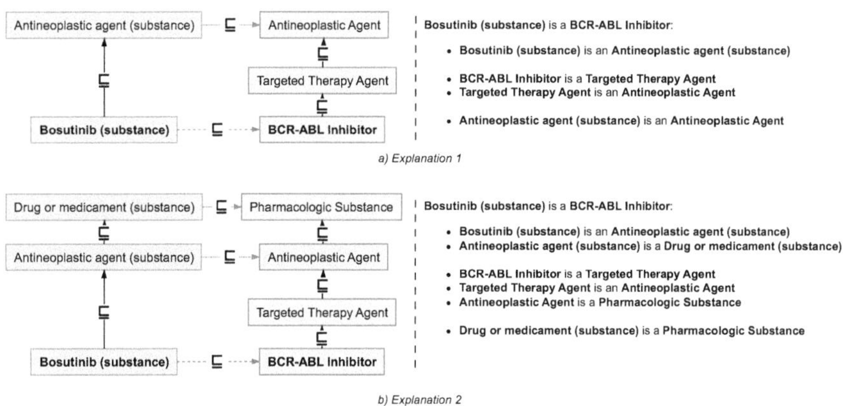

Fig. 2. Generated explanations for the use-case example.

Further sections present the methodology to generate these explanations.

3 Explaining Ontology Alignments via Abduction

We start by representing entity definitions as rooted trees, namely $\mathcal{EL}$ description trees [6].

Definition 1. *A description tree is a tuple* $T = \langle V, E, v_0, \Lambda \rangle$*, where* T *is a directed tree rooted in* v_0*, whose nodes* $v \in V$ *are labeled with* $\Lambda(v)$*, which can contain simple and complex concepts, and edges* $vrw \in E$ *are labeled with property names. The empty label corresponds with the top concept.*

To exemplify description trees, consider the concept definitions $C_1 \equiv A_1 \sqcap \exists r_1.A_2 \sqcap \exists r_2.A_3$ and $C_2 \equiv B_1 \sqcap \exists r_1.B_2 \sqcap \exists r_2.B_3 \sqcap r_3.B_4$. Their description trees are shown on Fig. 3.

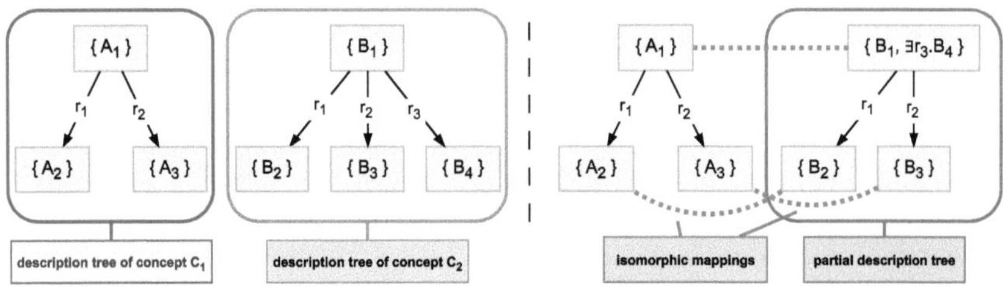

Fig. 3. Example description trees, partial description tree and an isomorphism.

Assuming that entities in a correspondence are correctly matched, to explain alignments we need to capture semantic relations between associated concepts. Abduction offers the theoretical tools to identify missing relations, which will be considered as part of the explanations for correspondences. To capture these semantic relations, we refer to description tree isomorphisms [25].

Definition 2. *Given a correspondence* $c = (e_1, e_2, \sqsubseteq)$ *between entities* $e_1 \in O_1$ *and* $e_2 \in O_2$*, and the description trees of the entities* e_1 *and* e_2*,* $T_{e_1} = \langle V_1, E_1, v_0, \Lambda_1 \rangle$ *and* $T_{e_2} = \langle V_2, E_2, w_0, \Lambda_2 \rangle$ *respectively, a mapping* $\varphi : T_{e_1} \mapsto T_{e_2}$ *is an isomorphism iff* $\varphi(v_0) = w_0$ *and:*

1) $vru \in E_1 \iff \varphi(v) r \varphi(u) \in E_2$*, and*
2) $\forall v \in V_1, w \in V_2$*, s.t.* $\varphi(v) = w$*,* $O_1 \cup O_2 \models \bigsqcap \Lambda_1(v) \sqsubseteq \bigsqcap \Lambda_2(w)$

To capture semantic relations we need to preserve the structure and integrity of concept definitions. This will provide semantic relations between simple concepts in the definitions. However, it is not unusual for semantic relations to exist between complex expression as well. The structures of entity definitions in these cases differ and in order to capture semantic relations between entire definitions, we need to abduct simple and complex concepts as well. Hence, if the structure of description trees cannot be preserved, i.e. they are not isomorphic, we need to propose modifications that will find subparts that are isomorphic, while retaining the original definitions of the matched entities. To do this, we introduce *partial description trees* and define the operation *contract*, which will enable us to embed complex expressions in the labels of description trees. They retain the original definitions of represented entities, but do not present the complete structure of the definitions. Instead, only parts relevant to upholding the assigned semantic relation between the entities in the correspondence are examined.

Definition 3. *Let* $T = \langle V, E, v_0, L \rangle$ *be a description tree,* $v, u \in V$*, such that* v *is the parent of* u*. The operation* contract$(T(u))$ *contracts the subtree* $T(u)$ *in* T *and updates the label of its parent node* v*. The new tree is denoted by* $T^* = \langle V^*, E^*, v_0, L^* \rangle$*, such that:*

- $V^* = V \setminus V(u)$,
- $E^* = E \setminus (E(u) \cup \{vru | v \in V\})$,
- $L^*(v) = L(v) \cup \{\exists r.C_{T(u)}\}$,

where $V(u)$ are the nodes and $E(u)$ are the edge sets of the subtree $T(u)$, $C_{T(u)}$ is the concept expression represented by $T(u)$, and r is the label of the edge connecting v to its child u.

A partial description tree is a description tree itself. The only difference is that node labels also contain complex expressions. This allows to embed non-isomorphic parts of trees as complex expressions in the node labels, whilst preserving the integrity of the original definition represented by the tree.

Figure 3 presents an isomorphism between a complete and a partial description tree. It can be easily verified that the description trees of C_1 and C_2 are not isomorphic. The reason for this is because the edge labels cannot be preserved, as the edge labeled with property r_3 cannot be mapped to any edge in the other tree. Thus, using the contract operation, the node is contracted and its expression is embedded in its parent. We now have a preserved structure between the trees and fully integrated the original concept definition in the node labels.

Description tree isomorphisms also preserve the structure and integrity of the (simple or complex) definitions of the entities they represent. This allows to identify relations between concepts within the definitions of the matched entities, which are crucial for the correspondence to satisfy the assigned semantic relation. Definition 2 captures these semantic relations w.r.t. some background knowledge via the semantic layer in point **2)**, and dependent on the assigned type of relation, this can be either subsumption or equivalence. In the case that the semantic layer is absent, isomorphisms allow to generate *hypotheses*, which depict explanations for why semantic relations between entities are not captured [25].

Definition 4. *Let $c = (e_1, e_2, \sqsupseteq)$ be a correspondence between entities $e_1 \in O_1$ and $e_2 \in O_2$. If $\varphi : T_{e_1} \mapsto T_{e_2}$ is an isomorphism, then an explanation for why $O_1 \cup O_2 \not\models e_1 \sqsupseteq e_2$ is a hypothesis defined as:*

$$\mathcal{H} = \{\sqcap L_{e_1}(v) \sqsupseteq \sqcap L_{e_2}(\varphi(v)) | O_1 \cup O_2 \not\models \sqcap L_{e_1}(v) \sqsupseteq \sqcap L_{e_2}(\varphi(v))\}, \tag{3}$$

such that $O_1 \cup O_2 \cup \mathcal{H} \models e_1 \sqsupseteq e_2$ and $O_1 \cup O_2 \cup \mathcal{H} \not\models e_1 \sqcap e_2 \sqsubseteq \bot$.

4 Computing Minimal Hypotheses

Since hypotheses are generated from isomorphisms, when the structures are not preserved the contract operation prunes the non-isomorphic parts and adds role restrictions to the set of abductions. Naturally, we want to obtain the maximal amount of preserved structure, so that we do not unnecessarily prune trees and introduce more complex abductions. Thus, we search for isomorphisms between partial description trees obtained by the minimal number of contract operations.

Definition 5. *A hypothesis $\mathcal{H}$ obtained from an isomorphism φ is considered minimal if there does not exist a hypothesis $\mathcal{H}'$ obtained from an isomorphism φ', such that φ is obtained with more* contract *operations than φ'.*

To compute minimal hypotheses, we introduce a heuristic-based approach. Given two description trees $T_1 = \langle V_1, E_1, v_0, \Lambda_1 \rangle$ and $T_2 = \langle V_2, E_2, w_0, \Lambda_2 \rangle$ the computation of the minimal partial description tree isomorphisms consists of three main steps: *i)* obtain all paths from the root to leaf nodes in both trees respectively, *ii)* rank all pairs of paths $(\pi_{v_l}^{v_0}, \pi_{w_l}^{w_0})$ where $\pi_{v_l}^{v_0} \in T_1$ and $\pi_{w_l}^{w_0} \in T_2$ w.r.t. the heuristic, and *iii)* map the nodes from paths that are ranked lowest.

The heuristic used to rank the solutions is defined as following:

$$h(\pi_{v_l}^{v_0}, \pi_{w_l}^{w_0}) = \frac{1}{J_w(\boldsymbol{\mathcal{R}}(T_1(v_1)), \boldsymbol{\mathcal{R}}(T_2(w_1))))^2} \cdot \frac{\mu(\boldsymbol{d}(\pi_{v_l}^{v_0}), \boldsymbol{d}(\pi_{w_l}^{w_0})) + \Delta_n}{\rho(\boldsymbol{r}(\pi_{v_l}^{v_0}), \boldsymbol{r}(\pi_{w_l}^{w_0}))}, \tag{4}$$

where: $\boldsymbol{r}(\cdot)$ and $\boldsymbol{r}(\cdot)$ are the *role vectors* of the paths, $\boldsymbol{d}(\cdot)$ and $\boldsymbol{d}(\cdot)$ are the *degree vectors* of the paths, $J_w(\cdot)$ is the weighted Jaccard similarity between the subtrees $T_1(v_1)$ and $T_2(w_1)$, $\mu(\cdot)$ is the similarity between the respective degree vectors, $\rho(\cdot)$ is the similarity between the respective role vectors, and Δ_n is the structural deviation between the subtrees $T_1(v_1)$ and $T_2(w_1)$.

We will now define each metric used in the heuristic. For each path represented as $\pi_{v_l}^{v_0} := v_0 \overset{r_1}{\rightsquigarrow} v_1 \overset{r_2}{\rightsquigarrow} \dots v_{l-1} \overset{r_l}{\rightsquigarrow} v_l$, i.e. a sequence of nodes in V and edges in E, we define two types of vectors, namely a *role vector* and a *degree vector*. A role vector is defined as:

$$\boldsymbol{r}(\pi_{v_l}^{v_0}) = \langle r_1, r_2, \dots, r_l \rangle, \tag{5}$$

where each element in the vector is the corresponding edge label in the path. The degree vector is defined as:

$$\boldsymbol{d}(\pi_{v_l}^{v_0}) = \langle deg^-(v_0), deg^-(v_1), \dots, deg^-(v_l) \rangle, = \langle d_1, d_2, \dots, d_l \rangle, \tag{6}$$

where each element in the vector is the out-degree of the corresponding node in the path. Next, we compute:

$$k = \min\{i \in \{1, \dots, \min(|\boldsymbol{r}(\pi_{v_l}^{v_0})|, |\boldsymbol{r}(\pi_{w_l}^{w_0})|)\} | \boldsymbol{r_i}(\pi_{v_l}^{v_0}) \not\equiv \boldsymbol{r_i}(\pi_{w_l}^{w_0})\}, \tag{7}$$

which is the first occurrence when labels in the role vectors are not equal. From Definition 2 point **1)**, if edge labels are not preserved between the trees, then they are not isomorphic and k is used to identify non-isomorphic parts between trees. These parts will be contracted and introduced as complex expressions in the labels of the partial description trees. We then evaluate the similarity of the role vectors, defined as:

$$\rho(\boldsymbol{r}(\pi_{v_l}^{v_0}), \boldsymbol{r}(\pi_{w_l}^{w_0})) = \begin{cases} 0, \text{ if } r_1 \not\equiv s_1 \\ 1, \text{ if } |\boldsymbol{r}(\pi_{v_l}^{v_0})| = |\boldsymbol{r}(\pi_{w_l}^{w_0})| = m \wedge r_i \equiv s_i \text{ for } 0 \leq i \leq m \\ \frac{k}{\max(|\boldsymbol{r}(\pi_{v_l}^{v_0})|, |\boldsymbol{r}(\pi_{w_l}^{w_0})|)}, \text{ otherwise} \end{cases} \tag{8}$$

and the similarity of degree vectors defined as:

$$\mu(\boldsymbol{d}(\pi_{v_l}^{v_0}), \boldsymbol{d}(\pi_{w_l}^{w_0})) = \sqrt{\sum_{i=1}^{k} (c_i - d_i)^2} + \delta(\boldsymbol{d}(\pi_{v_l}^{v_0}), \boldsymbol{d}(\pi_{w_l}^{w_0})), \tag{9}$$

where $\delta(\boldsymbol{d}(\pi_{v_l}^{v_0}), \boldsymbol{d}(\pi_{w_l}^{w_0})) = |\boldsymbol{d}(\pi_{v_l}^{v_0})| + |\boldsymbol{d}(\pi_{w_l}^{w_0})| - 2k$, and k is defined in Eq. (7).

A degree vector is a sequence of out-degrees (number of edges coming out from a node) of nodes contained in a path. The out-degrees are also a characteristic of the trees structures. Isomorphisms preserve node degrees and to identify if two paths contain nodes with similar branching we use the metric μ defined in Eq. (9). It measures the euclidean distance of the degree vectors up to a point where if nodes were mapped, the structures would be preserved. Moreover, the term δ signifies the amount of nodes that cannot be mapped in cases where edge labels do not allow for mappings. On the other hand, role vectors are a sequence of edge labels in a path. Even if μ measures that paths contain nodes with completely similar structures, by Definition 2, if the edge labels do not correspond, then an isomorphic mapping is not possible. Hence, the metric ρ presents an inverse condition when certain nodes can be mapped.

The above similarity metrics are used to compare the paths and identify the isomorphic and non-isomorphic parts between description trees. Because paths are preserved up to isomorphisms, by contracting the non-isomorphic subpaths the remaining nodes can be mapped and will preserve the structure of the partial description trees. To further determine the similarity between the paths, we use two additional metrics, namely the weighted Jaccard index between subtrees of description trees and a value measuring the structural deviation between subtrees.

The weighted Jaccard index is used to identify the similarity of edge labels between subtrees and takes into account the frequency of equally aligned edge labels. It is defined as:

$$J_w(\boldsymbol{\mathcal{R}}(T_1(v_1)), \boldsymbol{\mathcal{R}}(T_2(w_1))) = \frac{\sum_i \min(\boldsymbol{\mathcal{R}}_i(T_1(v_1)), \boldsymbol{\mathcal{R}}_i(T_2(w_1)))}{\sum_i \max(\boldsymbol{\mathcal{R}}_i(T_1(v_1)), \boldsymbol{\mathcal{R}}_i(T_2(w_1)))}, \tag{10}$$

where $\boldsymbol{\mathcal{R}}(T_1(v_1))$ and $\boldsymbol{\mathcal{R}}(T_2(w_1)))$ are vectors of the form $[\#r_1, \#r_2, \ldots, \#r_n]$, where $\#r_i$ is the number of times r_i occurs in the subtree. What this metric provides is how similar subtrees w.r.t. edge labels are. Once again, if two paths are structurally similar, but their nodes contain descendants that have non-related edge labels, the maximal number of isomorphic mappings would not be obtained. Together with ρ it carries an inverse correlation with the trees structural similarity and we emphasize this value in deciding whether two paths should be used to map nodes.

Finally, the structural deviation between entire subtrees is defined as:

$$\Delta_n = ||L_1||_F - ||L_2||_F + ||V(T_1(v_1))| - |V(T_2(w_1))||, \tag{11}$$

where $|V(T(v_1))|$ and $|V(T(w_1))|$ are the number of nodes in the subtrees, $||L_1||_F$ and $||L_2||_F$ are the Frobenius norms of the Laplacian matrices of $T(v_1)$ and $T(w_1)$. The Laplacian matrix of a tree is defined as $L = D - A$, where D is the degree matrix of a tree containing the out-degrees of nodes on the main diagonal, and A is the adjacency matrix.

We use the same idea as for the metric μ to evaluate the similarity of the subtrees of nodes contained in compared paths. As mentioned, even if paths contain similar structures, their descendants may represent completely different structures, which will influence the final solution, i.e. nodes that do not preserve the maximal structure could be mapped. The Laplacian matrices contain information on the structure the descendants in the subtrees of nodes in paths carry. Hence, we evaluate the similarity of entire subtrees by using the Frobenius norm on the Laplacian matrices, a way to measure the magnitude of a matrix:

$$||L||_F = \sqrt{\sum_{i=1}^{m}\sum_{j=1}^{n}|l_{ij}|^2} \tag{12}$$

The difference between cardinalities of node sets of subtrees provides an additional measure on how many nodes can be mapped and is included because isomorphisms preserve the number of mapped nodes.

The above metrics provides information on the structural and edge label similarity of the expressions represented by the subtrees. The heuristic in Eq. (4) is based on the idea of measuring the structural and role signature similarity of trees, as they present two crucial features of description tree isomorphisms. The formula weighs the sum of the structural similarity of paths and subtrees with the inverse of the weighted Jaccard index and role vector metrics to provide strict conditions on deciding which paths map the most amount of nodes that preserve the structure.

To exemplify the computation of isomorphisms and, subsequently, hypotheses, obtained from the minimal amount of contract operations, consider the description trees in Fig. 4, which depict the concepts $C_1 \equiv A_0 \sqcap \exists r_1.(A_1 \sqcap \exists r_1.A_2 \sqcap \exists r_2.A_3)$ and $C_2 \equiv B_0 \sqcap \exists r_1.(B_1 \sqcap \exists r_1.B_3 \sqcap \exists r_2.B_4) \sqcap \exists r_1.(B_2 \sqcap \exists r_2.B_5)$.

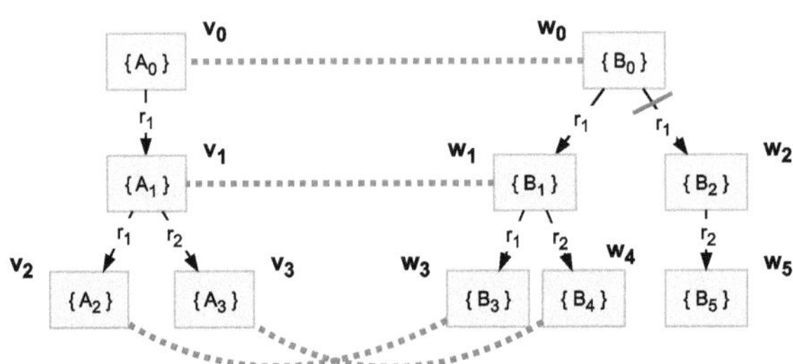

Fig. 4. Example isomorphism obtained from minimal number of contract operations.

The similarity metrics and heuristics evaluated for each pair of paths from T_{C_1} and T_{C_2} are shown in Table 1. The heuristics offer a quick way to filter out non-isomorphic parts of trees, contract them and return the isomorphisms with maximal number of nodes mapped. By choosing the pair of paths with minimal heuristics, we can obtain the isomorphism as shown in Fig. 4, i.e. $\varphi = (v_0w_0)(v_1w_1)(v_2w_3)(v_3w_4)$. All other combinations of paths will not map the maximal amount of nodes from both trees that preserve the structure and

Table 1. Evaluated similarity metrics and heuristics for trees T_{C_1} and T_{C_2}

π_1	π_2	$J(\pi_1)$ $[r_1, r_2]$	$J(\pi_2)$ $[r_1, r_2]$	$\boldsymbol{d_1}$	$\boldsymbol{d_2}$	$\boldsymbol{r_1}$	$\boldsymbol{r_2}$	k	J_w	ρ	μ	Δ_n	h
(v_0, v_1, v_2)	(w_0, w_1, w_3)	[2, 1]	[2, 1]	[1, 2, 0]	[2, 2, 0]	$[r_1, r_1]$	$[r_1, r_1]$	2	1	1	1	0	1
(v_0, v_1, v_3)	(w_0, w_1, w_4)	[2, 1]	[2, 1]	[1, 2, 0]	[2, 2, 0]	$[r_1, r_2]$	$[r_1, r_2]$	2	1	1	1	0	1
(v_0, v_1, v_3)	(w_0, w_2, w_5)	[2, 1]	[1, 1]	[1, 2, 0]	[2, 1, 0]	$[r_1, r_2]$	$[r_1, r_2]$	2	0.667	1	1.41421	2.03528	7.76135
(v_0, v_1, v_2)	(w_0, w_1, w_4)	[2, 1]	[2, 1]	[1, 2, 0]	[2, 2, 0]	$[r_1, r_1]$	$[r_1, r_2]$	1	1	0.5	5	0	10
(v_0, v_1, v_3)	(w_0, w_1, w_3)	[2, 1]	[2, 1]	[1, 2, 0]	[2, 2, 0]	$[r_1, r_2]$	$[r_1, r_1]$	1	1	0.5	5	0	10
(v_0, v_1, v_2)	(w_0, w_2, w_5)	[2, 1]	[1, 1]	[1, 2, 0]	[2, 1, 0]	$[r_1, r_1]$	$[r_1, r_2]$	1	0.667	0.5	5	2.03528	31.6587

will lead to unnecessary contraction of nodes and more complex abductions in hypotheses.

5 Implementation and Experiments

The algorithm is implemented in Python using the owlready2 module [3]. The general overview of the algorithm to explain ontology alignments is shown in Fig. 5.

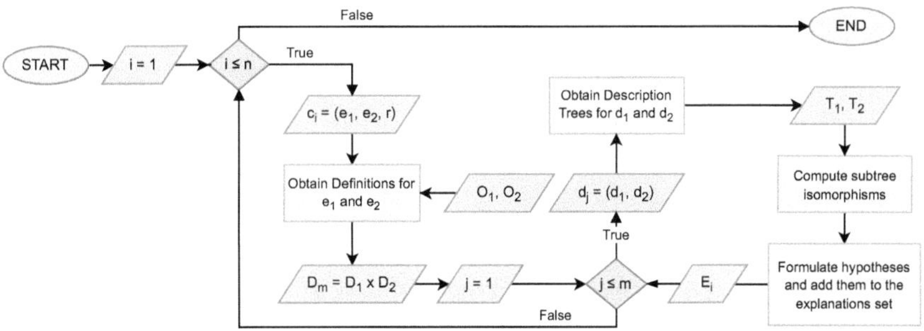

Fig. 5. Algorithm workflow.

There are three inputs that our approach requires, namely two ontologies and an alignment between them. The second input, i.e. the set of correspondences, can be obtained by any outside method, which can be either based on symbolic or subsymbolic approaches. Our algorithm needs only the set of determined correspondences and the ontologies which contain the knowledge of the domains of origin of the aligned entities. Our approach focuses on using the background knowledge and definitions within it. By using abduction, existing and missing relations between various definitions of aligned entities can be discovered. Thus, the algorithm in this paper is an out-of-the-box solution and complements existing ontology matching systems by providing explanations for their generated matches, and, subsequently, making them transparent and interpretable. From here, the set of minimal isomorphisms between the description trees are computed and hypotheses are formulated as described in Sect. 3. The implementation with the experiments and the generated explanations can be found in the repository https://github.com/Gochev-Ivan/Explaining-Ontology-Alignments-via-Abductive-Graph-Isomorphisms.

5.1 Experiments

We conducted practical experiments on computing abduction-based explanations of ontology matches on $\mathcal{EL}$ ontologies used in the Bio-ML track of the OAEI[1]. The experiments were conducted on a machine running Windows 10, CPU AMD Ryzen 7 4800H, RAM 16.0 GB. We specifically ran experiments on the alignment in two scenarios: SNOMED CT and NCIT, and SNOMED CT and FMA. The Deep Onto package from the Bio-ML track [19] was used to generate a set of correspondences between the ontologies and for each correspondence our method generated explanations. Based on similar evaluation approaches for abduction [27] [31], we report on the average, median and maximal values for the number of explanations generated for each correspondence ($\#\mathcal{E}$), size of explanations ($|\mathcal{E}|$), number of simple concepts in explanations ($\#A$), number of complex expressions in explanations ($\#C$), and the time (in seconds) needed to compute the explanations ($t[s]$) in Table 2.

Table 2. Experimental results.

Alignment	$\#\mathcal{E}$	$\lvert\mathcal{E}\rvert$	$\#A$	$\#C$	t[s]
snomed-ncit	4.59 — 4.00 — 21.00	1.00 — 1.00 — 2.25	2.90 — 2.00 — 10.00	0.58 — 0.00 — 3.00	0.01 — 0.00 — 0.30
snomed-fma	5.20 — 4.00 — 76.00	1.00 — 1.00 — 1.00	2.42 — 2.00 — 12.00	0.11 — 0.00 — 5.00	0.01 — 0.00 — 0.13

There are some cases in which a large amount of explanations occur, which depends on the complexity of concept definitions. However, there are only few of those cases. In addition, we noticed that not many complex expressions per explanation were abducted. This is due to a couple of reasons: *i)* concepts in the ontologies were not defined with a lot of nested expressions and did not contain a lot of restrictions, and *ii)* properties between the ontologies were not aligned and there were rarely cases in which isomorphisms were obtained without the contract operation. In future, we plan on expanding and evaluating the heuristics further, by adding an additional metric that will compare roles restricting concepts and will rank them and map accordingly to preserve concept expressions, as well as running systematic experiments on trees obtained from large complex expressions. Finally, the time in Table 2 represents computing all explanations for a correspondence and the method can achieve solutions in a feasible time. Nonetheless, we want to examine how our heuristic-based approach performs on much larger trees for individual cases also.

[1] https://krr-oxford.github.io/OAEI-Bio-ML/.

6 Conclusions

We presented an out-of-the-box methodology to generate post-hoc explanations for ontology alignments [32], by computing minimal abductions between graphical representations of concept expressions using isomorphisms. We defined an operation to obtain partial trees to identify structure preserving parts of descriptions and abduct simple and complex expressions, as well as a heuristic to obtain minimal hypotheses. In addition, the explanations contain underlying alignments that users can inspect and help them decide the next steps on how to update the initial alignment. We evaluated the performance of our approach on two existing alignments from the Bio-ML track of OAEI and the results show that our method can compute explanations in a feasible time.

Compared to similar methodologies, one advantage of ours is that we provide an out-of-the-box method that can be combined with any matching systems. Further, we propose an initial improvement of our previous work in [25], by providing a scalable approach that can filter non-minimal abductions in the solutions space. In addition, the (partial) description tree isomorphisms allow to perform abduction of both simple and complex concepts, which covers the area of complex ontology matching, and the transitive property that isomorphisms have can be potentially used for holistic ontology matching. Similarly, homomorphisms have been utilized for abduction in [27], which allow for abduction of atomic concepts only. Although our method is focused on ontology matching, both methods focus on explaining observations that are not entailed by background knowledge, and our method complements the one in [27] by allowing for abduction of complex concepts as well. Further, in [31] the abductive process is constrained by a set of predefined patterns that allow for abduction of both simple and complex concepts, whereas our method is only restricted by the signature of the ontologies included in the alignment process.

This is only an initial approach we presented for explaining ontology alignments. Currently, the method resorts to obtaining common subsumers to provide more clarifying definitions of matched entities. The explanations can further be improved. For example, the shortest paths to a common concept explaining the type of both entities in a correspondence can be found, and can be used to generate more concise explanations. In addition, intermediate connecting concepts can be found in the background ontologies to generate a sequence of semantic relations leading to the presumed alignments. Graphical visualization are needed of the ontology parts that represent the correspondences and hypothesis axioms as in Fig. 1 and an option to expand and contract the graph should be possible, so that each user can tailor the explanations to their preferences. Moreover, a limitation of our method is the dependency on the hierarchies of the background ontologies. In particular, our method requires properties between ontologies to be aligned, in order to find more simple semantic relations, which would in turn provide more interpretable explanations. To tackle this limitation, the following metrics could be used: *i)* a lexicographic metric that would define preferences when matching roles that are not already aligned, and *ii)* a structural similarity metric that would use the role hierarchy to measure the distance between roles.

A combination of these metric would provide more flexibility to our method by omitting the assumption of role alignments and directly align properties during the abductive process. The heuristics used to obtain minimal solutions can be further extended by a similarity metric between properties to rank and align them. Furthermore, the complexity of our method requires a more in-depth theoretical and empirical analysis. Finally, as part of our future work we plan on extending our methodology to more expressive DLs and conducting a user survey to evaluate the usefulness of the heuristic and how it influences the complexity of the generated explanations.

References

1. Nardi, D., Brachman, R.J.: An introduction to description logics. Description logic handbook **1**, 40 (2003)
2. Epp, S.S.: Discrete mathematics with applications. Brooks/Cole Publishing Co. (2010)
3. Jean-Baptiste, L.: Ontologies with python. Apress, Berkeley, CA (2021)
4. Cornet, R., de Keizer, N.: Forty years of SNOMED: a literature review. BMC Med. Inform. Decis. Mak. **8**, 1–6 (2008)
5. Fragoso, G., de Coronado, S., Haber, M., Hartel, F., Wright, L.: Overview and utilization of the NCI thesaurus. Comp. Funct. Genomics **5**(8), 648–654 (2004)
6. Baader, F., Küsters, R., Molitor, R.: July. Computing least common subsumers in description logics with existential restrictions. In IJCAI **99**, 96–101 (1999)
7. Euzenat, J., Shvaiko, P.: Ontology matching, vol. 18. springer, Heidelberg (2007)
8. Shvaiko, P., Euzenat, J.: Ontology matching: state of the art and future challenges. IEEE Trans. Knowl. Data Eng. **25**(1), 158–176 (2011)
9. Otero-Cerdeira, L., Rodríguez-Martínez, F.J., Gómez-Rodríguez, A.: Ontology matching: a literature review. Expert Syst. Appl. **42**(2), 949–971 (2015)
10. Euzenat, J., Shvaiko, P., Euzenat, J., Shvaiko, P.: Overview of matching systems. Ontology Matching, pp.201-283 (2013)
11. Thiéblin, E., Haemmerlé, O., Hernandez, N., Trojahn, C.: Survey on complex ontology matching. Semantic Web **11**(4), 689–727 (2020)
12. Lambrix, P., Tan, H.: SAMBO-A system for aligning and merging biomedical ontologies. J. Web Semantics **4**(3), 196–206 (2006)
13. Li, J., Tang, J., Li, Y., Luo, Q.: Rimom: a dynamic multistrategy ontology alignment framework. IEEE Trans. Knowl. Data Eng. **21**(8), 1218–1232 (2008)
14. Wang, Z.: AMD results for OAEI 2022. In OM@ ISWC (pp. 145-152) (2022)
15. Efeoglu, S.: Graphmatcher: a graph representation learning approach for ontology matching. (2024). arXiv preprint arXiv:2404.14450
16. Cotovio, P.G., et al.: Matcha-DL a tool for supervised ontology alignment. preprint (2024)
17. Cruz, I.F., Antonelli, F.P., Stroe, C.: AgreementMaker: efficient matching for large real-world schemas and ontologies. Proc. VLDB Endowment **2**(2), 1586–1589 (2009)
18. Gulić, M.A.R.K.O., Vrdoljak, B.O.R.I.S. and Ptiček, M.A.R.I.N.A.: CroMatcher 2.0: A Comprehensive Analysis of the Improved Ontology Matching System. IEEE Access (2025)
19. He, Y., et al.: DeepOnto: a Python package for ontology engineering with deep learning. Semantic Web **15**(5), 1991–2004 (2024)

20. Giunchiglia, F., Shvaiko, P., Yatskevich, M.: S-Match: an Algorithm and an Implementation of Semantic Matching. In: Bussler, C.J., Davies, J., Fensel, D., Studer, R. (eds.) ESWS 2004. LNCS, vol. 3053, pp. 61–75. Springer, Heidelberg (2004). https://doi.org/10.1007/978-3-540-25956-5_5
21. Dhamankar, R., Lee, Y., Doan, A., Halevy, A., Domingos, P.: IMAP: discovering complex semantic matches between database schemas. In: Proceedings of the 2004 ACM SIGMOD International Conference On Management of data (pp. 383-394) (2004)
22. Trojahn, C., Euzenat, J., Tamma, V., Payne, T.R.: Argumentation for reconciling agent ontologies. In: Semantic agent systems: Foundations and applications, pp. 89–111. Berlin, Heidelberg, Springer, Berlin Heidelberg (2011)
23. Shvaiko, P., Giunchiglia, F., Da Silva, P.P.: McGuinness, D.L., 2005, May. Web explanations for semantic heterogeneity discovery. In European Semantic Web Conference (pp. 303-317). Berlin, Heidelberg: Springer Berlin Heidelberg
24. Chai, X., Sayyadian, M., Doan, A., Rosenthal, A., Seligman, L.: Analyzing and revising data integration schemas to improve their matchability. Proc. VLDB Endowment **1**(1), 773–784 (2008)
25. Gocev, I., Meditskos, G., Bassiliades, N.: Towards explaining DL non-entailments by utilizing subtree isomorphisms, pp. 385–390. Springer Nature Switzerland, Cham (2022)
26. Koopmann, P.: January. Signature-based abduction with fresh individuals and complex concepts for description logics, In Description Logics (2021)
27. Haifani, F., Koopmann, P., Tourret, S., Weidenbach, C.: In: August. Connection-minimal abduction in EL via translation to FOL, pp. 188–207. Springer International Publishing, Cham (2022)
28. Di Noia, T., Di Sciascio, E., Donini, F.M.: Extending Semantic-Based Matchmaking via Concept Abduction and Contraction. In: Motta, E., Shadbolt, N.R., Stutt, A., Gibbins, N. (eds.) EKAW 2004. LNCS (LNAI), vol. 3257, pp. 307–320. Springer, Heidelberg (2004). https://doi.org/10.1007/978-3-540-30202-5_21
29. Wei-Kleiner, F., Dragisic, Z. and Lambrix, P.: Abduction framework for repairing incomplete EL ontologies: Complexity results and algorithms. In: Proceedings of the AAAI Conference on Artificial Intelligence (Vol. 28, No. 1) (2014)
30. Lambrix, P., Wei-Kleiner, F., Dragisic, Z. and Ivanova, V.: Repairing missing is-a structure in ontologies is an abductive reasoning problem. In Second International Workshop on Debugging Ontologies and Ontology Mappings (WoDOOM 2013), Montpellier, France, May 27, 2013 (pp. 33-44). Rheinisch-Westfaelische Technische Hochschule Aachen University (2013)
31. Du, J., Wan, H., Ma, H.: Practical TBox abduction based on justification patterns. In: Proceedings of the AAAI Conference on Artificial Intelligence (Vol. 31, No. 1) (2017)
32. Retzlaff, C.O., et al.: Post-hoc vs ante-hoc explanations: xAI design guidelines for data scientists. Cogn. Syst. Res. **86**, 101243 (2024)

Ontology-Aware Relation Prediction over EurostatKG: A Unified Neural Framework

Efstratios Skaperdas(✉) and Nick Bassiliades

School of Informatics, Aristotle University of Thessaloniki, 54124 Thessaloniki, Greece
{eskapera,nbassili}@csd.auth.gr

Abstract. Knowledge graphs about statistical data, such as the Eurostat Knowledge Graph (EurostatKG), present unique challenges for relation prediction due to their rich ontological schemas and strict semantic constraints. To address this, we reformulate relation prediction as a triple classification task and propose a unified, ontology-guided neural framework. Our approach integrates embedding-based models with graph-based semantic encoders (GCN, GAT), and introduces a hybrid loss function alongside a constraint-aware negative sampling strategy that respects ontology-defined domain and range axioms. Evaluated on EurostatKG, our method achieves strong results, including an F1-score of 0.96, PR-AUC of 0.99, Hits@5 of 0.618, and a Soft Hits@1 of 0.97. This work provides a principled integration of symbolic knowledge into neural models, while establishing a reproducible baseline for ontology-aware KG completion in complex, schema-rich domains.

Keywords: Triple Classification · Relation Prediction · Knowledge Graph Embeddings · Ontology-aware Learning · Statistical Knowledge Graphs

1 Introduction

Knowledge Graphs (KGs) offer structured, semantically rich representations of real-world entities and relations. They underpin a wide range of applications, including semantic search, recommendation systems, and automated reasoning. However, ensuring both completeness and semantic correctness at scale remains a persistent challenge—particularly in schema-rich domains.

A prominent example is the Eurostat Knowledge Graph (EurostatKG) [1], which encodes statistical data using formal ontologies. Relations in such graphs are governed by logical axioms (e.g., `rdfs:domain`, `rdfs:range`), and entities are extensively typed. As a result, the traditional relation prediction task—focused on ranking candidate triples—shifts toward validating whether a given triple (h, r, t) is both factually plausible and semantically compliant.

To address this, we reformulate relation prediction as a *triple classification* task that determines the validity of candidate triples under both factual and ontological constraints. Rather than proposing new architectures, we introduce a unified, ontology-guided neural framework that integrates embedding-based interaction models, graph-based semantic encoders (GCN/GAT), a hybrid loss function, and constraint-aware neg-

H. Takeda et al. (Eds.): IJCKG 2025, LNAI 16297, pp. 78–94, 2026.
https://doi.org/10.1007/978-981-95-5009-8_6

ative sampling. To our knowledge, this is among the first systems to tightly couple symbolic reasoning with neural modeling for relation prediction in large, schema-governed KGs.

Scope of the WN18RR Study. To complement EurostatKG, we also report results on WN18RR, a widely used ontology-agnostic benchmark. These experiments are intended solely to test the robustness of our architecture in settings without schema-level supervision.

Our Contributions are:

1. **A EurostatKG benchmark:** We introduce the first ontology-aware relation prediction benchmark for EurostatKG and release the accompanying code and data.
2. **Ontology-guided architectures:** We design neural triple classifiers that combine embeddings with type-aware GNNs enriched with `rdf:type` semantics.
3. **Semantic regularization:** We incorporate asserted `rdfs:domain` and `rdfs:range` axioms into training via a hybrid loss and constraint-aware negative sampling.
4. **Extensive evaluation:** We benchmark 10 model variants across classification and ranking metrics, including soft ranking to assess near-miss plausibility.
5. **A unified open-source pipeline:** We provide a modular ontology-aware framework for reproducible and semantically consistent relation prediction.

Scope of novelty. We do not introduce new embedding families; our novelty lies in the principled, end-to-end integration of established embeddings with ontology-aware GNNs, constraint-guided sampling, and a hybrid loss under a unified, reproducible pipeline.

2 Related Work

Embedding-Based Models for KG Completion. Relation prediction and triple classification are core tasks in knowledge graph (KG) completion. Early approaches learned continuous vector representations of entities and relations to infer missing triples. Translational models such as TransE [2] and its variants (TransH, TransR) treat relations as vector translations in embedding space. Bilinear models like DistMult [3] and ComplEx [4] use matrix factorization to capture symmetric and asymmetric relations. These models are efficient but typically treat triples in isolation, ignoring schema-level semantics and ontological constraints.

Graph-Based Models and Neural Architectures. Graph Neural Networks (GNNs) model structural dependencies in KGs. R-GCN [5] extends GCNs with relation-specific transformations, while GATs [6] weight neighbor features via attention. CompGCN [21] introduces compositional operators for message passing. Transformer-based methods like KG-BERT [22] and KEPLER [23] treat triples as text, enabling language-model-style pretraining. However, these architectures generally lack mechanisms to enforce symbolic constraints such as domain and range.

Ontology-Aware Representation Learning. Several works integrate formal semantics into KG learning, especially in biomedicine. Onto2Vec [7] embeds OWL axioms as co-occurrence patterns, while OPA2Vec and EL Embeddings [8] model richer description logic. RDF2Vec [24] captures structure via random walks but does not directly enforce schema constraints. In contrast, we explicitly operationalize axioms such as `rdfs:domain`, `rdfs:range`, and `rdf:type` within model training, tailored to statistical KGs like Eurostat. While related ideas have mainly been applied in biomedical KGs [8], we extend them to statistical domains and benchmark their effects in a modular, reproducible pipeline.

Ontology-Induced Embeddings and Semantic Losses. OWL2Vec [30] and OWL2Vec* [28] embed OWL axioms by transforming ontologies into sequences or graphs, capturing taxonomic and logical context. D'Amato et al. further show how OWL semantics can be injected into embedding objectives to regularize predictive tasks [29]. Our contribution is complementary: we combine ontology-aware message passing, constraint-guided negative sampling, and a semantic penalty into a single end-to-end triple *classification* pipeline.

Negative Sampling in KG Learning. Negative sampling is central to embedding training. Uniform corruption [2] and Bernoulli sampling [18] often generate trivial negatives. Recent strategies include type-consistent sampling [13], popularity bias [17], orthogonal sampling [25], and self-adversarial weighting (RotatE [9]). Yet most ignore ontological coherence. Our hybrid strategy blends uniform, type-violation, popularity-based, and flipped corruptions, while enforcing ontology-defined constraints.

Positioning. Surveys [19,26] stress the importance of tuning and modularity in KG models. Still, symbolic semantics are rarely integrated into end-to-end neural pipelines. Our framework addresses this by embedding ontological reasoning into all stages – representation, message passing, supervision, and evaluation – via GCN/GAT encoders, constraint-guided negative sampling, and a hybrid semantic loss. This design captures fine-grained plausibility in triple classification for schema-rich KGs like EurostatKG and provides a blueprint for principled ontology integration in KG learning.

3 Dataset and Preprocessing

This study builds upon the Eurostat Knowledge Graph (EurostatKG)[1] introduced in [1], modeling statistical concepts, datasets, and glossary articles published on the Eurostat and OECD portals. Due to its rich schema and extensive use of ontological constraints, EurostatKG provides a realistic and challenging testbed for relation prediction in ontology-guided settings.

3.1 EurostatKG Construction and Statistics

EurostatKG was constructed by scraping structured content from official Eurostat and OECD sources and mapping it to a custom OWL ontology [1]. For the experiments in this paper we use the full asserted graph as provided in Turtle form, i.e., we do not filter out datatype triples or `rdf:type` assertions. Entities without an asserted `rdf:type`

[1] EurostatKG available at: EurostatKG (dataset link).

are retained and assigned a dedicated "unknown–type" placeholder embedding so that message passing remains connected. All results are therefore computed on the asserted graph without RDFS/OWL materialization (no inferred types).

Key statistics for the asserted graph are as follows:

- **Triples and entities: 307,444** RDF triples, **59,938** unique entities, and **43** unique object properties.
- **Ontological schema:** The ontology defines **1,856** OWL classes, of which **1,435** are actively used in type assertions. Out of the 43 object properties, **27** have defined `rdfs:domain` and/or `rdfs:range` constraints.

3.2 Ontology Enrichment and Alignment

Building on the methodology introduced in [1], we enriched the KG using domain-specific logic and NLP-based techniques:

- Latent Dirichlet Allocation (LDA) and co-occurrence statistics were applied to generate `hasFrequentTerm` properties and relations between glossary articles.
- SPARQL CONSTRUCT queries were used to create links between Eurostat and OECD thematic resources.
- External alignments leveraged `rdfs:subClassOf`, `owl:equivalentClass`, and `skos: closeMatch` to map to widely used vocabularies such as DCAT, SKOS, DCMI, FOAF, Schema.org, and RDF Data Cube.

These enrichments were driven by real-world use cases from the NLP4StatRef project[2], including guided query interfaces and terminology linking, which provided iterative feedback for graph refinement.

3.3 Data Preparation and Negative Sampling

Let $\mathscr{T}$ denote the set of asserted triples from the Turtle file. We map each entity and relation IRI (including `rdf:type`) to integer IDs. For every entity, we build a type map from asserted `rdf:type` statements; entities lacking types receive a dedicated placeholder type index. These types are later used by the type-aware encoders and for soft constraint checking.

To train models capable of distinguishing between valid and invalid triples, we generate a set of negative triples $\mathscr{T}^-$ using a hybrid sampling strategy that blends structural, statistical, and ontological corruptions:

- **Uniform corruption:** For each true triple (s, p, o), we replace the object o with a randomly selected entity $o' \in \mathscr{E}$ such that $(s, p, o') \notin \mathscr{T}$. This serves as our baseline corruption method.

[2] NLP4StatRef: an EU-funded initiative on statistical reference and knowledge graphs. GitHub: https://github.com/eurostat/NLP4Stat.

- **Type-constraint violation:** For relations with defined `rdfs:domain` or `rdfs:range`, we generate triples that intentionally violate these constraints by selecting entities of incompatible types as the subject or object.
- **Popularity-biased noise:** Following [9], we bias draws toward frequent entities and relations to form additional hard negatives. This process can produce self-corruptions (e.g., (s, p, s)) among other frequent-entity variants.
- **Triple flipping:** We also reverse valid triples—(o, p, s)—as negatives without symmetry filtering.

Note that, under OWA, some unobserved but type-compatible triples may be true; our losses treat such cases softly via the semantic penalty rather than hard contradictions.

Note. Because we keep the full asserted graph (including datatype triples and `rdf:type` links), message passing operates on all edges, while semantic penalties consider only declared `rdfs:domain`/`rdfs:range` against the (possibly placeholder) asserted types. No RDFS/OWL materialization is performed.

Each positive triple is paired with one negative sample from each strategy, yielding a training set of approximately 1.2 million triples. This diverse corruption strategy ensures robustness against semantic, structural, and statistical noise.

4 Methodology

4.1 Design Rationale

Standard KG models typically rely on structural regularities or purely embedding-based representations, often overlooking the rich ontological semantics present in schema-aware knowledge graphs. The Eurostat Knowledge Graph, for instance, includes explicit typing (`rdf:type`) and logical constraints (e.g., `rdfs:domain`, `rdfs:range`) that can guide learning but are often neglected in link prediction models.

Our architecture integrates these symbolic signals through three core mechanisms:

- **Type-aware propagation**: Each entity's `rdf:type` assertions are embedded and incorporated into GNN-based encoders (e.g., GCN, GAT). This supports semantically informed message passing, where node representations reflect not only neighborhood structure but also ontological class membership. This idea is inspired by works such as Onto2Vec [7], but is implemented here within a fully differentiable, end-to-end framework.
- **Constraint-aware regularization**: Schema-level axioms (i.e., `rdfs:domain` and `rdfs:range`) are operationalized via two complementary mechanisms: (1) generation of type-violating negative samples during training, and (2) incorporation of a semantic penalty into the loss function for constraint violations. This builds upon prior efforts to integrate symbolic knowledge into neural models [8].
- **Hybrid supervision**: We combine three loss components—binary classification (BCE), margin-based ranking, and a semantic penalty—into a unified objective. This balances factual accuracy, ranking consistency, and ontological coherence.

These mechanisms are supported by a hybrid negative sampling strategy (Sect. 3.3), which blends type-violating, flipped, and popularity-biased corruptions to expose the model to diverse and realistic semantic noise.

Our overarching goal is to develop a framework that is not only accurate but also aligned with the symbolic structure underpinning formal knowledge graphs. The following subsections detail how these principles are instantiated within our model architecture and training pipeline.

4.2 Model Architecture

Our modeling framework follows a configurable, three-stage pipeline for triple classification, illustrated in Fig. 1. An input triple is embedded, optionally refined by ontology-aware GNNs, and classified by a scoring module.

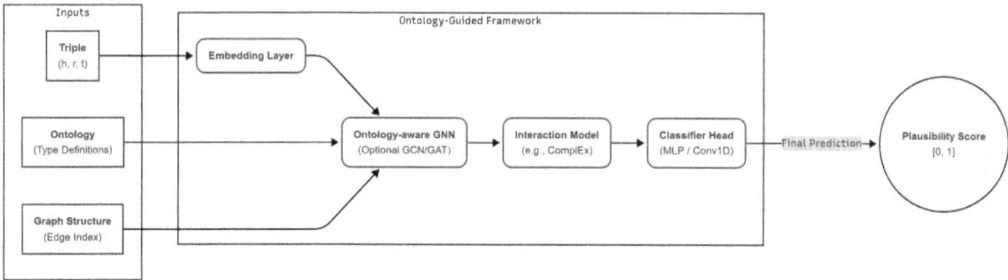

Fig. 1. The overall architecture of our ontology-guided framework.

- **Embedding Layer:** Entities and relations are first mapped to vector representations using trainable embeddings. We support a variety of interaction models for computing triple-level representations, including:
  - *Translational models* (e.g., TransE [2])
  - *Bilinear models* (e.g., DistMult [3], ComplEx [4])
  - *Compositional encodings* based on circular correlation (e.g., HolE [20])
- **Ontology-aware Graph-based Refinement:** Embeddings are optionally passed through schema-aware GNN layers. We implement:
  - A type-aware Graph Convolutional Network (GCN) that propagates neighborhood features modulated by `rdf:type` embeddings.
  - A gated Graph Attention Network (GAT) that uses semantic attention scores conditioned on ontology-derived type features, including domain and range class embeddings.

  These GNN layers enable the model to incorporate both topological and ontological context during embedding refinement.
- **Classifier Head:** The refined triple embedding is passed into one of two classifier architectures:

- A multi-layer perceptron (MLP) with residual connections and advanced activations (SiLU, GELU, and LeakyReLU used in rotation), enabling expressive nonlinear modeling.
- A Conv1D stack that processes triple embeddings as pseudo-sequences, capturing local compositional patterns across dimensions.

This modular design allows us to benchmark multiple model configurations and isolate the contribution of each architectural component. The framework also supports lightweight baseline variants that omit graph-based encoders, relying only on embedding interactions and a downstream classifier.

4.3 Hybrid Loss Function

To jointly enforce factual correctness, ranking consistency, and ontological validity, we design a hybrid loss function composed of three complementary objectives:

- **Binary Cross-Entropy Loss** ($\mathscr{L}_{\text{BCE}}$): Models triple classification as a binary decision, encouraging high confidence for valid triples and low confidence for corrupted ones.
- **Margin Ranking Loss** ($\mathscr{L}_{\text{rank}}$): Encourages margin-based separation between scores of positive and negative triples, aligning with standard link prediction objectives.
- **Semantic Violation Penalty** ($\mathscr{L}_{\text{sem}}$): Penalizes triples that violate the declared `rdfs:domain` and `rdfs:range` constraints of the ontology, promoting type-/consistent predictions.

The overall training objective is:

$$\mathscr{L}_{\text{total}} = \alpha \cdot \mathscr{L}_{\text{BCE}} + \beta \cdot \mathscr{L}_{\text{rank}} + \gamma \cdot \mathscr{L}_{\text{sem}}$$

where α, β, and γ are non-negative weighting coefficients. We tune them via grid search over $\{0.1, 0.5, 1.0, 2.0\}$ and set $\alpha = 1.0$, $\beta = 0.5$, and $\gamma = 0.5$ based on validation performance.

Semantic Violation Penalty. For each batch of candidate triples (h, r, t), we extract the expected domain and range classes for relation r from precomputed schema maps, and check whether h and t satisfy the typing constraints:

- A *domain violation* occurs when h lacks any `rdf:type` matching the declared `rdfs:domain` of r.
- A *range violation* occurs when t does not satisfy the `rdfs:range` constraint of r.

We operationalize domain/range constraints as soft penalties over asserted `rdf:type` statements; handling inferred types under the Open World Assumption (OWA) is orthogonal and left for future work.

Each violation contributes to a normalized batch-level penalty in $[0, 1]$, softly guiding the model toward ontology-compliant predictions.

Practical Note. If multiple `rdfs:domain/rdfs:range` axioms are present for a property, we use the first asserted class found in the graph. We do not currently expand `owl:unionOf` or other complex class expressions; supporting disjunctive domains/ranges is left to future work.

Implementation. All loss components are implemented within a unified PyTorch module. We compute scores for both positive and negative triples. Semantic constraints are enforced efficiently using a sparse entity-type matrix and preloaded domain/range mappings during training.

Semantic Constraints as Soft Logic. We model schema constraints as soft logical implications: $(h,r,t) \Rightarrow h \in \text{domain}(r),\ t \in \text{range}(r)$. Let $\mathcal{T}_h$ and $\mathcal{T}_t$ denote the known types of h and t. We define:

$$V_{\text{dom}}(h,r) = \mathbb{1}_{\text{domain}(r) \notin \mathcal{T}_h}, \quad V_{\text{range}}(r,t) = \mathbb{1}_{\text{range}(r) \notin \mathcal{T}_t}$$

The semantic penalty becomes:

$$\mathcal{L}_{\text{sem}} = \frac{1}{|\mathcal{B}|} \sum_{(h,r,t) \in \mathcal{B}} \left[V_{\text{dom}}(h,r) + V_{\text{range}}(r,t) \right]$$

This term softly enforces type constraints during training without requiring hard rejection of semantically imperfect triples.

4.4 Embedding Interaction Models

We evaluate six representative embedding models that span translational, bilinear, and compositional paradigms. Each model captures relational patterns in a distinct way within triples (h,r,t):

- **TransE** [2]: Models relations as vector translations in embedding space ($\mathbf{h} + \mathbf{r} \approx \mathbf{t}$). While simple, it effectively captures one-to-one and hierarchical relations.
- **TransH** [18]: Projects entities onto relation-specific hyperplanes before applying translation, enabling asymmetric and complex relation types.
- **DistMult** [3]: A bilinear model that scores triples via element-wise multiplication. It is efficient and suitable for symmetric relations, but lacks asymmetry modeling.
- **ComplEx** [4]: Extends DistMult to complex-valued embeddings, allowing the use of complex conjugation to model asymmetric and antisymmetric patterns.
- **RotatE** [9]: Encodes relations as rotations in complex space, capturing symmetry, inversion, and composition. It effectively models diverse relational structures.
- **HolE** [20]: Employs circular correlation to compactly represent compositional interactions between head and tail embeddings.

Each interaction model is integrated into our unified architecture, combined with optional ontology-guided graph encoders (GCN or GAT) and neural classifier heads (MLP or Conv1D). This design enables consistent evaluation across embedding paradigms and isolates the effects of architectural and semantic design choices.

Model Selection Rationale. Our focus on established embedding models is grounded in the findings of Ruffinelli et al. [19], who showed that, with proper tuning, classical methods such as TransE and ComplEx remain competitive with newer architectures. Rather than introducing novel embeddings, our contribution lies in enhancing these baselines with ontological semantics and constraint-aware, symbolic-structural integration tailored to schema-rich graphs like EurostatKG.

4.5 Ontology-Aware Graph Encoders

Intuitively, message passing updates a node by aggregating (and, in GAT, weighting) features from its neighbors; here we condition that aggregation on type semantics.

To enrich entity representations with schema-level semantics, we incorporate two graph-based encoders tailored to ontological contexts: a type-aware Graph Convolutional Network (GCN) and an ontology-guided Graph Attention Network (GAT). Both encoders are designed to propagate type semantics and enforce `rdfs:domain` and `rdfs:range` constraints during message passing.

Type-Aware GCN. We extend the standard GCN architecture by embedding OWL class types alongside entity embeddings. Each entity vector is concatenated with a gated transformation of its `rdf:type` representation. Neighborhood features are aggregated via mean pooling, followed by a linear projection, batch normalization, and dropout. Residual connections are applied when input and output dimensions match. This configuration enables the model to propagate type-specific information throughout the graph, promoting semantic coherence and supporting generalization across similarly typed entities. Entities without an asserted `rdf:type` receive an 'unknown' type embedding to preserve graph connectivity during message passing.

Ontology-Aware GAT. To capture relational asymmetries and semantic selectivity, we employ a multi-head GAT encoder where attention scores are conditioned on type semantics. Entity and type embeddings are fused through a gating mechanism, and attention coefficients are computed via learned projections over the fused representations. This allows the model to prioritize neighbors that are not only structurally connected but also ontologically relevant. Residual pathways further enhance gradient flow and support feature reuse.

Motivation. The GCN and GAT encoders offer complementary strengths in ontology-rich knowledge graphs. GCNs facilitate robust type propagation via neighborhood smoothing, while GATs provide fine-grained, adaptive attention over semantically aligned neighbors. In schema-driven domains like EurostatKG—where entities are richly typed and constrained by formal axioms—this hybrid design enables deeper semantic abstraction and more nuanced relational reasoning.

Both encoders are implemented as modular components in our framework and are jointly optimized alongside embedding models and classifier heads. This design enables our architecture to simultaneously capture structural dependencies, symbolic constraints, and type-based semantic signals.

4.6 Classification Heads

To determine the plausibility of a given triple (h,r,t), we evaluate two classifier architectures that operate on the feature vectors produced by the embedding interaction modules: a multi-layer perceptron (MLP) and a one-dimensional convolutional network (Conv1D).

MLP-Based Classifier. The MLP consists of five fully connected layers with progressively decreasing sizes: [2048, 1024, 512, 256, 128], followed by a final output layer. Each intermediate layer is paired with:

- *Layer Normalization* for stability on small batches,
- *Adaptive Dropout*, ranging from 0.5 to 0.2,
- *Advanced Activation Functions* (e.g., SiLU, GELU, and LeakyReLU used in rotation), and
- *Residual Connections* applied conditionally when dimensions align.

The input feature vector is a concatenation of h, r, t, and $h+r$ features, enabling the model to encode both factual plausibility and directional translation. A final sigmoid activation produces a plausibility score in the $[0,1]$ range.

Conv1D-Based Classifier. For models using Conv1D heads (e.g., DistMult + GAT + Conv1D), we treat the triple vector as a pseudo-sequential input and apply two convolutional layers:

- *Conv1D Layers:* Two stacked 1D convolutions with 128 and 256 filters, kernel size 3, and padding = 1.
- Each layer is followed by *Batch Normalization*, *ReLU* activation, and *Dropout* (rate = 0.5).
- The output is flattened and passed through a *dense projection layer* (512 units), and finally through a 1-unit linear layer with sigmoid activation.

This configuration captures localized compositional patterns across the triple vector while maintaining model compactness and interpretability.

Modularity. Both classifier heads are implemented as plug-and-play modules in our framework. The MLP supports expressive global reasoning over concatenated embeddings, while the Conv1D head provides compact modeling of local compositional interactions. This modularity allows for consistent evaluation and ablation studies across different embedding backbones and graph encoders.

5 Evaluation and Results

5.1 Experimental Setup

Data and Splits. We adopt a stratified 70/15/15 split of EurostatKG with a **1:3 positive-to-negative ratio**, pairing each positive with three negatives from our hybrid strategy (Sect. 3.3). This yields a realistic imbalanced classification task. For reference, we also report a **1:1 balanced baseline** using only uniform corruption. All splits preserve class ratios and are shuffled with a fixed seed.

Implementation Details. All models are implemented in PyTorch. Primary experiments are run on an NVIDIA A100 GPU (40GB VRAM), and a Titan V GPU is used for selected configurations. Hyperparameters such as embedding dimension, learning rate, and dropout rate are optimized on the validation set via grid search. The complete source code, pre-trained models, and experiment logs are publicly available for reproducibility.[3]

Evaluation Metrics. We report results using two families of evaluation metrics:

- **Classification Metrics:** Accuracy, Precision, Recall, F1-score, and PR-AUC are used to evaluate models as binary classifiers distinguishing valid from invalid triples.
- **Ranking Metrics:** Although our models are not optimized for ranking, we also report standard metrics – Hits@K (K=1,5,10) and mean reciprocal rank (MRR) – to assess how well models prioritize true triples over negatives.
  Additionally, we introduce a *Soft Hits@K* metric based on sigmoid probabilities. A ground-truth triple is counted as a soft hit if its score s_{true} lies within a small margin ε (default $\varepsilon = 0.01$) of any of the top-K scores, i.e., $\exists j \leq K: \ s_{\text{true}} \geq s_{(j)} - \varepsilon$. This metric captures semantically plausible near-misses in ontology-rich KGs, where relations may be interchangeable or hierarchically related. Soft Hits@K is reported strictly as an *auxiliary* diagnostic and always shown together with standard Hits/MRR. We emphasize that S.H@K does not by itself verify semantic similarity beyond score proximity; it should therefore be interpreted with caution. To support more rigorous use, our released code includes hooks to extend S.H@K with ontology- or embedding-based similarity checks, which we leave to future work.

5.2 Overall Performance Comparison

We evaluate ten representative model variants that combine diverse embedding strategies, ontology-aware graph encoders, and classifier heads. For brevity, we present a subset of configurations that best illustrate performance trends. Table 1 summarizes results across standard classification and ranking metrics. In addition to conventional evaluation, we report Soft Hits@1 (with $\varepsilon = 0.01$) to assess semantically plausible near-misses—capturing semantically tolerant ranking behavior in ontology-rich graphs.

To isolate the contribution of our classification head, we also report results for ontology-aware ranking-only variants. These models use the same embedding backbone (e.g., TransE, DistMult), and are trained with margin-based loss ($\mathscr{L}_{\text{rank}}$) and semantic penalty ($\mathscr{L}_{\text{sem}}$), but exclude binary classification loss ($\mathscr{L}_{\text{BCE}}$). Their performance, shown in parentheses in Table 1, offers a direct comparison to full hybrid models and highlights the benefit of classification supervision.

A full BCE-only ablation, which would isolate the incremental value of the ranking and semantic terms, is not included here due to scope constraints; we leave this analysis to future work and expose the toggle in our released code for reproduction.

[3] https://github.com/sskaperdas/eurostatkg-relation-prediction.

Table 1. Model performance on EurostatKG (classification and ranking). For ranking metrics, values in parentheses correspond to the performance of equivalent ranking-only baselines. Model abbreviations: TE = TransE, TH = TransH, DM = DistMult, C1D = Conv1D, O = ontology-aware (semantic penalty + constraint-guided negative sampling + type-aware GNNs).

ID	Model	F1	PR-AUC	H@1	H@5	MRR	S.H@1
Baseline Models							
M1	TE+ MLP	0.739	0.827	0.233 (0.02)	0.484 (0.08)	0.360 (0.09)	0.582 (0.02)
M2	HolE+ MLP	0.730	0.838	0.089 (0.139)	0.474 (0.219)	0.283 (0.204)	0.424 (0.139)
M3	TH+ C1D	0.490	0.500	0.060 (0.225)	0.171 (0.324)	0.150 (0.283)	0.151 (0.226)
M4	TH+ GCN+ C1D	0.577	0.529	0.221 (0.110)	0.270 (0.138)	0.269 (0.162)	0.532 (0.110)
M5	DM+ C1D	0.182	0.507	0.020 (0.02)	0.116 (0.124)	0.101 (0.106)	0.084 (0.174)
M6	DM+ GAT+ C1D	0.629	0.832	0.275 (0.02)	0.503 (0.113)	0.402 (0.099)	0.351 (0.034)
Proposed Ontology-Guided Models (Ours)							
M7	TE+ MLP+ O	**0.937**	**0.987**	0.342 (0.24)	0.614 (0.42)	0.476 (0.351)	0.800 (0.244)
M8	HolE+ MLP+ O	0.895	0.968	0.141 (0.209)	0.595 (0.400)	0.356 (0.330)	0.783 (0.212)
M9	TH+ GAT+ GCN+ O+ MLP	**0.973**	**0.998**	0.291 (0.116)	0.598 (0.218)	0.446 (0.198)	**0.969** (0.117)
M10	TH+ GAT+ O+ MLP	0.972	**0.998**	**0.309** (0.053)	**0.618** (0.246)	**0.467** (0.156)	0.964 (0.054)

1. **Ontology Guidance Significantly Boosts Classification.**
   All ontology-aware models outperform their baselines, especially in classification metrics. For instance, TransE improves from an F1 of 0.739 (M1) to **0.937** (M7), and PR-AUC rises from 0.827 to **0.987**. This affirms the effectiveness of semantic regularization, type-aware GNNs, and constraint-informed sampling.
2. **Specialization Trade-off: Ranking vs. Classification.**
   Full models are trained with both classification and ranking objectives, while ranking-only models rely solely on margin-based supervision. In some cases, ranking-only variants show marginally higher Hits@1 or MRR—reflecting sharper score separation. However, they consistently underperform on classification metrics, underscoring limited generalization to binary decision-making. Ontology-guided hybrids provide more robust and semantically consistent predictions, crucial for schema-rich KGs like EurostatKG.
3. **Graph Encoders Offer Consistent Gains.**
   Incorporating GCN or GAT layers improves performance across all metrics. For example, augmenting TH with a GCN (M3 vs. M4) raises F1 from 0.490 to 0.577. These encoders enhance type propagation and inject valuable structural context.
4. **Soft Hits@1 Captures Semantic Nuance.**
   In ontology-aware KGs, multiple relations may be semantically interchangeable or closely related. For instance, M9 yields a strict Hits@1 of 0.291, but a Soft Hits@1 of **0.969**—suggesting that many near-misses remain ontologically justifiable.
   Illustrative Example: If the ground truth triple is (`EU_GDP`, `skos:broader`, `National_GDP`), and the model predicts `hasParentConcept` with a nearly equiv-

alent score, the prediction still reflects correct hierarchical reasoning. Soft Hits@1 captures these cases, offering a more nuanced view of model performance in semantic contexts.

5.3 Generalizability on a Standard Benchmark

To assess the architectural robustness of our framework beyond schema-rich settings, we evaluate selected models on WN18RR—a widely used link prediction benchmark characterized by the absence of formal ontological structure. Unlike EurostatKG, WN18RR provides neither domain and range constraints nor comprehensive type assertions, making it an ideal testbed for evaluating how well our ontology-aware architecture generalizes in ontology-agnostic environments.

We adopt the same classification-based training protocol as in the EurostatKG experiments, while disabling ontology-specific mechanisms—such as constraint-based losses and type-guided message passing. This ensures a consistent evaluation of model architecture under settings where schema-level supervision is unavailable.

On Baseline Comparisons. The WN18RR evaluation is presented solely to examine the robustness of our architecture in an ontology-agnostic setting. Because WN18RR lacks ontological supervision, our semantic components are inactive. A thorough, like-for-like comparison against classical baselines on WN18RR is beyond the scope of this paper and is left for future work.

Table 2 presents results for four representative model configurations. Despite the lack of ontological structure, our models remain competitive across standard link prediction metrics. Notably, the most expressive variant (M9) performs best on all metrics, suggesting that enhancements introduced for schema-aware domains do not compromise generalization in lighter-weight benchmarks.

Caution and Scope. These results are not intended to validate the ontology-specific components of our architecture within WN18RR, as such mechanisms are inapplicable due to the absence of schema annotations. Rather, the goal is to assess whether our models degrade gracefully when transferred to structurally simpler benchmarks. This complements our findings on EurostatKG, where ontology-guided modeling yields substantial gains.

5.4 Qualitative Insights

We analyze representative test-set triples from EurostatKG to illustrate how ontology-guided models improve semantic plausibility, enforce ontological constraints, and offer robustness over baseline models. We report the ranking positions assigned by selected models—M1 (TransE + MLP), M3 (TransH + C1D), and M7–M10 (ontology-aware variants)—where a lower rank indicates higher confidence.

Table 2. Generalizability results on WN18RR for selected models using classification-based training. Abbreviations: TE = TransE, DM = DistMult, TH = TransH, C1D = Conv1D, O = Ontology-guided.

ID	Model	MRR	H@1	H@5	S.H@1
M1	TE + MLP	0.270	0.115	0.360	0.515
M6	DM + GAT + C1D	0.270	0.084	0.460	0.234
M7	TE + MLP + O	0.342	0.135	0.517	0.343
M9	TH + GAT + GCN + O + MLP	**0.468**	**0.315**	**0.632**	**0.840**

Category 1: Valid triples ranked higher by ontology-aware models. This category includes correct triples for which ontology-guided models significantly outperformed baselines:

- **(mar_pa_qm_el,** `level`**, 5)**—Rank: **2** (M9), 41 (M1), 20 (M10), 31 (M7), 40 (M3)
- **(tet00059,** `level`**, 4)**—Rank: **2** (M9), 37 (M1), **3** (M10), 18 (M7), 38 (M3)
- **(tin00099,** `level`**, 5)**—Rank: **1** (M9), 40 (M1), 25 (M10), 30 (M7), 40 (M3)

Category 2: Plausible but constraint-violating triples. These triples appear statistically plausible but violate ontological constraints (e.g., domain or range typing). Ontology-aware models correctly assign lower confidence:

- **(internet_in_households,** `hasTopic`**, SEArticle4068)**—Rank: 43 (M9), 35 (M1), 35 (M10), 40 (M7), **18** (M3)
- **(SEArticle8176,** `type`**, SEArticle8176)**—Rank: 43 (M9), 24 (M1), **8** (M10), 38 (M7), 41 (M3)
- **(SEArticle8987,** `type`**, SEArticle8987)**—Rank: 43 (M9), 18 (M1), **8** (M10), 38 (M7), 41 (M3)

Category 3: Deliberately invalid triples violating ontology rules. These examples were synthetically generated to violate `rdfs:domain` or `rdfs:range` axioms. Ontology-aware models consistently assign them low plausibility:

- **(paragraph5569_4519,** `content`**, The_statistical_population_consists...)**—Rank: 39 (M9), 39 (M1), **17** (M10), **16** (M7), 35 (M3)
- **(hlth_dsi080,** `term`**, hlth_dsi080)**—Rank: 42 (M9), **11** (M1), 43 (M10), 27 (M7), 38 (M3)
- **(paragraph5655_1000,** `content`**, Table_1_and_Figure...)**—Rank: 40 (M9), 37 (M1), 36 (M10), **17** (M7), 33 (M3)

Observations.

- **Ontology-aware models (M7–M10)** consistently elevated valid triples and demoted invalid or constraint-violating ones, demonstrating their semantic sensitivity.
- **GCN- and GAT-based encoders (M9, M10)** exhibited greater robustness by propagating semantic signals more effectively and recovering subtle patterns missed by embedding-only models.
- **Soft Hits@1 captures semantically near-correct predictions:** For example, models like M9 and M10 achieved Soft Hits@1 scores ≥ 0.96, suggesting strong plausibility even when strict top-1 accuracy was missed.
- **Baselines (M1, M3)** frequently over-ranked implausible or invalid triples and under-ranked valid ones, indicating a lack of constraint awareness.
- **TransH-based, ontology-aware models (M9, M10)** demonstrated the best trade-off between semantic compliance and generalization, owing to the synergy of embedding, graph encoding, and constraint-informed training.

6 Conclusion

We presented a unified ontology-guided framework for relation prediction in schema-rich KGs, using EurostatKG as a benchmark. Rather than proposing new architectures, we integrated established components—embedding models, type-aware GNNs, constraint-aware negative sampling, and a hybrid loss—into a cohesive pipeline that promotes semantic coherence and factual plausibility.

Experiments showed consistent gains in accuracy and robustness. The proposed soft ranking metrics complement strict ones by capturing near-miss plausibility, though they remain auxiliary diagnostics rather than replacements.

This work establishes a strong, reproducible baseline while opening several directions:

- **Richer logic:** Extend beyond domain/range to OWL features such as property chains, transitivity, and cardinality.
- **Benchmarks:** Expand evaluation with WN18RR and domains such as law, biomedicine, or scientific literature.
- **Embeddings:** Explore new paradigms within the ontology-aware pipeline.
- **Scalability:** Profile complexity, scale to larger KGs, and optimize type-aware message passing.
- **Post-processing:** Add symbolic consistency checks.
- **Asserted vs. inferred:** Move beyond asserted types to unions and inferred semantics under OWA.
- **Refined evaluation:** Enhance soft ranking with ontology- or embedding-based similarity to distinguish genuine near-misses from errors.

By embedding symbolic knowledge into neural learning, this work advances interpretable, trustworthy KG completion and fosters further hybrid neuro-symbolic research.

References

1. Vassiliades, A., Bassiliades, N., Meditskos, G., Spiliopoulos, K.: Building eurostat knowledge graph. In: KEOD (2022)
2. Bordes, A., Usunier, N., Garcia-Durán, A., Weston, J., Yakhnenko, O.: Translating embeddings for modeling multi-relational data. In: NeurIPS (2013)
3. Yang, B., Yih, W., He, X., Gao, J., Deng, L.: Embedding entities and relations for learning and inference in knowledge bases. In: ICLR (2015). arXiv:1412.6575
4. Trouillon, T., Welbl, J., Riedel, S., Gaussier, É., Bouchard, G.: Complex embeddings for link prediction. In: ICML (2016)
5. Schlichtkrull, M., Kipf, T., Bloem, P., van den Berg, R., Titov, I., Welling, M.: Modeling relational data with graph convolutional networks. In: ESWC (2018)
6. Veličković, P., Cucurull, G., Casanova, A., Romero, A., Liò, P., Bengio, Y.: Graph attention networks. In: ICLR (2018). arXiv:1710.10903
7. Smaili, F.Z., Gao, X., Hoehndorf, R.: Onto2Vec: joint vector-based representation of biological entities and ontology annotations. Bioinformatics (2018)
8. Kulmanov, M., Khan, M.A., Hoehndorf, R.: Deep learning with symbolic knowledge for biomedicine. Briefings Bioinform. (2021)
9. Sun, Z., Deng, Z.-H., Nie, J.-Y., Tang, J.: RotatE: knowledge graph embedding by relational rotation. In: ICLR (2019). arXiv:1902.10197
10. Kadlec, R., Bajgar, O., Kleindienst, J.: Knowledge base completion: baselines strike back. In: EACL (2017)
11. Guo, X., Wang, Q., Wang, B., Guo, L.: Jointly embedding knowledge graphs and logical rules. In: EMNLP (2016)
12. Rocktäschel, T., Singh, S., Riedel, S.: Injecting logical knowledge into embeddings. In: NAACL-HLT (2015)
13. Zhang, S., Tay, Y., Yao, L., Liu, Q., Wang, D.: Type-constrained representation learning in knowledge graphs. In: CIKM (2019)
14. Nguyen, D.Q., Nguyen, T.N., Phung, D.: A novel CNN-based embedding model. In: NAACL-HLT (2018)
15. Dettmers, T., Minervini, P., Stenetorp, P., Riedel, S.: Convolutional 2D knowledge graph embeddings. In: AAAI (2018)
16. Callahan, A., Cruz-Toledo, J., Ansell, P., Dumontier, M.: Bio2RDF release 2: coverage and provenance of life science linked data. In: ISWC (2011)
17. Sun, Z., Deng, Z.-H., Nie, J.-Y.: Self-adversarial negative sampling for knowledge graph embedding. In: ACL (2020)
18. Wang, Z., Zhang, J., Feng, J., Chen, Z.: Knowledge graph embedding by translating on hyperplanes. In: AAAI (2014)
19. Ruffinelli, D., Broscheit, S., Gemulla, R.: You CAN teach an old dog new tricks: on training knowledge graph embeddings. In: ICLR (2020)
20. Nickel, M., Rosasco, L., Poggio, T.: Holographic embeddings of knowledge graphs. In: AAAI (2016)
21. Vashishth, S., Sanyal, S., Nitin, V., Talukdar, P.: Composition-based graph convolutional networks. In: ICLR (2020)
22. Yao, L., Mao, C., Luo, Y.: KG-BERT: BERT for knowledge graph completion. arXiv:1909.03193 (2019)
23. Wang, Y., Liu, H., Wu, Y., Wang, W.-Y.: KEPLER: a unified model for knowledge embedding and pre-trained language representation. arXiv:1911.02168 (2019)
24. Ristoski, P., Paulheim, H.: RDF2Vec: RDF graph embeddings. In: ISWC (2016)

25. Tang, J., Hu, Z., Liu, M.: Orthogonal negative sampling for link prediction. In: NeurIPS (2020)
26. Paulheim, H.: Knowledge graph refinement: a survey. Semantic Web (2017)
27. Wang, Q., Mao, Z., Wang, B., Guo, L.: Knowledge graph embeddings: methods and applications. IEEE TKDE (2021)
28. He, J., Chen, J., Shao, C., Zhao, H., Wang, H., Liu, H.: OWL2Vec*: embedding of OWL ontologies. arXiv:2009.14654 (2020)
29. D'Amato, C., Rizzo, G., Staab, S., Zingoni, G.: Injecting background knowledge into embedding models for predictive tasks. In: ESWC (2021)
30. Holter, O.M., Chen, J., Jiménez-Ruiz, E., Antonyrajah, D., Horrocks, I.: Embedding OWL ontologies with OWL2Vec. In: ISWC Workshops (2019)

Semantic Annotation of Energy Data Using Ensemble Decision-Making with Large Language Models

Zhiyu Pan[1(✉)], Yongli Mou[2], Fengshuo Hao[1], Yuting Gao[1], Stefan Decker[2,3], and Antonello Monti[1,3]

[1] Institute for Automation of Complex Power Systems, RWTH Aachen University, Mathieustr. 10, 52074 Aachen, Germany
{zhiyu.pan,fengshuo.hao,yuting.gao,amonti}@eonerc.rwth-aachen.de
[2] Chair of Computer Science 5, RWTH Aachen University, Ahornst. 55, 52074 Aachen, Germany
{mou,decker}@dbis.rwth-aachen.de
[3] Fraunhofer Institute for Applied Information Technology FIT, 53757 Sankt Augustin, Germany

Abstract. Datasets in the energy domain often encounter significant challenges in achieving data interoperability due to semantically equivalent concepts being inconsistently named. Semantic annotation addresses this issue by standardizing concepts to enhance data interoperability and facilitate efficient data integration. Traditional annotation methods, however, rely heavily on static resources, which limit their ability to dynamically understand cross-domain context. Large Language Models (LLMs) have shown promise in capturing semantic similarities between terms, and their capabilities can be enhanced by incorporating ontological reasoning. In this paper, we propose a novel framework that leverages LLMs for ensemble decision-making to automate and improve semantic annotation. By combining LLM-based semantic understanding with structural reasoning from ontologies, our approach enhances reliability and annotation performance. We evaluate its effectiveness using both proprietary and open-source models on real data from industrial companies in the energy sector. The experimental results demonstrate superior annotation accuracy compared to baseline methods. This study highlights the potential of integrating ontological reasoning, LLMs, and ensemble decision-making to advance semantic data integration in the energy domain.

Keywords: Semantic Annotation · Large Language Models · Ensemble Decision-Making

1 Introduction

The energy sector today generates vast and rapidly growing volumes of data, from finely grained, time-stamped electricity usage logs at the building level, to

H. Takeda et al. (Eds.): IJCKG 2025, LNAI 16297, pp. 95–110, 2026.
https://doi.org/10.1007/978-981-95-5009-8_7

user profiles in XML, to unstructured textual reports and technical documents. This diversity in format and structure, coupled with the absence of universally adopted and consistently applied metadata standards, leads to deep semantic fragmentation: identical concepts may be labeled differently across datasets, key relationships remain implicit or inconsistently expressed, and domain-specific terms lack shared, machine-readable definitions [18]. As a result, integrating data from disparate sources becomes an error-prone, labor-intensive process, and automated systems struggle to interpret or combine information without manual intervention. Such semantic ambiguity not only hampers large-scale analytics and data exchange but also impedes collaboration across stakeholders in the energy ecosystem. Addressing these challenges requires the explicit encoding of meaning, clarifying what each term represents and how concepts relate, which is precisely the motivation for adopting ontology-based semantic annotation in the energy domain [2,13]. By anchoring data elements to shared vocabularies, semantic annotation provides the necessary metadata to ensure interoperability, enable reliable semantic search, and foster seamless data reuse.

In recent years, the emergence of large language models (LLMs) [12] has opened up new avenues for semantic annotation. These models are typically trained in a self-supervised manner on massive datasets and can be subsequently fine-tuned on relatively small, domain-specific corpora. Integrating pre-trained transformer models such as BERT or similar LLMs into the annotation workflow has shown promising results [19], substantially improving both the efficiency and accuracy of semantic annotation tasks. As the parameter sizes of LLMs grow, their capacity for natural language understanding and domain adaptation increases, laying a robust foundation for automated ontology-driven semantic annotation [17].

Despite these advances, current LLMs face notable limitations. One critical issue is the ability to understand complex language contexts [21]; large ontologies may exceed the token capacity of a single LLM, hindering its ability to fully comprehend and accurately annotate complex, domain-specific concepts. Moreover, relying solely on a single LLM can limit domain coverage and reasoning capabilities [22], as different models may exhibit varying strengths in distinct aspects of domain understanding. These challenges highlight the need for novel strategies that can mitigate token constraints and leverage the complementary strengths of LLM-based ensemble decision-making.

In this paper, we propose an LLMs-based ensemble annotation framework[1] that integrates multiple LLMs in a collaborative manner to achieve ontology-based semantic annotation. By dividing the ontology into manageable segments and assigning them to multiple LLMs, we improve interpretation ability for complex language contexts and the overall annotation accuracy. Our approach ensures that each agent focuses on a portion of the ontology, thereby enhancing coverage and reducing the impact of any single model's limitations.

The main contributions are summarized below:

[1] Source code is available from Github https://github.com/zhiyupan/SAED_llm/.

- We propose a novel semantic annotation framework that integrates general-purpose LLMs with domain-specific ontologies to automate the annotation of energy sector datasets.
- We introduce an ensemble-based strategy that leverages multiple LLMs to collaboratively interpret complex language contexts and improve domain coverage, overcoming the contextual limitations of single-model approaches.
- We curate and release new benchmark datasets[2] derived from real-world energy industry sources, comprising manually validated ground-truth annotations aligned with the energy domain ontology, to foster reproducible evaluation and accelerate future research.
- We demonstrate the effectiveness of our method through comprehensive evaluations using both new benchmark datasets and open-source models, achieving superior annotation accuracy compared to traditional baselines.

2 Related Work

In the energy sector, semantic annotation plays a crucial role in managing and interpreting vast amounts of heterogeneous data generated from various sources, including smart grids, renewable energy systems, and consumption monitoring devices. Ontologies have been developed to model energy-related concepts, enabling standardized data representation and interoperability.

Currently, most semantic annotation tools rely on fixed knowledge graphs (e.g., DBpedia, Wikidata) for entity and type lookups, without offering a user-driven way to upload domain-specific ontologies. For example, MantisTable [4] supports SPARQL-based entity linking against DBpedia/Wikidata but does not provide an interface to import custom OWL/RDF ontologies. As a result, users cannot seamlessly integrate domain-specific ontologies into the annotation workflow, and touted import capabilities often remain inaccessible or unreproducible. SiSEG [5] is an automatic annotation service in the energy domain that addresses the problem of heterogeneous data integration based on domain ontology. Based on different ontologies, different results of the annotation service are provided. For example, if the device is already defined in the ontology, the accuracy is 96.77%, otherwise the accuracy is only 27.27%. Automatic semantic annotation methods have traditionally relied on tools such as WordNet [15] or measures like cosine similarity [10] to determine the "semantic similarity" between data resources and ontology concepts. However, such tools often depend on manually constructed resources and predefined hierarchical relationships. This reliance on static resources limits their ability to dynamically understand context and generalize across complex, multidimensional semantic tasks. Consequently, these approaches are less suitable for handling large-scale, dynamic data with intricate semantic relationships [11].

[2] Datasets are available from Zenodo https://zenodo.org/records/15270008.

Recent advancements in deep learning techniques, have revolutionized automatic semantic annotation [7]. Techniques leveraging pre-trained language models [16], Transformer architectures [23], and LLMs [17] have demonstrated significant improvements in accuracy, scalability, and adaptability. LLMs represent a paradigm shift in semantic annotation by leveraging their ability to understand, generate, and manipulate human-like text on a scale. Pre-trained on vast amounts of text data, LLMs such as GPT, BERT, and their domain-specific variants have demonstrated exceptional capabilities in capturing contextual relationships, semantic nuances, and domain-specific knowledge [14]. Unlike traditional tools that rely on static ontologies or predefined hierarchies, LLMs dynamically learn and adapt to complex, multidimensional relationships within data [3].

The use of single large language models for semantic annotation presents limitations due to variations in training datasets, resulting in discrepancies in understanding domain-specific definitions. To address these challenges, we focus on Large Language Models-based Ensemble Decision-Making. This approach aims to enhance annotation efficiency, improve automation, and enable intelligent collaboration among agents, thereby addressing the inherent complexity and heterogeneity in semantic tasks [24].

3 Methodology

In this section, we detail the proposed methodology for automating the semantic annotation of energy data using an ensemble decision-making approach facilitated by LLMs. Our approach comprises two main parts: implementing an ontology-driven breadth-first search for class identification, and orchestrating an ensemble decision-making mechanism among multiple LLM instances to ensure accurate and reliable annotations.

3.1 Problem Statements

In this work, semantic annotation is the task of assigning semantic types to columns in tabular datasets by mapping them to classes in a domain ontology. While traditional database-to-ontology mappings often align columns with binary relations (e.g., attributes or properties between entities), in this work, we focus on mapping columns directly to classes. This approach is particularly suitable for energy datasets, where many columns represent independent, domain-specific concepts that align naturally with ontological classes, such as Fuel Type, Energy Source, or Region.

For example, consider a dataset of power plants with a column named “Fuel Type” containing values like “Coal”, “Natural Gas”, and “Solar”. Instead of mapping this to a property like “usesFuel”, we map it directly to the class “EnergySource” in the ontology. This allows the system to recognize “Coal” as an instance of “FossilFuel”, which is a subclass of “EnergySource”, enabling richer reasoning and improved data integration across datasets that use similar categorical labels.

Formally, let $\mathcal{T} = \{t_1, t_2, \cdots, t_n\}$ denote a set of tables, where each table t_i comprises a set of columns $\mathcal{C}_i = \{c_{i,1}, c_{i,2}, \cdots, c_{i,j}\}$. The task of semantic annotation involves mapping each column $c_{i,j}$ to a corresponding class or classes within a predefined ontology $\mathcal{O} = \{o_1, o_2, \cdots, o_m\}$, thereby standardizing the semantic representation across diverse datasets. Formally, the semantic annotation process can be represented as a mapping function:

$$\mathcal{M} : \mathcal{C} \to \mathcal{P}(\mathcal{O}), \quad \text{where } \mathcal{M}(c_{i,j}) = \begin{cases} \{o_s\}, & \text{a single ontology class,} \\ \{o_{s_1}, o_{s_2}, \cdots\} \subset \mathcal{O}, & \text{multiple ontology classes,} \\ \emptyset, & \text{no assigned class.} \end{cases} \tag{1}$$

where $\mathcal{C} = \bigcup_{i=1}^{n} \mathcal{C}_i$ is the set of all columns across all tables, $\mathcal{P}(\mathcal{O})$ is the power set of the ontology class set and $\mathcal{M}(c_{i,j})$ assigns an ontology class or multiple ontology classes or none to the column $c_{i,j}$. The diversity in column naming conventions and the complexity of energy domain terminologies pose significant challenges to achieving accurate semantic mappings. Columns representing semantically equivalent concepts may have varying or inconsistent names across different tables. There may exist columns $c_{i,j}$ and $c_{p,q}$ such that $c_{i,j} \neq c_{p,q}$ but $\mathcal{M}(c_{i,j}) = \mathcal{M}(c_{p,q})$. Formally, the objective of semantic annotation is to construct $\mathcal{M}$ using computational methods to assigns the most relevant ontological term o to each class c larger than a explicit or implicit threshold τ:

$$\mathcal{M}(c) = \{o \in \mathcal{O} \mid \mathcal{S}(c, o) \geq \tau\} \tag{2}$$

The scoring function $\mathcal{S} : \mathcal{C} \times \mathcal{O} \to [0, 1]$ measures similarity between c and o. To address these challenges, we propose a framework that integrates ontology-driven search mechanisms with an ensemble decision-making process among multiple LLMs. Figure 1 illustrates the architecture of the proposed framework. The framework consists of ontology-driven search mechanism and ensemble decision-making module.

3.2 Ontology-Driven Breadth-First Search

At the core of our semantic annotation approach lies the energy ontology. The ontology encapsulates the various classes and their interrelationships within the energy domain, providing a structured vocabulary for annotation.

The energy ontology classes are organized in a hierarchical structure, represented as a Directed Acyclic Graph (DAG). In such a graph, nodes represent ontology classes and edges represent hierarchical relationships, in our case, subclass-superclass relationships. Formally, the hierarchy is expressed as a DAG $\mathcal{G} = (\mathcal{O}, \mathcal{E})$, where $\mathcal{O}$ is the set of nodes (ontology classes). $\mathcal{E} \subseteq \mathcal{O} \times \mathcal{O}$ is the set of directed edges representing parent relation between classes. Figure 2 shows an example DAG where the root node represents an ontology class "Thing", and directed edges connect it to more specific subclasses, illustrating the hierarchical organization and specialization of ontology classes.

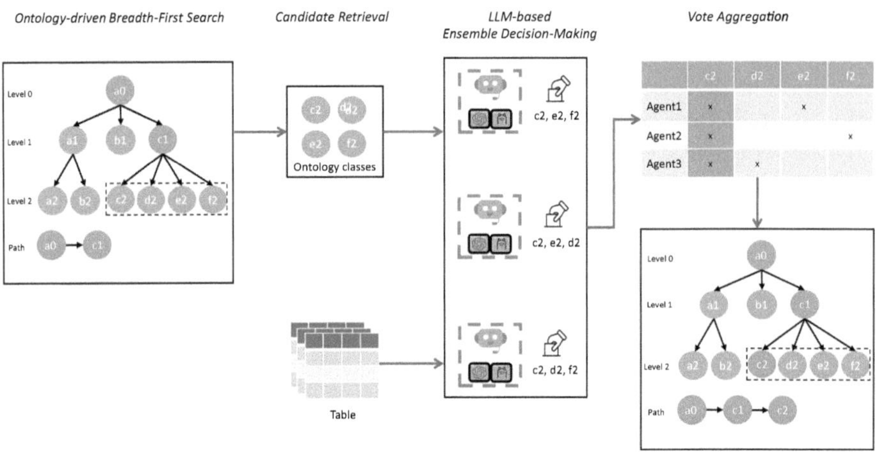

Fig. 1. Overview of the proposed framework integrating ontology-driven search with ensemble decision-making among multiple LLMs for semantic annotation.

To accurately map a table column to the most appropriate ontology class, we employ a Breadth-First Search (BFS) algorithm to traverse the ontology hierarchy systematically. The BFS algorithm ensures that the search begins from the most general classes and progresses towards more specific subclasses, facilitating the identification of the most semantically relevant class for annotation. This systematic approach ensures that the highest-level semantic context is considered before delving deeper into the ontology. In contrast to depth-first methods, which might quickly dive into a very specific branch and potentially overlook other broad, relevant classes, BFS guarantees a more balanced exploration across the ontology structure. This is especially important in semantic annotation, where maintaining a comprehensive understanding of all possible class relationships is crucial for accurately mapping table columns to the most semantically relevant ontology class. Algorithm 1 outlines the BFS process for class identification within the energy ontology. The algorithm systematically explores the ontology starting from the root class, utilizing LLM instances to evaluate the relevance of subclasses based on the context of the table.

The ontology-driven BFS algorithm (see Algorithm 1) starts with initialization: An empty queue is initialized to manage the BFS traversal. Each element in the queue is a tuple containing the current traversal level l_{current}, the current ontology class o_{current}, and the path taken to reach this class $path_{\text{current}}$. An empty list is prepared to store all possible paths that lead to candidate ontology classes meeting the similarity criteria. The algorithm begins by creating an initial path that contains only the root class o_r (e.g., "Thing"). This path is enqueued into Q with an initial traversal level of 0.

Then, the algorithm enters a loop that continues until the queue Q is empty, ensuring that all relevant classes within the traversal constraints are explored. The front element of the queue is dequeued for evaluation. If the current traver-

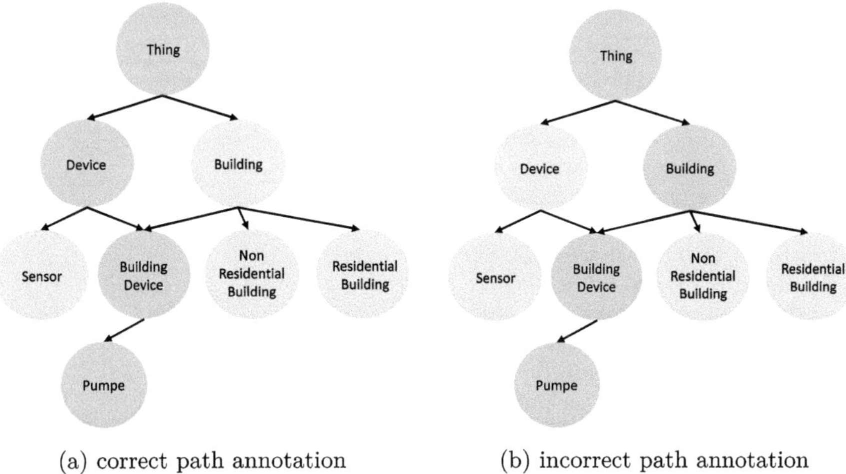

(a) correct path annotation (b) incorrect path annotation

Fig. 2. Hierarchical representation of ontology classes as a Directed Acyclic Graph (DAG), illustrating subclass relationships and specialization path annotation for the class within the ontology.

sal level l_{current} has reached or exceeded the maximum allowed traversal level $l_{\max}$, the current path is appended to P, and the algorithm proceeds to the next iteration. The algorithm retrieves all direct subclasses $\hat{\mathcal{O}}$ of the current class o_{current} from the ontology $\mathcal{O}$. If $\hat{\mathcal{O}}$ is empty, indicating that o_{current} has no subclasses, the current path is appended to P, and the algorithm continues to the next iteration.

Then, an Ensemble Decision-Making (EDM) module is employed (see Algorithm 2) with LLM instances to identify relevant ontology classes at the current level. The EDM module is described in Sect. 3.3. For each subclass o_{sub} identified by the LLM in $\hat{\mathcal{O}}$, a new path $path_{\text{new}}$ is created by appending o_{sub} to the existing $path_{\text{current}}$.

Once the queue Q is empty, indicating that all possible paths within the traversal constraints have been explored, the algorithm returns the list of paths P. Each path in P represents a potential mapping from the root class to a candidate ontology class relevant to the table column. The inclusion of the full path provides essential contextual and interpretative value beyond merely identifying the class itself. While the final ontology class is indeed the ultimate mapping target, the path encapsulates the hierarchical relationships and intermediate decisions made during the search. In complex ontologies, several classes might share similar characteristics or labels. For example, the correct path for annotating the ontology class "Pumpe" is illustrated in Fig. 2a. However, because "BuildingDevice" is defined as a subclass of both "Device" and "Building," it inherits properties from both, which can lead to erroneous semantic annotation paths (see Fig. 2b). Displaying the hierarchical path of each candidate class disambiguates these overlaps by pinpointing the specific branch of the ontology

Algorithm 1: Ontology-driven Breadth-First Search for Class Identification

Input: Ontology DAG $\mathcal{G} = (\mathcal{O}, \mathcal{E})$ with root class o_r, table name t_n, table header t_h, K rows of example data $t_d(K)$, column name c, maximum traversal level l_{max}, and a set of LLM agents $\mathcal{A} = \{A_1, \ldots, A_t\}$
Output: All possible paths to candidate ontology classes P

Initialize an empty queue Q;
Initialize an empty list of paths P;
Create the initial path with the root class: $path \leftarrow [o_r]$;
Enqueue $(0, R, path)$ into Q;
while *Q is not empty* **do**
  Dequeue the front element $(l_{\text{current}}, o_{\text{current}}, path_{\text{current}})$ from Q;
  if $l_{current} \geq l_{max}$ **then**
    Append $path_{\text{current}}$ to P;
    continue;
  Determine the direct subclasses of o_{current}:
    $\hat{\mathcal{O}} \leftarrow \{o \mid \exists o_{\text{current}} \in O : (o, o_{\text{current}}) \in \mathcal{E}\}$;
  if *$\hat{\mathcal{O}}$ is empty* **then**
    Append $path_{\text{current}}$ to P;
    continue;
  Identify classes at the traversal level using the EDM function:
    $O_{edm} \leftarrow EDM(t_n, t_h, t_d(K), \hat{\mathcal{O}}, \mathcal{A})$;
  for *each subclass $o_{sub} \in O_{edm}$* **do**
    Create a new path: $path_{\text{new}} \leftarrow path_{\text{current}} + [o_{\text{sub}}]$;
    Enqueue $(l_{\text{current}} + 1,\ o_{\text{sub}},\ path_{\text{new}})$ into Q;
return P;

from which the class originates. This added context helps ensure that the final mapping correctly aligns with the intended domain semantics.

3.3 Large Language Models-Based Ensemble Decision-Making

Our EDM module is designed to distribute the task of ontology class identification among multiple LLM instances, including orchestrating the distribution of ontology classes to LLM instances, the voting mechanism via LLMs, and the aggregation of votes to form consensus-based annotations. This ensemble approach leverages the collective intelligence of multiple LLM instances to mitigate individual model biases, handle context length limitations, and achieve consensus-based annotations. Given the direct subclasses $\hat{\mathcal{O}}$ of the current class , table context (name t_n, header t_h, and K sample rows $t_d(K)$), and a set of LLM agents $\mathcal{A} = \{A_1, \ldots, A_t\}$, the Algorithm 2 performs the following steps:

1. **Agent Distribution:** Distribute $\hat{\mathcal{O}}$ among the LLM agents using configurable parameters (avg_agents_per_class and avg_classes_per_agent).

Those parameters balance the workload and enhance consensus reliability, i.e., average classes per LLM specifies the average number of ontology classes each LLM instance is responsible for evaluating, average LLMs per class determines the average number of LLM instances assigned to evaluate each ontology class. This overlapping assignment ensures each candidate is evaluated by multiple agents while keeping individual workloads manageable.

2. **LLM Evaluation:** For each agent, construct a prompt that includes the table's metadata and its assigned subset of candidates. Each agent processes the prompt and returns a set of classes it deems relevant, effectively casting a vote for those classes. Each instance evaluates a subset of ontology classes based on its semantic understanding, and votes on the most appropriate mappings based on the table's context, specifically, the LLMs analyze the table name t_n, table header t_h, example data $t_d(K)$, and the current ontology class o_{current} to suggest relevant subclasses O.
3. **Vote Aggregation and Consensus:** For each candidate $c \in \hat{O}$, compute the vote ratio as

$$\text{ratio}[c] = \frac{\text{votes}[c]}{\text{seen}[c]},$$

where votes[c] is the number of positive votes and seen[c] is the number of agents that evaluated c. Candidates with a vote ratio exceeding a predefined consensus threshold are retained.

4 Evaluation

We evaluate the effectiveness of our proposed semantic annotation framework for the energy domain, comparing it against baseline approaches that utilize a single LLM instance. First, the dataset and metrics used for the evaluation are introduced. Secondly, the implementation details are described, and finally, an in-depth analysis and discussion of the results are given.

4.1 Datasets

We utilize a new benchmark dataset comprising an energy ontology and a collection of energy-related tables. Those tables are collected from industrial partners in the energy domain. This dataset is instrumental in assessing the framework's capability to accurately annotate columns with appropriate ontology classes. While we engaged with technical experts from these companies, we created the ground truth of this annotation task.

Tables. The benchmark dataset for annotation comprises tables from different city management departments, network operators, building management companies, and other related sources, representing a diverse range of information. The tables encompass various energy forms (e.g., wind, solar), energy consumption

Algorithm 2: EDM Ontology Class Identification

Input: t_n (table name), t_h (table header), $t_d(K)$ (K example data rows), $\hat{\mathcal{O}}$ (direct subclasses of o_{current}), $\mathcal{A} = \{A_1, \ldots, A_t\}$ (LLM agents), *avg_agents_per_class*, *avg_classes_per_agent*, *consensus_threshold*

Output: Relevant subclasses of $o_{current}$ meeting consensus threshold

Initialize task lists for each agent tasks$[A_i]$;
Initialize counters ($votes[c]$, $seen[c]$) for each subclass;
Initialize an empty *result*;

for *each class* $c \in \hat{\mathcal{O}}$ **do**
    $A_c \leftarrow$ select up to *avg_agents_per_class* agents with $|tasks[A]| < avg\_classes\_per\_agent$;
    for *each agent* $A_j \in A_c$ **do**
        tasks$[A_j] \leftarrow$ tasks$[A_j] \cup \{c\}$;
        $seen[c] \leftarrow seen[c] + 1$;

for *each agent* $A_i \in LLM$ **do**
    $Q_i \leftarrow$ generatePrompt(t_n, t_h, $t_d(K)$, tasks$[A_i]$);
    $response_i \leftarrow A_i$.query($Q_i$);
    $V_i \leftarrow \{\, c \in$ tasks$[A_i] \mid c$ is marked relevant in $response_i \,\}$;
    for *each class* $c \in V_i$ **do**
        $votes[c] \leftarrow votes[c] + 1$;

for *each class* $c \in \hat{\mathcal{O}}$ **do**
    if $\frac{votes[c]}{seen[c]} \geq consensus\_threshold$ **then**
        $result \leftarrow result \cup \{c\}$;

return *result*;

data, weather-related variables, and measurement types (e.g., meter data, aggregated summaries). The dataset spans three locations over a 12-month period and is multilingual (e.g., English, Portuguese), reflecting real-world diversity. In total, 48 tables are included, each comprising ten rows.

Ontology. The Building Energy Domain (BED) ontology[3] is employed as the reference ontology. It captures semantic information across the energy and built environment sectors, covering devices, measurements, user behavior, geospatial contexts, and environmental conditions. The ontology comprises approximately 602 classes and over 1,300 SubClassOf axioms, with hierarchical depth indicated by the distribution of classes across various levels (1 at level 0, 86 at level 1, 159 at level 2, 114 at level 3, 100 at level 4, 74 at level 5, 72 at level 6, and 88 at level 7). In total, there are 6,625 axioms, including 2,248 logical axioms and 1,456 declaration axioms.

[3] Building Energy Domain (BED) Ontology: https://w3id.org/def/bed.

4.2 Metrics

To quantitatively assess the performance of our semantic annotation framework, we employed a suite of metrics that evaluate the accuracy of the annotations. These metrics are categorized into precision, recall, and F1-score at both path and node levels using macro and micro averaging. Path-level metrics evaluate the accuracy of the entire annotation path from the root class to the target class within the ontology, while node-level metrics evaluate the accuracy of individual ontology class annotations without considering the entire path.

4.3 Implementation Details

In this section, we outline the implementation details, including baseline models and parameter settings.

Baseline Models. To contextualize the performance of our proposed framework, we compared it against the baseline models of single LLM annotation based on [9], which simply put the ontology content in the prompt and ask the LLM to classify table columns with the ontology classes. We also compared with single LLM annotation with Chain-of-Throught (CoT) prompting [20] that utilizes a single instance of GPT-4o-mini from OpenAI on Azure and Llama 3.1 8B from Meta for semantic annotation without the ensemble decision-making mechanism. Additionally, we conducted experiments with the traditional semantic annotation method SiSEG [5].

Parameter Settings. The ensemble decision-making framework was configured with the following parameters: average classes per LLM, average LLMs per class, and consensus threshold ratio set to 30, 3, and 0.8, respectively. To simplify the manual annotation process, we only perform two-level manual annotations as the ground truth; therefore, the maximum depth is limited to 2.

4.4 Results

This subsection presents the empirical findings of our evaluation, comparing the performance of the proposed ensemble decision-making framework against the baseline models across various metrics. The results are categorized into path-level and node-level analyses, as illustrated in Table 1 and Table 2, respectively.

Path-Level Result Analysis. Table 1 presents the performance metrics at the path level, evaluating the framework's ability to correctly annotate the entire annotation path from the root class to the target class within the ontology.

The ensemble decision-making (EDM) approach significantly enhances Precision compared to both the base and CoT variants of Llama 3.1 8B. Specifically, Llama 3.1 8B (EDM) achieves a Precision of 64.79%, a marked improvement from the base model's 37.01% and the CoT-enhanced model's 40.38%. This substantial

Table 1. Path-Level Performance Metrics of Semantic Annotation Models (Macro Metrics in bracket, CoT stands for Chain-of-Thought and EDM stands for Ensemble Decision-Making)

Model	Precision (%)	Recall (%)	F1 (%)
Llama 3.1 8B	37.01 (41.53)	90.58 (70.89)	52.55 (49.71)
Llama 3.1 8B (CoT)	40.38 (43.39)	87.20 (68.24)	55.20 (50.51)
Llama 3.1 8B (EDM)	64.79 (69.47)	100.00 (78.26)	78.63 (72.29)
GPT-4o-mini	40.58 (48.94)	100.00 (78.26)	57.74 (57.18)
GPT-4o-mini (EDM)	47.42 (56.30)	100.00 (78.26)	64.35 (62.69)

increase underscores the ensemble framework's effectiveness in accurately identifying complete annotation paths, thereby reducing false positives. Similarly, GPT-4o-mini (EDM) demonstrates an improvement in Precision from 40.58% to 47.42%, highlighting the framework's capability to enhance annotation accuracy across different LLM architectures.

In terms of Recall, both EDM-enhanced models attain a perfect score of 100.00%, ensuring that all relevant annotation paths are identified. This perfect recall is critical for comprehensive semantic annotation, as it guarantees that no pertinent ontology paths are omitted. The base models, while maintaining high Recall scores (90.58% for Llama 3.1 8B and 100.00% for GPT-4o-mini), do not achieve the same level of completeness as the EDM-enhanced counterparts.

The F1-Scores for EDM-enhanced models reflect a balanced enhancement in both Precision and Recall. Llama 3.1 8B (EDM) achieves an F1-Score of 78.63%, a significant improvement over the base model's 52.55% and the CoT-enhanced model's 55.20%. Similarly, GPT-4o-mini (EDM) attains an F1-Score of 64.35%, compared to 57.74% for the base GPT-4o-mini model. These results indicate that the ensemble framework effectively balances the trade-off between Precision and Recall, ensuring both accurate and comprehensive semantic annotations.

Additionally, the introduction of CoT in the Llama 3.1 8B model provides modest improvements in Precision (from 37.01% to 40.38%) and F1-Score (from 52.55% to 55.20%). However, these gains are overshadowed by the more substantial enhancements achieved through the EDM approach, emphasizing the superior impact of ensemble decision-making on path-level annotation performance.

Node-Level Result Analysis. Table 2 shows the performance of various approaches at the node level, assessing their ability to correctly annotate individual ontology classes.

The EDM approach demonstrates a substantial improvement in Precision compared to the base Llama 3.1 8B model. Specifically, Llama 3.1 8B (EDM) achieves a Precision of 68.44%, significantly higher than the base model's 43.79%. This enhancement indicates that the ensemble method effectively mitigates individual model biases, leading to more accurate class annotations. Similarly, GPT-

Table 2. Node-Level Performance Metrics of Semantic Annotation (Macro Metrics in bracket, CoT stands for Chain-of-Thought and EDM stands for Ensemble Decision-Making)

	Precision (%)	Recall (%)	F1 (%)
Llama 3.1 8B	43.79 (43.70)	90.75 (68.80)	59.08 (51.57)
Llama 3.1 8B (CoT)	45.36 (43.00)	87.33 (64.08)	59.71 (49.70)
Llama 3.1 8B (EDM)	68.44 (46.85)	100.00 (54.44)	81.27 (49.46)
GPT-4o-mini	49.77 (54.18)	100.00 (78.26)	66.46 (61.99)
GPT-4o-mini (EDM)	56.43 (57.48)	100.00 (75.42)	72.15 (63.38)
SiSEG	26.85 (20.82)	15.38 (19.09)	19.56 (17.79)

4o-mini (EDM) shows an increase in Precision from 49.77% to 56.43%, underscoring the efficacy of EDM in refining annotation precision across different LLM architectures.

In terms of Recall, both EDM-enhanced models attain a perfect score of 100.00%, indicating that they successfully identify all relevant classes in the dataset. This perfect recall highlights the ensemble framework's capability to ensure comprehensive coverage of ontology classes, which is critical for maintaining data interoperability. In contrast, the base models—Llama 3.1 8B and Llama 3.1 8B (CoT)—exhibit high Recall scores of 90.75% and 87.33%, respectively, though they do not achieve the same level of completeness as the EDM-enhanced counterparts.

The F1-Scores, which balance Precision and Recall, further illustrate the benefits of the ensemble approach. Llama 3.1 8B (EDM) attains an F1-Score of 81.27%, significantly outperforming the base model's 59.08%. Similarly, GPT-4o-mini (EDM) achieves an F1-Score of 72.15%, compared to 66.46% for the base GPT-4o-mini model. These improvements indicate that the ensemble framework not only enhances the accuracy of annotations but also ensures a balanced and comprehensive identification of relevant classes.

By contrast, SiSEG, a domain-specific, feature-based annotation service, performs markedly worse, achieving only 26.85% Precision, 15.38% Recall, and a 19.56% F1 score at the node level. Its dependence on exact ontology coverage and shallow fuzzy-matching heuristics prevents it from generalizing to unseen or symbol-rich inputs, resulting in substantially lower accuracy and coverage.

Furthermore, incorporating Chain-of-Thought (CoT) in the Llama 3.1 8B model yields a modest improvement in Precision (from 43.79% to 45.36%) and F1-Score (from 59.08% to 59.71%). However, the gains from CoT are relatively limited compared to the more substantial enhancements observed with the EDM approach. This suggests that while CoT contributes to performance, the EDM framework provides a more robust mechanism for improving annotation accuracy and reliability.

4.5 Discussion

The evaluation results robustly demonstrate the efficacy of our proposed EDM module in enhancing the accuracy and reliability of semantic annotations within the energy domain. By leveraging multiple LLM instances and an effective consensus mechanism, the EDM module consistently outperforms baseline models across both node-level and path-level metrics. This section delves into the interpretation of these results, discussing the strengths and limitations of the proposed approach.

Our approach is designed to leverage structured information from ontologies and LLMs to facilitate semantic annotation. It is applicable to all the domains that have well-defined ontology, (e.g. Biomedical: Ontologies like Gene Ontology [1], Geospatial, Social Sciences) to annotate the domain-specific data. Notably, we employ non-fine-tuned LLMs and apply them to a specialized domain (energy sector). The fact that our approach achieves reasonable performance despite the lack of fine-tuning highlights the inherent generalization capability of LLMs. We acknowledge that fine-tuning could further enhance the accuracy and adaptability across different domains. But this is not the focus of the current paper.

We observe an intriguing phenomenon where the Llama 3.1 8B (EDM) model exhibits higher micro Precision, Recall, and F1-Score compared to the base Llama 3.1 8B model. Concurrently, the macro metrics show a mixed outcome: Precision slightly increases, while Recall and F1-Score decrease, particularly at the node level. This discrepancy suggests that the framework performs exceptionally well on majority classes but may underperform on minority or less frequent classes. In datasets with imbalanced class distributions, this imbalance can lead to reduced performance for classes with fewer instances, impacting the overall semantic coverage.

A notable limitation of the EDM framework is the increased execution time required for annotations. Specifically, annotating 529 columns across 2 ontology hierarchical class levels using Llama 3.1 8B (EDM) necessitates approximately 300 min. Utilizing multiple LLM instances inherently demands greater computational resources, including processing power and memory. This resource-intensive approach may limit the framework's accessibility and scalability in environments with constrained computational capacities. Investigating resource-efficient ensemble methods (e.g. sparse or cascaded ensemble strategies [8]) or introducing lightweight models as proxies could help mitigate these constraints. Additionally, optimizing the annotation process through methods, such as parallel processing, model pruning [6], or leveraging specialized hardware accelerators, might further reduce the overall execution time and resource footprint.

5 Conclusion

In this study, we introduced a framework that integrates ontology-driven search mechanisms with an ensemble decision-making module based on multiple LLMs to enhance the semantic annotation of energy-related datasets. Addressing the inherent challenges posed by inconsistent or varying naming in energy data and the context length limitations, our framework effectively distributes ontology classes across multiple LLM instances, fostering consensus-based annotations that significantly outperform traditional single LLM approaches. The evaluation results demonstrate the framework's superiority, with EDM-enhanced models achieving substantial improvements in Precision, Recall, and F1-Score at both node and path levels.

However, the study also identified critical areas for improvement, including the need to balance performance across majority and minority classes, optimize annotation efficiency, and reduce computational resource demands. Addressing these challenges through adaptive consensus mechanisms, class-aware distribution strategies, and resource optimization will be explored in the future to further enhance the framework's effectiveness and applicability.

Acknowledgments. The research leading to these results has received funding from the Horizon Europe program under grant agreement No.101158432. This work was partially supported by the German Federal Ministry of Education and Research (BMBF) under the project WestAI (Grant no. 01IS22094D).

Disclosure of Interests. The authors have no competing interests to declare that are relevant to the content of this article.

References

1. Ashburner, M., et al.: Gene ontology: tool for the unification of biology. Nat. Genet. **25**(1), 25–29 (2000)
2. Booshehri, M., et al.: Introducing the open energy ontology: enhancing data interpretation and interfacing in energy systems analysis. Energy AI **5**, 100074 (2021)
3. Brown, T., et al.: Language models are few-shot learners. Adv. Neural. Inf. Process. Syst. **33**, 1877–1901 (2020)
4. Cremaschi, M., D'Adda, F., Nocco, S.: Mantistable UI: a web interface for comprehensive semantic table interpretation management
5. Haghgoo, M., Nazary Aghche Mazary, A., Monti, A.: Siseg-auto semantic annotation service to integrate smart energy data. Energies **15**(4), 1428 (2022)
6. Huang, W., Zhang, Y., Zheng, X., Chao, F., Ji, R.: Towards efficient automatic self-pruning of large language models. arXiv preprint arXiv:2502.14413 (2025)
7. Jovanović, J., Bagheri, E.: Semantic annotation in biomedicine: the current landscape. J. Biomed. Semant. **8**, 1–18 (2017)
8. Kolawole, S., Dennis, D., Talwalkar, A., Smith, V.: Revisiting cascaded ensembles for efficient inference. arXiv preprint arXiv:2407.02348 (2024)
9. Korini, K., Bizer, C.: Column type annotation using chatgpt. arXiv preprint arXiv:2306.00745 (2023)
10. Kulmanov, M., Smaili, F.Z., Gao, X., Hoehndorf, R.: Semantic similarity and machine learning with ontologies. Brief. Bioinform. **22**(4), bbaa199 (2021)
11. Liska, A., et al.: Streamingqa: a benchmark for adaptation to new knowledge over time in question answering models. In: International Conference on Machine Learning, pp. 13604–13622. PMLR (2022)
12. Min, B., et al.: Recent advances in natural language processing via large pre-trained language models: a survey. ACM Comput. Surv. **56**(2), 1–40 (2023)

13. Pan, Z., Gao, Y., Ponci, F., Monti, A.: Semi-automatic ontology development framework for building energy data management. IEEE Access (2023)
14. Pan, Z., Yang, M., Monti, A.: Schema matching based on energy domain pre-trained language model. Energy Inform. **6**(Suppl. 1), 22 (2023)
15. Sanfilippo, A.P., et al.: Ontological annotation with wordnet. Technical report, Pacific Northwest National Lab. (PNNL), Richland, WA (United States) (2006)
16. Suhara, Y., et al.: Annotating columns with pre-trained language models. In: Proceedings of the 2022 International Conference on Management of Data, pp. 1493–1503 (2022)
17. Tan, Z., et al.: Large language models for data annotation: a survey. arXiv preprint arXiv:2402.13446 (2024)
18. Taylor, P.: Amount of data created, consumed, and stored 2010-2020, with forecasts to 2025. Statista (2022). https://www.statista.com/statistics/871513/worldwide-data-created/. Accessed 24 Oct 2023
19. Tejani, A.S., Ng, Y.S., Xi, Y., Fielding, J.R., Browning, T.G., Rayan, J.C.: Performance of multiple pretrained BERT models to automate and accelerate data annotation for large datasets. Radiology. Artif. Intell. **4**(4), e220007 (2022)
20. Wei, J., et al.: Chain-of-thought prompting elicits reasoning in large language models. Adv. Neural. Inf. Process. Syst. **35**, 24824–24837 (2022)
21. Xue, F., Fu, Y., Zhou, W., Zheng, Z., You, Y.: To repeat or not to repeat: insights from scaling LLM under token-crisis. Adv. Neural Inf. Process. Syst. **36** (2024)
22. Zhang, Y., et al.: Way to specialist: closing loop between specialized LLM and evolving domain knowledge graph. arXiv preprint arXiv:2411.19064 (2024)
23. Zheng, S., et al.: Rethinking semantic segmentation from a sequence-to-sequence perspective with transformers. In: Proceedings of the IEEE/CVF Conference on Computer Vision and Pattern Recognition, pp. 6881–6890 (2021)
24. Zhou, X., He, J., Yang, C.: An ensemble learning method based on deep neural network and group decision making. Knowl.-Based Syst. **239**, 107801 (2022)

RDFGraphGen: An RDF Graph Generator Based on SHACL Shapes

Milos Jovanovik[1,2(✉)], Marija Vecovska[2], Maxime Jakubowski[1], and Katja Hose[1]

[1] Institute of Logic and Computation, TU Wien, Vienna, Austria
{maxime.jakubowski,katja.hose}@tuwien.ac.at
[2] Faculty of Computer Science and Engineering, Ss. Cyril and Methodius University in Skopje, Skopje, North Macedonia
milos.jovanovik@tuwien.ac.at, marija.vecovska@finki.ukim.mk

Abstract. Developing and testing modern RDF-based applications often requires access to RDF datasets with certain characteristics. Unfortunately, it is very difficult to publicly find domain-specific knowledge graphs that conform to a particular set of characteristics. Hence, in this paper we propose RDFGraphGen, an open-source RDF graph generator that uses characteristics provided in the form of SHACL (Shapes Constraint Language) shapes to generate synthetic RDF graphs. RDFGraphGen is domain-agnostic, with configurable graph structure, value constraints, and distributions. It also comes with a number of predefined values for popular schema.org classes and properties, for more realistic graphs. Our results show that RDFGraphGen is scalable and can generate small, medium, and large RDF graphs in any domain.

Keywords: Data Generator · Synthetic Data · Knowledge Graphs · RDF · SHACL

1 Introduction

The acceptance of RDF [1] as a standard for knowledge representation, publication on the Web [3], and internally in organizations [7], has led to the development of many tools, libraries, and applications that work with or use RDF as a data model. During the development lifecycle of each of these, benchmarking and testing are important steps, which validate the usage scenarios and provide metrics about their real-world usability [4]. These benchmarks and test scenarios often require application-specific or domain-specific RDF datasets with certain characteristics, that are not always readily available. One possible solution to this is synthetic RDF datasets—datasets containing entities and data that are artificial but follow the structure, distribution, and vocabulary expected from applications or systems. They fulfill the need for data in a specific format, for a specific domain, application, or system to be tested or benchmarked [9,10]. Depending on the specific needs, these synthetic RDF datasets can be small or consist of millions of different entities [8].

H. Takeda et al. (Eds.): IJCKG 2025, LNAI 16297, pp. 111–125, 2026.
https://doi.org/10.1007/978-981-95-5009-8_8

One of the more recent additions to the technology stack around RDF is the Shapes Constraint Language (SHACL) [2]. Its main purpose is defining structural constraints on RDF graphs, called *shapes*. SHACL is designed to validate RDF graphs to ensure their data quality. However, these shapes can also be used as *blueprints* for generating domain-agnostic RDF data. Recognizing this potential and aiming to address the need for an RDF data generator, this paper presents RDFGraphGen. RDFGraphGen is, to the best of our knowledge, the first *SHACL-based* RDF data generation tool. As opposed to the usual task of *validating* RDF data, we use SHACL shapes for *generating* RDF data.

Hence, in this paper we describe the design and implementation of RDFGraphGen—a domain-independent generator of RDF graphs based on SHACL shapes. SHACL shapes are interpreted as a description of the desired structure of an RDF graph, and in accordance with this view the corresponding RDF triples are generated. RDFGraphGen is domain-agnostic: the source SHACL shapes can be from any domain. The generator can generate an RDF graph with a specified scale factor that determines the number of generated entities, providing flexibility for end-users. RDF data generated by RDFGraphGen can be used for benchmarking, testing, quality control, or other similar tasks in the application lifecycle of systems using RDF. Additionally, RDFGraphGen is open-source[1] and available as a ready-to-use Python package[2], under the MIT license.

This paper is organized as follows. Section 2 reviews related work and existing tools. Section 3 describes the design and the underlying algorithms of RDFGraphGen. Section 4 illustrates its features, usability, and data generation through a practical example. Section 5 outlines the results of the performance evaluation of RDFGraphGen. Finally, Sect. 6 concludes with summary and future work.

2 Related Work

Generating synthetic RDF data is not a new topic as it has been of interest to the research community for quite a long time. In this section, we provide an overview of existing generators of synthetic RDF datasets.

Tab2KG [11] is a method that is used for interpretation of tables with previously unseen data and automatically infers their semantics to transform them into semantic data graphs. The Tab2KG algorithm transforms tabular data into a semantic data graph by automatically inferring the domain ontology and mapping the table columns to the ontology classes and properties, before transforming the rows of the table into RDF triples. In short, the tabular data is provided, and the ontology is adjusted to fit this data. In contrast, RDFGraphGen uses explicit rules and constraints—it takes a description of a target data graph as a SHACL shapes graph and generates entities according to this description, using

[1] RDFGraphGen on GitHub: https://github.com/etnc/RDFGraphGen/.

[2] RDFGraphGen on PyPi: https://pypi.org/project/rdf-graph-gen/.

random or structured values for the objects in the RDF triples of the generated entities.

GAIA [14] is a generic RDF data generator that allows users to generate RDF triples by conforming to an ontology. It is OWL-based and generates RDF objects based on any correctly defined OWL ontology. The generator correctly generates a user-defined number of objects following the OWL ontology but offers no way to constrain the objects' values beyond datatype. RDFGraphGen uses a SHACL shapes graph as a description of the entities that should be generated, allowing the user to describe the object's values in great detail. RDFGraphGen also allows using properties from multiple ontologies in a generated entity since it does not generate data based on a specific ontology. The ontology is implicitly defined in SHACL shapes via the entity classes and the specified properties.

GRR [5] is a system for generating random RDF data using SPARQL-like syntax to describe the desired ontology. GRR can generate entities using the desired ontology and allows the user to provide input for the objects' values in the triples. However, GRR offers no method for constraining these values beyond providing them beforehead, making the process much more complicated for the user. RDFGraphGen in contrast uses the input SHACL shapes graph to constrain the values of the objects in the generated RDF triples.

PyGraft [13] is another tool designed for the creation of highly customized, domain-independent RDF datasets and RDF schemas. It takes input RDF schemas and uses a set of rules and constraints to generate synthetic RDF triples that are compliant with the input structure. It has the ability to simulate realistic data distributions, including edge and node properties, while ensuring that the generated data adheres to the input ontology's constraints. PyGraft is scalable, which makes it suitable for large datasets, and it is capable of handling various RDF vocabularies to produce diverse datasets that closely mimic real-world RDF data. RDFGraphGen, on the other hand, relies on the level of details and constraints contained in the input SHACL shapes graph to generate the desired synthetic RDF knowledge graph.

GenACT [15] is a data generator for temporal and evolving RDF graphs in the domain of social media activity during academic events—more specifically, academic conference tweets (ACT). This approach tries to address the gap of missing real-world knowledge graphs with specific capabilities that are necessary for the use cases in this domain. In contrast, RDFGraphGen is domain-independent and tries to solve the same gap from a more generalized perspective.

3 RDFGraphGen

In this section, we describe the design of RDFGraphGen, we explain how the graph structure and the literal values are generated and how to use the tool to generate RDF data from SHACL shapes.

RDFGraphGen has a straightforward workflow (Fig. 1). Given a SHACL shapes graph as input, RDFGraphGen outputs RDF data. The size of the generated RDF graph is controlled by an input parameter: *scale-factor*. We also explain how the scale factor affects the number of entities and RDF triples.

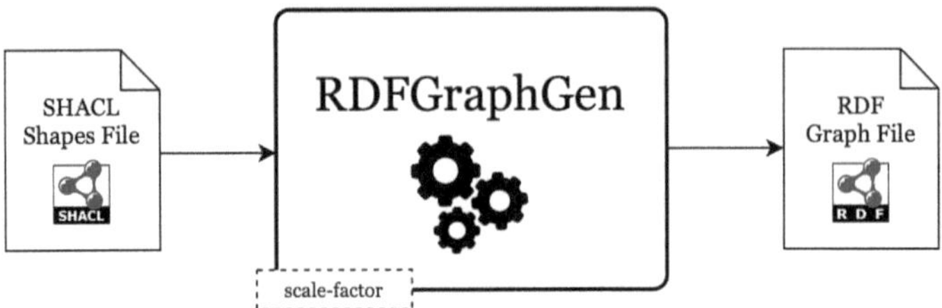

Fig. 1. RDFGraphGen Workflow Diagram

3.1 Graph Generation Algorithm

The goal is to generate realistic and useful RDF data. For the input shapes graphs, RDFGraphGen generates a data graph that conforms to them. The aim is to create an algorithm that uses the structure of the input SHACL shapes and generates data following this structure. The goal is to use as many details from the shapes graph in order to populate the generated RDF data graph, even if the shapes are contradictory. Generally, the task of generating conforming data graphs is non-trivial[3].

Preliminaries on SHACL. To provide sufficient background, we start by introducing SHACL [2]. In general, the main purpose of SHACL is to define a set of constraints on nodes called *shapes*, which in turn are used to define constraints on an RDF graph. The set of shapes is referred to as a *shapes graph*[4] while the data that is checked is called the *data graph.* When the data graph satisfies all the shapes from the shapes graph, we say it *conforms.* A node in the data graph that must satisfy a shape is called the *focus node.* Focus nodes for a given shape are specified by *target declarations.*

Conceptually, there are two different types of shapes: *node shapes*, which describe constraints on the focus nodes themselves, and *property shapes* describing constraints on a set of values reachable from the focus node by a specified property. Structurally, a node shape $s = (n, C, T)$ consists of three components: (1) a shape name n, typically an IRI, (2) a conjunction of constraint components C, and (3) a (possibly empty) set of target declarations T. When there are target declarations, the shape can be considered a constraint on the data graph. Similarly, a property shape $p = (n, C, T, E)$ consists of all the components of a node shape, with an additional element: the *property path* E. For our purposes, the property path is always a predicate path, i.e., an IRI.

A *constraint component* describes the specific conditions that a focus node must satisfy. Node shapes can be considered as a conjunction of constraint

[3] Because SHACL corresponds to a logic [6], generating data corresponds to finding a model for a given SHACL theory. Even for perceived "simple" shapes, the task might be best solved by resorting to first-order logical reasoners.

[4] It is called a *graph* because SHACL syntax is defined in RDF itself.

components—all constraint components must be satisfied for the focus node. In particular, shape-based constraint components[5] allow for creating a network of shapes that can be used to inform the structure of the generated data graph.

Generating the Graph. The algorithm for generating RDF data is primarily based on the set of constraint components. The shape name and the target declarations are only used as additional information. We assume that the shapes graph is well-formed[6] and non-recursive[7], i.e., that it is syntactically correct SHACL and there are no self-referential shapes.

The algorithm is designed as follows. Given an input SHACL shapes graph S, and the number of desired entities m, RDFGraphGen generates an RDF graph G that contains m entities per explicitly defined node shape s in S. Specifically, we generate m entities for each node shape s, such that the triple (s, rdf:type, sh:NodeShape) appears in the shapes graph S.

Algorithm 1 $\textsc{GenerateFromNodeshape}(S, s, e)$

$G = \emptyset$
$s = (s_n, C, s_T)$
for each constraint component $c \in C$ **do**
  if c is a property shape component **then**
    Let p be the property shape associated with c.
    $G', E := \textsc{GenerateFromPropshape}(S, p)$
    Let r the predicate path associated with p.
    $G := G \cup G' \cup \{(e, r, e') \mid e' \in E\}$
  else if c is a node constraint component **then**
    Let s' be the shape from S associated with c.
    $G := G \cup \textsc{GenerateFromNodeshape}(S, s', e)$
  else if c is a logical constraint component **then**
    $G := G \cup \textsc{GenerateLogical}(S, c, e)$
  end if
end for
return G

We focus our explanation on the generation of a single entity e for a node shape s. Recall that a shape contains a set of constraint components C. The procedure called $\textsc{GenerateFromNodeshape}(S, s, e)$ is described in Algorithm 1. It uses the shapes graph S, a node shape s, and an IRI e to generate triples describing entity e using s, possibly referring to other shapes from S, and outputs a synthetic RDF graph G. It looks at every supported constraint component and generates data accordingly. For node constraint components (`sh:node`), it simply generates some data for every component, for entity e. Similarly, for

[5] https://www.w3.org/TR/shacl/#core-components-shape.
[6] https://www.w3.org/TR/shacl/#dfn-well-formed.
[7] https://www.w3.org/TR/shacl/#shapes-recursion.

Algorithm 2 GENERATEFROMPROPSHAPE(S, p)

$G := \emptyset$
$p = (p_n, C, p_T, q)$
card $:= determineCardinality(C)$
if q is a supported schema.org predicate **then**
  return G, $createSchemaValues(q, \mathsf{card})$
else if $\exists c \in C$ such that c is a datatype constraint component **then**
  Let datatype be the datatype associated with c.
  return G, $createLiteralValues(\mathsf{datatype}, \mathsf{card})$
end if
Let $E := \{e_1, \ldots, e_{\mathsf{card}}\}$ be a set of newly created IRIs.
if $\exists c \in C$ such that c is a property shape component **then**
  Let $C' \subseteq C$ be the set of property shape components from C.
  Let S' be the set of all property shapes associated with C'.
  for each $s' \in S'$ **do**
    Let q be the predicate path associated with s'.
    $G', E' :=$ GENERATEFROMPROPSHAPE(S, s')
    $G := G \cup G' \cup \bigcup_{e \in E} \{(e, q, e') \mid e' \in E'\}$
  end for
else if $\exists c \in C$ such that c is a node constraint component **then**
  Let $C' \subseteq C$ be the set of node constraint components from C.
  Let $S' = \{s'_1, \ldots, s'_k\}$ be the set of all k node shapes associated with C'.
  Partition E into mutually disjoint sets $E_1, \ldots, E_k$
  for each $s'_i \in S'$ **do**
    $G = G \cup \bigcup_{e \in E_i}$ GENERATEFROMNODESHAPE(S, s'_i, e)
  end for
else if $\exists c \in C$ such that c is a logical constraint component **then**
  Let $C \subseteq C'$ be the set of logical constraint components from C.
  $G := \bigcup_{c' \in C'} \bigcup_{e \in E}$ GENERATELOGICAL(S, c', e)
end if
return G, E

Algorithm 3 GENERATELOGICAL(S, c, e)

$S' := \emptyset$
if c is an 'and' constraint component **then**
  S' is the set of shapes from S associated with c.
else if c is a 'or' constraint component **then**
  S' is a random, non-empty subset of the shapes in S associated with c.
else if c is a 'xone' constraint component **then**
  S' is a set containing exactly one random shape from S associated with c.
end if
return $\bigcup_{s' \in S'}$ GENERATEFROMNODESHAPE(S, s', e)

the supported logical constraint components `sh:and`, `sh:or`, and `sh:xone`, we select an appropriate subset of shapes for which data is generated w.r.t. e. (see Algorithm 3).

Example 1. Consider the shape <MyShape>:

```
<myShapeA> a sh:NodeShape ;
  sh:node <myShapeB>, <myShapeC> ;
  sh:or ( <myShapeD> <myShapeE> ) .
```

where we assume `<myShapeB>`, ...,`<myShapeE>` are defined. The generator creates the IRI e and calls GenerateFromNodeShape(S, s, e) for $s =$ `<myShapeA>` (initially), $s =$ `<myShapeB>` and $s =$ `<myShapeC>` (all node constraint components), and also, randomly, for $s =$ `<myShapeD>` or $s =$ `<myShapeE>` or both (logical constraints).

Finally, and most importantly, data is generated for every associated property shape (`sh:property`) p, handled by the procedure GenerateFromPropShape(S, p) described in Algorithm 2, which returns a set of new RDF terms E, and (a possibly empty) RDF graph G' representing triples that are recursively generated for entities from E. This is the most crucial procedure, as the property shapes indicate the structure of the generated RDF graph by describing the immediate neighborhood.

Example 2. Continuing from Example 1, we can add the following additions to the shape definition:

```
<myShapeA> sh:property <myPropShape> .
<myPropShape> sh:path <pred> ; sh:minCount 1 .
```

This means that the generator calls GenerateFromPropShape(S, p), where $p =$ (`<myPropShape>`, {`sh:minCount 1`}, $\emptyset$, `<pred>`)

The procedure works as follows. First, we determine how many values are generated for the property shape (indicated by *determineCardinality*(C)) based on the keywords `sh:minCount` and `sh:maxCount`. Essentially, a random integer is selected between the minimum and maximum cardinality. If there is an inconsistency (i.e., the minimal cardinality is greater than the maximal), the minimal indicated cardinality is followed.

Next, we distinguish between generating literal values, or new entities. If the predicate path is a schema.org predicate, we generate an appropriate value with the helper function *createSchemaValues*(q, `card`). Essentially, `card` indicates the number of values to be generated appropriate to q. Similarly, if the values are intended to be of a certain datatype (by default `xsd:string` values), these are handled accordingly by the helper function *createLiteralValues*(`datatype`, `card`). We elaborate on both helper functions in Sect. 3.2.

Finally, if we expect the values for the property to be other entities, e.g., referring to other shapes with `sh:node`, `sh:property`, or a logical constraint component, then the generation procedure is called recursively in accordance with the indicated structure of the shape.

Example 3. Recall `<myPropShape>` from Example 2. The generator recognizes that at least one value must be generated, so it will choose `card` ≥ 1, for example `card` = 2. Next, because `<pred>` is not a specially supported predicate path and because there is no datatype explicitly defined, it will generate two random strings $\texttt{str}_1$ and $\texttt{str}_2$, and adds the triples (e, `<pred>`, $\texttt{str}_1$) and (e, `<pred>`, $\texttt{str}_2$) to the output graph.

Size of the Generated RDF Graph. RDFGraphGen uses a scale factor to determine the size of the generated RDF dataset. The value of the parameter corresponds to the number of entities that will be generated from all top-level node shapes in the input SHACL graph. By *top-level* node shapes we mean node shapes that are either stand-alone and not related to other node shapes from the SHACL graph, or that only appear as subjects in the predicates pointing to other node shapes from the SHACL graph. The other node shapes, which solely appear as objects in predicates from other node shapes, are generated in numbers depending on the cardinality of the same predicate, as described in Algorithm 2.

Example 4. Consider the shape `<myShapeF>`:

```
<myShapeF> a sh:NodeShape ;
    sh:property [ sh:path <predicate> ;
        sh:node <myShapeG> ;
        sh:minCount 2 ;
        sh:maxCount 4 ] .
```

This means that `<myShapeF>` is a top-level shape, connected via `<predicate>` to the non top-level shape `<myShapeG>`.

If the scale-factor is set to 100, the generator will generate 100 entities based on `<myShapeF>`, and between 200–400 entities based on `<myShapeG>` (i.e., 2 to 4 per entity based on `<myShapeF>`), resulting in a total of 300–500 entities in the generated RDF graph.

3.2 Generating Literal Values

To generate *useful* RDF data, we generate meaningful literal values whenever possible. Recall that Algorithm 2 explicitly refers to schema.org for the creation of values, based on the predicate path of the property shape. The focus on schema.org is not a fundamental limitation: our implementation can easily be extended to include other vocabularies in the future.

If we need to generate a value that is not related to schema.org, then the values are generated based on the specified datatype (`sh:datatype`), and if none is given, we default to `xsd:string`.

Generating Schema.org Literals. When generating a new literal, the first step is to try and infer what would be its logical value, given the predicate and the RDF type of the entity as defined in the input SHACL shape. For example, if a triple is being generated for an entity of type `schema:Person` from the widely used schema.org vocabulary [12], where the predicate is `schema:firstName`, it is natural that the generated value of the object should be a human first name. In this case, the generator uses a set of first names to pick one randomly. Even further, if the `schema:Person` entity has a `schema:gender` value of either male or female, the generator will specifically use the list of male or female first names, accordingly, to pick one at random.

The generator contains sets of collected values for the most common properties from schema.org. These sets are used by RDFGraphGen to randomly select a value for such properties, when they appear in the SHACL shapes. More specifically, RDFGraphGen has collections of male and female names and surnames, job titles, addresses and street names, book titles, book genres, TV series titles, movie titles, movie genres, etc., that have been collected from online resources. This feature of RDFGraphGen can easily be extended in the future to include many more domains.

Additionally, the generator contains several rules about constructing values for other well-known schema.org predicates, e.g., a full name or an email address. In scenarios where the input SHACL shape uses `schema:name`, the value is constructed by concatenating a random (gendered) first name and a last name. When the predicate `schema:email` is used without a pattern, RDFGraphGen generates an email address in the form `firstname-lastname@gmail.com`.

If the predicate is unknown, it can still provide useful information to determine a suitable value. For example, if the name of the predicate is `birthDate`, from a random or unknown vocabulary, the term `date` suggests that the object should be a date. This will be caught by RDFGraphGen and a date value will be generated. Similar behavior is implemented for other predicates from unknown vocabularies, such as `telephone`, `phone`, `email`, etc.

Generating Generic Values. When the value cannot be picked from a predefined set of specific values, we generate it randomly. This random generation step uses all of the constraints components defined in the input SHACL shape, e.g., datatype, minimum length, maximum length, pattern, minimum and maximum value, etc.

This is what makes the RDFGraphGen generator general-purpose and domain-independent: it can generate synthetic RDF triples and graphs from any domain (life sciences, social networking, entertainment, linguistics, geography, etc.), regardless of whether it is explicitly familiar with it or not.

3.3 Usage

To make the RDFGraphGen generator more easily available for the community, we packaged it as a Python library and published it online on PyPi(see footnote 2). RDFGraphGen is available via the command: `pip install rdf-graph-gen`, which downloads and installs it locally on the user's machine. Afterwards, RDFGraphGen can be used via the command-line, using the command:

```
rdfgen input-shapes.ttl output-graph.ttl scale-factor
```

The command `rdfgen` takes three arguments as input:

- the file containing the input SHACL shapes graph, e.g., `input-shapes.ttl`,
- the file that the generated synthetic RDF graph is to be written to, e.g., `output-graph.ttl`, and
- the scale factor that determines the size of the generated RDF graph (`scale-factor`).

We optimized RDFGraphGen for generating large graphs—data is generated in batches that are serialized in the output file when the batch size is reached, in order to continuously free up main memory. We also implemented concurrency to parallelize and speed up data generation for large graphs.

The full RDFGraphGen code, along with the CSV files containing pre-defined sets of values, examples of SHACL shape files and generated graph examples, are publicly available on GitHub(see footnote 1). Using a CI/CD pipeline based on GitHub actions, we release a new version of the RDFGraphGen generator on PyPi when there are changes in the GitHub repository.

4 Example Use Case

To showcase how RDFGraphGen works for specific SHACL shapes, this section presents an example of using the `schema:Person` class from schema.org to represent people and their addresses as entities. Other examples, from various domains using different ontologies are available on the project's GitHub page(see footnote 1).

In this example, the input SHACL shapes graph (Listing 1) describes entities that are of type `schema:Person`. From the definition, each such entity is constrained to have a given name and a last name, or a full name; exactly one date of birth, which must be earlier than the potential death date; a gender with one of the available values; an email with a value constrained by a pattern; another using the `schema:email` property; a telephone number; a job title; an address, which has a specified complex type (it is an object, not a literal). Additionally, the address objects have a street address that is a string value; a postal code, which can be a string or an integer with values between 10,100–10,999.

Listing 1. Input SHACL Shapes Graph

```
ex:PersonShape a sh:NodeShape ;
    sh:targetClass schema:Person ;
    sh:xone ( [ sh:property [
                    sh:path schema:givenName ;
                    sh:datatype xsd:string ;
                    sh:minCount 1 ;
                ] ;
                sh:property [
                    sh:path schema:familyName ;
                    sh:datatype xsd:string ;
                    sh:minCount 1
```

```
                ]]
              [ sh:path schema:name ;
                sh:datatype xsd:string  ;
                sh:minCount 1 ] );
    sh:property [
        sh:path schema:birthDate ;
        sh:lessThan schema:deathDate ;
        sh:minCount 1 ;
        sh:maxCount 1 ] ;
    sh:property [
        sh:path schema:gender ;
        sh:in ( "female" "male" ) ;
        sh:maxCount 1 ] ;
    sh:property [
        sh:path ex:workMail ;
        sh:pattern "^[a-z0-9]+\\.[a-z0-9]+@work\\.[a-z]{2,3}$"
        ] ;
    sh:property [
        sh:path schema:email ;
        sh:maxCount 1 ] ;
    sh:property [ sh:path schema:telephone ] ;
    sh:property [ sh:path schema:jobTitle ] ;
    sh:property [
        sh:path schema:address ;
        sh:node ex:AddressShape ;
        sh:minCount 1 ] .

ex:AddressShape a sh:NodeShape ;
    sh:class schema:PostalAddress ;
    sh:property [
        sh:path schema:streetAddress ;
        sh:datatype xsd:string ; ] ;
    sh:property [
        sh:path schema:postalCode ;
        sh:or ( [ sh:datatype xsd:integer ;
                  sh:minInclusive 10100 ;
                  sh:maxInclusive 10999  ]
                [ sh:datatype xsd:string ] ) ] .
```

After running the RDFGraphGen generator on the input SHACL shapes file and setting the parameter `scale-factor` to 2, we get the generated synthetic RDF graph (Listing 2).

Listing 2. Output RDF Graph

```
<http://example.org/ns#Node100> a schema:Person ;
    ex:workMail "18b.dqjz8szqwne7nm@work.ekx" ;
    schema:address <http://example.org/ns#Node101> ;
    schema:birthDate "1965-07-07"^^xsd:date ;
    schema:deathDate "1989-07-07"^^xsd:date ;
```

```
6       schema:email "barton_aldridge@gmail.com" ;
7       schema:familyName "Aldridge" ;
8       schema:gender "male" ;
9       schema:jobTitle "bartender" ;
10      schema:givenName "Barton" ;
11      schema:telephone "647-466-552849" ;
12      sh:description ex:PersonShape .
13
14  <http://example.org/ns#Node102> a schema:Person ;
15      ex:workMail "sqq2.s7ojq@work.vab" ;
16      schema:address <http://example.org/ns#Node103> ;
17      schema:birthDate "1986-07-07"^^xsd:date ;
18      schema:email "sarajanebenjamin@gmail.com" ;
19      schema:gender "female" ;
20      schema:jobTitle "psychologist" ;
21      schema:name "Sarajane Benjamin" ;
22      schema:telephone "722-279-0247032" ;
23      sh:description ex:PersonShape .
24
25  <http://example.org/ns#Node101> a schema:PostalAddress ;
26      schema:postalCode 10481 ;
27      schema:streetAddress "no. 3 Lily st" ;
28      sh:description ex:AddressShape .
29
30  <http://example.org/ns#Node103> a schema:PostalAddress ;
31      schema:postalCode "6pl065qAPywG" ;
32      schema:streetAddress "no. 1 Gillette ave" ;
33      sh:description ex:AddressShape .
```

When generating the synthetic RDF graph, given this input, RDFGraphGen produces an output such as the one presented in Listing 2. When it generates the name of each person entity, only one option from the `sh:xone` constraint can be selected (Line 3, Listing 1). For the first person (Line 1, Listing 2), the generator has created separate predicates for the given name (Line 10, Listing 2) and the family name (Line 7, Listing 2). For the second person (Line 14, Listing 2), the generator has generated a single predicate for the full name (Line 21, Listing 2), and the full name includes both the given and the family name as a single value.

Furthermore, predicates related to dates have a date value for the object despite not explicitly containing a `sh:datatype` constraint in the description. The generator extracts information from the predicate names of `schema:birthDate` and `schema:deathDate` (Lines 17–18, Listing 1), determines that the objects' values in the generated RDF triples should be dates, and then generates them based on the constraints specified in the input SHACL shapes graph.

The email address values (`schema:email`) consist of the given name and the family name of the person (Lines 6 and 18, Listing 2). This has been specifically predefined in the generator when working with `schema:Person` entities and their email addresses. The other email address value (`ex:workMail`) has a definition that is not part of schema.org and therefore its value is constrained by the pattern in the SHACL graph (Line 27, Listing 1). The generator produces values that conform to the defined pattern (Lines 2 and 15, Listing 2).

The generated phone numbers follow a specific pattern. This is again based on the predicate name, `schema:telephone` (Line 32, Listing 1) and therefore the generator applies the predefined pattern constraint (Lines 11 and 22, Listing 2).

The address for each person is generated as a separate RDF entity (Lines 25 and 30, Listing 2) based on the SHACL shape in the input file, which constrains the address. Given that there are no explicit cardinality restrictions on the relation between people and their addresses, each person has exactly one address. So, given that `scale-factor` was set to 2 when starting the generator, the output contains a total of 4 entities: 2 person entities (as top-level shapes) and 2 address entities (as shapes directly connected to a top-level shape).

5 Performance Evaluation

To evaluate the performance of RDFGraphGen, we measured the processing time for generating some specific RDF datasets. The experiment suite is available on the project GitHub repository(see footnote 1).

Table 1. Performance Evaluation of RDFGraphGen

	Single Simple Shape		Single Complex Shape		Three Simple Shapes		Three Complex Shapes	
Scale Factor	RDF Triples	Time (s)	RDF Triples	Time (s)	RDF Triples	Time (s)	RDF Triples	Time (s)
1	2	4	9	4	8	4	13	4
10	20	4	89	4	80	4	135	4
100	200	4	862	4	800	5	1,401	5
1,000	2,000	5	8,477	5	8,000	5	13,939	5
10,000	20,000	8	84,975	8	80,000	8	139,969	9
100,000	200,000	10	850,313	24	800,000	16	1,400,109	30
1,000,000	2,000,000	35	8,504,410	164	8,000,000	102	14,004,068	244
10,000,000	20,000,000	287	85,042,729	1,608	80,000,000	1,352	140,044,876	2,335
100,000,000	200,000,000	2,871	850,497,903	17,312	800,000,000	9,888	1,371,272,422	22,734

We tested three dimensions: scale factor (from 1 to 100,000,000), number of shapes in the input SHACL graph (1 and 3), and SHACL shape complexity per input shape (from a single string predicate, up to 4 properties in varying datatypes, lengths, cardinalities, etc.). All experiments were run on a MacBook Pro with an Apple M4 chip, 10-core CPU, 24GB main memory and 120GB/s memory bandwidth.

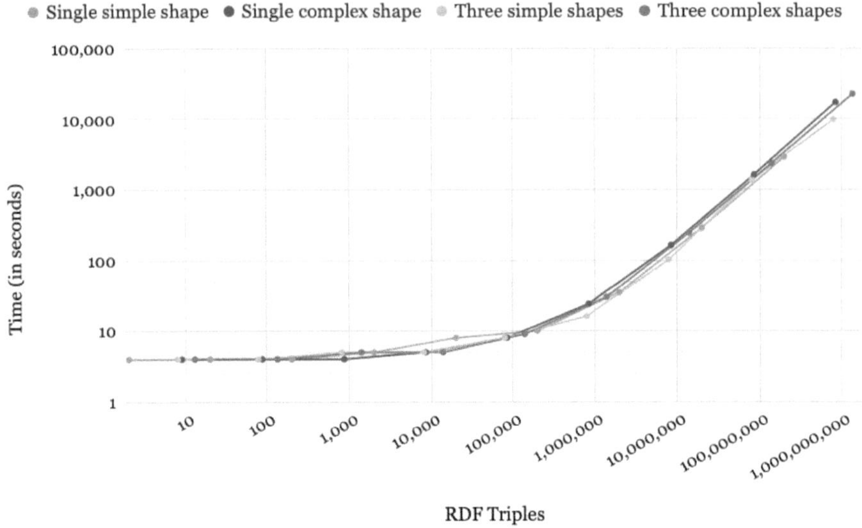

Fig. 2. Generation Time based on the Number of Generated RDF Triples

Our experiments, presented in Table 1, show that RDFGraphGen generates synthetic RDF graphs in a reasonably good timeframe. The generation time slightly depends on the complexity of the input SHACL shapes but is more strongly correlated with the number of RDF triples being generated (Fig. 2). The concurrent implementation of RDFGraphGen allows it to use all available CPUs on a given machine and scale efficiently. In our tests RDFGraphGen automatically ran with 10 processes in parallel.

6 Conclusion and Future Work

In this paper, we introduced RDFGraphGen—a general-purpose, domain-agnostic synthetic RDF graph generator based on SHACL shapes. To the best of our knowledge, this is the first RDF generator based on SHACL constraints. The synthetic RDF knowledge graphs generated by RDFGraphGen can be used in many different scenarios in the software development cycle: application testing, algorithm testing, application benchmarking, software quality control, training of machine learning models, etc.

RDFGraphGen is domain-independent: the generated RDF graph contains data from the domain that is described in the source SHACL shapes file. Additionally, RDFGraphGen has the ability to use pre-collected values for schema.org classes and predicates and can be extend to include values for any ontology. This allows for the generator to be fine-tuned and adapted to the needs of specific use cases and domains.

In the future, we plan to extend the support for human-friendly literal values for an increased number of ontologies, classes, and properties in a manner similar to what we did with a subset of schema.org. To make our finding reproducible, and enable others to easily use RDFGraphGen, we have published it as open-source, under the MIT license.

Acknowledgments. The work in this paper was partially financed by the Faculty of Computer Science and Engineering, Ss. Cyril and Methodius University in Skopje, and the European Union's Horizon Europe Research and Innovation Program under grant agreement no 101136244 (TARGET).

References

1. RDF 1.1. https://www.w3.org/TR/rdf11-concepts/
2. Shapes Constraint Language (SHACL). https://www.w3.org/TR/shacl/
3. Web Data Commons - RDFa, Microdata, Embedded JSON-LD, and Microformats Data Sets - October 2024. http://webdatacommons.org/structureddata/2024-12/stats/stats.html
4. Angles, R., et al.: The linked data benchmark council: a graph and RDF industry benchmarking effort. ACM SIGMOD Rec. **43**(1), 27–31 (2014)
5. Blum, D., Cohen, S.: Generating RDF for application testing. In: 9th International Semantic Web Conference ISWC 2010, p. 105 (2010)
6. Bogaerts, B., Jakubowski, M., Van den Bussche, J.: SHACL: a description logic in disguise. In: Logic Programming and Nonmonotonic Reasoning, pp. 75–88 (2022)
7. Buchgeher, G., Gabauer, D., Martinez-Gil, J., Ehrlinger, L.: Knowledge graphs in manufacturing and production: a systematic literature review. IEEE Access **9**, 55537–55554 (2021)
8. Duan, S., Kementsietsidis, A., Srinivas, K., Udrea, O.: Apples and oranges: a comparison of RDF benchmarks and real RDF datasets. In: SIGMOD, pp. 145–156 (2011)
9. Georgala, K., et al.: MOCHA2018: the mighty storage challenge at ESWC 2018. In: Semantic Web Challenges, pp. 3–16 (2018)
10. Georgala, K., Spasić, M., Jovanovik, M., Petzka, H., Röder, M., Ngomo, A.C.N.: MOCHA2017: the mighty storage challenge at ESWC 2017. In: Semantic Web Challenges, pp. 3–15 (2017)
11. Gottschalk, S., Demidova, E.: Tab2KG: semantic table interpretation with lightweight semantic profiles. Semantic Web **13**(3), 571–597 (2022)
12. Guha, R.V., Brickley, D., Macbeth, S.: Schema.org: evolution of structured data on the web. Commun. ACM **59**(2), 44–51 (2016)
13. Hubert, N., Monnin, P., d'Aquin, M., Monticolo, D., Brun, A.: PyGraft: Configurable Generation of Synthetic Schemas and Knowledge Graphs at Your Fingertips. arXiv (2024)
14. Raynaud, T., Amir, S., Haque, R.: A generic and high-performance RDF instance generator. Int. J. Web Eng. Technol. **11**(2), 133–152 (2016)
15. Singh, G., et al.: GenACT: an ontology-based temporal web data generator. In: International Conference on Conceptual Modeling, pp. 317–336. Springer (2024)

RANA: Conflict Resolution for Knowledge Graphs with Better Representation Learning

Zihao Song, Huang Peng, Weixin Zeng, and Xiang Zhao(✉)

National Key Laboratory of Big Data and Decision, National University of Defense Technology, Changsha, China

3236274662@qq.com, {phmail,zengweixin13,xiangzhao}@nudt.edu.cn

Abstract. Knowledge fusion combines multi-source data to address the inherent incompleteness of knowledge graphs (KGs), with conflict resolution serving as a critical technique for resolving knowledge conflicts during the fusion process. However, current advanced methods often suffer from inadequate representation learning for KGs, which results in suboptimal performance. To tackle these challenges, we propose RANA, a framework that combines self-adversarial negative sampling and learnable weight matrices for learning better representation of KGs, thus benefiting conflict resolution. RANA uses a graph neural network (GNN) based encoder with a self-adversarial negative sampling mechanism to enhance representation learning quality and improve the model's ability to identify noise. Furthermore, learnable parameters were introduced to project entity representations, enabling better capture of interactions between different attributes and entities, thus improving the accuracy of representation learning for attributed KGs. By integrating source credibility with the high-quality representations, RANA utilizes a truth inference module to update the KG, thus achieving effective conflict resolution. Experimental results demonstrate the effectiveness of RANA in conflict resolution tasks, achieving an F1-score of 0.715 and provide deeper insights into the proposed method.

Keywords: knowledge graphs · knowledge fusion · conflict resolution · self-adversarial negative sampling

1 Introduction

Knowledge graphs (KGs), as graph-structured knowledge bases, employ structured semantic representations of entities, relationships, and attributes, which enhance the interconnectedness of heterogeneous datasets and provide a semantic foundation for advanced intelligent services across multiple domains [1]. In applications like recommendation systems, KGs have shown the ability to transform abstract knowledge structures into computable semantic networks [2, 3]. However, in open environments, knowledge continuously evolves through dynamic changes and updates. To maintain the timeliness and completeness of KGs, it is essential to integrate heterogeneous data from multiple sources through knowledge fusion [4]. Distinct from static knowledge fusion

H. Takeda et al. (Eds.): IJCKG 2025, LNAI 16297, pp. 126–141, 2026.
https://doi.org/10.1007/978-981-95-5009-8_9

approaches like entity alignment [32, 33], when collecting and processing knowledge from various origins, knowledge conflicts may arise. Thus, conflict resolution techniques play a crucial role. They can identify and retain correct knowledge, ensure the credibility of KGs, and prevent errors in downstream applications.

Current knowledge conflict resolution approaches, including probabilistic fusion frameworks, Bayesian models, and syntactic fusion methods, predominantly rely on symbolic information from KGs [5–7]. Some methods suggest incorporating the structural and semantic information of triples into the conflict resolution process and employing deep representation learning [11]. Nevertheless, these methods often underperform when dealing with rapidly updated and diversely expressed information, attributed to their insufficient learning of knowledge graph information.

On the other hand, approaches to open knowledge graph completion (open KGC) have also been leveraged for conflict resolution. These methods exploit representation learning on existing KGs to infer new facts in open settings, implicitly assessing the veracity of candidate triples and thereby identifying inconsistencies [21]. Recent work encodes entities from their names and descriptions [22] and designs scoring functions to evaluate triple plausibility [30]. Notably, the current state of the art method, Nari [29], discovers new facts by coupling a negative-sampling strategy with source-reliability analysis. Although open KGC techniques can capture rich structural and semantic KG information, their effectiveness is often undermined by inadequate negative-sample learning, yielding suboptimal embeddings. Moreover, most existing approaches focus almost exclusively on relational triples, neglecting attribute triples and leaving ample scope for improving representation quality.

To sum up, the main limitations of current conflict resolution methods include: (1) insufficient negative samples learning. (2) overlook of effective modeling and learning for attribute triples. These limitations hinder their ability to effectively model attributed knowledge graphs and consequently result in suboptimal performance.

To address these challenges, we build upon the latest advances and propose a novel framework, RANA. Specifically, we introduce two key innovations: a self-adversarial negative sampling mechanism that dynamically generates harder negatives, thereby enhancing the model's ability to identify noise; and a learnable weight matrix that projects entity representations into an attribute-aware space, effectively capturing the complex interactions between entities and their attributes. By incorporating this matrix into the scoring function, RANA achieves more accurate modeling of attribute triples and significantly improves representation learning for attributed knowledge graphs. Furthermore, RANA integrates source credibility with the high-quality representations produced by the model to further optimize overall performance.

The main contributions of this article can be summarized as:

1. A conflict resolution framework RANA is proposed, which combines self-adversarial negative sampling, a learnable weight matrix and multi-source reliability reasoning to accurately resolve knowledge conflict resolution.
2. We learn attribute triples by introducing learnable parameters, capturing the interaction between entities and attributes and enhancing the accuracy of representation.
3. Experiments have proved that the conflict resolution method of our study is more effective compared with the existing conflict resolution methods.

2 Related Works

In this section, we first introduce several representative conflict resolution methods in the database field and knowledge graph conflict resolution. Then, we discuss the recent progress in open knowledge graph completion.

2.1 Conflict Resolution

Traditional conflict resolution methods have primarily focused on the database domain, which can be mainly categorized into three types: iterative methods, optimization-based methods, and probabilistic graphical model-based methods [5]. Iterative methods [13] begin by fixing the credibility of data sources to predict the true value and iteratively update the credibility until convergence. Optimization-based methods transform the conflict resolution problem into an optimization problem. Probabilistic graphical model-based methods treat observed values as conditional probability distributions to evaluate the credibility of data sources. For example, the TruthFinder algorithm [14] integrates iterative and probabilistic graphical model methods to assess credibility and identify truth values. Tian et al. [31] designed the TFAR optimization framework, which uses Bayesian probabilistic graphical models and Gibbs sampling inference algorithms to estimate the reliability of attributes and objects. These methods have been effective in resolving conflicts among databases.

With the advent of the knowledge graph era, conflict resolution turns to addressing conflict among KGs. Traditional conflict resolution methods, due to their disregard for data structure, can be directly applied to knowledge graph conflict resolution. However, this disregard also means they fail to leverage the unique properties of knowledge graphs, leading to suboptimal performance. Some early approaches address knowledge fusion conflicts through rule-based methods and programmable algorithms, such as probabilistic fusion frameworks, Bayesian models, and syntactic fusion methods. Bryl et al. [10] proposed utilizing metadata from Wikipedia's revision history to learn a conflict resolution model. Dong et al. [24] introduced the probabilistic knowledge fusion method KnowledgeVault, which relies on knowledge extractors to transform triples into vectors and leverages high-quality knowledge graphs as prior knowledge to assess triple correctness. Belth et al. [11] adopted a graph summary learning method to identify abnormal knowledge in the graph by generating, ranking, filtering, and refining candidate rules. However, there are only a limited number of relevant methods in this area, and few new developments have emerged in recent years, necessitating more advanced conflict resolution techniques.

2.2 Open Knowledge Graph Completion

In recent years, various frameworks for open KGC have been introduced. Early frameworks utilized pre-existing KGs to train classifiers for recognizing external relation triples. However, these approaches had limitations such as insufficient processing of diverse triple types and heavy reliance on prior knowledge [15, 16]. Shah et al. [21] extended traditional KGC methods by learning a transformation from text embedding space to KG embedding space for identifying unseen entities through their names and descriptions. Shi et al. [22] used a convolutional neural network to encode entities with their names and descriptions, and designed a score function to measure triples. Niu et al. [30] further introduced a multiple interaction attention mechanism to better utilize information in entity descriptions. Nevertheless, these methods mainly focus on relational facts, neglecting attribute triples and only able to identify facts similar to known ones. Peng et al. [29] proposed a method that integrates representation learning with source evaluation, but the quality of negative sampling in their representation learning is suboptimal.

3 Methodology

In this section, we first define the conflict resolution problem and describe the entire framework of RANA. Then, we elaborate on our solution in detail, including the structure of the GNN-based encoder, the steps of the training stage and the inference stage.

3.1 Problem Formulation

In the problem of conflict resolution, we consider triples composed of relations and attributes. Formally represent the knowledge graph as $KG = \{E, R, A, V, T_r, T_a\}$, where E, R, A, V denote the set of entities, relations, attributes and values. $T_r \subset E \times R \times V$ is the collection of relation triples and $T_a \subset E \times A \times V$ is the collection of attribute triples. The fact is the triple in the KG, denoted as $f = (h, r, t) \in T_r \cup T_a$. The claim is a triple extracted from open resources, denoted as $C = \{(h, r, t, s) | h \in E, r \in R \cup A, s \in S\}$, where S is the collection of all sources. That is to say, a claim contains a triple and its source.

Assume that the head entity and the relationship (or attribute) are contained in the KG, but the tail entity (or value) may not be contained. The correct claims are called truths. The KG is often referred to as prior KG because it provides a source of prior knowledge for claim identification. Correspondingly, prior knowledge specifically involves the facts contained in the KG. Given a $KG = \{E, R, A, V, T_r, T_a\}$ and set of claims C extracted from a set of sources S, the goal of conflict resolution is to identify a subset of the true claims $C' \subseteq C$.

The overall framework of RANA is shown in the Fig. 1. RANA learns embeddings of KGs through a GNN-based encoder, and combines self-adversarial negative sampling with an inference module to assess the credibility and truthfulness of claims, thereby achieving continuous KG updates.

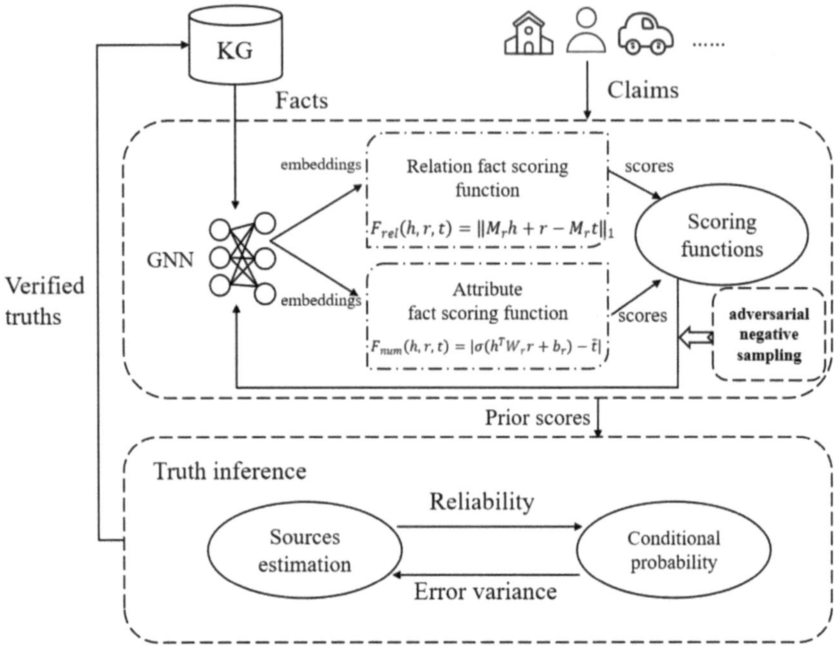

Fig. 1. The overall architecture of RANA

3.2 GNN-Based Encoder

Drawing on Huang et al.'s TKGC [25] approach in their paper, we use a GNN to capture the structural features of the KG. Given that two-hop neighbors often help identify missing entities and values, we design a two-layer GNN to implicitly capture neighbor information. Each entity's representation is calculated as follows:

$$zi^{(l+1)} = \sigma\left[\frac{1}{|\mathcal{N}_i|}\sum_{j\in\mathcal{N}_i} \boldsymbol{W}^l z_j^l\right] \tag{1}$$

where z_i^l denotes the implicit representation of entity e_i at layer l, and $\sigma(\cdot)$ represents the sigmoid function. $\mathcal{N}_i$ indicates the neighbor entities of e_i, and $\boldsymbol{W}^l$ is the learnable weight matrix of the GNN at layer l. To mitigate potential information distortion in multi-layer GNNs, we use residual connections. This involves concatenating the initial information with the output of the final GNN layer, thereby preserving the original information:

$$z_i^{final} = \mathrm{Concat}\left(z_i^0, \mathrm{Res}\left(z_i^0, z_i^L\right)\right) \tag{2}$$

$$\mathrm{Res}\left(z_i^0, z_i^L\right) = z_i^0 + \sigma\left(\boldsymbol{W}^{final} z_i^L + z_i^0\right) \tag{3}$$

where z_i^0 represents the input to the GNN, namely the initial entity embedding of e_i. $\mathrm{Concat}(\cdot,\cdot)$ denotes the vector concatenation operation, $\boldsymbol{W}^{final}$ is the weight matrix, and L refers to the total number of layers.

3.3 Training Stage

Preprocessing. To handle heterogeneous attribute value types, such as strings, numbers, and dates, we categorize the attribute facts into two groups based on their value types. Specifically, string-based values are treated as entities and fed into a pre-trained BERT model, with the hidden states of the final layer projected into the entity embedding space to yield dense entity representations. Date values are converted into integers; for instance, a calendar date is mapped to the number of days elapsed since 1 January 2000. All other discrete types are treated as strings and processed accordingly. Consequently, downstream components only need to reason over relational facts and numerical attribute facts.

Negative-Aware Relational Fact Learning. To evaluate relational facts, we have designed a relational scoring function to assess their plausibility. The function outputs a non-negative real number, with lower values indicating a higher likelihood of the fact being true. Here, we employ TransR [28], a translation-based embedding method, as the scoring function. For a given relational fact (h, r, t), the scoring function is defined as follows:

$$F_{rel}(h, r, t) = \|\boldsymbol{M}_r\boldsymbol{h} + \boldsymbol{r} - \boldsymbol{M}_r\boldsymbol{t}\|_1 \tag{4}$$

where $\boldsymbol{h}$, $\boldsymbol{r}$, and $\boldsymbol{t}$ denote the representation of h, r, and t, respectively. $\boldsymbol{M}_r$ is the transformation matrix for relation r, mapping entity representations to the semantic space of relation r. $\|\cdot\|_1$ denotes the norm of $L1$.

However, the encoder may overfit as the prior knowledge graph only provides true facts, leading to poor identification of false claims. To address this, we propose a loss function based on self-adversarial negative sampling:

$$Loss_{rel} = \sum_{h,r}\Big(\sum_{t \in T_{h,r}} F_{rel}(h, r, t) + \beta \cdot \sum_{t' \in T_{h,r}'} p(h, r, t') \cdot [\theta - F_{rel}(h, r, t')]_+\Big) \tag{5}$$

$$p(h, r, t) = \frac{\exp(F_{rel}(h, r, t))}{\sum_{\tau \in T_{h,r}'} \exp(F_{rel}(h, r, \tau))} \tag{6}$$

For each relation r, positive samples $T_{h,r}$ are extracted from the knowledge graph, and negative samples $T_{h,r}^{'}$ are generated by considering all possible tail entities $t^{'}$ and weighting their contributions based on sampling probabilities. The model's scoring function $F_{rel}(h, r, t)$ evaluates both sample types. The loss is calculated as the sum of positive sample scores and the sum of margin-based losses for negative samples, with the margin defined by hyperparameter θ. $p(h, r, t)$ denotes the weighting factor assigned to different negative samples (h, r, t). In this way, $p(h, r, t)$ enables the model to dynamically focus on more misclassifiable negative samples during training. It helps the model target these confusing samples, thereby improving its ability to distinguish between true and false facts. β is a hyperparameter, which balances the loss from positive and negative samples by adjusting the weight of negative-sample loss in the total loss. The optimizer adjusts model parameters to minimize this loss function. This self-adversarial negative sampling approach enhances the model's ability to distinguish true facts from false claims, reducing overfitting.

Numerical Fact Learning. Similar to Nari [29], we design a numerical scoring function, which is essentially a regression loss function with non-linear transformations. Given a numerical fact (h, r, t), we first compute $\tilde{t}$, which normalizes a for attribute r using method *max-min.*

$$\tilde{t} = \frac{t - \min Y_r}{\max Y_r - \min Y_r} \tag{7}$$

where $Y_r = \{t | \exists h \in E, (h, r, t) \in T_a\}$ represents the set of all possible values of attribute r, and the scoring function is defined as:

$$F_{num}(h, r, t) = \left| \sigma\left(\boldsymbol{h}^T \boldsymbol{W}_r \boldsymbol{r} + b_r\right) - \tilde{t} \right| \tag{8}$$

where $\boldsymbol{W}_r$ is the learnable weight matrix for attribute r, which is used to project entity representations. This allows the model to better capture interactions between different attributes and entities. b_r is the bias for r. The loss for all numerical facts is the sum of all scores:

$$Loss_{num} = \sum_{(h,r,t) \in T_a} F_{num}(h, r, t) \tag{9}$$

Training Loss Function. Finally, the total loss is defined as:

$$Loss_{tot} = Loss_{rel} + Loss_{num} \tag{10}$$

3.4 Inference Stage

The goal of the reasoning phase is to identify factual information from claims extracted from open sources. Once the training phase is complete, the encoder can map entities and relations to the object space. A fact-scoring function can then assess triple compatibility with the KG. However, the KG may be incomplete, and external claims might contain previously unseen facts that are hard to identify using the prior KG. Therefore, we propose a truth-inference algorithm based on a GNN encoder and scoring function to uncover potential truths absent from the KG.

Source Reliability Estimation. Similar to methods in some truth inference papers, we assume observations come from a normal distribution. Formally, given claim $(h, r, t, s), t \sim N(t^*, (u_r \epsilon_s)^2)$, where t^* is the latent truth, u_r is a fixed scaling factor for relation r, and ϵ_s represents the reliability of source s. A smaller ϵ_s indicates a more reliable source s. Also, the range of t^* depends on h and r, so we have:

$$\sum_{r,t \in Y_s} \frac{d(t, t^*)^2}{(u_r \epsilon_s)^2} \sim \chi^2(|Y_s|) \tag{11}$$

where $Y_s = \{(r, t) | (h, r, t, s) \in C\}$ is the set of all relations (or attributes) and corresponding tail entities (or values) extracted from source s. $d(\cdot, \cdot)$ denotes the distance function between tail entities and values. If t is an entity, option $d(t, t^*) = \|M_r t - M_r t^*\|$

is selected; whereas if t is a digital number, option $d(t, t^*) = \|t - t^*\|$ is chosen. Note that for a given h and r, there may be multiple truths t^*. In such cases, we select the nearest t^* from the KG for the computation of $d(t, t^*)$.

Following the approach used in CATD [26], we calculate the upper bound of $(1-\alpha)$ confidence interval as the estimation ϵ_s^2 for source reliability:

$$\epsilon_s^2 \approx \sum_{r,t \in Y_s} \frac{d(t, t^*)^2}{u_r^2 \chi^2\left(\frac{\alpha}{2}, |Y_s|\right)} \tag{12}$$

Likelihood Calculation. Following the initial estimation for open source, we can compute a credibility probability $P(t|h, r, s)$ for each claim. As previously assumed, t is related to t^*, whose value range depends on h and r. Thus, the credibility probability can be expressed as:

$$P(t|h, r, s) = \sum_{t^* \in T^*_{h,r}} P\left(t|t^*, h, r, s\right) P\left(t^* \middle| h, r, s\right) \tag{13}$$

$$= \sum_{t^* \in T^*_{h,r}} P\left(t|t^*, s\right) P\left(t^* \middle| h, r\right) \tag{14}$$

where $T^*_{h,r}$ represents the set of all possible true values t^*. Although prior KG can provide candidates for $T^*_{h,r}$, $T^*_{h,r}$ is often indeterminate and incomplete. To address this, we approximate $T^*_{h,r}$ using the corresponding fact set and all observed values from each source, $T^*_{h,r} \approx \left\{t' \middle| (h, r, t', s) \in C \bigwedge (h, r, t') \in T_r \bigcup T_a\right\}$. For $P(t|t^*, s)$, it can be computed via a normal probability distribution function and distance function $d(\cdot, \cdot)$. In the case of $P(t|h, r)$, we reuse the fact scoring function $exp(-F(h, r, t^*))$ to calculate the probability with the incorporation of prior knowledge.

Truth Verification. By setting a threshold p, we ultimately determine the truth set $C' = \{c = (h, r, t, s) \in C | P(t|h, r, s) > p\}$. Subsequently, over multiple epochs, the KG can undergo enrichment, and the quality of the sources can be re-estimated. This iterative process allows for the continuous improvement of both the KG and the assessment of source reliability.

4 Experiments

In this section, we first present the experimental settings along with relevant details. Then, experiments are carried out to validate the effectiveness of the proposed method. Next, ablation studies are used to assess the impact of the learnable weight matrix and self-adversarial negative sampling. Additionally, the model's performance is evaluated across different levels of prior knowledge. Finally, a case study is provided to demonstrate the model's performance.

4.1 Experiment Settings

Dataset. Since subsets of DBpedia/Wikidata and resources such as OpenEA are not inherently multi-source collections, we utilize a dataset compiled by OKELE [17], comprising triples extracted from various web sources. The dataset gathers 10 popular entity classes from Freebase [18], with 1,200 entities in each class, and a total of 191,759 triples. Table 1 presents the dataset's statistics. In our experiments, we gather at least 1-hop and 2-hop facts for each entity to form the prior KG, while the remaining triples are treated as claims.

Table 1. Dataset statistics.

Classes	#Relational triples	#Attribute triples	Classes	#Relational triples	#Attribute triples
actor	64983	330	film	45233	576
album	5897	155	food	23041	842
book	10766	499	mountain	2720	623
building	2823	361	ship	1805	852
drug	26432	1002	software	2322	487

Evaluation Metrics. We use precision (P), recall (R), and F1-score as evaluation metrics, where higher values indicate better performance. Precision is the ratio of correctly identified true claims to all identified claims. Recall is the ratio of correctly identified true claims to all actual true claims. The F1-score, which is the harmonic mean of precision and recall, balances these two metrics.

Implement Details. Our experiments were performed on a server equipped with an Intel Core i7-12700F CPU, 16 GB of memory, and an NVIDIA GeForce RTX 3050 Ti GPU. For hyperparameters, we set the margin $\gamma = 10$, confidence $\alpha = 0.05$, and probability threshold $p = 0.5$. Initial entity representations were generated using a pre-trained BERT model. The entity representation dimension was set to $d_e = 100$, and the relation representation dimension to $d_r = 300$. We optimized model parameters using the Adam algorithm with a learning rate of $l_r = 0.01$. Training was conducted for $EPOCHS_t = 50$ epochs, and inference for $EPOCHS_i = 20$ epochs.

Comparative Models. We compare our model with several representative open KGC and truth inference models. These baseline models can be divided into two groups. The first group consists of methods that recognize facts from multiple sources but do not consider the features of the KG:

- TruthFinder [19], which estimates source reliability and identifies truths through Bayesian analysis for data fusion.
- Latent Credibility Analysis (LCA) [20], which constructs a strongly principled probabilistic model to capture the credibility of information sources.

- Latent Truth Model (LTM) [9], which proposes a probabilistic graphical model capable of automatically inferring true records and assessing source quality in an unsupervised manner.
- Multi-truth Bayesian Model (MBM) [23], which presents an integrated Bayesian approach to address the multi-truth-finding problem.
- OKELE [18], which constructs a probabilistic graphical model to infer true facts about long-tail entities from open sources.

The second group comprises open KGC methods that harness prior knowledge to identify facts:

- Open World Extension (OWE) [21], which offers an extension for embedding-based knowledge graph completion models, enabling open-world link prediction.
- ConMask [22], which learns embeddings from entity names and text descriptions to link unseen entities to the KG.
- KnowledgeVault [24], which uses supervised machine learning to integrate prior knowledge from existing repositories with distinct information sources.
- TKGC [25], which presents a trustworthy method leveraging existing KG facts to infer truths from open sources.
- Nari [29], which enhances open KGC performance by combining negative-perception representation learning with multi-source reliability reasoning to account for source quality.

Additionally, we propose a simple, training-free baseline that leverages ChatGPT to assess the veracity of each claim by directly querying the model for a true or false judgment.

4.2 Overall Performance

Table 2 presents the comparison results, from which we can draw the following conclusions: (1) Our model attains the best overall performance among all competitors. Although it trails the strongest baseline (Nari) in precision by only a razor-thin margin, it achieves a markedly higher recall, yielding a 0.062-point gain in F1-score, the primary composite metric that balances precision and recall, thereby corroborating the effectiveness of self-adversarial negative sampling and learnable weight matrices. (2) Among the first group of methods, OKELE successfully models long-tail distributions and assesses source quality but neglects KG semantic information. LTM yields the worst results, primarily due to its reliance on assumptions about prior distributions. (3) The second group of methods, which overlook explicit source estimation, perform unsatisfactorily. Additionally, while OWE, KnowledgeVault, and ConMask identify external facts based on existing KGs, they do not distinguish between relation facts and attribute facts, leading to inferior performance compared to TKGC. (4) ChatGPT achieves precision, recall, and F1 scores in the 0.4–0.5 range, all of which are surpassed by RANA. Although ChatGPT encapsulates extensive commonsense knowledge, it is incapable of acquiring prior knowledge from an existing knowledge graph or explicitly assessing source quality.

Table 2. Comparison of overall performance. The best scores are highlighted in bold and the second-best scores are underlined.

Models	Precision	Recall	F1-score
TruthFinder [19]	0.279	0.374	0.320
LCA [20]	0.364	0.404	0.383
LTM [9]	0.262	0.394	0.315
MBM [23]	0.340	0.539	0.417
OKELE [17]	0.459	0.485	0.472
OWE [21]	0.351	0.421	0.383
KnowledgeVault [24]	0.385	0.455	0.417
ConMask [22]	0.376	0.443	0.407
TKGC [25]	0.524	0.491	0.507
ChatGPT	0.417	0.477	0.445
Nari [29]	**0.728**	0.591	0.653
RANA	0.722	**0.709**	**0.715**
RANA w/o lwm	0.660	0.720	0.686
RANA w/o san	0.682	0.602	0.638

4.3 Ablation Study

Self-Adversarial Negative Sampling. Unlike existing conflit resolution methods, RANA actively synthesizes hard negative triples during training to improve false-claim detection. Concretely, the sampler enumerates every possible tail entity, re-weights the loss by the current posterior probability of each candidate, and forces the model to concentrate on the most misleading negatives. We ablate this strategy in the RANA w/o san variant, which instead draws negatives uniformly as in ConMask. As Table 2 shows, removing self-adversarial sampling decreases the F1 score by 0.077 and sharply reduces precision, demonstrating that negatives generated via adversarial sampling provide indispensable supervisory signal for recognizing spurious facts. The substantial degradation confirms that self-adversarial sampling is significantly superior to uniform negative sampling for discovering hard negatives.

Learnable Weight Matrix. Eliminating the learnable weight matrix, RANA w/o lwm drops F1 from 0.715 to 0.686: precision falls from 0.722 to 0.660 while recall rises from 0.709 to 0.720, indicating that the model recovers more true claims but at the cost of numerous false positives. The weight matrix projects entity representations into a space where subtle interactions between attributes and entities are disentangled; it also re-distributes model capacity so that parameters focus on the most discriminative directions. Without this module, the encoder loses its ability to capture fine-grained interaction signals, blurring the decision boundary between true and false facts. These results underscore the critical role of the learnable weight matrix in enhancing both model expressiveness and interaction modeling.

4.4 Sensitivity Analysis

Proportion of Prior Knowledge. We retain KG facts at varying proportions to analyze the impact of prior knowledge. Specifically, we randomly retain facts from 10% to 100%, in 10% increments, and run the model independently for each proportion. As shown in Fig. 2(a), the model exhibits low precision and recall in the absence of prior knowledge, making truth identification nearly random. Upon introducing prior knowledge, the metrics generally rise with the proportion of retained facts. For instance, precision rises by 20%, recall by 40%, and F1-score by 35%. This highlights how prior knowledge aids in external fact identification. Overall, incorporating prior knowledge boosts model performance by over 20%.

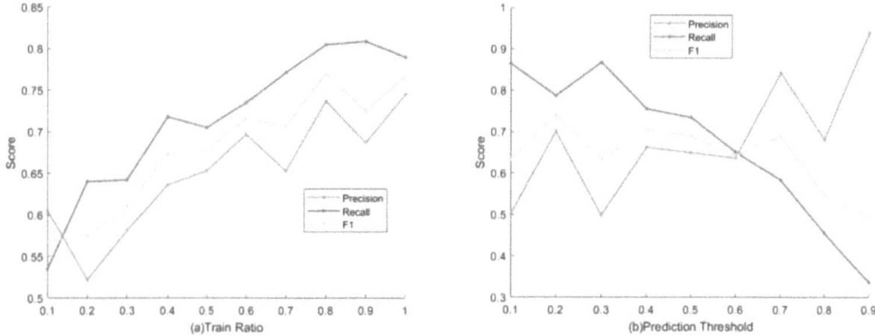

Fig. 2. Sensitivity analysis

Threshold. To test the impact of the threshold, we ran experiments by varying the threshold p. Figure 2(b) shows that as p increases, the model becomes more conservative in identifying true claims. At low thresholds, the model misclassifies many false claims as true, leading to low precision but high recall. As the threshold rises, the model's criteria for identifying true claims become stricter, increasing precision but decreasing recall. The F1-score peaks around a threshold of 0.5, indicating this value best balances precision and recall. When the threshold exceeds 0.5, the F1-score drops significantly, showing that too high a threshold makes the model overly conservative and causes it to miss many true claims.

4.5 Case Study

We further demonstrate the comparison of the KG before and after completion. As shown in Table 3, we illustrate a set of facts and claims about three entities and two relations, which correspond to three types of food and data on ingredients and content. Initially, our model identifies claims similar to existing facts in the KG. It then assigns a high credibility value to source A due to its accurate provision of two claims, leading to the correct identification of the remaining claims from source A. Notably, the KG lacks information on "Cachapa" containing "sodium." However, "Cachapa" and "Tomato and egg soup" share the same relations and tail entities, which may cause the encoder to generate similar embeddings for them. This facilitates the discovery of missing facts, such as ("Cachapa", Nutrient, "sodium").

To locate RANA's remaining flaws, we sampled low-F1 triples from OKELE for manual review and identified two main issues: poor adaptability to certain data types and temporal-shift false negatives. For example, ("Barack Obama", "presidentOf", "United States") was wrongly rejected in 2022 because RANA uses static embeddings without temporal modeling. For future work, we plan to append a timestamp as the fourth dimension to each fact to enable dynamic fact tracking.

Table 3. Comparison of food KG before and after completion

KG before	("Tomato and egg soup", Nutrient, "sodium")
	("Tomato and egg soup", quantity, "165mg")
	("Far Breton", Nutrient, "sugars")
	("Far Breton", quantity, "34g")
	("Cachapa", Nutrient, "potassium")
	("Cachapa", quantity, "32mg")
Claims	("Tomato and egg soup", quantity, "165mg", source A)
	("Tomato and egg soup", quantity, "serves", source B)
	("Far Breton", quantity, "34g", source A)
	("Far Breton", quantity, "33.5g", source A)
	("Far Breton", quantity, "sichuan", source B)
	("Cachapa", Nutrient, "potassium", source A)
	("Cachapa", Nutrient, "sodium", source A)
	("Cachapa", Nutrient, "ingredients", source B)
KG after	("Tomato and egg soup", Nutrient, "sodium")
	("Tomato and egg soup", quantity, "165mg")
	("Far Breton", Nutrient, "sugars")
	("Far Breton", quantity, "34g")
	("Far Breton", quantity, "33.5g")
	("Cachapa", Nutrient, "potassium")
	("Cachapa", Nutrient, "sodium")
	("Cachapa", quantity, "32mg")

5 Conclusion

Conflict resolution is a key technology in the fusion process and can enhance the accuracy of the fused knowledge graph. However, current methods often underperform due to inadequate representation learning of knowledge graphs. To address these issues, we propose a novel solution that combines self-adversarial negative sampling and learnable weight matrices to boost conflict resolution performance. By considering all possible tail entities and weighting losses by their probabilities, the model's ability to distinguish true from false facts is enhanced. The learnable weight matrix projects entity representations, aiding the capture of complex interactions between attributes and entities. Experimental results show our approach achieves superior performance in conflict resolution tasks. This advancement is crucial for real-world applications like intelligent recommendation systems and accurate information retrieval, which rely on identifying reliable information from diverse sources. For limitations of RANA, data in the real world does not always align with the assumptions made in this paper. Additionally, our approach does not handle temporal-related data. Future research directions include developing more sophisticated conflict resolution mechanisms and incorporating temporal dynamics to enhance the model's robustness and accuracy in dynamic, complex environments.

Acknowledgments. This work was partially supported by NSFC under grant No. 62302513.

Disclosure of Interests. The authors have no competing interests to declare that are relevant to the content of this article.

References

1. Ehrlinger, L., Wöß, W.: Towards a definition of knowledge graphs. In: Martin, M., Cuquet, M., Folmer, E. (eds.) Joint Proceedings of the Posters and Demos Track of the 12th International Conference on Semantic Systems SEMANTiCS2016 and the 1st International Workshop on Semantic Change & Evolving Semantics (SuCCESS'16) co-located with the 12th International Conference on Semantic Systems (SEMANTiCS 2016), Leipzig, Germany, September 12–15, 2016, Vol. 1695 of CEUR Workshop Proceedings. CEUR-WS.org, pp. 1–4 (2016)
2. Guo, J., Zhou, Y., Zhang, P., Song, B., Chen, C.: Trust-aware recommendation based on heterogeneous multi-relational graphs fusion. Inf. Fusion **74**, 87–95 (2021). https://doi.org/10.1016/J.INFFUS.2021.04.001
3. Chen, X., Jia, S., Xiang, Y.: A review: Knowledge reasoning over knowledge graph. Expert Syst. Appl. **141** (2020). https://doi.org/10.1016/J.ESWA.2019.112948
4. Nguyen, H.L., Vu, D., Jung, J.J.: Knowledge graph fusion for smart systems: a survey. Inf. Fusion **61**, 56–70 (2020). https://doi.org/10.1016/J.INFFUS.2020.03.014
5. Dutta, A., Meilicke, C., Ponzetto, S.P.: A probabilistic approach for integrating heterogeneous knowledge sources. In: Proceedings of the European Semantic Web Conference (ESWC), pp. 286–301. Anissaras, Crete, Greece, May 25–29 (2014)
6. Hu, X., Hu, J., Peng, Y., et al.: Conflict resolution methods in semantic knowledge fusion. Shanghai Jiao Tong Univ. J. **43**(11), 1730–1733 (2009)
7. Zhao, X., Jia, Y., Li, A., et al.: Multi-source knowledge fusion: a survey. World Wide Web J. **23**(4), 2567–2592 (2020)
8. Gao, Y., Ge, C., Guo, Y., et al.: A review of data integration technologies for relational data and knowledge graphs. J. Softw. 2204–2230 (2022)

9. Zhao, B., Rubinstein, B.I.P., Gemmell, J., Han, J.: A Bayesian approach to discovering truth from conflicting sources for data integration. Proc. VLDB Endow. **5**(6), 550–561 (2012). https://doi.org/10.14778/2168651.2168656
10. Bryl, V., Bizer, C.: Learning conflict resolution strategies for cross-language wikipedia data fusion. In: Proceedings of the 23rd International World Wide Web Conference (WWW), pp. 1129–1134. Seoul, Republic of Korea, April 7–11 (2014)
11. Belth, C., Zheng, X., Vreeken, J., et al.: What is normal, what is strange, and what is missing in a knowledge graph: unified characterization via inductive summarization. In: Proceedings of the ACM Web Conference 2020 (WWW), pp. 1115–1126. Taipei, Taiwan, China, April 20–24 (2020)
12. Dutta, A., Meilicke, C., Ponzetto, S.P.: A Probabilistic approach for integrating heterogeneous knowledge sources. In: Proceedings of the European Semantic Web Conference (ESWC), pp. 286–301. Anissaras, Crete, Greece (2014)
13. Li, Y., Gao, J., Meng, C., et al.: A survey on truth discovery. ACM SIGKDD Explorations Newsl **17**(2), 1–16 (2016)
14. Yin, X., Han, J.S.: Truth discovery with multiple conflicting information providers on the web. In: Proceedings of the 13th ACM Conference on Knowledge Discovery and Data Mining (KDD), pp. 1048–1052. San Jose, California, USA, August 12–15 (2007)
15. Abboud, R., Ceylan, I., Lukasiewicz, T., Salvatori, T.: Boxe: a box embedding model for knowledge base completion. Adv. Neural. Inf. Process. Syst. **33**, 9649–9661 (2020)
16. Balazevic, I., Allen, C., Hospedales, T.M.: Tucker: tensor factorization for knowledge graph completion. In: Inui, K., Jiang, J., Ng, V., Wan, X. (eds.) Proceedings of the 2019 Conference on Empirical Methods in Natural Language Processing and the 9th International Joint Conference on Natural Language Processing, EMNLP-IJCNLP 2019, Hong Kong, China, November 3–7, 2019. Association for Computational Linguistics, pp. 5184–5193 (2019). https://doi.org/10.18653/V1/D19-1522
17. Cao, E., Wang, D., Huang, J., Hu, W.: Open knowledge enrichment for long-tail entities. In: Huang, Y., King, I., Liu, T., van Steen, M. (eds.) WWW '20: The Web Conference 2020, Taipei, Taiwan, April 20–24, 2020. ACM / IW3C2, pp. 384–394 (2020). https://doi.org/10.1145/3366423.3380123
18. Bollacker, K.D., Evans, C., Paritosh, P.K., Sturge, T., Taylor, J.: Freebase: a collaboratively created graph database for structuring human knowledge. In: Wang, J.T. (ed.) Proceedings of the ACM SIGMOD International Conference on Management of Data, SIGMOD 2008, Vancouver, BC, Canada, June 10–12, 2008. ACM, pp. 1247–1250 (2008). https://doi.org/10.1145/1376616.1376746
19. Yin, X., Han, J., Yu, P.S.: Truth discovery with multiple conflicting information providers on the web. IEEE Trans. Knowl. Data Eng. **20**(6), 796–808 (2008). https://doi.org/10.1109/TKDE.2007.190745
20. Pasternack, J., Roth, D.: Knowing what to believe (when you already know something). In: Huang, C., Jurafsky, D. (eds.) COLING 2010, 23rd International Conference on Computational Linguistics, Proceedings of the Conference, 23–27 August 2010, Beijing, China. Tsinghua University Press, pp. 877–885 (2010)
21. Shah, H., Villmow, J., Ulges, A., Schwanecke, U., Shafait, F.: An open-world extension to knowledge graph completion models. In: The Thirty-Third AAAI Conference on Artificial Intelligence, AAAI 2019, The Thirty-First Innovative Applications of Artificial Intelligence Conference, IAAI 2019, The Ninth AAAI Symposium on Educational Advances in Artificial Intelligence, EAAI 2019, Honolulu, Hawaii, USA, January 27-February 1, 2019. AAAI Press, pp. 3044–3051 (2019). https://doi.org/10.1609/AAAI.V33I01.33013044
22. Shi, B., Weninger, T.: Open-world knowledge graph completion. In: McIlraith, S.A., Weinberger, K.Q. (eds.) Proceedings of the Thirty-Second AAAI Conference on Artificial Intelligence, (AAAI-18), the 30th innovative Applications of Artificial Intelligence (IAAI-18), and

the 8th AAAI Symposium on Educational Advances in Artificial Intelligence (EAAI-18), New Orleans, Louisiana, USA, February 2–7, 2018. AAAI Press, pp. 1957–1964 (2018). https://doi.org/10.1609/AAAI.V32I1.11535
23. Wang, X., Sheng, Q.Z., Fang, X.S., Yao, L., Xu, X., Li, A.: An integrated bayesian approach for effective multi-truth discovery. In: Bailey, J., Moffat, A., Aggarwal, C.C., Rijke, M. de, Kumar, R., Murdock, T.K., Sellis, T., Yu, J.X. (eds.) Proceedings of the 24th ACM International Conference on Information and Knowledge Management, CIKM 2015, Melbourne, VIC, Australia, October 19–23, 2015. ACM, pp. 493–502 (2015). https://doi.org/10.1145/2806416.2806443
24. Dong, X.L., et al.: From data fusion to knowledge fusion. Proc. VLDB Endow. **7**(10), 881–892 (2014). https://doi.org/10.14778/2732951.2732962
25. Huang, J., et al.: Trustworthy knowledge graph completion based on multi-sourced noisy data. In: Laforest, F., Troncy, R., Simperl, E., Agarwal, D., Gionis, A., Herman, I., Medini, L. (eds.) WWW '22: The ACM Web Conference 2022, Virtual Event, Lyon, France, April 25–29, 2022. ACM, pp. 956–965 (2022). https://doi.org/10.1145/3485447.3511938
26. Li, Q., et al.: A confidence-aware approach for truth discovery on long-tail data. Proc. VLDB Endow. **8**(4), 425–436 (2014). https://doi.org/10.14778/2735496.2735505
27. Devlin, J., Chang, M., Lee, K., Toutanova, K.: BERT: pre-training of deep bidirectional transformers for language understanding. In: Burstein, J., Doran, C., Solorio, T. (eds.) Proceedings of the 2019 Conference of the North American Chapter of the Association for Computational Linguistics: Human Language Technologies, NAACL-HLT 2019, Minneapolis, MN, USA, June 2–7, 2019, Volume 1 (Long and Short Papers). Association for Computational Linguistics, pp. 4171–4186 (2019). https://doi.org/10.18653/V1/N19-1423
28. Lin, Y., Liu, Z., Sun, M., Liu, Y., Zhu, X.: Learning entity and relation embeddings for knowledge graph completion. In: Bonet, B., Koenig, S. (eds.) Proceedings of the Twenty-Ninth AAAI Conference on Artificial Intelligence, January 25–30, 2015, Austin, Texas, USA. AAAI Press, pp. 2181–2187 (2015). https://doi.org/10.1609/AAAI.V29I1.9491
29. Huang, P., Weixin, Z., Jiuyang, T., Wang, M., Huang, H., Zhao, X.: Open knowledge graph completion with negative-aware representation learning and multi-source reliability inference. Inf. Fusion **115**, 102729 (2025)
30. Niu, L., et al.: Open-world knowledge graph completion with multiple interaction attention. World Wide Web **24**(1), 419–439 (2021). https://doi.org/10.1007/s11280-020-00847-2
31. Tian, H., Sheng, W., Shen, H., Wang, C.: Truth finding by reliability estimation on inconsistent entities for heterogeneous data sets. Knowl.-Based Syst. **187**, 104828 (2020). https://doi.org/10.1016/j.knosys.2019.06.036
32. Zhao, X., et al.: An experimental study of state-of-the-art entity alignment approaches. IEEE Trans. Knowl. Data Eng. **34**(6), 2610–2625 (2020)
33. Zeng, W., et al.: On entity alignment at scale. VLDB J. **31**(5), 1009–1033 (2022)

Semantic Web and Software Agents – A Forgotten Wave of Artificial Intelligence?

Tapio Pitkäranta[1(✉)] and Eero Hyvönen[1,2]

[1] Department of Computer Science, Aalto University, Espoo, Finland
tapio.pitkaranta@iki.fi

[2] Helsinki Centre for Digital Humanities (HELDIG), University of Helsinki, Helsinki, Finland

Abstract. The history of Artificial Intelligence (AI) is a narrative of waves, rising optimism followed by crashing disappointments. This paper argues that such a perspective overlooks a crucial wave of AI that seems to be forgotten: the rise of the Semantic Web, which is based on knowledge representation, logic, and reasoning, and its interplay with intelligent Software Agents. The Semantic Web aimed to transform the WWW into an ecosystem where AI could reason, understand, and act but is seldom mentioned in the context of AI after ChatGPT and Large Language Models (LLM) reignited the AI enthusiasm of today. By analyzing its academic impact through bibliometric data, we highlight the Semantic Web's role in AI history and its untapped potential for modern Software Agent development. This analysis not only deepens our understanding of AI's cyclical evolution but also offers key insights for integrating emerging technologies.

Keywords: Artificial Intelligence · Software Agents · Semantic Web · WWW

1 Introduction

The notion of AI was introduced in 1956 at a workshop at Dartmouth College by early pioneers such as John McCarthy, Marvin Minsky, and others. Since then, the history of AI has experienced recurring cycles of enthusiasm and subsequent disillusionment: "AI summers" and "winters" [31]. These cycles are associated with two main approaches to creating AI: symbolic ("white box") approaches, based on explicit knowledge representation, logic, reasoning, and search, and sub-symbolic ("black box") approaches, based on neural networks, machine learning, and statistical models.

Following the first AI summer in the 1960s, the first AI winter occurred in the 1970s after it was argued in [35] that sub-symbolic neural networks (then based on single-layer perceptron) would never be useful for solving real-world tasks. As a response, the symbolic approach, including knowledge-based expert systems

H. Takeda et al. (Eds.): IJCKG 2025, LNAI 16297, pp. 142–157, 2026.
https://doi.org/10.1007/978-981-95-5009-8_10

and logic programming, fuelled the next AI summer in the 1980s. However, challenges in, e.g., knowledge representation and acquisition [34] soon led to another AI winter. Research on multi-layer sub-symbolic neural models and learning systems continued despite—or perhaps because of—the shortcomings of the knowledge-based approach. By expanding machine learning models through deep learning, employing new training algorithms such as backpropagation, introducing novel neural architectures and non-linear activation functions, leveraging Big Data from the Web and available databases for training, and by using new efficient AI chips for parallel neural computations AI has entered an unprecedented summer in the 2020s [23]. A prominent research topic today, aimed at addressing challenges in neural systems such as hallucinations, is the combination of symbolic and sub-symbolic approaches in neuro-symbolic (hybrid) AI systems [5,21].

The early 2000s are often labelled as an AI winter. However, this narrative overlooks significant developments from that era, including the emergence of the Semantic Web, which gained momentum in 2001 [3,4]. Although rooted in knowledge representation, logic, reasoning, and ontologies—core subjects of traditional symbolic AI—this branch of research is typically not classified as part of AI. For example, in the authoritative ACM Computing Classification System[1] of Computer Science, AI is a major first-level category under Computing Methodologies while Semantic Web is only mentioned under the minor category under Information Systems. According to the original Semantic Web vision outlined in 2001 [3], the Semantic Web is centered around intelligent agents on the Web. The category of Intelligent Agents can be found as a minor subcategory under Artificial Intelligence. Even review papers on the Semantic Web do not focus on the AI or Software Agent research conducted during that period within the Semantic Web domain. For example, Hitzler [20] provides a comprehensive review of the Semantic Web, detailing its evolution, core principles, and ongoing challenges. The paper highlights how the Semantic Web extends the traditional Web by incorporating structured and linked data to enhance machine interoperability. Key technologies such as RDF (Resource Description Framework), OWL (Web Ontology Language), and SPARQL are discussed, emphasizing their roles in enabling semantic interoperability. In addition, the paper explores real-world applications in domains such as AI, knowledge graphs, and data integration. Hitzler also addresses challenges such as scalability, reasoning complexity, and data quality, underscoring the need for continued research and development to fully realize the Semantic Web's potential.

The rapid advancement of Generative Artificial Intelligence (GAI) and the widespread adoption of Large Language Models (LLMs) based on deep learning models [23,48] laid the foundation for applications such as ChatGPT leading to a surge in research and commercial interest. More recently, the focus has expanded beyond standalone LLMs to the integration of Software Agents and agentic capabilities, where models exhibit autonomy, goal-directed behavior, and enhanced reasoning abilities [43,59]. These agentic systems, which incorporate

[1] ACM Computing Classification System: https://dl.acm.org/ccs.

planning, memory, and interaction mechanisms, are now being extensively studied to augment the functionality of LLMs, enabling them to operate dynamically in complex environments, solve multi-step problems, and interact more effectively with users and digital ecosystems.

It appears that for some reason the Semantic Web is a forgotten wave of AI although the current AI trend is very much happening on the Web. Why? To study and analyze this question, this paper reviews previous trend analyses related to the Semantic Web and AI and presents statistical analyses of academic literature on these topics.

2 AI Summers and Winters

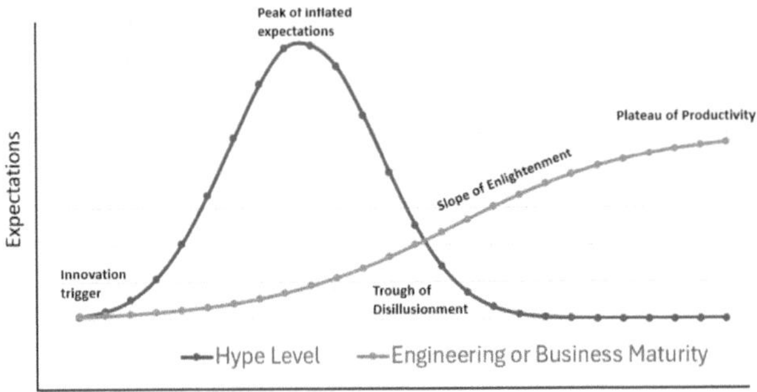

Fig. 1. Hype Cycle: Hype Level, Engineering or Business Maturity and combination Hype Cycle [47]

The hype cycle model, introduced by Gartner Inc. in 1995, [47] illustrates the evolution of technological expectations over time. As shown in Fig. 1 [47], the model is formed by merging two distinct curves: a human-centric expectation curve, represented by the dashed blue line, and a classical technology maturity S-curve, represented by the dotted green line. The first equation describes the hype level, which follows a bell-shaped curve driven by excitement, social contagion, and heuristic decision making. This curve captures the tendency of people to overestimate the potential of a technology, leading to an initial peak of inflated expectations, followed by a sharp decline when real-world limitations become evident. The second equation, the S-curve of technological maturity, represents the gradual advancement of a technology over time. Initially, progress is slow due to limited understanding and minimal performance gains from early investments. However, as knowledge accumulates and improvements accelerate, the technology experiences rapid development until it eventually stabilizes at a plateau defined by its inherent limitations. Together, these two curves form the hype cycle (solid gray line), which models the typical trajectory of emerging technologies, highlighting both the initial overhype and the eventual stabilization at a realistic level of maturity.

Although the hype-cycle is typically illustrated as a single expectation curve, it actually overlays the S-curve of technology diffusion; how these two curves interact determines both the tempo and the amplitude of successive boom-and-bust phases, yet this relationship remains only sketchily articulated in the literature. Consequently, existing research offers limited guidance on when firms and investors should initiate, scale, or exit their involvement in order to capture upside while containing downside risk. Closing this gap will require analytical frameworks that fuse expectation dynamics with adoption kinetics and strategic positioning along the hype-cycle.

AI is not a single invention but a shifting portfolio of techniques from symbolic reasoning and expert systems to machine learning, deep learning, and today's large-scale generative models. Each element in this portfolio follows its own trajectory, so the broad history of AI is best understood as the superposition of many overlapping hype cycles rather than one monolithic wave. Take *software agents* as an illustrative example. Introduced in the early 1990s, agent technology ascended to widespread media enthusiasm, encountered technical and market barriers, resurfaced with the rise of web services, and is now experiencing renewed interest through autonomous AI assistants. In other words, the same underlying concept has already traversed the Gartner hype-cycle several times.

Viewed in aggregate, these technology-specific curves explain why the field repeatedly climbs to a *Peak of Inflated Expectations*, descends into a *Trough of Disillusionment*, and, where genuine utility is present, climbs the *Slope of Enlightenment* to a durable *Plateau of Productivity*. Recognising AI as a constellation of evolving technologies helps clarify why optimism and disappointment can coexist—and why practical value often emerges only after multiple passes through the cycle.

In our research, we synthesized 16 different studies of AI history[2] To quantify "summer" and "winter", a grading scale from +10 to -10 was used, normalizing each illustration to this same scale. This means that all figures receive a maximum value of +10 when they visually reach their peak and -10 when they visually reach their lowest point. This grading serves as an indication of "temperature," verbally expressed as "AI summer" (hot) or "AI winter" (cold). The aggregated illustration is presented in Fig. 2. This figure also indicates that the period of active Semantic Web research is generally considered an AI winter. Furthermore, none of the studies mention either the topic *Semantic Web* or *Software Agents* as representing any form of AI boom in historical AI illustrations.

AI history illustrations known to the authors of this study depict the period from 1995 to 2010 as an AI winter rather than a time of substantial research advancements. These accounts consistently emphasize stagnation and minimal innovation, portraying this period as one characterized by diminished funding, waning public interest, and a general lack of transformative breakthroughs in

[2] The references to these studies are available in this arxiv version of our paper [42].

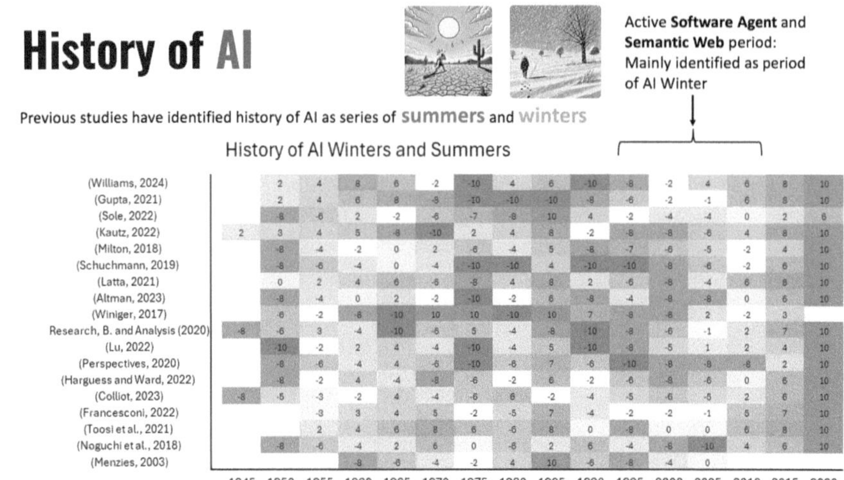

Fig. 2. History of Artificial Intelligence as illustrated by previous studies. Period of active Semantic Web development is identified typically as AI Winter period

AI. While some research activities continued during this time, the prevailing perception is that AI did not experience any major surge or revolution.

None of the AI history illustrations reviewed in our study explicitly mention Software Agents or the Semantic Web. This absence suggests that these topics were not widely recognized or highlighted as pivotal elements within the broader narrative of AI development during this period. Their exclusion reinforces the notion that this era was marked more by stagnation than by the emergence of significant new paradigms or technologies.

3 Semantic Web and Its Vision of AI Software Agents

Software agent technology has gone through a hype cycle a couple of times during its history. The following quote [40] is from 1996: "Agents are here to stay, not least because of their diversity, their wide range of applicability and the broad spectrum of companies investing in them. As we move further and further into the information age, any information-based organization which does not invest in agent technology may be committing commercial hara-kiri". The authors of this quote wrote another article a few years later [39] stating that the assumptions were too high. If anything, companies that invested in agent technology in the 1990s were perhaps the ones to commit a form of hara-kiri.

In December 2024 Microsoft CEO Satya Nadella stated that AI-driven software agents are fully revolutionizing business-to-business (B2B) applications. AI-driven agents will take over business logic and enable new AI-powered applications, placing software agents at the heart of the current AI hype [36]: "People want more AI-native business apps. It means the logic tier can be orchestrated by AI and AI Agents in a very seamless way."

Obviously, noting that agent technology has had inflated expectations in the past does not prove anything about the future. This time, it might be different, as the overall technology infrastructure has evolved significantly since the 1990s. We also have to remember that, for example, neural networks have experienced several hype cycles before the major breakthrough in the 2010s.

Software Agents Before the Semantic Web. Long before the Web era, the conceptual foundations for software agents was established by research in *Distributed Artificial Intelligence* (DAI) and the *Actor model* [19]. DAI cast intelligence as an *emergent* phenomenon arising from multiple, loosely-coupled problem solvers that cooperate across a network, thereby elevating autonomy and negotiation to primary design concerns [6,17]. In parallel, the Actor model supplied a concrete computational metaphor of independent, concurrently executing entities that communicate solely through asynchronous message passing. Together these two traditions contributed the twin conceptual pillars of **distribution** and **concurrency**, which later researchers distilled into the now-standard agent properties of social ability and proactive–reactive behaviour [54].

The 1990s witnessed the emergence of software agents as a concept in computer science, driven by the need for autonomous and intelligent systems in increasingly complex environments. This decade saw the consolidation of core agent characteristics, including autonomy, proactivity, reactivity, and social ability [40,55]. The rapid expansion of the World Wide Web around 1994 underscored the potential for agents to operate within this new digital space [29]. Foundational research explored diverse aspects of agent technology, from theoretical underpinnings and formal methods to practical applications for information management and personalized assistance [29,55].

The development of various agent architectures, such as reactive, deliberative, and hybrid models, further illustrated the field's commitment to creating sophisticated autonomous entities [53]. Researchers also focused on defining the boundaries of agency, distinguishing agents from conventional programs through taxonomies and formal definitions [16]. The intellectual roots of software agents were traced back to Distributed Artificial Intelligence (DAI) and Carl Hewitt's Actor model from 1977, highlighting the long-standing interest in distributed and interacting intelligent systems [19,40,54]. The explicit taxonomies and formal definitions produced during this period played a pivotal role in *standardising* the concept of agency. By establishing a common vocabulary, they enabled the Foundation for Intelligent Physical Agents (FIPA) to publish its initial specifications—most notably the *FIPA Agent Communication Language* (FIPA ACL) and a reference architecture that has since underpinned almost every industrial-grade multi-agent platform [14]. These efforts, together with subsequent W3C initiatives (e.g. the *Agent Markup Language* submission), ensured that later Semantic-Web agents could interoperate across organisational and domain boundaries on the basis of well-defined semantics rather than ad-hoc conventions.

The Vision of the Semantic Web and Intelligent Agents. Tim Berners-Lee, James Hendler, and Ora Lassila, in their seminal 2001 paper "The Semantic Web" [3], envisioned an evolution of the World Wide Web into a more intelligent and interconnected system. This Semantic Web would be an infrastructure where information has explicitly defined meaning, facilitating human-computer cooperation and enabling machines to understand and process web data through a universal framework of metadata and ontologies. This involved enriching web resources with semantic markup using languages like RDF and OWL, making content machine-processable [32]. The goal was a transition from unstructured documents to interconnected data, enhancing knowledge sharing, reuse, and integration [45], aligning directly with AI's goals of machine understanding and reasoning [3]. As of the time of writing, this highly influential paper has garnered over 30 000 citations.

Intelligent agents were conceived as crucial actors within this vision, able to automatically access, interpret, and reason about the semantically marked-up information [18]. These agents would act on behalf of users, performing tasks such as information retrieval, service discovery, automated decision-making (e.g., scheduling meetings or making purchases), and seamless interaction across different platforms [33]. Personal assistants, guided by rules, ontologies, and user profiles, were highlighted as prime examples of potential applications [26].

Realising this vision depends on the *ability of autonomous agents to converse*. The FIPA ACL specification meets this requirement by supplying a performative-based message envelope whose semantics are defined in a modal logic of communicative acts [15]. When the `content` field of such messages is expressed in RDF or OWL, the resulting utterances are simultaneously interpretable at the *data* and the *speech-act* levels, thereby enabling richly contextualised dialogues among heterogeneous Web agents [49]. Accordingly, ACLs form the critical bridge between the Semantic Web's data layer and the multi-agent community's interaction layer.

The paper's emphasis on AI Software Agents, and the backgrounds of Hendler and Lassila in intelligent agents and knowledge representation, underscore the Semantic Web's connection to advancing autonomous systems capable of reasoning, decision-making, and communication [3]. By leveraging standardized semantics, agents could interpret and act upon information with greater sophistication. This vision linked the future of the Semantic Web to developments in AI and positioned Software Agents as a critical component in realizing the Web's full potential. While the full vision remains unrealized, it remains a cornerstone in the intersection of Semantic Web technologies and AI Software Agent development.

Agent Communication Languages (ACLs), such as the Foundation for Intelligent Physical Agents (FIPA) ACL [12], were recognized as essential for facilitating interaction and information exchange between autonomous agents. Furthermore, the *DAML-S* ontology—renamed *OWL-S* after the release of OWL— [30] was created to add a machine-interpretable semantic layer to conventional Web-service descriptions. It exposes three mutually supportive components: a *Service*

Profile that advertises what the service does, a *Process Model* that specifies how the service can be executed, and a *Grounding* that binds abstract inputs and outputs to concrete message exchanges. Together these components allow an intelligent agent to match its goals to advertised profiles and thus discover relevant services, to use the process model and grounding information to construct and send the correct messages for direct invocation, and to chain multiple process models into coherent workflows for automatic composition of complex tasks. The development of DAML-S/OWL-S therefore represents a focused effort to supply the semantic infrastructure—rooted in ontologies, RDF, and OWL—along with the inter-agent communication mechanisms required for truly autonomous agents on the Semantic Web.

Software Agents in the LLM Age. The advent of large language models (LLMs) after the advent of the Transformer [48] has dramatically reshaped the landscape of software agent development, opening up new possibilities and challenges. Pre-LLM, agents often relied on meticulously crafted rules, ontologies, and knowledge bases, limiting their adaptability and generalizability [44,54]. While these approaches provided a degree of control and explainability, they struggled to handle the complexity and ambiguity of real-world scenarios. LLMs, trained on massive datasets, offer a fundamentally different approach. They possess an unprecedented ability to understand and generate natural language, reason, and even exhibit some degree of common sense, albeit imperfectly [7,41,51]. This capability allows LLM-powered agents to interact with users and environments in a much more natural and flexible way, reducing the need for extensive manual programming and domain-specific knowledge engineering.

The integration of LLMs with agent frameworks has led to the emergence of "LLM-powered agents" or "LLM agents," capable of performing a wider range of tasks than their predecessors. The simple definition for an LLM Agent is an artificial entity with prompt specifications (initial state), conversation trace (state), and ability to interact with the environments such as tool usage (action) [8,38]. These agents can leverage LLMs for various functions, including natural language understanding (NLU), natural language generation (NLG), planning, decision-making, and even tool use [50,57,58]. For instance, an LLM agent can be prompted to break down a complex task into smaller, manageable sub-tasks, search the web for relevant information, interact with APIs to execute actions, and summarize the results for the user [46,59]. Frameworks like LangChain [9] and AutoGen [56] provide tools and abstractions to facilitate the development and deployment of these agents, enabling developers to chain together LLM calls, external tools, and custom logic. This marks a significant shift from symbolic AI to a more data-driven, emergent approach to agent development.

Despite the significant advancements, LLM agents also present unique challenges. One major concern is the potential for "hallucinations," where the LLM generates plausible-sounding but factually incorrect or nonsensical information [22]. This can lead to unreliable agent behavior and requires careful mitigation strategies, such as grounding the LLM's responses in external knowledge sources or employing techniques like retrieval-augmented generation (RAG) [27]. Other challenges include ensuring the safety and ethical behavior of agents, managing the computational cost of running large models, and addressing biases inherent in the training data [2,52]. Furthermore, the "black box" nature of LLMs can make it difficult to understand the reasoning behind an agent's actions, hindering debugging and limiting trust in critical applications [11].

The ongoing research in LLM-powered agents focuses on addressing these limitations and exploring new capabilities with the emerging multi-agent systems (MAS). This includes developing methods for improving the factual accuracy and reliability of LLMs, designing more robust and explainable agent architectures, and creating effective evaluation metrics to assess agent performance across diverse tasks [24,37]. The convergence of LLMs and software agents represents a significant step towards building more general-purpose, adaptable, and intelligent systems, with the potential to transform how we interact with computers and automate complex tasks. However, responsible development and deployment, with a keen awareness of the inherent challenges, are crucial to realizing the full potential of this technology.

Could Semantic Web Principles Complement LLMs? Unlike the high-dimensional token spaces in which LLMs operate, an ontology offers a *canonical, machine-interpretable vocabulary* of entities, relations and axioms. Mapping generated spans to URI-identified concepts therefore gives each symbol an unambiguous referent, providing the first step toward faithful explanations [10]. OWL and RDF guarantee sound, monotonic inference but require careful manual modelling, limit expressivity to stay decidable, adopt an Open World Assumption and incur worst-case exponential reasoning cost. LLMs sit at the opposite end of the spectrum: they learn broad coverage automatically, answer in closed-world style with constant-time decoding, and tolerate noisy data—yet offer no built-in notion of truth, identity or provenance. Combining the two thus pairs the symbolic rigour of OWL/RDF with the statistical reach of LLMs.

Pattern 1 – Ontology-Constrained Decoding. During generation the model is prompted or fine-tuned so that every entity mention must be resolvable to an ontology concept; illegal continuations are pruned. This drastically reduces hallucinations and yields provenance for every token. A practical example is the plant-disease classifier of [1], where OWL reasoning is used online to check that the captioned visual attributes entail a unique disease class. **Pattern 2 – knowledge-graph RAG.** Retrieval-augmented generation in which the retriever indexes a knowledge graph rather than free text grounds answers in triples whose semantics are explicit. Recent systems such as SUBGRAPHRAG show that even small models achieve state-of-the-art accuracy *and* produce subgraph-level rationales that users can inspect [28]. **Pattern 3 – neuro-symbolic post-hoc explanation.** After an LLM produces an answer, a reasoning engine aligns the answer with the ontology, derives minimal entailment paths, and verbalises them as "because-clauses." Confalonieri and Guizzardi identify reference modelling, commonsense reasoning and knowledge-refinement as the three main roles ontologies play in such explanatory pipelines [10].

4 Academic Significance and Statistics

In this section, we use Google Scholar and other bibliometric tools to evaluate the significance of the research wave surrounding the Semantic Web. By analyzing citation counts, publication trends, and related metrics, we assess the impact and reach of Semantic Web research within the academic community.

To assess the significance of the Semantic Web and its associated AI trends compared to other common AI research themes, we conducted Google Scholar searches. The Semantic Web was undoubtedly a notable research wave, as evidenced by 3.41 million Google Scholar search results (March 2025). While broader topics like "Artificial Intelligence" (7.25M) and "Machine Learning" (5.47M) have higher volumes, the Semantic Web's specific focus on structured, meaningful data integration, along with its connection to related areas such as Artificial Intelligence (1.66M) and Agents (1.19M), underscores its significant impact. Despite its narrower scope, the Semantic Web established foundational principles that continue to influence modern AI and data technologies, solidifying its importance in the research landscape.

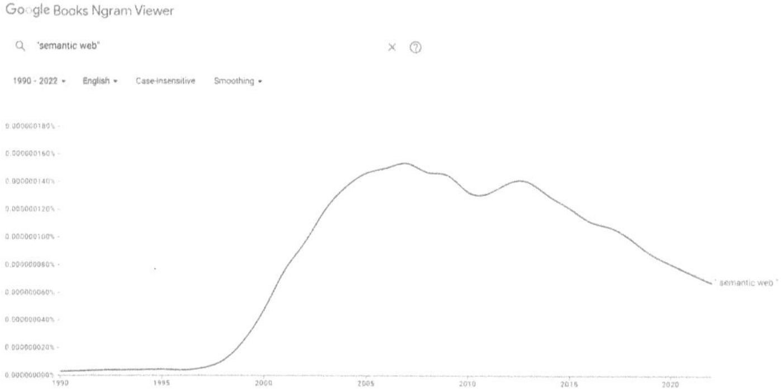

Fig. 3. Google Books: number of hits using "Semantic Web" as key words (Data not available after 2022)

Figure 3 presents a screenshot from the Google Books Ngram Viewer, illustrating the historical prominence of the term Semantic Web in English-language books from 1990 to 2022. The graph shows a significant rise in usage beginning in the early 2000s, coinciding with the publication of the seminal 2001 paper by Tim Berners-Lee, James Hendler, and Ora Lassila, which introduced the concept. Usage peaks between 2005 and 2010, reflecting heightened academic and industrial interest in its potential to transform data structuring and access on the Web. However, the decline after this peak suggests a waning emphasis on the Semantic Web in scholarly and public discourse, possibly due to challenges in realizing its full vision or a shift toward emerging technologies such as machine learning and deep learning. This trend underscores the cyclical nature of technological interest, aligning with broader discussions of the hype cycle.

Search: "Software Agents"

Search from	Search to	Google Scholar: hits	Google Books: frequency
1970	1974	7	0
1975	1979	12	0
1980	1984	8	5,00E-09
1985	1989	39	1,00E-08
1990	1994	463	5,00E-08
1995	1999	6 320	3,00E-07
2000	2004	15 700	4,50E-07
2005	2009	16 700	4,00E-07
2010	2014	16 300	2,50E-07
2015	2019	15 400	2,00E-07
2020	2024	13 700	2,20E-07

Fig. 4. Search results with "Software Agents" with five year time intervals Google Scholar (hits) and Google Books (frequencies)

One of the most promising future approaches is integrating software agents with large language models (LLMs) to enable not only discussions but also the execution of actions and commands by applications. To analyze the origins and activity trends of software agent research, we compiled statistics from Google Scholar and Google Books Ngram Viewer (cf. in Fig. 4). Google Scholar searches indicate that the term "Software Agents" appeared with minimal activity in the 1970s and 1980s, with only a handful of mentions. Significant growth occurred in the 1990s, reaching 463 results between 1990 and 1994. The field experienced exponential growth after 2000, peaking between 2005 and 2015, with over 16,000 search results in each five-year period. Activity has slightly declined in recent years, with 15,400 results from 2015 to 2020 and a projected 13,700 from 2020 to 2024, indicating sustained but slightly reduced interest in the field.

Figure 4 highlights the historical prominence of the term "Software Agents" in English-language books from 1990 to 2019, showing a significant rise in usage between 2000 and 2010. This trend suggests that Software Agent research received substantial attention during this period, coinciding with the heightened focus on the Semantic Web. The overlap is noteworthy, as the original Semantic Web vision in 2001 envisioned a machine-readable web where Software Agents could perform complex tasks on behalf of users. The simultaneous growth in interest in Software Agents and the Semantic Web underscores their interconnectedness, with Software Agents positioned as a key application of Semantic Web technologies. This period represents a peak in academic and industrial enthusiasm for intelligent, autonomous systems capable of leveraging structured web data, aligning with broader trends in artificial intelligence research.

The illustration in Fig. 5 depicts the number of Google Scholar search hits for the general terms Software Agent and specific combination "Software Agents" from 1990 to 2010, highlighting their academic prevalence over time. The data indicates a steady increase in publications for both terms, with Software Agent experiencing continuous growth and peaking around 2010 at approximately 413,000 hits. In contrast, the more specific term "Software Agents" saw rapid growth until about 2005, reaching a peak of approximately 5,230 hits before experiencing a slight decline. This suggests that while interest in the broader concept of software agents continued to grow, research using the specific phrase "Software Agents" peaked mid-decade before stabilizing or declining.

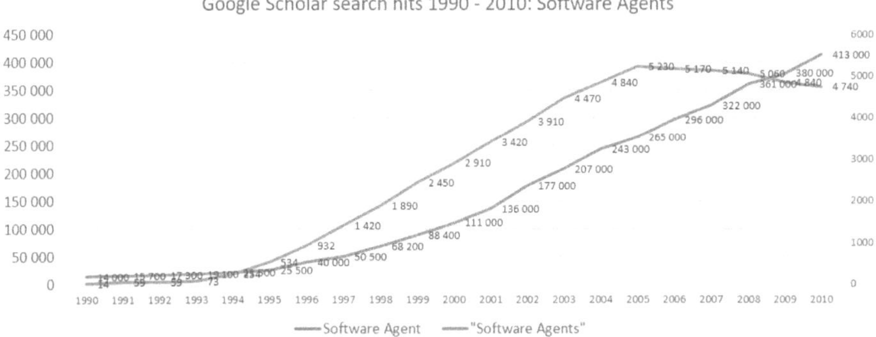

Fig. 5. Citation Trends in Software Agent Research in Google Scholar

5 Discussion

AI history is marked by cycles of optimism and setbacks. This paper reviewed prior studies of AI history, concluding that the Semantic Web and Software Agents represent a neglected phase in AI's evolution. Software Agents are currently a hot topic, with roots in Semantic Web research. A key question arises: Why wasn't the Semantic Web recognized as a distinct AI wave? The original Semantic Web vision in 2001 sought to create a machine-readable web enabling intelligent reasoning and automation. Software Agents played a central role, designed to operate autonomously, interact with structured data, and execute user tasks. During the early 2000s, Semantic Web research saw a surge in academic publications and funding, as evidenced by our citation analysis. One possible explanation, as suggested by Ora Lassila in his keynote talk [25], is that Artificial Intelligence was perceived negatively at the time, leading researchers to distance themselves from the term. Instead, they used alternatives terms, such as Software Agents and Knowledge Representation. A similar trend occurred in the early 1980s when Geoffrey Hinton and other researchers avoided using the term Neural Networks due to its controversial status in AI research [13].

Despite initial momentum, the Semantic Web failed to become a mainstream AI paradigm. One primary reason was the dominance of symbolic reasoning and logic-based AI at the time, which contrasted with the statistical and data-driven approaches that later fueled AI's success. Software Agents lacked access to the powerful natural language processing (NLP) and deep learning techniques that characterize modern AI applications. Although the Semantic Web did not achieve its grand vision of intelligent software agents, its principles are resurfacing in contemporary AI, particularly through Large Language Models (LLMs) and AI-driven Software Agents.

Recent advancements highlight the renewed relevance of Semantic Web concepts such as knowledge representation, reasoning, and autonomous agents. Knowledge graphs, an evolution of Semantic Web technologies, are widely used by companies like Google and Microsoft to enhance AI-driven search and recommendation systems. Moreover, modern Software Agents are now integrated with LLMs, enabling them to interpret unstructured text, execute commands, and autonomously interact with various data sources—effectively realizing the Semantic Web's vision through alternative means.

A growing emphasis on AI explainability, data provenance, and structured knowledge integration suggests that Semantic Web technologies could find new applications in the evolving AI landscape. Hybrid neuro-symbolic AI approaches that combine statistical learning with symbolic reasoning may leverage the Semantic Web's structured data to enhance AI reasoning and interpretability.

The Semantic Web's history offers valuable lessons on AI's cyclical evolution. The shift from logic-based AI to deep learning was not a rejection of earlier methods but rather an adaptation to technological constraints and industry priorities. As AI progresses, integrating structured knowledge with modern techniques could yield more powerful, interpretable, and autonomous systems. Recognizing the Semantic Web as part of AI's history provides a more comprehensive view of technological evolution. Instead of seeing it as a failed wave, we can understand it as a crucial step in AI's development—one whose principles may yet play a key role in the next generation of intelligent systems.

The key contributions in this paper were 1) to review and synthesize previous illustrations of AI history and 2) to identify the Semantic Web and Software Agents as a forgotten wave of AI. The Semantic Web represents a significant yet often underappreciated chapter in the evolution of artificial intelligence. Emerging in the early 2000s, its vision of a machine-readable web, capable of empowering AI-driven Software Agents, promised transformative applications ranging from intelligent automation to advanced data integration. While the grand vision articulated by Tim Berners-Lee, James Hendler, and Ora Lassila remains unrealized in its entirety, its influence on structuring and interpreting web data continues to resonate in modern technologies, including knowledge graphs, ontologies, and interoperability standards. We revisited the academic and industrial impact of the Semantic Web, using bibliometric analysis and historical context to argue for its rightful place within AI's broader narrative of hype cycles, AI winters, and eventual breakthroughs. In particular, the overlooked connection between the Semantic Web and AI Software Agent research from 2000 to 2010 highlights its contributions to advancing intelligent systems. Despite challenges in achieving widespread adoption and the subsequent shift in focus toward machine learning and deep learning paradigms, the Semantic Web established foundational principles that remain relevant to contemporary AI development.

Recognizing the contributions of the Semantic Web enriches our understanding of AI's history and underscores the importance of examining past technological waves to draw lessons for the future. By reintegrating this forgotten wave into AI's historical discourse, we gain a more comprehensive perspective on the cyclical nature of innovation and the interplay between vision and practicality in shaping technological progress.

References

1. Amara, J., König-Ries, B., Samuel, S.: Enhancing explainability in multimodal large language models using ontological context (2024). https://arxiv.org/abs/2409.18753
2. Bender, E.M., Gebru, T., McMillan-Major, A., Shmitchell, S.: On the dangers of stochastic parrots: can language models be too big? In: Proceedings of the 2021 ACM Conference on Fairness, Accountability, and Transparency, pp. 610–623 (2021)
3. Berners-Lee, T., Hendler, J., Lassila, O.: The Semantic Web. The Scientific American (2001)
4. Berners-Lee, T.: Semantic web wave. In: Presentation at SIIA (2003). Accessed 26 Jan 2024
5. Bhuyan, B.P., Ramdane-Cherif, A., Tomar, R., Singh, T.P.: Neuro-symbolic artificial intelligence: a survey. Neural Comput. Appl. **36**, 12809–12844 (2024). https://doi.org/10.1007/s00521-024-09960-z
6. Bond, A.H., Gasser, L.: Readings in Distributed Artificial Intelligence. Morgan Kaufmann (2014)
7. Brown, T., Mann, B., Ryder, N., et al.: Language models are few-shot learners. In: Advances in Neural Information Processing Systems, vol. 33, pp. 1877–1901 (2020)
8. Cemri, M., Pan, M.Z., Yang, S., et al.: Why do multi-agent LLM systems fail? (2025). https://arxiv.org/abs/2503.13657
9. Chase, H.: Langchain (2022). https://github.com/langchain-ai/langchain. Accessed 26 Jan 2024
10. Confalonieri, R., Guizzardi, G.: On the multiple roles of ontologies in explainable AI (2023). https://arxiv.org/abs/2311.04778
11. Doshi-Velez, F., Kim, B.: Towards a rigorous science of interpretable machine learning (2017). https://arxiv.org/abs/1702.08608
12. FIPA: FIPA communicative act library specification. Foundation for Intelligent Physical Agents (2002). http://www.fipa.org/specs/fipa00037/. Accessed 26 Jan 2024
13. Ford, M.: Architects of Intelligence: The Truth About AI from the People Building it. Packt Publishing, Birmingham (2018)
14. Foundation for Intelligent Physical Agents (FIPA): FIPA '97 Specification, Version 1.0 – Part 2: Agent Communication Language. Standard specification, FIPA, Geneva, Switzerland (1997)
15. Foundation for Intelligent Physical Agents (FIPA): FIPA ACL Message Structure Specification. Standard Specification SC00061G, FIPA, Geneva, Switzerland (2002)
16. Franklin, S., Graesser, A.: Is it an agent, or just a program?: A taxonomy for autonomous agents. In: Third International Workshop on Agent Theories, Architectures, and Languages (ATAL 1996), pp. 21–35. Springer (1996)

17. Gasser, R., Huhns, M.N.: Distributed Artificial Intelligence: Volume II, vol. 2. Morgan Kaufmann (2014)
18. Hendler, J.: Agents and the semantic web. IEEE Intell. Syst. **16**(2), 30–37 (2001)
19. Hewitt, C.: Viewing control structures as patterns of passing messages. Artif. Intell. **8**(3), 323–364 (1977)
20. Hitzler, P.: A review of the semantic web field. Commun. ACM **64**(2), 76–83 (2021)
21. Hitzler, P., Sarker, M.K. (eds.): Neuro-Symbolic Artificial Intelligence: The State of the Art. IOS Press (2022). https://doi.org/10.3233/FAIA210348
22. Ji, Z., Lee, N., Frieske, R., et al.: Survey of hallucination in natural language generation. ACM Comput. Surv. **55**(12), 1–38 (2023). https://doi.org/10.1145/3571730
23. Kelleher, J.D.: Deep Learning. The MIT Press Essential Knowledge Series. The MIT Press, Cambridge (2019)
24. Laskin, M., Wang, L., Oh, J., et al.: In-context reinforcement learning with algorithm distillation (2022). https://arxiv.org/abs/2210.14215
25. Lassila, O.: Semantic web and AI: can we finally realize the vision? Keynote Address at the 23rd International Semantic Web Conference (ISWC 2024) (2024). slides: https://www.lassila.org/publications/2024/lassila-iswc2024-keynote.pdf, Video: https://videolectures.net/videos/iswc2024_lassila_web_and_ai
26. Lassila, O., Swick, R.R.: Enabling the web of data with RDF. Bull. IEEE Comput. Soc. Tech. Committee Data Eng. **24**(4), 6–10 (2001)
27. Lewis, P., Perez, E., Piktus, A., et al.: Retrieval-augmented generation for knowledge-intensive NLP tasks (2020). https://arxiv.org/abs/2005.11401
28. Li, M., Miao, S., Li, P.: Simple is effective: the roles of graphs and large language models in knowledge-graph-based retrieval-augmented generation (2025). https://arxiv.org/abs/2410.20724
29. Maes, P.: Agents that reduce work and information overload. Commun. ACM **37**(7), 30–40 (1994)
30. Martin, D., et al.: OWL-S: semantic markup for web services. In: W3C Member Submission, vol. 22, p. 2004 (2004)
31. McCorduck, P.: Machines Who Think: A Personal Inquiry into the History and Prospects of Artificial Intelligence. AK Peters Ltd (2004)
32. McGuinness, D.L., Van Harmelen, F.: Owl web ontology language overview. W3C Recommendation **10**(10), 2004 (2004)
33. McIlraith, S.A., Son, T.C., Zeng, H.: Semantic web services. IEEE Intell. Syst. **16**(2), 46–53 (2001)
34. Mettrey, W.: An assessment of tools for building large knowledge-based systems. AI Mag. **8**(4) (1987)
35. Minsky, M., Papert, S.: Perceptrons: An Introduction to Computational Geometry. MIT Press (1969)
36. Nadella, S.: BG2 pod series: December 2024 episode. YouTube (2024). https://youtu.be/9NtsnzRFJ_o?t=2811. Accessed Mar 2025
37. Nakano, R., Hilton, J., Balaji, S., et al.: Webgpt: browser-assisted question-answering with human feedback (2022). https://arxiv.org/abs/2112.09332
38. Naveed, H., et al.: A comprehensive overview of large language models (2024). https://arxiv.org/abs/2307.06435
39. Nwana, H.S., Ndumu, D.T.: A perspective on software agents research. Knowl. Eng. Rev. **14**(2), 1–18 (1999). http://agents.umbc.edu/introduction/hn-dn-ker99.pdf

40. Nwana, H.S.: Software agents: an overview. Knowl. Eng. Rev. **11**(3), 205–244 (1996)
41. Ouyang, L., Wu, J., Jiang, X., et al.: Training language models to follow instructions with human feedback. Adv. Neural. Inf. Process. Syst. **35**, 27730–27744 (2022)
42. Pitkäranta, T., Hyvönen, E.: Semantic web and software agents – a forgotten wave of artificial intelligence? (2025). https://arxiv.org/abs/2503.20793
43. Reed, S., Zolna, K., et al.: A generalist agent. DeepMind (2022)
44. Russell, S.J., Norvig, P.: Artificial Intelligence: A Modern Approach, 3rd edn. Prentice Hall (2010)
45. Shadbolt, N., Hall, W., Berners-Lee, T.: The semantic web revisited. IEEE Intell. Syst. **21**(3), 96–101 (2006)
46. Shinn, N., Cassano, F., Berman, E., Gopinath, A., Narasimhan, K., Yao, S.: Reflexion: language agents with verbal reinforcement learning (2023). https://arxiv.org/abs/2303.11366
47. Steinert, M., Leifer, L.: Scrutinizing gartner's hype cycle approach. In: Picmet 2010 Technology Management for Global Economic Growth, pp. 1–13. IEEE (2010)
48. Vaswani, A., et al.: Attention is all you need. In: Advances in Neural Information Processing Systems, vol. 30 (2017)
49. Vreeswijk, G.A.: Liberalizing protocols for argumentation in multi-agent systems. In: Proceedings of the Fourth International Joint Conference on Autonomous Agents and Multiagent Systems, pp. 1259–1260 (2005)
50. Wang, L., Ma, C., Feng, X., et al.: A survey on large language model based autonomous agents. Front. Comput. Sci. **18**(6) (2024). https://doi.org/10.1007/s11704-024-40231-1
51. Wei, J., Wang, X., Schuurmans, D., et al.: Chain-of-thought prompting elicits reasoning in large language models. Adv. Neural. Inf. Process. Syst. **35**, 24824–24837 (2022)
52. Weidinger, L., et al.: Ethical and social risks of harm from language models (2021). https://arxiv.org/abs/2112.04359
53. Weiss, G.: Multiagent Systems: A Modern Approach to Distributed Artificial Intelligence. MIT Press (1999)
54. Wooldridge, M.: An Introduction to Multiagent Systems. Wiley (2009)
55. Wooldridge, M., Jennings, N.R.: Intelligent agents: theory and practice. Knowl. Eng. Rev. **10**(2), 115–152 (1995)
56. Wu, Q., Bansal, G., Zhang, J., et al.: Autogen: enabling next-gen LLM applications via multi-agent conversation (2023). https://arxiv.org/abs/2308.08155
57. Xi, Z., Chen, W., Guo, X., et al.: The rise and potential of large language model based agents: a survey (2023). https://arxiv.org/abs/2309.07864
58. Yang, S., Nachum, O., Du, Y., Wei, J., Abbeel, P., Schuurmans, D.: Foundation models for decision making: problems, methods, and opportunities (2023). https://arxiv.org/abs/2303.04129
59. Yao, S., Zhao, J., Yu, D., et al.: React: synergizing reasoning and acting in language models (2023). https://arxiv.org/abs/2210.03629

MugraTAG: Multi-granularity Representation Learning on Text-Attributed Graphs

Shu Zhao[1,3,4], Ying Zha[1,3,4], Ziwei Du[1,3,4](✉), Huanjing Zhao[2,3,4](✉), Xingsheng Lu[1,3,4], Ao Liu[1,3,4], Jie Chen[1,3,4], and Zhen Duan[1,3,4]

[1] School of Computer Science and Technology, Anhui University, Hefei 230601, Anhui, China
duziwei1225@163.com

[2] Tsinghua University, Beijing 100084, China
hwangyeong@163.com

[3] Key Laboratory of Intelligent Computing and Signal Processing, Ministry of Education, Anhui University, Hefei 230601, Anhui, China

[4] Information Materials and Intelligent Sensing Laboratory of Anhui Province, Anhui University, Hefei 230601, Anhui, China

Abstract. Representation learning on Text-Attributed Graphs (TAGs) is crucial for deeply analyzing complex network data in real-world scenarios. Current approaches which utilize Pretrained Language Model (PLMs) and Graph Neural Networks (GNNs) mainly analyze the intrinsic correlation between semantic and structural information at a single granularity. However, semantic and structural features inherently exhibit multi-granularity characteristics, which are essential for effectively modeling TAGs. In this paper, we propose a multi-granularity representation learning method for Text-Attributed Graphs (MugraTAG) to achieve effective integration of semantic and structural information at various granularities. Firstly, we construct a multi-granularity TAG via a graph coarsening strategy, leveraging Large Language Models (LLMs) to integrate and inherit structural and semantic attributes across granularities. Secondly, we design the embedding method to explicitly capture intra-granularity structural and semantic features. Finally, we perform inter-granularity fusion by integrating node representations across granularities, yielding unified node embeddings enriched with hierarchical structural and semantic insights. Experimental results on five publicly available TAG datasets demonstrate that our proposed multi-granularity integration framework achieves better performance compared to various baselines, particularly achieving a 10.35% average improvement on PubMed dataset.

Keywords: Text-Attributed Graphs · Representation Learning · Multi-granularity · Large Language Models

H. Takeda et al. (Eds.): IJCKG 2025, LNAI 16297, pp. 158–172, 2026.
https://doi.org/10.1007/978-981-95-5009-8_11

1 Introduction

Text-Attributed Graphs (TAGs) are widely available in real-world scenarios [1]. Each node in TAGs is associated with a text description. A paper citation graph is a typical application scenario for TAGs, where the nodes represent academic papers and the node attributes comprise textual content such as the titles and abstracts. The core of representation learning on TAGs lies in analyzing the intrinsic correlations between semantic information of textual attribute and structural connections. The semantic information of nodes is generally captured by PLMs [2–4], while structural relationships are captured by GNNs [5,27,28].

As illustrated in Fig. 1, PLM-based and GNN-based methods are two traditional paradigms of learning on TAGs. Firstly, the textual content of each node is directly input into the PLMs without considering topological information. Secondly, GNN-based methods pre-encode the individual node's textual content as non-contextualized shallow embeddings [17], and then feed feature vectors into GNNs [5,27,28]. Recent methods increasingly shift focus to co-training paradigms which integrate PLMs and GNNs through cascaded [6, 7,31,34] or nested [1,10,11] architectures, effectively consolidating textual semantics and graph topology for node representation. Additionally, LLMs [13,14] are also incorporated into TAG representation learning [7–9,15,32–34] with their superior semantic understanding and contextual awareness. However, these methods typically analyze semantic-structural relationships at a single fine-grained level, overlooking coarse-grained information that could be crucial for capturing richer contextual semantics among neighboring nodes. Moreover, while multi-granularity graph representations [20,21] offer the potential to encode higher-level semantics, most existing approaches rely on pre-processed textual features, thereby missing the opportunity to fully exploit the rich information embedded in raw texts.

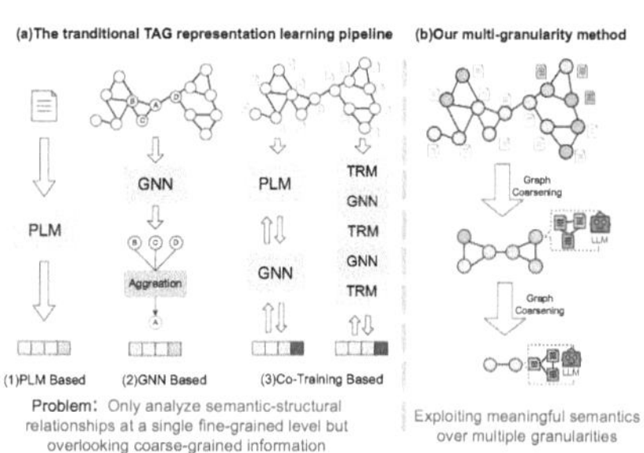

Fig. 1. (a) Most existing works leverage processed text features for analysis at a single fine-granularity. (b) Our method utilizes raw texts directly rather than processed text features and introduces a coarsening strategy to capture contextual information.

When analyzing TAGs over multiple granularities, two key challenges arise: (1) How to preserve semantic information when building multi-granularity TAGs? During graph coarsening, coarser hyper-nodes should inherit the common properties of the corresponding fine-grained node sets, forming a hierarchical inheritance relationship. In a text-attributed graph, not only the topological structure but also the textual attributes must be consistently inherited at coarser grain sizes, which adds complexity to building hierarchical multi-granularity TAG structures. (2) How to fuse intra-grained and inter-grained relationships? While the multi-granularity structure captures richer relationships among nodes at different grain sizes, each node at the original granularity may exhibit distinct relational properties cross over those coarser granularities. Integrating these multi-granularity relationships back into a unified representation remains a critical challenge.

In this paper, we propose an effective multi-granularity framework for TAGs, called MugraTAG, to address two key challenges. Our approach leverages the capabilities of LLMs to capture meaningful semantics across multiple granularities, thereby fully exploring the contextual relationships between a node and the textual descriptions of its neighbors. To address the first challenge, we design a prompt-summary-expert method that effectively captures shared contextual information, guided by the topological structure obtained through community partitioning. For the second challenge, we build mapping connections across different granularities to align cross-granular relationships, while preserving the structural and semantic information within each level. Dedicated TAG models are applied at each granularity to jointly extract and integrate hierarchical information. The main contributions of this work are as follows:

- We design multi-granularity TAG structures by combining LLM-based textual attribute coarsening with heuristic community partitioning, thereby establishing hierarchical inheritance of both semantic and structural information cross over different granularities.
- We develop an integration architecture that combines LLM and GNN to effectively quantify and merges node relationships at multiple granularities, fusing inter-grained and intra-grained representations.
- Experiments on five public TAG datasets demonstrate that our proposed MugraTAG method achieves better performance compared to various baselines, demonstrating its effectiveness in TAG representation, while maintaining enriched semantic information.

2 Related Work

2.1 Language Models on TAGs

Current works explore various techniques for representation learning on TAGs. A categorization of the existing methods is presented below.

Initial TAG representation learning first encodes the textual information of nodes as non-contextualized shallow embeddings [17], and then feeds node features as inputs to graph-based learning algorithms, e.g., GCN [5], SAGE [27]. Researchers explore deep embedding techniques by fine-tuning PLMs such as BERT [2] and DeBERTa [4] without incorporating graph topology. Following existing surveys [11], we categorize these two paradigms as **GNN-based** and **PLM-based methods**. Several recent studies attempt to integrate PLMs and GNNs, allowing the model to utilize both textual and structural information. Integration strategies include cascading approaches such as GIANT [31], UniGraph [34], P2TAG [6], nested approaches such as GraphFormers [1], and

GLEM [10]. For instance, GIANT [31] introduces a graph-aware self-supervised framework that fine-tunes language models using a neighborhood prediction task, improving node feature extraction from raw text. GLEM [10] proposes an EM-based co-training scheme to alternately optimize a LM and a GNN. Following [11], we fall these pipelines into **Co-Training Based methods**. In recent years, LLMs such as ChatGPT [13], PaLM [18], LLaMA [14], have demonstrated their effectiveness in generating higher quality text features. TAGs representation learning are beginning to make significant progress. Existing methods explore some ways to integrate LLMs with graphs. For instance, some approaches prompt LLMs to produce extensive additional knowledge [7,15,32] or generate final predictions [9,16,33], to boost initial fine-grained embeddings quality of TAGs. In this paper, we follow GLBench [12] referring to this new paradigm as **GraphLLM methods**. TAPE [7] designs an LLM-to-LM interpreter to advance fine-grained understanding of TAGs. Additionally, LLaGA [33] focuses on generating graph content from LLMs to augment graph learning tasks. However, they analyze the potential semantic-structural relationships within a single granularity, disregarding the coarse-grained information, which may limit their intelligence to fully capture contextual semantics between neighbors.

2.2 Multi-granularity Graph Representations

Multi-granularity graph representations are obtained by coarsening the graph into different granularities [19–23]. Coarsening is a graph summarization [22] technique in which nodes and edges of the original graph are fused to form a coarse-grained graph [23]. Semantic and structural features inherently exhibit multi-granularity characteristics, which are essential for effectively modeling TAGs. Recent efforts have also explored multi-granularity strategies to improve graph representation learning. HANE [21] proposes a hierarchical attributed network embedding framework that constructs progressively coarser graphs by fusing topology and attributes. TACO [19] proposes a topology-aware graph coarsening framework that stores information from previous tasks as a reduced graph. However, previous works usually use features that have been processed, do not focus on the original textual descriptions of nodes when performing coarsening, which may result in semantic loss. Moreover, traditional graph coarsening methods are also difficult to quantify the complex semantic information of node texts effectively.

3 Preliminaries

In this section, we introduce the notation of TAGs and the formalization of some concepts related to LLMs.

3.1 Text-Attributed Graphs

Given a TAG $\mathcal{G} = (\mathcal{V}, \mathcal{E}, \mathcal{T})$ also expressed as $\mathcal{G} = (\mathcal{A}, \mathcal{T})$, where $\mathcal{V}$ represents a set of nodes and $\mathcal{E}$ represents the set of edges. $\mathcal{T} = \{T^i \mid i \in \mathcal{V}\}$ denotes the textual description associated with each node v_i. Let $\mathcal{A} \in \mathbb{R}^{|V| \times |V|}$ denote the adjacency matrix, where each entry $A(u, v)$ represents the edge between the node u and v with weight.

3.2 Large Language Models

Large language models (LLMs) with powerful context generative capability, demonstrate excellent performance on various natural language processing (NLP) tasks. LLMs show excellent semantic comprehensive and summary capacity based on a natural language prompt. Additionally, LLMs also have the ability to encode textual content in sentences for text representation learning. Formally, given a node-associated text input x, we denote the summary output and embedding representation obtained from the LLM as:

$$X_g = LLM_{\text{Prompt}}(x), Q_g = LLM_{\text{Embedding}}(x) \tag{1}$$

where X_g refers to the generation-based output guided by task-specific prompts $\text{LLM}_{\text{Prompt}}$, and Q_g refers to the feature embedding extracted from the model $\text{LLM}_{\text{Embedding}}$ respectively.

4 Method

In this section, we display the process of MugraTAG in three major stages: (1) LLM-based multi-granularity TAG construct strategy that considers both semantic and structural information. (2) Build corresponding TAG embedding models to capture intra-granularity features independently. (3) Inter-grained and intra-grained representations integration (see Fig. 2).

4.1 Graph Coarsening

In this phase, we aim to generate multi-granularity TAG while preserving both structural relationships and original semantic information. Given a TAG $\mathcal{G}_0 = (\mathcal{V}_0, \mathcal{E}_0, \mathcal{T}_0)$ (or $\mathcal{G}$), the purpose is to iteratively construct coarser graphs $\mathcal{G}_1 = (\mathcal{V}_1, \mathcal{E}_1, \mathcal{T}_1)$, $\mathcal{G}_2 = (\mathcal{V}_2, \mathcal{E}_2, \mathcal{T}_2)$, $\ldots$, $\mathcal{G}_L = (\mathcal{V}_L, \mathcal{E}_L, \mathcal{T}_L)$ iteratively, where$|\mathcal{V}_0| > |\mathcal{V}_1| > \cdots > |\mathcal{V}_L|$, $|\mathcal{E}_0| > |\mathcal{E}_1| > \cdots > |\mathcal{E}_L|$. The initial graph $\mathcal{G}_0$ is the finest and the last graph $\mathcal{G}_L$ is the coarsest. The graph coarsening of TAG is much more challenging because not only the topological structure but also the textual attributes must be consistently inherited at coarser granularities, which adds complexity to building multi-granularity TAG structures. To tackle the first challenge, we introduce a coarsening strategy that considers both graph topology and textual semantics, denoted as structure coarsening (SC) and text coarsening (TC), respectively.

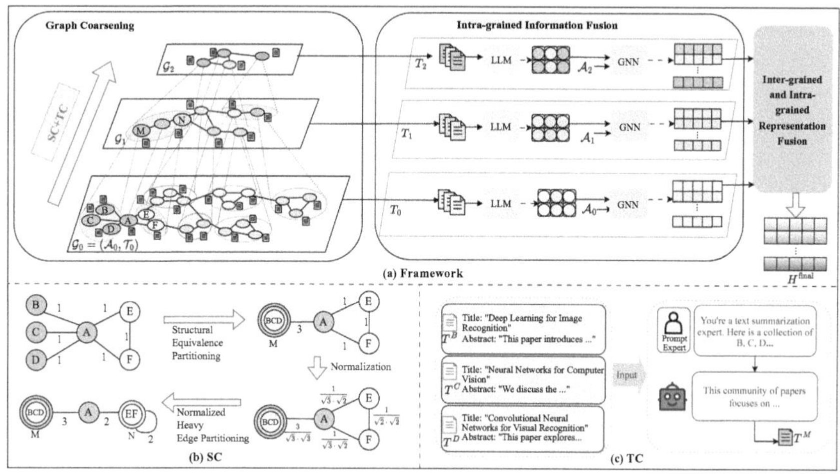

Fig. 2. The MugraTAG framework with two rounds of coarsening is applied.

Structure Coarsening (SC). Structure coarsening module contains node granulation and edge granulation. One or more nodes in the fine-granularity graph $\mathcal{G}_l$ are merged into a hyper-node in the coarser-granularity graph $\mathcal{G}_{l+1}$. The process of forming node sets of hyper-nodes is called community partitioning. Here, we use a hybrid of two partitioning methods. Firstly, use structural equivalence partitioning [20]. Specifically, if node u and node v have same neighborhoods, consider them as structurally equivalent nodes and send them into the same group. Secondly, use normalized heavy edge partitioning to compute the normalized adjacency matrix, facilitating the computation of similarity.

$$A_{\text{norm}} = D^{-\frac{1}{2}} A D^{-\frac{1}{2}} \tag{2}$$

where A is the original adjacency matrix, and D is the degree matrix. $d_i = \sum_j A_{ij}$ represents the degree of the i-th node. Then, we sort nodes based on their degree, match low-degree nodes first as they have fewer matching options. For each unmatched node u, find its neighbors and select the neighbor v with the highest normalized adjacency weight $A_{\text{norm}}(u, v)$.

After community partitioning, the node set $\mathcal{V}_l$ of $\mathcal{G}_l$ is divided into P groups $S_1, S_2, \ldots, S_P$, where each group S_p is a subset of the node set $\mathcal{V}_l$, and the groups are disjoint. Especially, some nodes may not be matched to other nodes, and they are treated as a separate group. Each fine-grained node group S_p corresponds to a hyper-node $v_p^{(l+1)}$ of the coarsened node set $\mathcal{V}_{l+1}$:

$$\mathcal{V}_{l+1} = \{v_p^{(l+1)} \mid S_p \subseteq \mathcal{V}_l, p = 1, 2, \ldots, P\}. \tag{3}$$

The edge granulation process forms the edge set $\mathcal{E}_{i+1}$ of $\mathcal{G}_{i+1}$ according to the node set $\mathcal{V}_{i+1}$ of $\mathcal{G}_{i+1}$ and the edge set $\mathcal{E}_i$ of $\mathcal{G}_i$. If node w in S_p and node m in S_q are connected at $\mathcal{G}_i$, then there exists an edge between the corresponding hyper-node of S_p and S_q at $\mathcal{G}_{i+1}$.

Text Coarsening (TC). After completing the structural community partitioning, it is a key to maximize the preservation of the original textual semantics in fine-grained nodes within each community. LLMs have demonstrated strong capabilities in natural language understanding and generation. On this basis, we apply LLMs to generate semantic abstraction during graph coarsening. Specifically, we import a prompt-summary-expert method to summarize rich fine-grained node texts which correspond to the same hyper-node in a unified way, generate representative textual descriptions of hyper-nodes, thereby providing richer semantic information for subsequent graph representation learning. In order to fully explore the contextual relationships between a node and its community neighbors, we utilize a prompt that includes the titles and abstracts of papers. The general structure of the prompt-summary-expert is as follows:

> ... **Abstract:** [paper abstract]
> **Title:** [paper title] ...
> **Question:** [You're a text summarization expert. Here is a collection of text descriptions from paper titles and abstracts. Please integrate the information and generate a concise summary that captures the main themes or common features without missing any key details.]
> **Answer:**

The textual information of a hyper-node $v_i^{(l+1)}$ is obtained by summarizing the texts of all fine-grained nodes in S_p.

$$T_{l+1}^{p} = \mathrm{LLM}_{summarize}\left(\{T_l^i \mid v_i^{(l)} \in S_p\}\right) \tag{4}$$

$LLM_{summarize}$ represents prompt statements that contain texts that are sent into the LLM. This ensures that the text of the fine-granularity graph is effectively compressed while retaining key semantic details.

The steps above demonstrate a round of the coarsening pyramid. By using the coarsen method repeatedly, we construct Multi-granularity TAG.

4.2 Intra-grained Information Aggregation

In this section, we develop an architecture that integrates LLM and GNN to effectively quantify intra-grained representations.

Node Embeddings Initialization with LLM. After obtaining TAGs at different granularities, we process the text attributes of nodes in each granularity respectively. Specifically, we use an LLM to encode the node texts and convert them into embedding vectors to fully explore the semantic information of node texts at each granularity. This process not only preserves the key semantic features in the original text, but also helps to enhance the expressive capability of the graph representation in the multi-granularity semantic space, thus providing more effective semantic support for downstream tasks. We adopt the LLM embedding model to extract semantic information such as BGE-M3 [24].

$$z_i = \mathrm{LLM}_{Embedding}(T^i) \tag{5}$$

where T^i represents the textual information of node i, and z_i is the text vector representation of node i obtained by the LLM.

Structure-Aware Node Embeddings Learning. After obtaining node text vectorization at each granularity of TAGs, we employ a self-supervised learning strategy to achieve meaningful node embeddings. Specifically, we adopt a reconstruction-based objective that encourages the learned embeddings to preserve both the structural information and node semantics. Given an input feature matrix $\mathbf{X} = [z_1, z_2, \ldots, z_N]^\top$, $\mathbf{X} \in \mathbb{R}^{N \times d}$ and graph $G = (V, E)$, the encoder generates structure-aware embeddings $\mathbf{H} = [h_1, h_2, \ldots, h_N]^\top \in \mathbb{R}^{N \times d}$. This is typically achieved via a graph convolution operation defined as:

$$\mathbf{H} = \sigma\left(\hat{\mathbf{A}}\mathbf{X}\mathbf{W}\right) \tag{6}$$

where $\hat{\mathbf{A}} \in \mathbb{R}^{N \times N}$ is the normalized adjacency matrix with self-loops, $\mathbf{W} \in \mathbb{R}^{d \times d'}$ is a learnable weight matrix, and $\sigma(\cdot)$ denotes a activation function such as ReLU. The reconstruction loss function is:

$$\mathcal{L}_{\text{recon}} = -\sum_{(u,v) \in E} \log \sigma(h_u^\top h_v) - \sum_{(u,\tilde{v}) \in \tilde{E}} \log \sigma(-h_u^\top h_{\tilde{v}}) \tag{7}$$

where $E \subseteq V \times V$ represents the set of observed edges, and $\tilde{E} \subseteq (V \times V) \setminus E$ denotes negative edges sampled from non-existent connections. For each node u, $h_u \in \mathbb{R}^d$ is its structure-aware embedding obtained from the encoder. The inner product $h_u^\top h_v$ measures similarity between nodes, which is converted to a probability by the sigmoid function $\sigma(\cdot)$. For negative samples, we use $\tilde{v}$ to denote a node that is not connected to u, and $h_{\tilde{v}}$ is its corresponding embedding. The loss encourages embeddings of connected nodes to be similar, while separating the embeddings of unconnected nodes.

4.3 Inter-grained and Intra-grained Representation Fusion

While the multi-granularity structure captures richer relationships among nodes at different granularities, each node at the original granularity may exhibit distinct relational properties cross over those coarser granularities. Effectively integrating these multi-granularity relationships back into a unified representation remains a critical challenge. To tackle the second challenge, we proceed to integrate inter-grained and intra-grained information to enhance the node representations. For each node in the original graph, we identify its corresponding coarse representations over multiple layers based on the mapping relationships generated during the graph coarsening process. The final representation of a node is constructed by fusing its original embedding with its corresponding coarsen-grained node embeddings. This fusion strategy allows the model to leverage both intra-grained local features and inter-grained semantic information, thereby improving the node representations for downstream tasks such as classification.

Let $\mathbf{h}_i^{(l)}$ denote the embedding of node i at the l-th coarsened graph. The final multi-granularity embedding $\mathbf{h}_i^{\text{final}}$ of node i is defined as:

$$\mathbf{h}_i^{\text{final}} = \left[\mathbf{h}_i^{(0)}, \mathbf{h}_i^{(1)}, \mathbf{h}_i^{(2)}, \ldots, \mathbf{h}_i^{(L)}\right] \tag{8}$$

where L is the number of coarsening rounds. Let $\mathbf{H}^{\text{final}} \in \mathbb{R}^{N \times d}$ denote the final embedding matrix composed of all nodes. For the evaluation, we employ the linear probing.

5 Experiments

5.1 Datasets

To evaluate the performance of MugraTAG, we employ five publicly available TAG datasets: Cora [25], Wiki-CS [26], PubMed [7], Photo [11] and History [11]. We use Cora and PubMed TAG version released publicly by TAPE [7]. For all datasets considered, we conform to the standard train/validation/test data splits used in prior works. In our experiments, we opt to leverage raw texts directly rather than processed text features so as to retain the textual semantics. The statistical details of datasets can be found in Table 1.

Table 1. Statistics of TAG datasets used in this work.

Dataset	#Nodes	#Edges	#Classes	Raw Texts
Cora	2,708	5,429	7	paper titles and abstracts
Wiki-CS	11,701	215,863	10	Wikipedia entry names and contents
PubMed	19,717	44,338	3	paper titles and abstracts
History	41,551	358,574	12	title and description of the respective books
Photo	48,362	500,928	12	electronics-related products and user reviews

5.2 Baselines

To assess the effectiveness of our proposed method, we compare MugraTAG with 15 baselines categorized into four types: (i) Graph-agnostic methods, (ii)Traditional GNN-based method, (iii) LM-based methods, (iv) Recent GraphLLM works designed for TAGs. Briefly, graph-agnostic methods include MLP. Traditional GNN-based models include SAGE [27], GCN [5], GAT [28], SGFormer [29], NodeFormer [30]. LM-based methods include BERT [2], Sentence-BERT [3]. Recent works designed for TAGs include GIANT [31], GLEM [10], LLaGA [33], OFA [32], UniGraph [34], GraphAdapter [9], PATTON [8].

5.3 Experimental Details

In MugraTAG, we function exclusively via API access to an LLM. We adopt GLM-4-Flash[1] to generate text information for hyper-node. As for LLM embedding method, we select BGE-M3 [24]. We use *Accuracy* and $Macro-F1$ to evaluate the model performance of node classification task. The number of training epochs is in the range of {100, 500}. The learning rate is chosen from the set {1e-2, 5e-3}. In our experiments, we follow specific dataset splits and utilized random seeds for reproducibility. For baselines, we use the results reported in the corresponding papers [12,35].

Table 2. Experimental results of node classification accuracy.

Methods	Cora	Wiki-CS	PubMed	History	Photo
MLP	62.29	68.41	62.88	64.62	61.21
GCN [5]	82.47	76.33	77.36	81.93	78.58
SAGE [27]	83.69	79.56	83.22	82.12	80.35
GAT [28]	82.92	78.21	74.36	<u>82.85</u>	82.08
SGFormer [29]	82.36	76.56	78.92	78.98	80.12
NodeFormer [30]	81.55	75.47	76.49	79.60	78.51
BERT [2]	79.02	77.98	76.74	72.97	68.82
SentenceBERT [3]	78.09	77.72	75.48	74.10	69.02
GIANT [31]	85.52	75.81	76.89	-	81.27
GLEM [10]	<u>85.60</u>	78.92	-	-	81.47
LLaGA [33]	81.25	<u>80.74</u>	<u>86.54</u>	82.55	-
OFA [32]	75.72	78.02	75.26	81.43	<u>84.46</u>
UniGraph [34]	84.23	-	80.11	-	81.22
MugraTAG	**85.79**	**81.55**	**87.74**	**83.11**	**84.79**

5.4 Performance Analysis

From Table 2, we draw several meaningful conclusions across all five datasets. First, graph-agnostic method like MLP pales in comparison to those GNN-based methods (i.e., GCN, SAGE, GAT), confirming that structural information is crucial. Moreover, methods that utilize standalone language models (i.e., BERT, SentenceBERT) generally achieve lower performance on textual graph tasks when compared to co-training based GraphLLM methods (i.e., GIANT, GLEM, UniGraph), indicating the effectiveness of fusing semantic understanding from LLMs with topological awareness captured by GNNs. Finally, our proposed

[1] https://bigmodel.cn/dev/howuse/llm/GLM-4-FlashX-250414.

method, **MugraTAG**, consistently achieves state-of-the-art performance across the majority of datasets. Notably, on PubMed dataset, MugraTAG achieves a 10.35% improvement compared to baseline models. This demonstrates that our multi-granularity design, which integrates inter- and intra-grained textual and structural information, effectively captures multi-grained semantics and contextual dependencies essential for node classification.

Similar experimental conclusions can be drawn from the Macro-F1 results as shown in Table 3. GraphLLM methods show notable improvements across the majority of datasets. For instance, on Wiki-CS, GIANT surpasses the best performing PLM-based model BERT by 3.56% in terms of Macro-F1. Lastly, our method outperforms all GraphLLM methods, demonstrating that our multi-granularity method leads to more discriminative node representations.

5.5 Sensitive Analysis

The Number of Coarsen Rounds. We evaluate the performance of the MugraTAG framework under different numbers of coarsening rounds k. Starting from $k = 0$, we incrementally increase k up to 5. Figure 3 illustrates how the Accuracy score for node classification as we vary the number of coarsening rounds k. Each experiment is independently implemented for five times. We summarize the following insights: For the smaller coarsening rounds (i.e., $k = 1$), our MugraTAG method steadily improves representations quality over the $k = 0$ (i.e., without applying the multi-granularity framework) across all datasets. For instance, on Cora, the Accuracy represents an increase of approximately 2.42%. Even if the number of coarsening rounds is high ($k = 4$ for History, $k = 4$ for Cora/PubMed/Photo), MugraTAG also offers a qualitative improvement over the original granularity. We hypothesize that the observed

Table 3. Experimental results of Macro-F1.

Methods	Cora	Wiki-CS	PubMed
GCN [5]	<u>80.65</u>	77.63	79.19
SAGE [27]	79.35	77.05	76.91
GAT [28]	79.00	77.40	74.36
BERT [2]	67.53	75.74	64.93
SentenceBERT [3]	67.59	75.11	67.33
GIANT [31]	80.13	<u>78.67</u>	76.05
GraphAdapter [9]	70.66	66.49	73.19
OFA [32]	74.20	74.97	75.60
LLaGA [33]	72.50	70.90	68.82
PATTON [8]	67.97	77.72	<u>79.25</u>
MugraTAG	**83.07**	**78.74**	**87.12**

qualitative improvement is because representations capture inter-grained and intra-grained semantics.

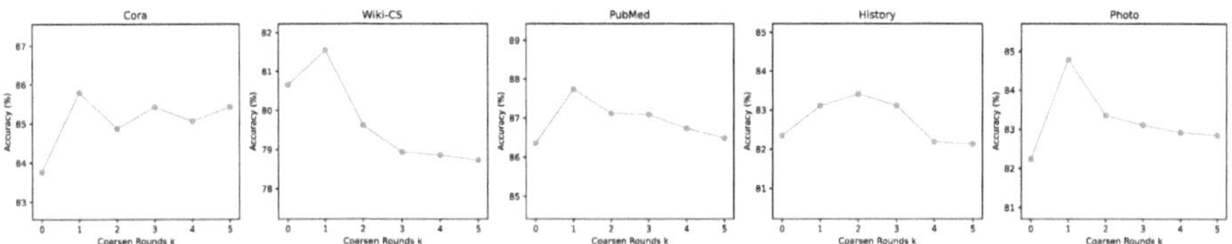

Fig. 3. Changes in performance as the number of coarsening rounds in MugraTAG increases. Note that coarsening rounds = 0 refers to the original representation without applying the multi-granularity framework.

To quantify the scale variation of graphs at different granularity, we introduce the *Granulated Ratio*. *Granulated Ratio* refers to the ratio between the number of nodes and edges in the granulated graph and those in the original graph. Specifically, we denotes the nodes' *Granulated Ratio* as NG_R while the edges' *Granulated Ratio* as EG_R. As illustrated in Fig. 4, we report these ratios over varying granularity for five datasets. Both node and edge counts diminish progressively with each round of coarsening. However, excessive compression may lead to information loss. Thus, choosing an appropriate coarsening depth (e.g., 1–2 rounds) is essential for maintaining representation quality. More importantly, hierarchical coarsening enables to capture inter-grained and intra-grained structural and semantic dependencies.

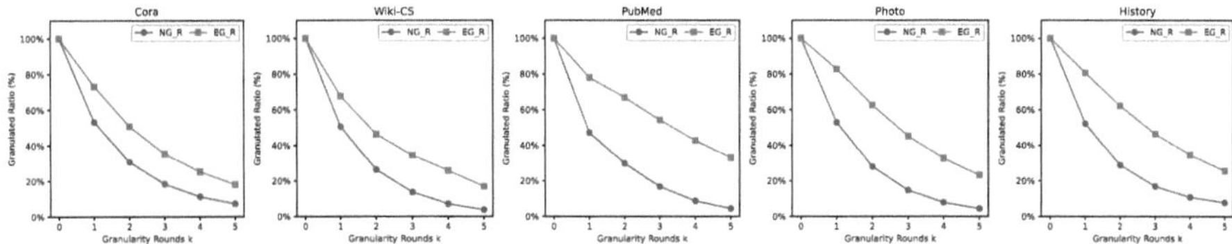

Fig. 4. The *Granulated_Ratio* on all datasets. X axis is the number of coarsen rounds k, we set $k = 0, 1, 2, 3, 4, 5$. Y axis is the *Granulated_Ratio*. The blue line denotes the nodes *Granulated_Ratio*(NG_R) and the red line represents the edges' *Granulated_Ratio*(EG_R).

5.6 Ablation Study

In this section, we investigate the contribution of individual components: (1) w/o raw texts, which replaces raw texts with processed text features; (2) w/o multi-gra, which without applying the multi-granularity framework; (3) w/o GNN, which removes the GNN module. As illustrated in Table 4, our MugraTAG method consistently achieves superior performance in all comparisons. Experimental results highlight the importance of structural information, original semantic, and multi-granularity architecture in improving representation quality. Specifically, compared to the best-performing ablated variants, MugraTAG achieves notable absolute improvements of +12.54% on Cora (vs. w/o GNN), +6.62% on PubMed (vs. w/o raw texts). Additionally, MugraTAG achieves improvements of +2.03% on Cora (vs. w/o multi-gra).

Table 4. Experimental results of ablation study. We report the mean accuracy with a standard deviation across 5 different random seeds.

Variants	Cora	Wiki-CS	PubMed	History	Photo
w/o raw texts	78.78	78.58	80.85	79.82	81.65
w/o multi-gra	83.76	80.65	86.36	82.35	82.24
w/o GNN	73.25	79.50	83.52	81.87	83.38
MugraTAG	**85.79**	**81.55**	**87.47**	**83.41**	**84.79**

6 Conclusion and Future Work

In this paper, we present MugraTAG, an effective multi-granularity integration framework to integrate multi-granularity node representations. Address the dual challenges of constructing multi-granularity structures by combining LLM-based textual attribute coarsening with heuristic community partitioning, and fusing cross-granularity relationships effectively. We develop a methodology that integrates LLM and GNN to effectively quantify and fuse node relationships at multiple granularities, fusing inter-grained and intra-grained representations. In future work, we will explore more adaptive fusion mechanisms, such as attention to better combine representations from different granularities. We also note that the current method has limitations in scalability and generality. LLM-based summarization increases computational cost and limits scalability to large graphs. In general, we will continue exploring the integration of LLMs with graph learning in diverse domains.

Acknowledgments. Our work is supported by the National Natural Science Foundation of China (62476003), Anhui Province Excellent Scientific Research and Innovation Team (2024AH010004), Anhui Provincial Natural Science Foundation - Water Science Joint Fund (2408055US006), the University Synergy Innovation Program of Anhui Province (GXXT-2023-050), and SMP-Zhipu.AI Large Model Cross-Disciplinary Fund (SMP-Zhipu20240210). We also acknowledge the support from Zhipu AI-Anhui University Joint Research Center, and the High-Performance Computing Platform of Anhui University.

References

1. Yang, J., et al.: GraphFormers: GNN-nested transformers for representation learning on textual graph. In: Neural Information Processing Systems, pp. 28798–28810 (2021)
2. Devlin, J., Chang, M. W., Lee, K., Toutanova, K.: Bert: pre-training of deep bidirectional transformers for language understanding. In: North American Chapter of the Association for Computational Linguistics: Human Language Technologies, pp. 4171–4186 (2019)
3. Reimers, N., Gurevych, I.: Sentence-bert: sentence embeddings using siamese bert-networks. In: Empirical Methods in Natural Language Processing, pp. 3980–3990 (2019)
4. He, P., Liu, X., Gao, J., Chen, W.: Deberta: decoding-enhanced bert with disentangled attention. In: International Conference on Learning Representations (2021)
5. Kipf, T.N., Welling, M.: Semi-supervised classification with graph convolutional networks. In: International Conference on Learning Representations (2017)
6. Zhao, H., et al.: Pre-training and prompting for few-shot node classification on text-attributed graphs. In: Knowledge Discovery and Data Mining, pp. 4467–4478 (2024)
7. He, X., Bresson, X., Laurent, T., Perold, A., LeCun, Y., Hooi, B.: Harnessing explanations: LLM-to-LM interpreter for enhanced text-attributed graph representation learning. In: International Conference on Learning Representation (2024)
8. Jin, B., et al.: Patton: language model pretraining on text-rich networks. In: Annual Meeting of the Association for Computational Linguistics, pp. 7005–7020 (2023)
9. Huang, X., et al.: Can GNN be good adapter for LLMs? In: International World Wide Web Conference, pp. 893–904 (2024)
10. Zhao, J., et al.: Learning on large-scale text-attributed graphs via variational inference. In: International Conference on Learning Representation (2023)
11. Yan, H., et al.: A comprehensive study on text-attributed graphs: benchmarking and rethinking. In: Neural Information Processing Systems, pp. 17238–17264 (2023)
12. Li, Y., et al.: Glbench: a comprehensive benchmark for graph with large language models. In: Neural Information Processing Systems, pp. 42349–42368 (2024)
13. Brown, T., et al.: Language models are few-shot learners. In: Neural Information Processing Systems, pp. 1877–1901 (2020)
14. Touvron, H., et al.: LlaMA: Open and Efficient Foundation Language Models. arXiv preprint arXiv:2302.13971 (2023)
15. Fang, Y., Fan, D., Zha, D., Tan, Q.: GAugLLM: improving graph contrastive learning for text-attributed graphs with large language models. In: Knowledge Discovery and Data Mining, pp. 747–758 (2024)
16. Ye, R., Zhang, C., Wang, R., Xu, S., Zhang, Y.: Natural language is all a graph needs. In: European Chapter of the Association for Computational Linguistics, pp. 1955–1973 (2024)
17. Miaschi, A., Dell'Orletta, F.: Contextual and non-contextual word embeddings: an indepth linguistic investigation. In: Representation Learning for NLP, pp. 141–151 (2020)
18. Chowdhery, A., et al.: PaLM: scaling language modeling with pathways. In: International Conference on Learning Representations (2022)
19. Han, X., Feng, Z., Ning, Y.: A topology-aware graph coarsening framework for continual graph learning. In: Neural Information Processing Systems, pp. 132491–132523 (2024)

20. Liang, J., Gurukar, S., Parthasarathy, S.: Mile: a multi-level framework for scalable graph embedding. In: Association for the Advancement of Artificial Intelligence, pp. 361–372 (2021)
21. Zhao, S., Du, Z., Chen, J., Zhang, Y., Tang, J., Yu, P.S.: Hierarchical representation learning for attributed networks. IEEE Trans. Knowl. Data Eng. **35**, 2641–2656 (2021)
22. Liu, Y., Safavi, T., Dighe, A., Koutra, D.: Graph summarization methods and applications: a survey. ACM Comput. Surv. **51**(3), 1–34 (2019)
23. Dickens, C., Huang, E., Reganti, A., Zhu, J., Subbian, K., Koutra, D.: Graph coarsening via convolution matching for scalable graph neural network training. In: International World Wide Web Conference, pp. 1502–1510 (2024)
24. Chen, J., Xiao, S., Zhang, P., Luo, K., Lian, D., Liu, Z.: M3-embedding: multi-linguality, multi-functionality, multi-granularity text embeddings through self-knowledge distillation. In: Association for Computational Linguistics, pp. 2318–2335 (2024)
25. McCallum, A.K., Nigam, K., Rennie, J., Seymore, K.: Automating the construction of internet portals with machine learning. Inf. Retrieval **3**, 127–163 (2000)
26. Mernyei, P., Cangea, C.: Wiki-cs: A Wikipedia-based Benchmark for Graph Neural Networks. arXiv preprint arXiv:2007.02901 (2020)
27. Hamilton, W., Ying, Z., Leskovec, J.: Inductive representation learning on large graphs. In: Neural Information Processing Systems (2017)
28. Velickovic, P., Cucurull, G., Casanova, A., Romero, A., Lio, P., Bengio, Y.: Graph attention networks. In: International Conference on Learning Representation (2018)
29. Wu, Q., et al.: Simplifying and empowering transformers for large-graph representations. In: Neural Information Processing Systems, pp. 64753–64773 (2023)
30. Wu, Q., Zhao, W., Li, Z., Wipf, D.P., Yan, J.: Nodeformer: a scalable graph structure learning transformer for node classification. In: Neural Information Processing Systems, pp. 27387–27401 (2022)
31. Chien, E., et al.: Node feature extraction by self-supervised multi-scale neighborhood prediction. In: International Conference on Learning Representation (2022)
32. Liu, H., et al.: One for all: towards training one graph model for all classification tasks. In: International Conference on Learning Representation (2024)
33. Chen, R., Zhao, T., Jaiswal, A., Shah, N., Wang, Z.: LLaGA: large language and graph assistant. In: International Conference on Machine Learning, pp. 7809–7823 (2024)
34. He, Y., Sui, Y., He, X., Hooi, B.: UniGraph: learning a unified cross-domain foundation model for text-attributed graphs. In: Knowledge Discovery and Data Mining, pp. 448–459 (2025)
35. Zhu, X., et al.: LLM as GNN: Graph Vocabulary Learning for Text-attributed Graph Foundation Models. arXiv preprint arXiv:2503.03313 (2025)

HECATE at the Crossroads: Guiding Astrophysics into the Semantic Web

Sophia Sideri[1(✉)], Iosif Oikonomakis[1], Petros Tsirmakos[2], Giorgos Flouris[1], Vasilis Efthymiou[1,2], Dimitrios Rompogiannakis[3,4], Elias Kyritsis[3,4], Ioanna Leonidaki[3,4], and Andreas Zezas[3,4]

[1] Institute of Computer Science, FORTH, Heraklion, Greece
{sophisid,sifisoik,fgeo}@ics.forth.gr

[2] Department of Informatics and Telematics, Harokopio University, Athens, Greece
{it21693,vefthym}@hua.gr

[3] Physics Department, University of Crete, Heraklion, Greece
ph4502@edu.physics.uoc.gr, {ekyritsis,ioanna,azezas}@physics.uoc.gr

[4] Institute of Astrophysics, FORTH, Heraklion, Greece

Abstract. Massive volumes of raw astrophysical data are collected every day, and subsequently processed into public catalogues used for scientific research. While deep learning has accelerated astrophysics in recent years, the field has yet to benefit from the Semantic Web's advances in data representation and management. As a result, most of those catalogues are typically stored as flat CSV/ASCII files in a tabular format, limiting their interoperability and reuse. In this work, we introduce a semantically enriched astrophysical catalogue, in the form of an ontology-backed Knowledge Graph (KG). Our KG is based on a custom, lightweight ontology (DOCBO), and was populated by transforming the Heraklion Extragalactic Catalogue (HECATE) into RDF using two alternative mapping approaches, X3ML and RML, whose characteristics and trade-offs are discussed and compared. Both the mapping rules and the resulting Knowledge Graphs are openly available to foster reuse, interoperability and community feedback. This work, which is part of the interdisciplinary PARSEC project, constitutes a first proof-of-concept effort towards a more generic methodology for representing astrophysical knowledge of various subdomains (e.g., Supernova Remnants and transients) using semantic technologies, thereby forming and automatically populating semantically-enriched KGs based on appropriately defined ontologies.

Keywords: Knowledge Graph Construction · Astrophysical Data · Data Mappings

1 Introduction

Telescopes, artificial satellites and other astrophysical instruments constantly probe the sky, gathering astronomical data and resulting to large-scale multi-

H. Takeda et al. (Eds.): IJCKG 2025, LNAI 16297, pp. 173–187, 2026.
https://doi.org/10.1007/978-981-95-5009-8_12

wavelength surveys. The resulting data are pre-processed and organised in *astrophysical catalogues* [20,24,30], which contain information about celestial sources such as galaxies, stars and transient events.

Astrophysical research crucially relies on the availability of such catalogues to enable cross-domain research tasks [6]. However, the exponentially growing volume of astronomical data highlights the need for a new approach in astronomical catalogue curation, as these surveys (and the resulting catalogues) constitute enormous datasets that contain valuable information, but often lack the semantic structure needed to fully support reuse, integration, and automated exploration appropriate for downstream tasks [1].

In practice, catalogue curation remains a manual and time consuming task. As the volume of incoming data increases, along with the different data formats, the task of catalogue curation poses major challenges. Catalogues are currently stored as static tables in CSV or other textual formats, or, in the best scenario, as a single-table relational database [12]. As a result, these resources risk becoming outdated, difficult to maintain, and hard to exploit in new research contexts [27].

Semantic technologies and Knowledge Graphs (KGs) [2,13] could provide a solution to this problem, by representing the data as entities and relationships, aligned with a domain-specific ontology that defines concepts and their properties. This model could be used to perform data querying in more expressive and semantically meaningful ways, and can allow efficient data searching, querying, reasoning, data analysis, pattern identification, provenance, trust and credibility analyses, catalogue maintenance and enrichment, data reuse, catalogue interlinking, and others. Nowadays, the astrophysical community have introduced ontologies for astronomical objects and metadata [16,17], but large-scale catalogue conversion into RDF-based KGs remains rare, due to the scale, complexity, and flatness of the existing datasets.

The development and application of a machine-assisted methodology for constructing and curating semantically enriched astrophysical catalogues in the form of KGs is the main objective of the PARSEC project[1]. In this paper, we focus on the subproblem of automatically constructing a domain-specific KG based on the data contained in the Heraklion Extragalactic Catalogue (HECATE) [19], an astrophysics catalogue which contains information about extragalactic sources. We present a methodology for transforming this catalogue into a semantic KG, called **HECAT3KG**, using a custom lightweight ontology named **DOCBO**, tailored to the domain. For the population, we explored different mapping frameworks, in particular the RML [8] and X3ML [21] mapping languages.

To enhance reproducibility, the resulting resources are publicly available on Zenodo[2], whereas the code and mappings are available on Github[3]. All resources are released under open licenses (CC BY 4.0 for data and Apache-2.0 for code).

---

[1] Propelling Astrophysics Research via Semantically Enriched Catalogues - https://isl.ics.forth.gr/PARSEC/.

[2] DOCBO: https://zenodo.org/records/16281112, HECAT3KG: https://doi.org/10.5281/zenodo.15379419.

[3] https://github.com/sophisid/HECAT3KG.

The rest of the paper is organized as follows. In Sect. 2, we review related work on astrophysical catalogues and semantic mapping frameworks. Section 3 describes the ontology, some competency queries used for its validation, and the mappings used for its population. Section 4 provides a comparison among the two mapping approaches that we used. Finally, Sect. 5 concludes the paper and outlines directions for future work.

2 Background and Related Work

In this section, we provide some background knowledge regarding HECATE, the astrophysics catalogue that we use throughout this paper, and then describe and compare against existing efforts for KGs in astrophysics, as well as mapping languages for transforming (semi-)structured data into RDF.

2.1 HECATE

The Heraklion Extragalactic Catalogue[4] (HECATE) [19] describes 204,733 galaxies in the local Universe (i.e., with distance less than 200 Mpc). Galaxies are individually selected from the HyperLEDA astronomical database, mainly based on radial velocity ($< 14,000$). For this reason, the methodology adopted for constructing the catalogue is specifically tailored to increase the quality of the data for nearby galaxies and to facilitate cross-matching procedures.

The main features of HECATE are the calculation of robust distances based on the combination of available redshift-independent indicators and redshifts, along with homogenized size information, and the computation and homogenization of star-formation rates based on infrared indicators.

HECATE provides various characteristics for each of the featured galaxies, e.g., position, morphological information, distance, star-formation rate, stellar mass, gas-phase metallicity, nuclear classification and others. The wealth of this information can be particularly useful for a wide-range of studies such as: the exploration of nearby galaxies and their properties; the selection of galaxy samples for future surveys; the identification of the host galaxies of multi-wavelength source populations and the characterization of their connection with their host's parameters; and the application of astrophysical priors in the search for electromagnetic counterparts of gravitational-wave sources.

As mentioned, HECATE is essentially a large table, each line of which describes one galaxy. The columns of this table contain measurements or other information about different physical properties of the respective galaxy. From an ontological representation perspective, HECATE contains four kinds of columns:

1. *Identification columns*, which contain galaxy identification information taken from other catalogues, thereby helping determining the galaxy that the respective line in the catalogue corresponds to.

[4] https://hecate.ia.forth.gr/.

2. *Measurement columns*, which contain the measurements regarding the galaxies' attributes and physical properties.
3. *Error and confidence interval columns* about (some of) the measurements that appear in the measurement columns; these are very important for the astrophysicists and are related to the accuracy of the instruments and methodologies used to determine a measurement.
4. *Flag columns*, which provide metadata about other data columns, such as the reliability and context of measurements (e.g., a flag could indicate whether a measurement is reliable or not).

2.2 KGs for Astrophysics

Creating and maintaining astrophysical catalogues is a long-standing problem for astrophysics research, currently being accomplished mainly by manual effort, either by catalogue curators who detect and ingest the data, as in the case of NASA/IPAC Extragalactic Database (NED) [24] and HyperLEDA [20], or by researchers and scientific journals that notify the catalogues of interest about their new findings, as in the case of SIMBAD [30]. Some astrophysics studies of wide impact (e.g., [11]) were, in essence, the manual work of several months, combining astrophysics literature and catalogues.

However, such catalogues are usually in semi-structured, flat tabular formats, expressed in plain ASCII, CSV or MRT files (the latter are similar to CSV). The transformation of existing scientific findings into a machine-readable and reusable format, as recommended by the linked data principles [5], will improve the catalogues' expressiveness, facilitate their reuse, maintenance and update, and allow added-value services to be implemented on top of them, such as catalogue integration and interlinking, expressive provenance recording, etc. As a result, this transition would significantly raise the value of catalogues and help scientists in their data-intensive research.

Efforts to address these issues through ontologies and KGs have been proposed. For example, ASON [17], models astrophysical services and objects, emphasizing service interoperability. Similarly, the International Virtual Observatory Alliance (IVOA) standards [16] provide vocabularies for celestial objects and metadata, enabling semantic integration across datasets. However, IVOA's ontology maintenance has been discontinued, limiting its applicability for ongoing projects. These approaches, while valuable, tend to be overly expressive and complex, making them difficult to apply efficiently on flat and large-scale datasets such as HECATE. In our case, such ontologies would significantly increase the number of generated triples, which could hinder scalability in future use cases, especially in the context of streaming data.

2.3 Mapping Languages

Transforming semi-structured data into RDF is a fundamental step for creating a KG. Mapping languages allow the expression of rules that determine how to represent the source data as RDF triples, in accordance to the target ontology.

Over the past years, many mapping frameworks have been developed to translate the data from formats like CSV, XML or relational databases into semantic representations. Among the most used tools are X3ML [21] and RML [8], each of which serves specific purposes and has been used in various projects requiring RDF materialization.

X3ML [21] is a declarative, XML-based language designed for clear, readable mappings. It was originally created to support mappings in the cultural heritage domain [22], where the complexity of the data often requires a more flexible and understandable set of transformation rules. X3ML has been used in various projects on diverse domains [4,10,26], in environments relying on semantic interoperability for connecting and reusing data among different disciplines. One of the key strengths of X3ML is its focus on human-readable mappings that can be written, reviewed, and maintained by both technical and domain experts. The X3ML framework has also been extended with tools like instance and label generator, and is supported by the 3M Mapping Memory Manager [18] interface, which simplifies collaboration between engineers and domain experts, like astrophysicists.

The *RDF Mapping Language (RML)* [8] is widely used, as it is a superset of the standard knowledge graph construction engine, R2RML [25], which is a W3C recommendation. While R2RML is restricted to relational databases and relies on SQL queries, RML has been extended to support additional data formats such as CSV, JSON and XML, offering flexibility while retaining the expressivity of the mapping syntax. However, RML mappings can become quite complex and are often difficult for non-technical users to understand and manage. To address RML's complexity and performance, recent works such as the modular redesign of the RML ontology [14] and the development of scalable processors like Morph-KGC [3] and SDM-RDFizer [8] have been proposed. These efforts aim to improve expressiveness, performance and scalability, particularly for large heterogeneous datasets. Comparative studies of mapping frameworks such as RML, R2RML, and YARRRML [7,15,29] have shown that RML excels in automation and scalability.

3 Designing and Populating the Ontology

3.1 Ontology Design and Motivation

Although standard ontologies to represent scientific measurements and observational data exist, they tend to be complex, overly general, or not suited for the scale and flat structure of HECATE. To address this, we developed a domain-specific ontology, DOCBO (Domain Ontology for Celestial Bodies and Observations), having in mind the structure and semantics of HECATE.

The DOCBO ontology (see Fig. 1) provides a lightweight but expressive model to describe astrophysical records expressed in tabular formats, i.e., the format of most astrophysical data. It was developed using Protégé [23], and contains 15 classes, 18 object properties and 15 data properties. As shown in Fig. 1, the ontology models:

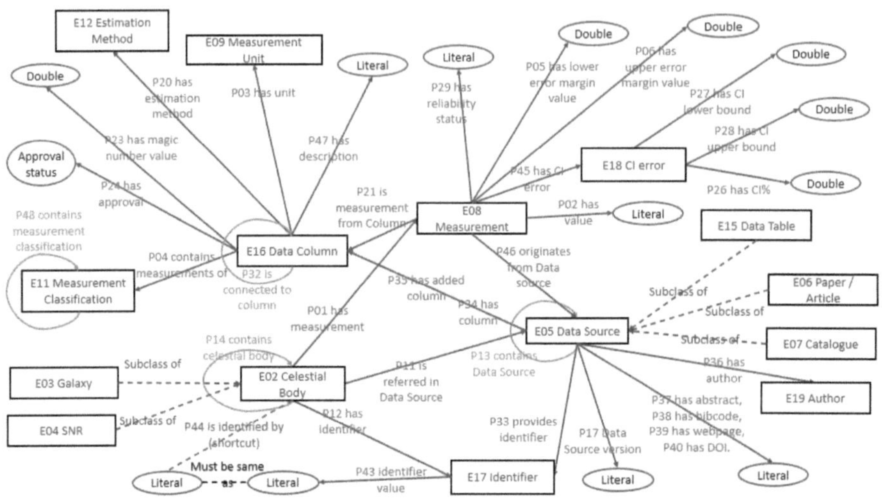

Fig. 1. The DOCBO ontology.

- Celestial bodies (`E02 Celestial Body`), namely objects in the universe (galaxies in the case of HECATE). Note that celestial bodies can contain other celestial bodies (a galaxy can contain a star system, a star system can contain a planet, etc.).
- Identifiers (`E17 Identifier`), which can be used to identify and disambiguate celestial bodies that often have different identifiers in different data sources.
- Data sources (`E05 Data Source`), which are used to record provenance information, i.e., the data table, article, or catalogue from which the information was derived.
- Measurements (`E08 Measurement`), which are used to record the actual values (measurements) of different physical quantities found in the table. Various metadata on the measurements can be recorded (such as error margins, measurement reliability, measurement unit, estimation method etc.), some of which are recorded using intermediate classes for proper modelling.

These entities capture the semantics of HECATE entries while allowing future ontology alignment with other relevant ontologies in future work. A more detailed documentation of these classes and properties is provided in the accompanying Zenodo resource, found in https://zenodo.org/records/16281112.

3.2 Competency Queries

To assess the usefulness of DOCBO, a number of competency queries were developed, capturing various features and types of searches that astrophysicists typically perform. All these queries are supported by the current version of DOCBO. Due to space considerations, only a partial list of these competency queries is included here, expressed in natural language.

1. Return all available photometric data and magnitudes for a given galaxy.
2. Give all sources with ages > 103yrs or diameter > 50pc or distance < 3Mpc.
3. For a given galaxy, how many/which publications present shock velocity/proper motion measurements and which are those values?
4. Return all galaxies with sizes < 0.5 degrees from SDSS and distances between 100 and 200 Mpc.
5. Return all galaxies with photometry g < 20.0 mag, SNR > 3.0 and activity classification AGN, based on spectroscopy.
6. Return all pairs of galaxies having distances less than 10 arcsec
7. Return all systems of galaxies that contain three objects or more, have distances less than 10 arcsec, and g-band photometry that differs more than 1.0 mag at the 90% confidence.

3.3 Mappings for Ontology Population

To populate the HECAT3KG Knowledge Graph with instances from HECATE (which contains more than 200,000 records and each more than 100 columns stored in a single CSV), we examined two different mapping frameworks: X3ML [21] and RML [8]. Both approaches can be used to transform the source data into RDF triples, based on the target ontology described above, DOCBO. Each mapping approach offers different strengths and trade-offs in terms of readability, maintainability, and performance, and we compare their effectiveness in handling large-scale astrophysical data (see Sect. 4). Interestingly, both mapping approaches required a preprocessing step to split the initial dataset into several ones that are processed individually. Below, we present the two approaches, describing the mapping workflow, preprocessing steps, and their main strengths and limitations.

X3ML Mappings. For the X3ML mappings, we adopted a declarative approach which helped collaboration with the domain experts. X3ML divides the mapping process into two parts: (i) the *Mapping Definition Document*, which defines how source data elements translate to ontology concepts and properties, and (ii) the *Generator Policy Document*, which defines templates on how new Unique Resource Identifiers (URIs) and literals are going to be generated. The transformation was performed using the X3ML Engine[5].

Our initial approach was to create individual mapping rules, and to transform the values for each column in the dataset; this is very much like the way a domain expert would approach the data for the first time. To support this, we transformed the CSV into an XML format, where each row became a `<ROW/>` element, and each column was a nested element named after its column header (e.g., `<RA/>`, `<DEC/>`, for right ascension and declination, respectively). We then treated each XML element and attribute independently, resulting in one-to-one mappings. This approach is simple and straightforward, but also computationally

[5] https://www.ics.forth.gr/isl/x3ml-toolkit.

inefficient. With more than 200,000 entities and over 100 rules, the mapping engine encountered performance bottlenecks.

To tackle the above issues, we implemented a second approach that introduces a preprocessing step that transformed the CSV into a semantically organized XML structure. This structure followed a schema that reflected the domain semantics and exhibited mapping efficiency. For instance, identifiers were grouped under the same XML node <identifiers>, positional data under <degrees>, orientation data in <rotation>, measurement uncertainties in <errors> and confidence intervals under <confidence_intervals>. Flags and classifications were included as standalone elements, except from flux data from IRAS[6] and distance-related flags, which were placed in <flux> and <DIST> respectively. The remaining column categories were placed in a generic block, named <gen>.

The key benefits of X3ML are that it uses an intuitive, human-readable structure, making it easy for both engineers and astrophysicists to review and validate the mappings. Additionally, X3ML allows tracking how the RDF data was produced, by the mapping rules that produced it. On the other hand, X3ML is XML-based, which limits its applicability when working with different data source formats, requiring a preprocessing step. This also causes efficiency issues when dealing with many flat data instances, as was the case with HECATE.

RML Mappings. For RML, we utilized the RMLMapper[7] engine to transform the HECATE CSV into RDF, leveraging RML's native support for CSV, JSON, XML and relational databases. RML defines declarative mappings as Turtle files, specifying how CSV columns correspond to ontology classes and properties. However, the flat structure of HECATE CSV and RML's lack of native conditional logic for CSV inputs, necessitated a preprocessing step to ensure semantic accuracy and minimize redundant triples.

The preprocessing enriches the CSV by adding 10 auxiliary columns to support mappings. These include error margins, reliability status, flux quality descriptions from IRAS flux flags, metallicity status, source labels, and distance estimation methods. For example, **RFLAG** values are mapped to descriptions like "Reliable"/"Unreliable" (for the values 1 or 0 respectively). Additionally, the **F_ASTROM** field, which indicates astrometric uncertainty of the galaxy's position, is used to compute an error margin using the formula $2.78 \times 10^{-4+F\_ASTROM}$. However, the preprocessing does not filter rows with invalid measurements, resulting to redundant triples. While this increased the CSV size, it facilitated precise RDF triple generation.

To manage the mapping for the ~ 100 columns, we developed a tool to automate rule generation for measurement columns. In a subsequent step, we refined the URI generation using custom functions to ensure consistency and semantic clarity. The resulting URIs follow a clear and reusable naming template like: `https://parsec.forth.gr/celestial_body`

[6] https://lambda.gsfc.nasa.gov/product/iras/.
[7] https://github.com/RMLio/rmlmapper-java.

`/galaxy_{PGC}/column/{column}/measurement/...` . In this template, `PGC` refers to a Principal Galaxy Catalogue (PGC) Identifier, which is a unique numerical ID assigned to each galaxy, while `column` represents the name of a specific column in the dataset (such as `r1`, `metal`, etc.). These placeholder values are dynamically replaced with the actual ones during the URI generation process.

Another challenge is RML's limited ability to filter empty or invalid values, leading to redundant RDF triples for cases like missing measurements. However, in some cases, it was not possible to identify or label all invalid values during preprocessing—either because the condition was too complex to encode, or due to missing metadata in the source. This results in a larger KG, but it does not affect the semantics of the output KG, or the results of the queries that we want to perform over it. Additionally, RML required around 100 mapping rules, i.e., one mapping per column, increasing the maintenance complexity.

On the other hand, RML excelled in performance using the RMLMapper, even when producing the same or bigger KG compared to X3ML. Its versatility to handle diverse data formats makes it suitable for future integration. However, it tends to be more technical.

3.4 Developed Resources

Both the DOCBO ontology and the two KGs (HECAT3KG-X3ML and HECAT3KG-RML, which were derived from the HECATE dataset using the X3ML and RML mappings respectively), adhere to the FAIR principles [31]. The ontology and the produced KGs are *Findable* through their publication on Zenodo with persistent identifiers (DOIs). They are *Accessible* on Zenodo and Github. The KGs are *Interoperable*, as they are linked to external datasets (astrophysical catalogues) through their identifiers. The use of RDF and a custom ontology ensures *Reusability* by providing a machine-readable, semantically rich data.

4 Comparison of Mapping Approaches

As mentioned, the two mapping approaches we used generated different KGs, HECAT3KG-X3ML and HECAT3KG-RML; this was due to the distinct characteristics of the X3ML and RML mapping frameworks. However, it should be noted that both KGs preserve the semantics of the HECATE dataset, mapping the records of all galaxies to DOCBO classes and properties. In other words, the aforementioned variations do not affect the correctness of query results, but they do affect the storage and maintenance process.

The characteristics of these two KGs and of the mapping approaches employed are summarized in Tables 1 and 2, and explained in more details below. It is important to note that our comparison should not be used to infer generic conclusions about the two approaches. Indeed, our work is not meant to be a general comparison of the two mapping languages (RML and X3ML), as such a

Table 1. X3ML Engine vs RML Mapper: Performance Metrics

Metric	X3ML Engine	RML Mapper
Execution Time	~18 h	~50 min
Output Size	~5 GB	~9 GB
Number of Triples	~44 Million	~55 Million

Table 2. X3ML vs RML Language: Features and Characteristics

Feature	X3ML	RML
Input Formats	XML	CSV, XML, JSON, RDBs
Mapping Rules	10	157
Mapping Intuitiveness	High	Low
Column Merging	Yes	With preprocessing
Conditionals	Yes	Partly, with preprocessing
URI Consistency	Yes	With preprocessing

comparison would require a more thorough theoretical and experimental setting that would take full advantage of their capabilities. Instead, in this paper we only report on our experiences and findings for our particular use case and our implementation of those approaches.

Execution Time. In our original application of the mapping rules over HECATE, we experienced a very slow execution time. Upon further tests, we realized that the problem was related to the fact that the input file with all the galaxy entities was too large to be processed efficiently. The problem appeared in both mapping languages, but was much more pronounced for the X3ML language.

To address this problem, we split the original input file into smaller chunks that were processed independently. For X3ML, we split the original file into smaller chunks of 100 galaxies each, resulting to a total of ~2,000 smaller files. As RML was fairly efficient, we used a more coarse-grained splitting, resulting to ~200 files containing 1,000 galaxies each. In both cases, after the migration of each smaller chunk in an RDF file, all the output files were merged and uploaded into the target triple store, resulting to the expected RDF graph.

The times reported in Table 1 were measured under the above process. As seen in the table, X3ML ran in ~ 18 h, whereas RML in just ~ 50 min. This was despite the fact that X3ML uses less rules and produces a smaller KG (see Table 1). The main reason for this finding is that RML mappings are processed using engines optimised for speed, like RMLMapper [9], leading to low execution times. On the other hand, X3ML is designed with other goals in mind, namely good human-readability and easiness of collaboration, rather than execution speed. Moreover, for our use case, X3ML required the intermediate CSV to XML transformation, which introduced an additional overhead.

These results highlight performance trade-offs. Future applications on larger or even streaming astrophysical catalogues may introduce further scalability challenges. Utilizing incremental or distributed processing of the transformation could help alleviate this problem.

Size of Resulting KGs. HECAT3KG-X3ML and HECAT3KG-RML differ vastly with respect to their size. As Table 1 shows, X3ML produced a much smaller file (with size of about 5GB), containing 44 Million triples; on the other hand, RML produced a file size of about 9GB, with 55 Million triples. The main reason for this was the conditional limitations in RML with CSV files (see below), particularly for the uncertainties (errors) in measurements and the related flags, which caused RML to produce more triples than needed, and eventually resulting to a much bigger output file.

Input Formats. An important difference between the two approaches is that they support different input formats (see Table 2). X3ML requires XML data, which raises the need for a preprocessing step whenever the data are in another format (as in our use case), whereas RML supports a variety of input types, including CSVs, eliminating the need for format conversion. We opted not to transform the CSV into XML for the RML approach, as this would undermine its design for handling different formats.

Intuitiveness and Size of the Mappings. When creating the mapping rules, X3ML was found to be more intuitive and user-friendly, especially for domain experts without technical knowledge or background. Using the 3M Editor [28], as well as the instance and label generators, astrophysicists were able to actively participate in the mapping process. By leveraging the semantic structure we introduced in the preprocessing step, we managed to reduce the X3ML mapping rules to 10, which increases the intuitiveness and simplifies the maintenance of the mappings. For comparison, RML required the definition of a separate rule for nearly every column, leading to 157 mappings that are more complex and difficult to maintain (see Table 2). Moreover, the syntax of RML is more technical and less accessible for non-experts.

Column Merging. With regards to data mapping functionalities, X3ML leverages XPath, thus allowing the creation of data triples from information found in two or more different columns without the need for any preprocesing. This proved very handy during the creation process for the mapping rules. On the other hand, RML does not support this functionality in CSVs; to achieve the same effect, an additional temporary column had to be created (containing the combined information) and used for the mapping (see Table 2).

Conditionals. X3ML's XPath conditional logic helps skip empty or invalid values, producing fewer redundant triples. In contrast, RML resulted in more triples for missing values in the CSV, like measurement flags and error margins (see Table 2). The preprocessing step tried to mitigate this by labeling the invalid values, but the main issue persisted, leading to the size difference between HECAT3KG-RML and HECAT3KG-X3ML that is shown in Table 1.

URI Consistency. Another important aspect for our use case was URI consistency, which refers to the ability of the mapping language to generate a uniform and intuitive set of URIs whose naming makes sense for a human reader and can be used to track the origin of the information in the source data. We should note here the importance of URI consistency with respect to the pipeline we followed for the processing of HECATE: the splitting of the original XML file into many, and the subsequent merging of the resulting outputs to a single RDF file, would not produce a correct result without this feature, as cross-referencing among different chunks would fail, resulting to wrong mappings. URI consistency is supported natively by X3ML through its URI generator policies, ensuring consistent and traceable URIs even when data is processed incrementally. On the contrary, RML does not support URI generation policies, so we had to manually create appropriate URI naming schemes during the preprocessing step to avoid inconsistencies in the execution of our pipeline (see Table 2).

Summary of Findings. In summary, the two approaches had significant trade-offs related to their semantics, performance and usability for the described use case. RML excels performance-wise and can handle different data formats, making it suitable for heavy integration scenarios. On the other hand, X3ML can create rich and expressive semantics while allowing the mapping process to be more intuitive and more easily graspable by domain experts.

5 Conclusions

In this paper, we described our methodology for transforming the HECATE astrophysical catalogue into a semantic Knowledge Graph (KG), HECAT3KG, using a custom ontology (DOCBO) and two mapping frameworks (X3ML and RML). In astrophysics, large amounts of data are generated constantly, and stored in a tabular form using unstructured or semi-structured formats (ASCII, CSV) in large text files (called catalogues). The aim of this work, which is part of the more ambitious PARSEC project, was to provide a proof-of-concept methodology to migrate one of those catalogues, namely HECATE [19], into a KG. Our more general objective is to show how structured formats, like KGs, can be used to enhance the value of astrophysical catalogues by allowing the development of added-value services (e.g., related to maintenance and curation, provenance and trust management, advanced querying and navigation, etc.).

Our contributions include: (i) the DOCBO ontology tailored to celestial objects and measurements, (ii) the HECAT3KG knowledge graph constructed using X3ML and RML, and (iii) comparative insights into the mapping frameworks' usability and performance.

The generated resources are publicly available in Zenodo[8] and Github[9] under open licenses. While X3ML facilitated collaboration with domain experts, RML offered better performance and scalability. Our methodology has already been reused in an ongoing effort to build a KG for Supernova Remnants (SNRs), showing its applicability to other domains.

For future work, we plan to align DOCBO with existing standards (such as CIDOC CRM extensions) and other astrophysical ontologies and vocabularies (e.g., DCAT, Measurement Ontology, IVOA), defining a long-term maintenance and update strategy for the ontology and knowledge graph.

Acknowledgement. This work was partially funded by the "Theodore Papazoglou" FORTH Synergy Grants, through the PARSEC project. We would like to thank David Chaves-Fraga for his valuable guidance in R2RML/RML tools.

References

1. Accomazzi, A., Eichhorn, G., Kurtz, M.J., Grant, C.S., Murray, S.S.: The NASA astrophysics data system: architecture. Astron. Astrophys. Suppl. Ser. **143**(1), 85–109 (2000)
2. Antoniou, G., Groth, P., Harmelen, F.V., Hoekstra, R.: A Semantic Web Primer, 3rd edn. MIT Press (2012)
3. Arenas-Guerrero, J., Alobaid, A., Navas-Loro, M., Pérez, M.S., Corcho, Ó.: Boosting knowledge graph generation from tabular data with RML views. In: ESWC-23. LNCS, vol. 13870, pp. 484–501. Springer (2023)
4. Basset, A., Los, W.: Biodiversity e-science: lifewatch, the European infrastructure on biodiversity and ecosystem research. Plant Biosyst. Int. J. Dealing Aspects Plant Biol. **146**(4), 780–782 (2012)
5. Bizer, C., Heath, T., Berners-Lee, T.: Linked data - the story so far. In: Seneviratne, O., Hendler, J.A. (eds.) Linking the World's Information - Essays on Tim Berners-Lee's Invention of the World Wide Web, ACM Books, vol. 52, pp. 115–143 (2023)
6. Budavári, T., Szalay, A.S.: Probabilistic cross-identification of astronomical sources. Astrophys. J. **679**(1), 301–309 (2008)
7. Dimou, A.: R2RML and RML comparison for RDF generation, their rules validation and inconsistency resolution. CoRR abs/2005.06293 (2020)
8. Dimou, A., Sande, M.V., Colpaert, P., Verborgh, R., Mannens, E., de Walle, R.V.: RML: a generic language for integrated RDF mappings of heterogeneous data. In: Proceedings of the Workshop on Linked Data on the Web, vol. 1184 (2014)

[8] https://doi.org/10.5281/zenodo.15379419 and https://zenodo.org/records/16281112.

[9] https://github.com/sophisid/HECAT3KG.

9. Dimou, A., Vander Sande, M., Colpaert, P., Verborgh, R., Mannens, E., Van de Walle, R.: RML: a generic language for integrated RDF mappings of heterogeneous data. LDOW **1184** (2014)
10. Doerr, M., Theodoridou, M.: Mapping archaeological databases to CIDOC-CRM. In: Proceedings of the 40th Annual Conference on Computer Applications and Quantitative Methods in Archaeology (CAA) (2011)
11. Green, D.: SNR catalogue (online resource). https://www.mrao.cam.ac.uk/surveys/snrs/snrs.paper.html. Accessed 30 Apr 2025
12. Hanisch, R.: Data standards for the international virtual observatory. Data Sci. J. **5**, 168–173 (2006)
13. Hitzler, P., Krötzsch, M., Rudolph, S.: Foundations of Semantic Web Technologies. Chapman & Hall/CRC (2009)
14. Iglesias-Molina, A., et al.: The RML ontology: a community-driven modular redesign after a decade of experience in mapping heterogeneous data to RDF. In: ISWC 2023, vol. 14266, pp. 152–175 (2023)
15. Iglesias-Molina, A., Chaves-Fraga, D., Dasoulas, I., Dimou, A.: Human-friendly RDF graph construction: which one do you choose? In: ICWE 2023, vol. 13893, pp. 262–277. Springer (2023)
16. IVOA: IVOA vocabulary. https://www.ivoa.net/rdf/object-type/2020-10-06/object-type.html. Accessed 30 Apr 2025
17. Karray, F., Ahmed, M.B., Kacem, A.H.: ASON: an OWL-S based ontology for astrophysical services. Comput. Geosci. **94**, 71–80 (2016)
18. Konsolaki, K., Samaritakis, G.: 3M: mapping memory manager (2015). https://www.ics.forth.gr/3m-mapping-memory-manager. Accessed 26 Apr 2025
19. Kovlakas, K., et al.: The heraklion extragalactic catalogue (HECATE): a value-added galaxy catalogue for multimessenger astrophysics. Mon. Not. R. Astron. Soc. **506**(2), 1896–1915 (2021)
20. Makarov, D., Prugniel, P., Terekhova, N., Courtois, H., Vauglin, I.: HyperLEDA. III. the catalogue of extragalactic distances. A&A **570**, A13 (2014)
21. Marketakis, Y., Minadakis, N., Samaritakis, G., Doerr, M.: X3ML mapping framework for information integration in cultural heritage and beyond. Int. J. Digit. Libr. **18**(4), 301–319 (2017)
22. Minadakis, N., et al.: X3ML framework: an effective suite for supporting data mappings. In: Ronzino, P. (ed.) Proceedings of the Workshop on Extending, Mapping and Focusing the CRM. CEUR Workshop Proceedings, vol. 1656, pp. 1–12 (2015)
23. Musen, M.A.: Protégé team: the protégé project: a look back and a look forward. AI Matters **1**(4), 4–12 (2015)
24. NASA/IPAC: NASA/IPAC extragalactic database (NED). http://ned.ipac.caltech.edu. Accessed 30 Apr 2025
25. Pinkel, C., Binnig, C., Haase, P., Martin, C., Sengupta, K., Trame, J.: How to best find a partner? An evaluation of editing approaches to construct R2RML mappings. In: Presutti, V., d'Amato, C., Gandon, F., d'Aquin, M., Staab, S., Tordai, A. (eds.) ESWC 2014. LNCS, vol. 8465, pp. 675–690. Springer, Cham (2014). https://doi.org/10.1007/978-3-319-07443-6_45
26. Remy, L., et al.: Building an integrated enhanced virtual research environment metadata catalogue. Electron. Libr. **37**(6), 929–951 (2019)
27. Richards, J.W., et al.: On machine-learned classification of variable stars with sparse and noisy time-series data. Astrophys. J. **733**(1) (2011)
28. Samaritakis, G., Konsolaki, K.: 3meditor: web application for editing/viewing X3ML mapping files (2019). https://github.com/isl/3MEditor. Accessed 26 Apr 2025

29. Warren, P., Mulholland, P., Daga, E., Asprino, L.: Path-based and triplification approaches to mapping data into RDF: user behaviours and recommendations. Semantic Web **15**(6), 2479–2505 (2024)
30. Wenger, M., et al.: The SIMBAD astronomical database. the CDS reference database for astronomical objects. Astron. Astrophys., Suppl. Ser. **143**, 9–22 (2000)
31. Wilkinson, M.D., et al.: The fair guiding principles for scientific data management and stewardship. Sci. Data **3**(1), 1–9 (2016)

Answering MetaQueries over RDFS Ontologies Under the SPARQL Metamodeling Semantics Entailment Regime

Diego Calvanese[1(✉)], Gianluca Cima[2], Julien Corman[1], Roberto Maria Delfino[2], Maurizio Lenzerini[2], Lorenzo Marconi[2], Antonella Poggi[2], and Ognjen Savkovic[1]

[1] Free University of Bozen-Bolzano, Bolzano, Italy
{Calvanese,Corman,Savkovic}@inf.unibz.it
[2] Sapienza University of Rome, Rome, Italy
{Cima,Delfino,Lenzerini,Marconi,Poggi}@diag.uniroma1.it

Abstract. We study (unions of) conjunctive metaqueries over ontologies expressed in the Description Logics counterpart of RDFS equipped with metamodeling capabilities. This generalization to metaquerying goes beyond the extensively investigated query answering problem over ontologies, as now queries may also contain metavariables and TBox atoms, aligning with the classical "Ask-and-Tell" paradigm in knowledge representation. For the semantics of queries, we adopt the SPARQL Metamodeling Semantics Entailment Regime, a recently introduced regime that generalizes the commonly used SPARQL Direct Semantics Entailment Regime by relaxing the typing constraints. This new regime supports a more flexible querying mechanism over ontologies involving metaclasses and metaproperties, as query variables can appear in positions of different types (i.e. as individual, class, or property). As a result, it enables the formulation of meaningful metaqueries related to metamodeling. We show that this query answering task can be reduced to the evaluation of a linear Datalog program. This Datalog encoding allows us to establish tight complexity bounds for the associated decision problem, in combined complexity, ontology complexity and data complexity.

Keywords: Metamodeling · Metaquerying · RDFS · SPARQL Entailment Regime · Datalog · Computational Complexity

1 Introduction

Description Logic (DL) ontologies enable the structured and symbolic representation of knowledge pertaining to a specific domain of interest. A DL ontology (or simply ontology in what follows) typically consists of two complementary components: an intensional part and an extensional part. The first one, called

H. Takeda et al. (Eds.): IJCKG 2025, LNAI 16297, pp. 188–203, 2026.
https://doi.org/10.1007/978-981-95-5009-8_13

TBox, is a set of axioms expressed in some formal language that define the intensional knowledge of the modeled domain. The latter, called ABox, is a set of logical ground atoms that provides the extensional knowledge of the domain.

Query answering is arguably one of the most studied tasks for extracting information from ontologies, where a query is a declarative expression defining the information an end user wants to retrieve. But with very few exceptions (see related work in Sect. 2), this vast body of literature on ontology-mediated query answering [6] and ontology-based data access [35] does not consider *metamodeling* and *metaquerying*.

As pointed out in [24], metamodeling (also known as *multi-level modeling* [4]) allows to model domain knowledge by using *metaclasses* and *metaproperties* in the ontology, which are, respectively, classes and properties that can predicate over classes and properties (in addition to individuals).

Example 1. According to scientific classification, living beings are categorized in progressively more specific taxonomic ranks, namely domains (the most generic), kingdoms, phyla, classes, orders, families, genera, and finally, species (the most specific). Every species (and the same holds for the other ranks) can be thought of as a set of individuals. Therefore, it is natural to treat a concept like "species", "genus", etc. as a metaclass (in this example, a class whose instances are in turn classes). In the following, we partially model the above domain of interest through the ontology $\mathcal{O} = \langle \mathcal{T}, \mathcal{A} \rangle$, with TBox $\mathcal{T} = \{\mathsf{Warthog} \sqsubseteq_c \mathsf{Animal}, \mathsf{Lion} \sqsubseteq_c \mathsf{Animal}, \mathsf{predatorOf} \sqsubseteq_p \mathsf{preyOf}^-\}$, where $\sqsubseteq_c$ and $\sqsubseteq_p$ refer to inclusions between classes and properties, respectively, and ABox $\mathcal{A} = \{\mathsf{predatorOf}(\mathsf{Scar}, \mathsf{Pumbaa}), \mathsf{Warthog}(\mathsf{Pumbaa}), \mathsf{Lion}(\mathsf{Scar}), \mathsf{Kingdom}(\mathsf{Animal}), \mathsf{Species}(\mathsf{Warthog}), \mathsf{Species}(\mathsf{Lion})\}$. Note that Lion and $\mathsf{Warthog}$ are not only used as predicates (e.g. in $\mathsf{Lion} \sqsubseteq_c \mathsf{Animal}$ or $\mathsf{Lion}(\mathsf{Scar})$), but are also predicated over (e.g. in $\mathsf{Species}(\mathsf{Lion})$). Note also that the statements that predicate over classes (e.g. $\mathsf{Species}(\mathsf{Lion})$) are part of the ABox.

The current W3C Recommendation for representing ontologies on the Semantic Web is `OWL 2`, and the de-facto query language for expressing queries over `OWL 2` ontologies is `SPARQL 1.1` [20] (in what follows, simply `SPARQL`). The core construct of `SPARQL` queries is the *basic graph pattern* (*BGP*), which can be seen as a conjunction of atoms. Each atom in a BGP corresponds to an axiom expressible in the language of the queried ontology, but its terms may include variables other than `IRI`s. This is coherent with the classical knowledge representation *Ask-and-Tell* framework [25], whose guiding principle is to "ask anything that can be told to a knowledge base". As previously noted, this differs from the majority of works in the query answering literature, where queries can typically contain only ABox axioms and variables are restricted to individual positions.

Example 2. In the scenario modeled in Example 1, one might ask which taxa Scar's prey belongs to, limited to the animal kingdom (and sub-taxa thereof):

$$q_1(X) = \exists Y.\mathsf{preyOf}(Y, \mathsf{Scar}) \land X(Y) \land X \sqsubseteq_c \mathsf{Animal}$$

Interestingly, the syntax of the OWL 2 profiles [28], such as OWL 2 QL, provide support for metamodeling through OWL 2 *punning*, which is the capability of an IRI (*Internationalized Resource Identifier*) to play simultaneously the role of an individual, class, or property. Although such languages allow for metamodeling at least from a syntactic point of view, the official semantics, called *Direct Semantics* [29], imposes that occurrences of the same IRI in different positions are treated as if they were occurrences of different IRIs.

As a result of this choice, the *SPARQL Direct Semantics Entailment Regime* (*DSER*) [17] (the most typical SPARQL Entailment Regime for OWL 2 QL ontologies) enforces the so-called *typing constraints*, which forbid a variable in a SPARQL query to appear both in an individual position (i.e. as an argument of a predicate atom) and in predicate position (i.e. as a class or a property). This means that, under DSER, we cannot join variables denoting classes with those denoting individuals, thus preventing the specification of interesting metaqueries related to metamodeling.

Example 3. According to DSER, the set of answers to $q_1(X)$ of Example 2 over the ontology of Example 1 is $\{\mathsf{Warthog}, \mathsf{Animal}\}$. To filter out taxa that may be too generic (like Animal), one can restrict the assignments for X to species:

$$q_2(X) = \exists Y.\mathsf{preyOf}(Y, \mathsf{Scar}) \wedge X(Y) \;\; \wedge X \sqsubseteq_c \mathsf{Animal} \wedge \mathsf{Species}(X)$$

Unfortunately, such a query is not allowed in DSER, because the typing constraints prohibit X from appearing both in class and in individual position.

To remedy this limitation, [23] introduced a novel semantics, called *Metamodeling Semantics* (*MS*) in [24], which relies on the notion of interpretation structures with a second-order flavor, and is a conservative extension of the Direct Semantics. Building on this new semantics, [11] proposed the SPARQL *Metamodeling Semantics Entailment Regime* (*MSER*), a novel SPARQL Entailment Regime that generalizes DSER by *(i)* adopting the MS for OWL 2 QL ontologies and *(ii)* relaxing the typing constraints.

Example 4. According to MSER, the set of answers to the query $q_2(X)$ of Example 3 over the ontology of Example 1 is $\{\mathsf{Warthog}\}$.

In this paper, we focus on the fragment of OWL 2 QL corresponding to the DL *DL-Lite*$_{\mathtt{RDFS}}$ [34] extended with *OWL punning*, which we call *MDL-Lite*$_{\mathtt{RDFS}}$. As an example, the ontology illustrated in Example 1 is a *MDL-Lite*$_{\mathtt{RDFS}}$ ontology. We recall that *DL-Lite*$_{\mathtt{RDFS}}$ corresponds to the DL counterpart of RDFS, and, as already pointed out in [34], it is very similar to the DL *RDFS(DL)* [14]. This popular DL has been the subject of several recent studies , including tasks related to abstraction [10,13], as well as query answering over ontologies formulated in this language [34] and slight extensions of it [9,12].

To the best of our knowledge, however, the topic of metaqueries over ontologies expressed in this language equipped with metamodeling capabilities (i.e. over

MDL-Lite$_{\texttt{RDFS}}$ ontologies) is still unexplored. In this paper, we focus on unions of conjunctive metaqueries (UCMQs), which correspond to unions of SPARQL Basic Graph Patterns. Both queries in Example 2 and Example 3 are UCMQs.

Contributions. Our main contributions are twofold. First, we show that the problem of answering UCMQs over *MDL-Lite*$_{\texttt{RDFS}}$ ontologies under MSER can be reduced to the evaluation of a Datalog program. In particular, we provide a fixed set of Datalog rules that simulates inference under MSER and, for an *MDL-Lite*$_{\texttt{RDFS}}$ ontology $\mathcal{O}$ and UCMQ q, we show how to encode $\mathcal{O}$ as a set Datalog facts and q as another set of rules, such that the evaluation of the resulting program returns exactly the answers to q over $\mathcal{O}$.

Second, we exploit this program to provide tight complexity bounds for the decision problem associated to answering a UCMQ over an *MDL-Lite*$_{\texttt{RDFS}}$ ontology. Specifically, we show that this problem is NP-complete in combined and query complexity (which is already the complexity of conjunctive query answering [1]), NLogSpace-complete in ontology and TBox complexity, and in AC^0 in ABox (a.k.a. data) complexity.

We conclude this section by mentioning that a Datalog reduction from the problem of answering UCMQs over OWL 2 QL ontologies under MSER was already provided in [11]. So, in principle, since *MDL-Lite*$_{\texttt{RDFS}}$ is a fragment of OWL 2 QL, we could use that encoding. However, two observations are in order: *(i)* to the best of our knowledge, there is no proof of completeness for the Datalog program defined in [11]; and *(ii)* tight complexity bounds were not provided in this work.

2 Related Work

Metamodeling techniques have been applied in various domains, including model-driven development [5] and conceptual modeling [7]. In the Knowledge Representation and Reasoning research area, several works [22,26,27,30] investigate the satisfiability problem (and, in some cases, also the subsumption problem) of ontologies expressed in languages that support a higher-order syntax. Notably, inspired by HiLog [8,27] proposes a metamodeling semantics for OWL 2 DL enriched with the metamodeling features of OWL 2. However, these works do not consider query answering.

The problem of answering SPARQL queries over ontologies expressed in (fragments of) OWL 2 has been an active research topic in recent years [2,3,18,19,21,31]. However, although these works address the case of metaqueries thanks to SPARQL's native support for them, the ontology languages adopted in these papers lack metamodeling capabilities.

Closer to the present paper are the investigations on answering metaqueries over OWL 2 QL ontologies in [11,16,24]. Similarly to [27], the semantics for ontologies considered in [16] is inspired by HiLog and thus differs from the MS [24] adopted in this paper. On the other hand, [11] uses the same semantics for ontologies (MS) and for queries (MSER) adopted in this paper. As already said in the introduction, this work provides a Datalog encoding for answering metaqueries over OWL 2 QL ontologies under MSER. Very recently, this encoding has been implemented in various Datalog tools and its practical effectiveness has been evaluated over the LUBM and MODEUS ontologies [32,33]. However, we remind the reader that the exact complexity of the associated query answering problem remains open, as [11] only establishes upper bounds derived from the complexity of the evaluation of Datalog queries.

Finally, we mention that [24] also addresses the problem of answering UCMQs over `OWL 2 QL` ontologies. However, the semantics adopted for queries is a variant of MSER that is more closely aligned with classical logic. In particular, under the semantics for queries adopted in [24], a Boolean UCMQ q is true over an `OWL 2 QL` ontology $\mathcal{O}$ if, for every model $\mathcal{I}$ of $\mathcal{O}$, there exists an assignment for the existential variables of q that makes the resulting quantifier-free sentence true in $\mathcal{I}$. However, this assignment may differ from one model to another. In contrast, under MSER, q is true over $\mathcal{O}$ if a *same* assignment satisfies this condition in every model of $\mathcal{O}$. In other words, the semantics for queries adopted in [24] assigns to the union ($\vee$) and the existential quantification ($\exists$) operators the classical logical meaning, while MSER is closer in spirit to DSER (the official W3C semantics for interpreting queries over `OWL 2` ontologies). Indeed, we recall that MSER extends DSER to properly handle both metamodeling (by adopting the MS) and metaquerying (by removing the typing constraints) with `SPARQL`.

3 Preliminaries

We assume basic knowledge about first-order logic, databases and Description Logics. In what follows, we refer to countably infinite and disjoint sets $\mathbb{V}$ and $\mathbb{I}$ of symbols, used for variables and *Internationalized Resource Identifiers* (`IRI`s), respectively. We also use $\bar{t}$ to denote a tuple of symbols and, when clear from the context, we may use the same notation to refer to the set of symbols occurring in this tuple.

An *MDL-Lite*$_{\texttt{RDFS}}$ ontology is a set of axioms, expressed in *MDL-Lite*$_{\texttt{RDFS}}$, i.e., the language that extends *DL-Lite*$_{\texttt{RDFS}}$ [34] by allowing the same expression to denote an individual, a class, or a property. As usual in DLs, an ontology $\mathcal{O}$ can be seen as a pair $\mathcal{O} = \langle \mathcal{T}, \mathcal{A} \rangle$, where $\mathcal{T}$ and $\mathcal{A}$, respectively called *TBox* and *ABox*, are sets of intensional and extensional assertions called *axioms*, expressing knowledge about elements of the domain of interest.

Syntactically, the set of *well-formed expressions over* $\mathbb{I}$ $Exp(\mathbb{I})$ is the union of the set of *role expressions* $Exp_p(\mathbb{I}) = \mathbb{I} \cup \{e^- \mid e \in \mathbb{I}\}$, and the set of *concept expressions* $Exp_c(\mathbb{I}) = \mathbb{I} \cup \{\exists e \mid e \in Exp_p(\mathbb{I})\}$. An *MDL-Lite*$_{\texttt{RDFS}}$ *TBox* $\mathcal{T}$ is a finite set of axioms of the following form:

$$b \sqsubseteq_c a \quad \text{(concept inclusion)} \qquad\qquad p_1 \sqsubseteq_p p_2 \quad \text{(role inclusion)}$$

where $a \in \mathbb{I}$ is an `IRI`, $b \in Exp_c(\mathbb{I})$ is a concept expression, and $p_1, p_2 \in Exp_p(\mathbb{I})$ are role expressions. An *MDL-Lite*$_{\texttt{RDFS}}$ *ABox* $\mathcal{A}$ is a finite set of axioms of the form $a(e)$ or $r(e_1, e_2)$, with $e, e_1, e_2, a, r \in \mathbb{I}$. We use $\Sigma(\mathcal{O}) \subset \mathbb{I}$ to denote the set of IRIs occurring in and ontology $\mathcal{O}$ (also called the *signature* of $\mathcal{O}$).

The semantics of *MDL-Lite*$_{\texttt{RDFS}}$ ontologies is based on the so-called *Metamodeling Semantics* (*MS*) [24], which we tailor here for this language. An *MS-interpretation* (or simply *interpretation*) $\mathcal{I}$ is a tuple $\mathcal{I} = \langle \Delta, \cdot^C, \cdot^P, \cdot^{\mathcal{I}} \rangle$, where:

- Δ is a non-empty set of *objects*, called the *domain* of $\mathcal{I}$;
- $\cdot^C : \Delta \to \mathcal{P}(\Delta)$ is a partial function mapping an object o to a set o^C of objects;
- $\cdot^P : \Delta \to \mathcal{P}(\Delta \times \Delta)$ is another partial function mapping an object o to a binary relation o^P over Δ;
- $\cdot^\mathcal{I} : Exp(\mathbb{I}) \to \Delta$ is a total function, called *interpretation function*, mapping each expression in $Exp(\mathbb{I})$ to an object $o^\mathcal{I} \in \Delta$ such that *(i)* for every $p \in \mathbb{I}$, if $\cdot^P$ is defined for $p^\mathcal{I}$, then $((p^-)^\mathcal{I})^P = \{(o', o) \mid (o, o') \in (p^\mathcal{I})^P\}$; *(ii)* for every $p \in Exp_p(\mathbb{I})$, if $\cdot^P$ is defined for $p^\mathcal{I}$, then $((\exists p)^\mathcal{I})^C = \{o \mid \exists o'.(o, o') \in (p^\mathcal{I})^P\}$.

In the above definition, $\cdot^C$ (resp. $\cdot^P$) applied to a concept o forms the extension of o as a concept (resp. role). If the function is undefined for o, then o is intuitively not a concept (resp. role) in the world represented by $\mathcal{I}$.

We now provide the notion of satisfaction of an *MDL-Lite*$_{\texttt{RDFS}}$ axiom ϕ with respect to an MS-interpretation $\mathcal{I} = \langle \Delta, \cdot^C, \cdot^P, \cdot^\mathcal{I} \rangle$ (denoted by $\mathcal{I} \models \phi$):

- if $\phi = b \sqsubseteq_c a$, then $\mathcal{I} \models \phi$ if $\cdot^C$ is defined for $b^\mathcal{I}$ and $a^\mathcal{I}$, and $(b^\mathcal{I})^C \subseteq (a^\mathcal{I})^C$;
- if $\phi = p \sqsubseteq_p r$, then $\mathcal{I} \models \phi$ if $\cdot^P$ is defined for $p^\mathcal{I}$ and $r^\mathcal{I}$, and $(p^\mathcal{I})^P \subseteq (r^\mathcal{I})^P$;
- if $\phi = a(e)$, then $\mathcal{I} \models \phi$ if $\cdot^C$ is defined for $a^\mathcal{I}$, and $e^\mathcal{I} \in (a^\mathcal{I})^C$;
- if $\phi = r(e_1, e_2)$, then $\mathcal{I} \models \phi$ if $\cdot^P$ is defined for $r^\mathcal{I}$, and $(e_1^\mathcal{I}, e_2^\mathcal{I}) \in (r^\mathcal{I})^P$.

We use $\mathcal{I} \models \mathcal{T}$ (resp. $\mathcal{I} \models \mathcal{A}$) to denote the fact that $\mathcal{I}$ satisfies all the axioms occurring in a TBox $\mathcal{T}$ (resp. ABox $\mathcal{A}$). We say that $\mathcal{I}$ is a *model* of an *MDL-Lite*$_{\texttt{RDFS}}$ ontology $\mathcal{O} = \langle \mathcal{T}, \mathcal{A} \rangle$ (denoted by $\mathcal{I} \models \mathcal{O}$) if $\mathcal{I} \models \mathcal{T}$ and $\mathcal{I} \models \mathcal{A}$. We may also write $\mathcal{O} \models \phi$ to indicate that $\mathcal{I} \models \phi$ holds for every model $\mathcal{I}$ of $\mathcal{O}$.

A *query atom* is an expression having the form of an *MDL-Lite*$_{\texttt{RDFS}}$ axiom, with the difference that variables may also occur among its terms. A *Conjunctive MetaQuery* (CMQ) $q(\bar{X}; \bar{Y})$ over an ontology $\mathcal{O}$ is an expression of the form $\exists \bar{Y}.\varphi$, where $\bar{X}$ and $\bar{Y}$ are tuples of variables in $\mathbb{V}$, and φ is a conjunction of atoms with variables in $\bar{X} \cup \bar{Y}$. When convenient, we simply write $q(\bar{X})$ or q, and we call *arity* of $q(\bar{X})$ the arity of $\bar{X}$. A *Union of* CMQs (UCMQ) over $\mathcal{O}$ is a finite disjunction of CMQs over $\mathcal{O}$ with the same arity, called its disjuncts. We say that a query is *Boolean* if it has arity 0.

For the semantics of (U)CMQs, we adopt the SPARQL *Metamodeling Semantics Entailment Regime* (*MSER*) [11], which defines the answers to a query similarly to the SPARQL Direct Semantics Entailment Regime (DSER) [17], except that it uses MS for logical entailment.

Let $q(\bar{X}) = \exists \bar{Y}.\alpha_1 \wedge \cdots \wedge \alpha_n$ be a CMQ, and let $\bar{d}$ and $\bar{e}$ be two tuples of IRIs of the same arity as $\bar{X}$ and $\bar{Y}$, respectively. We denote by $q(\bar{X}/\bar{d}; \bar{Y}/\bar{e})$ the conjunction of ground atoms $\phi_1, \ldots, \phi_n$, where each ϕ_i is obtained from α_i by replacing each variable $X \in \bar{X}$ and $Y \in \bar{Y}$ with the corresponding IRI in $\bar{d}$ and $\bar{e}$, respectively. Note that, by construction, every ϕ_i in $q(\bar{X}/\bar{d}; \bar{Y}/\bar{e})$ is an *MDL-Lite*$_{\texttt{RDFS}}$ axiom. Then, we say that a tuple $\bar{d}_a$ of IRIs is an *answer to* $q(\bar{X})$ *over* $\mathcal{O}$ if there is a tuple $\bar{e}_a$ of IRIs such that $\mathcal{O} \models \phi_i$ holds for every $\phi_i \in q(\bar{X}/\bar{d}_a; \bar{Y}/\bar{e}_a)$. If q is a UCMQ, then $\bar{d}_a$ is an answer to q over $\mathcal{O}$ if it is an answer to any disjunct of q. For a Boolean UCMQ $q()$, we say that q *is entailed by* $\mathcal{O}$, written $\mathcal{O} \models q$, if the empty tuple is an answer to q over $\mathcal{O}$.

Hereafter, we will use the terms *class* and *property* in place of the DL notions of "concept" and "role", respectively, as these best align with the knowledge graph terminology.

Throughout this paper, we will also use Datalog. A Datalog *atom* may have constants and/or variables as arguments. For instance, the atom $\mathtt{subClass}(a, X)$ contains a constant a and a variable X. A Datalog *fact* is an atom without variables, while a Datalog *rule* is an expression of the form $\alpha_0 \leftarrow \alpha_1, \dots, \alpha_m$, where $m \geq 1$, each α_i is an atom and all variables in α_0 appear in some α_j, for $j \geq 1$. The *head* of the rule is α_0 and $\alpha_1, \dots, \alpha_m$ is its *body*. A Datalog *program* is a finite set of facts and rules. In a Datalog program, a predicate is called *intensional* if it appears in the head of some rule(s), and *extensional* otherwise. A Datalog program is called *linear* if the body of each rule contains at most one intensional predicate. We use a standard characterization of the evaluation of a Datalog program Π (see, e.g., [1]), based on the notion of *least model* Π^* of Π, i.e. the least (w.r.t. set inclusion) set of facts that contains all facts in Π and satisfies all its rules.

4 Datalog Encoding

In this section, we introduce a linear Datalog program Π that encodes an ontology $\mathcal{O}$ and has linear size. We show that the least model Π^* of Π coincides (modulo some simple encoding of DL axioms as Datalog facts) with the set of formulas that are logically implied by $\mathcal{O}$. This will allow us to derive results about the complexity of query answering, in Sect. 5.

Encoding axioms. Before defining the program Π, we need to clarify how we encode DL axioms as Datalog facts (i.e. grounded atoms). First, to each expression $\eta \in \mathit{Exp}(\mathbb{I}) \setminus \mathbb{I}$ (meaning that η is a composite role or concept), we associate a Datalog fact $\mathsf{expr}(\eta)$ that specifies how η is constructed. Precisely, we define $\mathsf{expr}(r^-)$ and $\mathsf{expr}(\exists r)$ as the Datalog facts $\mathtt{inverse}(r^-, r)$ and $\mathtt{domain}(\exists r, r)$, respectively. Then, a *MDL-Lite*$_{\mathtt{RDFS}}$ axiom α is encoded as the set

$$\mathsf{enc}(\alpha) = \{\mathsf{atom}(\alpha)\} \cup \{\mathsf{expr}(\eta) \mid \eta \text{ appears in } \alpha\}$$

of Datalog facts, where $\mathsf{atom}(\alpha)$ is defined as follows:

MDL-Lite$_{\mathtt{RDFS}}$ axiom α	Datalog fact $\mathsf{atom}(\alpha)$
$a(e)$	$\mathtt{classInst}(e, a)$
$r(e_1, e_2)$	$\mathtt{propInst}(e_1, e_2, r)$
$a \sqsubseteq_c b$	$\mathtt{subClass}(a, b)$
$r \sqsubseteq_p s$	$\mathtt{subProp}(r, s)$

Soundness and completeness. With this encoding, we can clarify what it means for our program Π to be correct, namely

$$\mathcal{O} \models \alpha \quad \text{iff} \quad \mathsf{enc}(\alpha) \subseteq \Pi^*,$$

for any *MDL-Lite*$_{\texttt{RDFS}}$ axiom α over $\Sigma(\mathcal{O})$, where Π^* is the least model of Π.

The rules of our program Π (which we will introduce below) all have an intuitive meaning. For instance, the first rule of Fig. 1 states that if a is a subclass of b and e is an instance a, then e is also an instance of b. In other words, these rules only derive (atoms representing) axioms that are indeed implied by $\mathcal{O}$. As a consequence, correctness (as we just defined it) trivially holds in one direction (*soundness*), namely

$$\mathsf{enc}(\alpha) \subseteq \Pi^* \text{ implies } \mathcal{O} \models \alpha$$

The proof of the converse property (*completeness*) is more involved:

$$\mathcal{O} \models \alpha \text{ implies } \mathsf{enc}(\alpha) \subseteq \Pi^*$$

This property means that every *MDL-Lite*$_{\texttt{RDFS}}$ axiom (semantically) implied by $\mathcal{O}$ can be derived via the Datalog program Π. So throughout this section, while introducing Π, we will also prove that it is complete.

ABox axioms. We start by introducing a sub-program Π^{ABox} of Π, in charge of inferring all ABox axioms implied by $\mathcal{O}$. In order to ensure that Π is linear, we use two predicates `subClassExt` and `subPropExt` to encode the TBox axioms that are (syntactically) present in $\mathcal{T}$. Precisely, for an axiom α, we use $\mathsf{encExt}(\alpha)$ for the set of atoms identical to $\mathsf{enc}(\alpha)$, but where the predicate `subClass` (resp. `subProp`), if present, is replaced with `subClassExt` (resp. `subPropExt`).

Our subprogram Π^{ABox} includes the set $\mathcal{D}_{\mathcal{O}}$ of Datalog facts that encode the axioms of $\mathcal{O}$, i.e.:

$$\mathcal{D}_{\mathcal{O}} = \bigcup_{\alpha \in \mathcal{O}} \mathsf{encExt}(\alpha) \subseteq \Pi^{ABox}$$

In addition to $\mathcal{D}_{\mathcal{O}}$, the program Π^{ABox} contains the 7 rules listed in Fig. 1.

Next, we observe that $\mathsf{enc}(\alpha) = \{\mathsf{atom}(\alpha)\}$ for any ABox axiom α, so that completeness for this fragment can be reformulated as

$$\mathcal{O} \models \alpha \text{ implies } \mathsf{atom}(\alpha) \in (\Pi^{ABox})^* \tag{1}$$

To show that this property holds, we focus on the set $(\Pi^{ABox})^*_{\mathsf{facts}}$ of Datalog atoms that represent ABox axioms (a.k.a. "facts") in the least model of this program, i.e.:

$$(\Pi^{ABox})^*_{\mathsf{facts}} = \{\alpha \mid \mathsf{atom}(\alpha) \in (\Pi^{ABox})^* \text{ and } \alpha \text{ is an ABox axiom}\}$$

$$\begin{array}{rl}
\texttt{classInst}(E,B) \leftarrow & \texttt{classInst}(E,A), \texttt{subClassExt}(A,B)\\
\texttt{classInst}(E_1,A) \leftarrow & \texttt{propInst}(E_1,E_2,R),\ \texttt{domain}(W,R),\ \texttt{subClassExt}(W,A)\\
\texttt{classInst}(E_2,A) \leftarrow & \texttt{propInst}(E_1,E_2,R),\ \texttt{inverse}(S,R), \texttt{domain}(W,S),\ \texttt{subClassExt}(W,A)\\
\texttt{propInst}(E_1,E_2,S) \leftarrow & \texttt{propInst}(E_1,E_2,R),\ \texttt{subPropExt}(R,S)\\
\texttt{propInst}(E_2,E_1,S) \leftarrow & \texttt{propInst}(E_1,E_2,R),\ \texttt{inverse}(R',R),\ \texttt{subPropExt}(R',S)\\
\texttt{propInst}(E_2,E_1,S) \leftarrow & \texttt{propInst}(E_1,E_2,R),\ \texttt{inverse}(S',S),\ \texttt{subPropExt}(R,S')\\
\texttt{propInst}(E_1,E_2,S) \leftarrow & \texttt{propInst}(E_1,E_2,R),\ \texttt{inverse}(R',R),\ \texttt{inverse}(S',S),\ \texttt{subPropExt}(R',S')
\end{array}$$

Fig. 1. Rules of the program Π^{ABox}, in charge of inferring ABox axioms.

This set $(\Pi^{ABox})^*_{\mathsf{facts}}$ can intuitively be viewed as a first-order structure. Or more formally, under the MS, as any interpretation $\mathcal{I} = \langle \Delta, \cdot^C, \cdot^P, \cdot^\mathcal{I} \rangle$ for $\Sigma(\mathcal{O})$ such that *(i)* $\cdot^\mathcal{I}$ is injective, *(ii)* $o \in (a^\mathcal{I})^C$ iff $a(e) \in (\Pi)^*_{\mathsf{facts}}$ and $e^\mathcal{I} = o$, and *(iii)* $(o_1, o_2) \in (r^\mathcal{I})^P$ iff $r(e_1, r_2) \in (\Pi)^*_{\mathsf{facts}}$, $e_1^\mathcal{I} = o_1$ and $e_2^\mathcal{I} = o_2$.

By construction, for an ABox axiom α,

$$\mathcal{I} \models \alpha \text{ implies } \mathsf{atom}(\alpha) \in (\Pi^{ABox})^* \tag{2}$$

In addition, we will show below that $\mathcal{I}$ is a model of $\mathcal{O}$. These two properties are sufficient to prove completeness (a.k.a. Property (1)) for our fragment. Indeed, if $\mathcal{O} \models \alpha$, then $\mathcal{I} \models \alpha$ (since $\mathcal{I}$ is a model of $\mathcal{O}$). Together with (2), this implies that $\mathsf{atom}(\alpha) \in (\Pi^{ABox})^*$.

So to complete the proof, we only need to show that $\mathcal{I}$ is a model of $\mathcal{O}$, i.e. that it satisfies every axiom $\beta \in \mathcal{O}$. Due to space limitations, we focus on one type of axiom as an illustration, namely the case where β is of the form $\exists r \sqsubseteq_c a$ with $a, r \in \mathbb{I}$ (the proof for the other types of axioms is either simpler or analogous). By construction, $\texttt{subClassExt}(\exists r, a) \in \mathcal{D}_\mathcal{O}$, and $\mathsf{expr}(\exists r) = \{\texttt{domain}(\exists r, r)\} \subseteq \mathcal{D}_\mathcal{O}$. So from the second rule of Fig. 1, $\texttt{propInst}(e_1, e_2, r) \in (\Pi^{ABox})^*_{\mathsf{facts}}$ implies $\texttt{classInst}(e_1, a) \in (\Pi^{ABox})^*_{\mathsf{facts}}$ for any e_1, e_2. Viewing $(\Pi^{ABox})^*_{\mathsf{facts}}$ as the interpretation $\mathcal{I}$, this implication translates to $(\exists r^\mathcal{I})^C \subseteq (a^\mathcal{I})^C$, i.e. $\mathcal{I} \models \beta$.

TBox Axioms. Next, we extend the program Π^{ABox} to a program Π that also derives TBox axioms, which gives us a complete system for *MDL-Lite*$_{\mathrm{RDFS}}$. We use a technique that reduces the inference of such axioms to the inference of some ABox atoms, thus relying on the results of the previous section.

For each IRI $e \in \Sigma(\mathcal{T})$, we introduce three fresh constants (not in $\mathbb{I}$) that we call *witnesses* for e, denoted with w_e, w_e^s and w_e^o (for "subject" and "object"). Then we extend our program with the Datalog facts $\texttt{classWit}(w_e, e)$ and $\texttt{classInst}(w_e, e)$, which indicate that w_e is the class witness for e and an instance of e. We proceed similarly with the other two witnesses, adding to Π the atoms $\texttt{subjWit}(w_e^s, e)$, $\texttt{objWit}(w_e^o, e)$, $\texttt{propWit}(w_e^s, w_e^o, e)$ and $\texttt{propInst}(w_e^s, w_e^o, e)$. We also add to Π the set of inference rules listed in Fig. 2, which rely on witnesses to infer TBox axioms.

We can now extend our proof of completeness to *MDL-Lite*$_{\mathrm{RDFS}}$. We saw in the previous section that Π^{ABox} is complete for ABox axioms. Since $\Pi^{ABox} \subseteq \Pi$,

$$
\begin{aligned}
\texttt{subClass}(A, B) &\leftarrow \texttt{classWit}(E, A), \texttt{classInst}(E, B)\\
\texttt{subClass}(W, A) &\leftarrow \texttt{domain}(W, R), \texttt{subjWit}(E, R), \texttt{classInst}(E, A)\\
\texttt{subClass}(W, A) &\leftarrow \texttt{inverse}(R, S), \texttt{domain}(W, S), \texttt{objWit}(E, R), \texttt{classInst}(E, A)\\
\texttt{subProp}(R, S) &\leftarrow \texttt{subjWit}(E_1, R), \texttt{objWit}(E_2, R), \texttt{propInst}(E_1, E_2, S)\\
\texttt{subProp}(R, S') &\leftarrow \texttt{subjWit}(E_1, R), \texttt{objWit}(E_2, R), \texttt{inverse}(S', S), \texttt{propInst}(E_2, E_1, S)\\
\texttt{subProp}(R', S) &\leftarrow \texttt{subjWit}(E_1, R), \texttt{objWit}(E_2, R), \texttt{inverse}(R', R), \texttt{propInst}(E_2, E_1, S)\\
\texttt{subProp}(R', S') &\leftarrow \texttt{subjWit}(E_1, R), \texttt{objWit}(E_2, R), \texttt{inverse}(R', R), \texttt{inverse}(S', S),\\
&\quad\ \texttt{propInst}(E_1, E_2, S)
\end{aligned}
$$

Fig. 2. Rules in charge of inferring TBox axioms.

this property trivially extends to Π. So we can focus on proving that Π is complete for TBox axioms.

First, we note that for any TBox axiom α over $\Sigma(\mathcal{O})$, $\mathcal{D}_{\mathcal{O}} \subseteq \Pi$ contains the set $\{\mathsf{expr}(\eta) \mid \eta \text{ appears in } \alpha\}$. So, again, completeness can be reformulated as

$$\mathcal{O} \models \alpha \text{ implies } \mathsf{atom}(\alpha) \in \Pi^* \tag{3}$$

As above, we use $(\Pi)^*_{\mathsf{facts}}$ to refer to the set of Datalog facts that encode ABox axioms and belong to the least model Π^* of Π, and we leverage the fact that $(\Pi)^*_{\mathsf{facts}}$ can be viewed as an interpretation $\mathcal{I}$. We will also write $(\mathcal{D}_{\mathcal{O}})_{\mathsf{facts}}$ for the set of ABox atoms in $\mathcal{D}_{\mathcal{O}}$, i.e. all $\mathsf{atom}(\beta) \in \mathcal{D}_{\mathcal{O}}$ s.t. β is an ABox axiom.

As we saw in the previous section, in order to prove (3), it is sufficient to show that $\mathcal{I}$ is a model of $\mathcal{O}$, and that for any TBox axiom α,

$$(\Pi)^*_{\mathsf{facts}} \models \alpha \text{ implies } \mathsf{atom}(\alpha) \in \Pi^* \tag{4}$$

The proof that $\mathcal{I}$ is a model of $\mathcal{O}$ is identical to the one above, due to the fact that $\Pi^{ABox} \subseteq \Pi$. As for Property (4), for space limitations, we once again focus on one syntactic type of TBox axioms as an illustration, namely the case where α is of the form $\exists r \sqsubseteq_c a$ for $r, a \in \mathbb{I}$ (the proof for the other cases is either simpler or analogous).

Since $\mathcal{O} \models \alpha$ (by assumption) and $\mathcal{I}$ is a model of $\mathcal{O}$, we know that $\mathcal{I} \models \alpha$, or in other words, that every $(s, o) \in (r^{\mathcal{I}})^P$ satisfies $s \in (a^{\mathcal{I}})^C$. In Datalog terms (leveraging the correspondence between $\mathcal{I}$ and $(\Pi)^*_{\mathsf{facts}}$), this means that $\texttt{propInst}(s, o, r) \in (\Pi)^*_{\mathsf{facts}}$ implies $\texttt{classInst}(s, a) \in (\Pi)^*_{\mathsf{facts}}$. In particular, since $\texttt{propInst}(w^s_r, w^o_r, r) \in (\mathcal{D}_{\mathcal{O}})_{\mathsf{facts}} \subseteq (\Pi)^*_{\mathsf{facts}}$, this implies

$$\texttt{classInst}(w^s_r, a) \in (\Pi)^*_{\mathsf{facts}} \subseteq \Pi^* \tag{5}$$

Besides, by construction, the atoms $\texttt{subjWit}(w^s_r, a)$ and $\texttt{domain}(\exists r, r)$ are both in $\Pi \subseteq \Pi^*$. Together with (5) and the second rule of Fig. 2, this implies

$$\texttt{subClass}(\exists r, a) = \mathsf{atom}(\alpha) \in \Pi^*,$$

which is what we wanted to prove.

This result, together with the structure of the program Π, allows us to derive an important (albeit unsurprising) property of our logic. For a TBox axiom α, the atom $\mathsf{atom}(\alpha)$ belongs to Π^* iff $\mathsf{atom}(\alpha)$ has been derived by means of one of the rules of Fig. 2. Then, a simple syntactic analysis of our program shows that the ABox $\mathcal{A}$ cannot play any role in the derivation of such atoms (in particular, in Fig. 2, atoms with predicate `classInst` or `propInst` can only be instantiated with witnesses). As a consequence, all TBox axioms implied by $\mathcal{O}$ follow from the TBox $\mathcal{T}$ alone, i.e.:

Lemma 1. *For a MDL-Lite*$_{\texttt{RDFS}}$ *ontology* $\mathcal{O} = \langle \mathcal{T}, \mathcal{A} \rangle$ *and TBox axiom* α,

$$\mathcal{O} \models \alpha \textit{ implies } \mathcal{T} \models \alpha$$

Filtering out Auxiliary Constants. The Datalog program Π that we introduced is complete (in the sense defined above), but not sound, because Π^* may contain atoms of the form $\texttt{classInst}(w_{e_1}, e_2)$ and $\texttt{propInst}(w^s_{e_1}, w^o_{e_1}, e_2)$, which are technically not implied by $\mathcal{O}$, since witnesses are fresh. We made this choice to simplify our presentation and proof of completeness. However, soundness can easily be recovered by eliminating such atoms from Π^*.

To do this, we first modify the definition of $\mathsf{atom}(\alpha)$, in such a way that it now uses a fresh predicate $\texttt{classInst}'$ (resp. $\texttt{propInst}'$, $\texttt{subClass}'$, $\texttt{subProp}'$) in place of `classInst` (resp. `propInst`, `subClass`, `subProp`). For instance the definition of $\mathsf{atom}(a(e))$ becomes $\texttt{classInst}(e, a)$, instead of $\texttt{classInst}(e, a)$.

Then we introduce a predicate `IRI`, in charge of collecting the signature of $\mathcal{O}$, and we use $\mathcal{D}_\Sigma$ for the set $\{\texttt{IRI}(e) \mid e \in \Sigma(\mathcal{O})\}$. To complete the construction, we add to Π this set of atoms $\mathcal{D}_\Sigma$, as well as four additional rules, one for each of the fresh predicates, which retains only constants from $\Sigma(\mathcal{O})$, thanks to the predicate `IRI`. For instance, for $\texttt{classInst}'$, the rule is

$$\texttt{classInst}(E, A) \leftarrow \texttt{classInst}(E, A), \texttt{IRI}(E), \texttt{IRI}(A)$$

5 Query Answering

In this section, we show how answering an UCMQ $q(\bar{X})$ over an *MDL-Lite*$_{\texttt{RDFS}}$ ontology $\mathcal{O}$ under MSER amounts to evaluating a Datalog query over the program defined in Sect. 4. Then we study the complexity of this problem.

Query answering. We first generalize the definition of $\mathsf{atom}(\alpha)$ to atoms that may contain variables (the definition is identical otherwise). For instance, if $\alpha = a \sqsubseteq X$, then $\mathsf{atom}(\alpha) = \texttt{subClass}(a, X)$. Next, to each disjunct $q'(\bar{X}) = \exists \bar{Y}.\alpha_1 \wedge \cdots \wedge \alpha_n$ of q, we associate the Datalog rule with head $\mathsf{q}'(\bar{X})$ and body $\bigcup_{i \in [1..n]} \mathsf{enc}(\alpha_i)$. We extend Π with each of these rules, as well as one rule for q, with head $\mathsf{q}(\bar{X})$ and body $\{\mathsf{q}'(\bar{X}) \mid q' \text{ is a disjunct of } q\}$. Let $\Pi_{\mathcal{O},q}$ be the resulting program. Because Π is complete, a valuation $\bar{d}$ for $\bar{X}$ is an answer to q over $\mathcal{O}$ iff the atom $\mathsf{q}(\bar{d})$ is in $(\Pi_{\mathcal{O},q})^*$.

Decision problem. As is conventional in the database literature, in order to study the complexity of query answering, we focus on Boolean queries, i.e. we investigate the following decision problem:

QUERYANSWERING $(q, \mathcal{O})$
Input: Boolean UCMQ q, *MDL-Lite*$_{\texttt{RDFS}}$ ontology $\mathcal{O}$
Decide: $\mathcal{O} \models q$

Ontology Complexity, Membership. Let n denote the size of (the encoding of) $\mathcal{O}$. Because our program Π is sound and complete (in the sense of Sect. 4), one can use the following procedure to decide $\mathcal{O} \models q$. First, compute Π out of $\mathcal{O}$, in $O(n)$. Then for each disjunct $q'(\bar{X}; \bar{Y})$ of q, and for each valuation $(\bar{d}, \bar{e})$ for $(\bar{X}, \bar{Y})$, decide whether $\mathsf{enc}(\phi) \subseteq \Pi^*$ holds for every ground atom $\phi \in q(\bar{X}/\bar{d}; \bar{Y}/\bar{e})$.

The amount of extra space needed to iterate over all such valuations does not depend on $\mathcal{O}$ (but only on $\bar{X}$ and $\bar{Y}$). Then we observe that the size of the set $\mathsf{enc}(\phi)$ is fixed (it contains up to three atoms). Deciding whether a fact is in the least model of a linear Datalog program is known to be in NLOGSPACE (see [15], Theorem 6.8). Since Π is linear (and has size in $O(n)$), this allows us to conclude that QUERYANSWERING is in NLOGSPACE in the size of $\mathcal{O}$.

ABox (a.k.a. Data) Complexity, Membership. We now use n to denote the size of (the encoding of) $\mathcal{A}$, and rely on the same procedure as in the previous section to provide an upper bound on the cost of QUERYANSWERING in terms of n.

The amount of time needed to iterate over all valuations for $(\bar{X}, \bar{Y})$ is in $O(n)$, and the amount of time needed to iterate over each disjunct q' of q, each atom $\phi \in q(\bar{X}/\bar{d}; \bar{Y}/\bar{e})$ and each atom $\alpha \in \mathsf{enc}(\phi)$ is in $O(1)$. If α is an extensional atom, then $\alpha \in \Pi^*$ can be decided in $O(n)$, by iterating over Π. If α is intensional with predicate `subClass` or `subProp`, then from Lemma 1, $\alpha \in \Pi^*$ can be decided in $O(1)$. The last possible case is if α has predicate `classInst` or `propInst`. We show that in this case, $\alpha \in \Pi^*$ can be decided by answering a Boolean First-Order (FO) query over the encoding $\mathcal{D}_{\mathcal{O}}$ of $\mathcal{O}$, and that this query can be computed in $O(1)$ (and therefore also has size in $O(1)$). This implies that QUERYANSWERING is in AC^0 in ABox complexity.

Let β be the atom identical to α, but where `classInst` (resp. `propInst`) is replaced with `classInst` (resp. `propInst`). Then by construction, $\alpha \in \Pi^*$ iff $\beta \in \Pi^*$. And from (1), we know that $\beta \in \Pi^*$ iff $\beta \in (\Pi^{ABox})^*$. So we can focus on the subprogram Π^{ABox}, whose rules are in Fig. 1. These rules can be partially grounded (with constants) for all variables but E, E_1, and E_2, by inspecting only atoms of $\mathcal{D}_{\mathcal{O}}$ with predicate `inverse`, `domain`, `subClassExt` or `subPropExt`. The number of such atoms is independent of $\mathcal{A}$ (it only depends on $\mathcal{T}$), so this partial grounding can be performed in $O(1)$. For instance, a partial grounding of the first rule may be $\texttt{classInst}(E, b) \leftarrow \texttt{classInst}(E, a), \texttt{subClassExt}(a, b)$, where a and b are constants.

Next, each rule in the resulting set Ψ of partially grounded rules can be finitely"unfolded", meaning that recursively replacing each intensional atom by

ABox complexity	in AC^0
TBox complexity	NLogSpace-complete
ontology complexity	NLogSpace-complete
query complexity	NP-complete
combined complexity	NP-complete

Fig. 3. Complexity of QueryAnswering.

its definitions in Ψ (and renaming variables accordingly) yields a finite set of non-recursive rules. This can be easily seen by observing that *(i)* the only rules in Ψ whose body introduces variables (i.e. not present in the head) are instantiations of the second and third rules of Fig. 1, *(ii)* the only intensional predicate in these rules is `propInst` and *(iii)* the rules of Ψ with `propInst` as predicate do not introduce variables (their bodies only use E_1 and E_2).

For $\psi \in \Psi$, let $\mathsf{unfold}(\psi)$ denote the unfolding of ψ. If β is of the form $\texttt{classInst}(e, a)$, then $\beta \in (\Pi^{ABox})^*$ iff $\beta \in (\mathcal{D}_\mathcal{O} \cup \Psi_\beta)^*$, where $\Psi_\beta = \bigcup\{\mathsf{unfold}(\psi) \mid \psi \in \Psi \text{ and } \psi \text{ has head predicate } \texttt{classInst}(E, a)\}$, and similarly for $\beta = \texttt{propInst}(e_1, e_2, r)$ (with head predicate $\texttt{propInst}(E_1, E_2, r)$). Because Ψ_β is non-recursive, deciding $\beta \in (\mathcal{D}_\mathcal{O} \cup \Psi_\beta)^*$ is equivalent to answering a Boolean FO query (precisely, a union of conjunctive queries) over $\mathcal{D}_\mathcal{O}$, and this query only depends Ψ and β, which are independent of $\mathcal{A}$ (Fig. 3).

TBox Complexity, Hardness. We provide a reduction from the *graph reachability problem*, which is known to be NLogSpace-complete, to QueryAnswering. Given a directed graph G with nodes V and edges E, and a pair (s, t) of nodes in V, graph reachability consists in deciding whether there exists a path from s to t in G. Viewing E as a binary relation over V, this is equivalent to deciding whether (s, t) is in the transitive closure E^* of E. W.l.o.g., we can assume that E is a superset of the identity relation over V.

We represent each node $v \in V$ as a distinct atomic concept c_v, and leverage the fact that the relation $\sqsubseteq_c$ is (semantically) transitive. Precisely, given $G = \langle V, E\rangle$ and $s, t \in V$ as input, we define the UCMQ $q() = c_s \sqsubseteq_c c_t$ and the ontology $\mathcal{O}^G = \langle \mathcal{T}^G, \emptyset\rangle$, where $\mathcal{T}^G = \{c_{v_1} \sqsubseteq_c c_{v_2} \mid (v_1, v_2) \in E\}$.

Now let $R = \{(c_{v_1}, c_{v_2}) \mid c_{v_1} \sqsubseteq_c c_{v_2} \in \mathcal{T}^G\}$, and let R^* denote the transitive closure of R. Immediately from the semantics of *MDL-Lite*$_{\mathsf{RDFS}}$, $\mathcal{O} \models c_{v_1} \sqsubseteq_c c_{v_2}$ iff $(c_{v_1}, c_{v_2}) \in R^*$, for any $v_1, v_2 \in V$. And by construction, $(c_{v_1}, c_{v_2}) \in R^*$ iff $(v_1, v_2) \in E^*$. Therefore $\mathcal{O} \models q$ iff $(s, t) \in E^*$.

This reduction requires a constant amount of extra space, and the sizes of q and $\mathcal{A}$ are fixed (i.e. independent of the size of G). This allows us to conclude that QueryAnswering is NLogSpace-hard in the size of $\mathcal{T}$.

Query and Combined Complexity. Finally, we show that QueryAnswering is NP-complete in both query and combined complexity. NP-hardness for query (and hence combined) complexity is inherited from the well-known complexity of Boolean CQ entailment over an ABox. Membership in combined (and hence query) complexity in NP is also straightforward: we can use a non-deterministic version of the algorithm used above for NLogSpace membership (in the size of $\mathcal{O}$): guess a disjunct $q'(\bar{X};\bar{Y})$ of q, and a valuation $(\bar{d},\bar{e})$ for $(\bar{X},\bar{Y})$ such that $\mathsf{enc}(\phi) \subseteq \Pi^*$ holds for every ground atom $\phi \in q(\bar{X}/\bar{d};\bar{Y}/\bar{e})$. Then check that this holds for each such ϕ. The number of such checks is linear in the size of q'. and the cost of each of them is independent on the size of q, and, as we have seen above, (less than) polynomial in the size of $\mathcal{T}$ and $\mathcal{A}$.

6 Conclusion

In this paper, we addressed metaquerying on *MDL-Lite*$_{\mathtt{RDFS}}$ (the metamodeling counterpart of *DL-Lite*$_{\mathtt{RDFS}}$) ontologies under MSER semantics. Specifically, we focused on unions of conjunctive metaqueries, and we provided a reduction from the metaquery answering problem to the evaluation of a Datalog query over a (positive) Datalog program that encodes in a sound and complete way the knowledge represented by the ontology. Besides, the rules of this program are independent of the input ontology. The fact that this program is linear allowed us to established a tight NLogSpace bound for TBox and ontology complexity, while showing that data complexity remains in AC^0, and that query and combined complexity are NP-complete. Our approach also allows immediate implementations using off-the-shelf Datalog engines.

We plan to extend our work in several directions, possibly simultaneously. First, an interesting extension would be to consider the first-order logic semantics (rather than the SPARQL semantics) for query answering. Second, both the ontology and the query language could be enriched by allowing, e.g., concept and role disjointnesses, inequalities, negated atoms, and concept (resp. role) expressions in the right-hand side of concept (resp. role) inclusions.

Acknowledgments. This work has been supported by the PNRR MUR project PE0000013-FAIR.

References

1. Abiteboul, S., Hull, R., Vianu, V.: Foundations of Databases. Addison-Wesley (1995)
2. Ahmetaj, S., Fischl, W., Pichler, R., Simkus, M., Skritek, S.: Towards reconciling SPARQL and certain answers. In: WWW 2015, pp. 23–33 (2015)
3. Arenas, M., Gottlob, G., Pieris, A.: Expressive languages for querying the semantic web. ACM Trans. Database Syst. **43**(3), 13:1–13:45 (2018)
4. Atkinson, C., Kühne, T.: The essence of multilevel metamodeling. In: Gogolla, M., Kobryn, C. (eds.) UML 2001. LNCS, vol. 2185, pp. 19–33. Springer, Heidelberg (2001). https://doi.org/10.1007/3-540-45441-1_3
5. Atkinson, C., Kühne, T.: Model-driven development: a metamodeling foundation. IEEE Softw. **20**(5), 36–41 (2003)

6. Bienvenu, M.: Ontology-mediated query answering: Harnessing knowledge to get more from data. In: IJCAI 2016, pp. 4058–4061 (2016)
7. de Carvalho, V.A., Almeida, J.P.A., Fonseca, C.M., Guizzardi, G.: Multi-level ontology-based conceptual modeling. Data Knowl. Eng. **109**, 3–24 (2017)
8. Chen, W., Kifer, M., Warren, D.S.: HILOG: A foundation for higher-order logic programming. J. Logic Program. **15**(3), 187–230 (1993)
9. Cima, G., Console, M., Delfino, R., Lenzerini, M., Poggi, A.: Answering conjunctive queries with safe negation and inequalities over RDFS knowledge bases. In: AAAI 2025, pp. 14824–14831 (2025)
10. Cima, G., Console, M., Lenzerini, M., Poggi, A.: Monotone abstractions in ontology-based data management. In: AAAI 2022, pp. 5556–5563 (2022)
11. Cima, G., De Giacomo, G., Lenzerini, M., Poggi, A.: On the SPARQL metamodeling semantics entailment regime for OWL 2 QL ontologies. In: WIMS 2017, pp. 10:1–10:6 (2017)
12. Cima, G., Lenzerini, M., Poggi, A.: Answering conjunctive queries with inequalities in DL-Lite$_{\mathcal{R}}$. In: AAAI 2020, pp. 2782–2789 (2020)
13. Cima, G., Poggi, A., Lenzerini, M.: The notion of abstraction in ontology-based data management. Artif. Intell. **323**, 103976 (2023)
14. Cuenca Grau, B.: A possible simplification of the semantic web architecture. In: WWW 2004, pp. 704–713 (2004)
15. Dantsin, E., Eiter, T., Gottlob, G., Voronkov, A.: Complexity and expressive power of logic programming. ACM Comput. Surv. **33**(3), 374–425 (2001)
16. De Giacomo, G., Lenzerini, M., Rosati, R.: Higher-order description logics for domain metamodeling. In: AAAI 2011, pp. 183–188 (2011)
17. Glimm, B.: Using SPARQL with RDFS and OWL entailment. In: RW 2011, pp. 137–201. Lecture Notes in Computer Science, Springer (2011)
18. Glimm, B., Kazakov, Y., Kollia, I., Stamou, G.B.: Lower and upper bounds for SPARQL queries over OWL ontologies. In: AAAI 2015, pp. 109–115 (2015)
19. Gottlob, G., Pieris, A.: Beyond SPARQL under OWL 2 QL entailment regime: Rules to the rescue. In: IJCAI 2015, pp. 2999–3007 (2015)
20. Harris, S., Seaborne, A.: SPARQL 1.1 query language. W3C Recommendation (2013). available at https://www.w3.org/TR/sparql11-query/
21. Kontchakov, R., Rezk, M., Rodríguez-Muro, M., Xiao, G., Zakharyaschev, M.: Answering SPARQL Queries over Databases under OWL 2 QL Entailment Regime. In: Mika, P., et al., (eds.) ISWC 2014. LNCS, vol. 8796, pp. 552–567. Springer, Cham (2014). https://doi.org/10.1007/978-3-319-11964-9_35
22. Kubincová, P., Kluka, J., Homola, M.: Expressive description logic with instantiation metamodelling. In: KR 2016, pp. 569–572 (2016)
23. Lenzerini, M., Lepore, L., Poggi, A.: A higher-order semantics for metaquerying in OWL 2 QL. In: KR 2016, pp. 577–580 (2016)
24. Lenzerini, M., Lepore, L., Poggi, A.: Metamodeling and metaquerying in OWL 2 QL. Artif. Intell. **292**, 103432 (2021)
25. Levesque, H.J.: Foundations of a functional approach to knowledge representation. Artif. Intell. **23**, 155–212 (1984)
26. Martinez, M., Rohrer, E., Severi, P.: Complexity of the description logic ALCM. In: KR 2016, pp. 585–588 (2016)
27. Motik, B.: On the properties of metamodeling in OWL. J. Logic Comput. **17**(4), 617–637 (2007)
28. Motik, B., Cuenca Grau, B., Horrocks, I., Wu, Z., Fokoue, A., Lutz, C.: OWL 2 Web Ontology Language profiles (second edition). W3C Recommendation (2012). available at http://www.w3.org/TR/owl2-profiles/

29. Motik, B., Patel-Schneider, P.F., Cuenca Grau, B.: OWL 2 Web Ontology Language direct semantics (second edition). W3C Recommendation (2012). available at https://www.w3.org/TR/owl2-direct-semantics/
30. Motz, R., Rohrer, E., Severi, P.: The description logic SHIQ with a flexible meta-modelling hierarchy. J. Web Sem. **35**, 214–234 (2015)
31. Poggi, A.: On the SPARQL direct semantics entailment regime for OWL 2 QL. In: DL 2016. CEUR Workshop Proc. vol. 1577 (2016)
32. Qureshi, H.M., Faber, W.: Efficient OWL2QL meta-reasoning using asp-based hybrid knowledge bases. In: ICLP 2024. EPTCS, vol. 416, pp. 188–200 (2024)
33. Qureshi, H.M., Faber, W.: Evaluating datalog tools for meta-reasoning over OWL 2 QL. Theory Pract. Logic Program. **24**(2), 368–393 (2024)
34. Rosati, R.: The limits of querying ontologies. In: ICDT 2007. Lecture Notes in Computer Science, vol. 4353, pp. 164–178. Springer (2007)
35. Xiao, G., Calvanese, D., Kontchakov, R., Lembo, D., Poggi, A., Rosati, R., Zakharyaschev, M.: Ontology-based data access: A survey. In: IJCAI 2018, pp. 5511–5519 (2018)

User Access Control in Policy-Protected Virtual Knowledge Graphs

Divya Baura[1(✉)] and Diego Calvanese[2]

[1] Umeå Universitet, Umeå, Sweden
divya.baura@umu.se
[2] Free University of Bozen-Bolzano, Bolzano, Italy
diego.calvanese@unibz.it

Abstract. *Virtual Knowledge Graph* (VKG) is a well-established framework in which users can access a relational data source through an ontology and declarative mappings. VKG systems traditionally assume uniform access rights for all users, an assumption that does not always hold in real-world scenarios involving diverse user roles and sensitive information requiring protection. *Controlled Query Evaluation* (CQE) provides a privacy-preserving framework by enforcing policies that define confidential information and implementing censors to prevent policy violations. However, it does not account for differences in user privileges during query answering. To address this gap, we extend the *Policy-Protected VKG* (PPVKG) framework, which embeds CQE policies into VKG mappings, by enabling role-sensitive query answering. Specifically, we incorporate *Role-Based Access Control* (RBAC) into PPVKG, by associating to each user role a specific set of policies, and ensuring that during query evaluation, only the policies relevant to the user's role are applied. We validate our RBAC enhanced PPVKG approach using the MIMIC-III critical-care database, mapped to the *Fast Healthcare Interoperability Resources* (FHIR) ontology. Our experiments, conducted with the open-source VKG system Ontop, demonstrate effective policy enforcement with RBAC.

Keywords: Virtual Knowledge Graph · Controlled Query Evaluation · Policy-protected VKG · Role-Based Access Control

1 Introduction

The Virtual Knowledge Graph (VKG) paradigm [16,20] enables seamless integration and querying of heterogeneous relational databases through an ontology layer, typically expressed in OWL 2 QL [15], using declarative mappings (e.g., R2RML [7]) that connect database schema elements to ontology classes and properties. This allows users to pose high-level semantic queries over the ontology without directly engaging with the underlying data sources.

Despite its strengths, VKG inherits critical privacy and inference risks, since TBox axioms may implicitly reveal sensitive information through logical entailments. *Controlled Query Evaluation (CQE)* addresses this by modeling confidentiality requirements as policies and applying censoring functions that evaluate

H. Takeda et al. (Eds.): IJCKG 2025, LNAI 16297, pp. 204–219, 2026.
https://doi.org/10.1007/978-981-95-5009-8_14

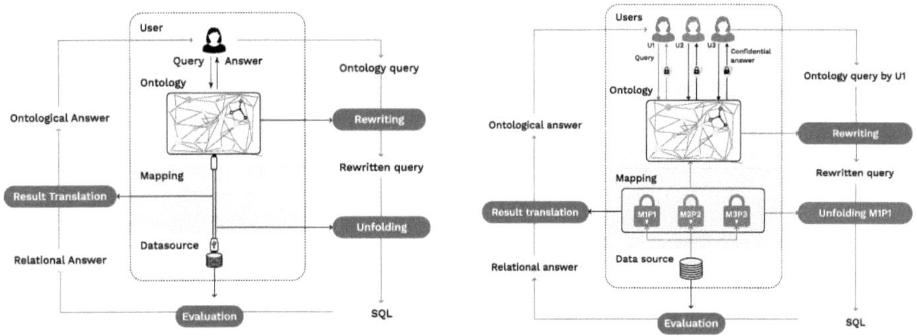

Fig. 1. Query processing in VKG (left) vs. RBAC-PPVKG (right)

and restrict query answers [3,11,14]. Extending this concept, *Policy-Protected VKG* (PPVKG) integrates CQE in VKG [2,6]. PPVKG specifies policies that are first-order denial constraints. These policies are then embedded into VKG mapping assertions, resulting in new mapping assertion that enforce policy compliance during the rewriting process. However, both VKG and PPVKG assume a uniform access model, granting identical data visibility to all authenticated users regardless of their operational context or responsibilities. This one-size-fits-all paradigm proves inadequate in sensitive domains like healthcare, where granular access differentiation based on professional roles is essential not only to prevent data leakage but also to comply with stringent regulations like the *General Data Protection Regulation* (GDPR) [19].

Role-Based Access Control (RBAC) [9,18] provides the necessary conceptual framework to address these limitations by enforcing the privacy principle of least privilege, restricting users to the minimum level of access required for their organizational roles or responsibilities. In this work, we show that mappings can be enriched with RBAC, thereby ensuring that query results are dynamically filtered based on user roles and data exposure is minimized. We have implemented constrained RBAC, enforcing session-specific separation-of-duty rules and limiting role activations per session, in compliance with the NIST RBAC standard's more secure models. This integration enables enterprise-grade features such as role-differentiated views and dynamic session-aware policy enforcement. The resulting RBAC-enhanced PPVKG framework combines static privacy policies with runtime, context-aware access control, making it suitable for deployment in complex domains like healthcare and finance, where both confidentiality and compliance are paramount. As illustrated in Fig. 1, our approach activates role-specific mapping assertions during query processing. For example, when User 1 (specified with red color) queries the ontology, only policy augmented mappings applicable to their role with the specific policy (e.g., M1P1) are activated.

In the rest of the paper, after introducing technical preliminaries on VKG and PPVKG (Sect. 2), we define the notion of RBAC (Sect. 3) and introduce PPVKG with RBAC functionalities (Sect. 4). We evaluate our implementation of role-base policy embedding into mappings on the MIMIC-III clinical dataset [13], demonstrating that role-aware query answering delivers fine-grained access control with minimal performance overhead (Sect. 5).

Our implementation and experimental evaluation is available on GitHub[1].

2 Preliminaries

2.1 Virtual Knowledge Graph

In VKG, users can query a data source through an ontology TBox linked to the data source via declarative mappings [16,20]. We formalize VKG using the notion of *VKG specification*, which is a triple $\mathcal{V} = \langle \mathcal{T}, \mathcal{S}, \mathcal{M} \rangle$, where $\mathcal{T}$ is a TBox, $\mathcal{S}$ a relational database (DB) schema, and $\mathcal{M}$ a mapping between $\mathcal{T}$ and $\mathcal{S}$. The TBox captures knowledge at the intensional level about the classes (denoting sets of individuals) and properties (denoting binary relations between individuals or values) of the domain of interest. It is expressed in the lightweight description logic (DL) *DL-Lite*$_\mathcal{R}$ [5], which is the formal counterpart of the ontology language OWL 2 QL, standardized by the W3C [15]. For the formal details about *DL-Lite*$_\mathcal{R}$ and OWL 2 QL we refer to [5,15]. The *mapping* $\mathcal{M}$ is a finite set of *mapping assertions* from $\mathcal{S}$ to $\mathcal{T}$, each of the form $\varphi(\boldsymbol{x}) \rightsquigarrow \psi(\mathsf{IRI}(\boldsymbol{x}))$. Here, $\varphi(\boldsymbol{x})$ denotes a FOL (or SQL) source query over $\mathcal{S}$ with answer variables $\boldsymbol{x}$, while $\psi(\mathsf{IRI}(\boldsymbol{x}))$, referred to as the *target* of the mapping, is an atom whose predicate is a class or property of $\mathcal{T}$ over the variables in $\boldsymbol{x}$ and so-called *IRI-templates* $\mathsf{IRI}(\boldsymbol{x})$. Each IRI-template $\mathsf{iri}(\boldsymbol{x})$ in $\mathsf{IRI}(\boldsymbol{x})$ is a term that concatenates string values and the answer variables in $\boldsymbol{x}$, and is used to construct values (i.e., literals) and object identifiers (i.e., IRIs) from the DB values returned by $\varphi(\boldsymbol{x})$. A concrete mapping language that provides such mappings is R2RML, standardized by the W3C [7]. In our examples we provide mappings in the mapping language of the Ontop system (see Sect. 2.3), where the source query is expressed in SQL, the target atom is specified in RDF syntax as a *triple template* (in which the SQL answer variables of the IRI-templates are enclosed in `{...}`), and multiple mapping assertions with the same source query might be grouped together, resulting in a target consisting of multiple RDF triple templates.

Given a VKG specification $\mathcal{V} = \langle \mathcal{T}, \mathcal{S}, \mathcal{M} \rangle$ and a DB instance $\mathcal{D}$ for $\mathcal{S}$, the pair $\langle \mathcal{V}, \mathcal{D} \rangle$ is called a *VKG instance*. The *retrieved ABox* for $\langle \mathcal{V}, \mathcal{D} \rangle$, denoted $\mathsf{ret}(\mathcal{V}, \mathcal{D})$, consists of all facts $\psi(\mathsf{IRI}(\boldsymbol{t}))$, where $\varphi(\boldsymbol{x}) \rightsquigarrow \psi(\mathsf{IRI}(\boldsymbol{x}))$ is a mapping assertion in $\mathcal{M}$, and $\boldsymbol{t} \in \mathsf{eval}(\varphi(\boldsymbol{x}), \mathcal{D})$ is a tuple of constants in the evaluation of $\varphi(\boldsymbol{x})$ over $\mathcal{D}$. Notice that the fact $\psi(\mathsf{IRI}(\boldsymbol{t}))$ contains IRIs of the form $\mathsf{iri}(\boldsymbol{t})$ constructed from the answer tuple $\boldsymbol{t}$ using the IRI template $\mathsf{iri}(\boldsymbol{x})$ in the mapping target. Hence, $\mathsf{ret}(\mathcal{V}, \mathcal{D})$ is an ABox over constants in the set $\Delta_{\mathcal{O},\mathcal{D}}$, consisting of: *(i)* all DB values in $\mathcal{D}$ and *(ii)* all possible IRIs $\mathsf{iri}(\boldsymbol{t})$ constructed from some tuple $\boldsymbol{t}$ of values in $\mathcal{D}$ and some IRI template $\mathsf{iri}(\boldsymbol{x})$ in a mapping assertion in $\mathcal{M}$.

A *model* of the VKG instance $\langle \mathcal{V}, \mathcal{D} \rangle$ is defined as a model of the DL knowledge base $\langle \mathcal{T}, \mathsf{ret}(\mathcal{V}, \mathcal{D}) \rangle$. We denote the set of models of $\langle \mathcal{V}, \mathcal{D} \rangle$ by $\mathsf{Mod}(\mathcal{V}, \mathcal{D})$,

[1] https://github.com/divyabaura/PPVKG-role-based-policies.

and we say that $\langle \mathcal{V}, \mathcal{D} \rangle$ is *inconsistent* if $\mathsf{Mod}(\mathcal{V}, \mathcal{D}) = \emptyset$. Moreover, $\langle \mathcal{V}, \mathcal{D} \rangle \models \alpha$, indicating that $\langle \mathcal{V}, \mathcal{D} \rangle$ entails a sentence α, holds if α is true in every model in $\mathsf{Mod}(\mathcal{V}, \mathcal{D})$. Here, we adopt the *standard name assumption*, i.e., given a VKG instance $\langle \mathcal{V}, \mathcal{D} \rangle$, we consider interpretations over a fixed domain Δ containing $\Delta_{\mathcal{O},\mathcal{D}}$ such that all values in $\Delta_{\mathcal{O},\mathcal{D}}$ are interpreted as themselves.

We provide an example of VKG mappings in a medical domain, on which we will build in the rest of the paper.

Example 1. We assume to have in the ontology a class `:Patient` with data property `:Patient.gender`, and a class `:Prescription` with object property `:Prescription.subject` relating a prescription to the patient who is the subject of the prescription. In the data source we have a table `Person`, which is mapped to the class `:Patient` and the `:Patient.gender` data property, and a table `DrugExposure`, which is mapped to the class `:Prescription` and the `:Prescription.subject` object property, as follows:

```
mappingId m1
target :Patient/{person_id} a :Patient ; :Patient.gender {gender}^^xsd:string .
source SELECT person_id, gender FROM Person P

mappingId m2
target :Prescription/{drug_exposure_id} a :Prescription ;
         :Prescription.subject :Patient/{person_id} .
source SELECT drug_exposure_id, person_id FROM DrugExposure
```

$\triangleleft$

2.2 Policy-Protected VKG

Policy-Protected VKG (PPVKG) extends the traditional VKG paradigm by embedding data protection policies directly into VKG mapping definitions [2,6]. While standard VKG enables querying heterogeneous sources through an ontological semantic layer, PPVKG makes use of policies to specify information that should be protected. Specifically, policies in PPVKG are expressed as denial constraints and the disclosure of protected information, i.e., information that causes a violation of the denial constraint, is prevented by dynamically rewriting queries in such a way that protected information is suppressed from query results. This is achieved by compiling the policy rules into the declarative VKG mappings. In this way, a policy-aware VKG is transformed into a standard VKG that transparently enforces the policies.

PPVKG builds upon the principles of *controlled query evaluation* (CQE), which is a privacy-preserving framework for query answering in the presence of ontologies [3,11,14], and it extends the VKG framework to incorporate CQE. To formally introduce the PPVKG setting, we first define a *denial (assertion)* as a first-order logic (FOL) sentence of the form $\forall \boldsymbol{x}.\varphi(\boldsymbol{x}) \rightarrow \bot$, where $\exists \boldsymbol{x}.\varphi(\boldsymbol{x})$ is a Boolean conjunctive query (BCQ). Given a set $\mathcal{V}$ of FOL sentences (e.g., a TBox) and a denial δ, we say that $\mathcal{V} \cup \{\delta\}$ is *consistent* if $\mathcal{V} \not\models \exists \boldsymbol{x}.\varphi(\boldsymbol{x})$.

Following [6], we define a *PPVKG specification* as a 4-tuple $\mathcal{E} = \langle \mathcal{T}, \mathcal{S}, \mathcal{M}, \mathcal{P} \rangle$, where $\langle \mathcal{T}, \mathcal{S}, \mathcal{M} \rangle$ is a VKG specification and $\mathcal{P}$ is a *policy*, i.e., a finite set of denials over the signature of $\mathcal{T}$, such that $\mathcal{T} \cup \mathcal{P}$ is consistent. The semantics of a PPVKG specification is the same as that of the underlying VKG specification, and all other notions (source DB $\mathcal{D}$, instance $\langle \mathcal{E}, \mathcal{D} \rangle$, retrieved ABox $\mathsf{ret}(\mathcal{E}, \mathcal{D})$, and set $\mathsf{Mod}(\mathcal{E}, \mathcal{D})$ of models) naturally extend to PPVKG.

For a query language $\mathcal{L}$ (e.g., the language **CQ** of conjunctive queries or **GA** of ground atoms), let $\mathcal{L}(\mathcal{T})$ denote the restriction of $\mathcal{L}$ to the predicates in $\mathcal{T}$, and $\mathcal{L}_{\mathcal{D}}$ the formulas in $\mathcal{L}$ mentioning only constants in $\mathcal{D}$. An *optimal censor* for $\mathcal{E}$ in $\mathcal{L}$ is a function $\mathsf{cens}(\cdot)$ that, for each source DB $\mathcal{D}$ for $\mathcal{E}$, returns a set $\mathsf{cens}(\mathcal{D}) \subseteq \mathcal{L}_{\mathcal{D}}$ such that *(i)* $\langle\langle \mathcal{T}, \mathcal{S}, \mathcal{M} \rangle, \mathcal{D} \rangle \models \varphi$, for each $\varphi \in \mathsf{cens}(\mathcal{D})$, and *(ii)* $\mathcal{T} \cup \mathcal{P} \cup \mathsf{cens}(\mathcal{D})$ is consistent. $\mathcal{L}$ is called the *censor language*. We are interested in *optimal* censors, i.e., those that return as many formulas as possible. The set of all optimal censors in $\mathcal{L}$ for a PPVKG specification $\mathcal{E}$ is denoted $\mathcal{L}\text{-}\mathsf{OptCens}_{\mathcal{E}}$.

To obtain a notion of censor that allows for embedding a policy into the mapping, [6] define censors that approximate censors for $\mathcal{E}$ in **GA** [14].

Definition 2 (IGA censor [6]**).** Given a PPVKG specification $\mathcal{E} = \langle \mathcal{T}, \mathcal{S}, \mathcal{M}, \mathcal{P} \rangle$, the *intersection GA (IGA)* censor for $\mathcal{E}$ is the function $\mathsf{cens}_{IGA}(\cdot)$ such that, for every DB instance $\mathcal{D}$ for $\mathcal{S}$, $\mathsf{cens}_{IGA}(\mathcal{D}) = \bigcap_{\mathsf{cens} \in \mathbf{GA}\text{-}\mathsf{OptCens}_{\mathcal{E}}} \mathsf{cens}(\mathcal{D})$. ◁

Thus, an IGA censor, when applied to a DB instance $\mathcal{D}$ for the source schema $\mathcal{S}$ of $\mathcal{E}$, returns the intersection of the sets of ground atoms computed by all optimal censors. In [2], an algorithm called EncodeMapping is presented, which generates a new VKG mapping $\mathcal{M}'$ that embeds a policy $\mathcal{P}$ into a given mapping $\mathcal{M}$. When the VKG engine processes queries posed over the ontology using $\mathcal{M}'$, the obtained answers comply with $\mathcal{P}$ according to an IGA censor.

Example 3 (Example 1 cont'd). Let us now consider two policies p_1 : $\forall x.\forall y.\mathit{Patient.gender}(x, y) \land \forall z.\mathit{Patient.address}(x, z) \rightarrow \bot$, which ensures the confidentiality of the combination of a patient's gender and address, and p_2 : $\forall x.\mathit{Prescription}(x) \land \forall y.\mathit{Prescription.subject}(x, y) \rightarrow \bot$, which ensures the confidentiality of patient ids for which a prescription is made. Let us assume that the `Person` table contains also a `location_id` attribute that is mapped to the *Patient.address* data property. By compiling these policies into m_1 and m_2, we obtain the following policy-protected mapping assertions:

```
mappingId mp1
target :Patient/{person_id} a :Patient ;  :Patient.gender {gender}^^xsd:string .
source SELECT person_id, gender FROM Person WHERE location_id IS NULL

mappingId mp2
target :Prescription/{drug_exposure_id} a :Prescription ;
        :Prescription.subject :Patient/{person_id} .
source SELECT drug_exposure_id, person_id FROM DrugExposure WHERE person_id IS NULL
```

Intuitively, the condition `location_id IS NULL` in *mp₁* ensures that the gender location can be returned only for patients whose `location_id` is not known. Similarly, mp_2 ensures that prescriptions can only be disclosed when the `person_id` is not known. ◁

2.3 The VKG System Ontop

We utilize the state-of-the-art open-source VKG system Ontop [4,21], which supports query answering over a VKG instance through query transformation and sophisticated optimization techniques. An Ontop installation operates on a VKG specification $\mathcal{V} = \langle \mathcal{T}, \mathcal{S}, \mathcal{M} \rangle$, where $\mathcal{T}$ is an OWL 2 QL TBox, $\mathcal{S}$ a relational schema with constraints, and $\mathcal{M}$ an R2RML [7] mapping. Ontop supports also a proprietary mapping language that is more user-friendly and fully interoperable with R2RML, and in our examples we have used such language.

Ontop efficiently answers SPARQL queries [12] over a VKG instance $\langle \mathcal{V}, \mathcal{D} \rangle$, where $\mathcal{D}$ is a DB instance for $\mathcal{S}$, using the OWL 2 QL entailment regime [10]. It does so by implementing *query answering by rewriting.* This process involves pre-processing $\mathcal{T}$, $\mathcal{S}$, and $\mathcal{M}$ for optimization purposes. At runtime, Ontop transforms a given SPARQL query into an equivalent SQL query and optimizes it using mapping information and database constraints in $\mathcal{S}$. The resulting SQL query gets then executed by the underlying DB engine.

3 Role Based Access Control

Role-Based Access Control (RBAC) is a well-established authorization model that assigns permissions based on organizational roles rather than individual identities. This abstraction simplifies policy management by aligning access rights with job responsibilities, reducing administrative overhead and mitigating risks from over-privileged users [18]. In the RBAC paradigm, three core entities interact: *users* (human or system agents), *roles* (functional positions within an organization), and *permissions* (authorizations to perform operations on protected resources).

The foundational variant of RBAC, known as *Flat RBAC* or *Core RBAC*, mandates many-to-many relations between users and roles and between roles and permissions, it also requires that users can activate multiple roles concurrently. Building upon this, *Hierarchical RBAC* introduces role inheritance, where senior roles automatically inherit permissions from their juniors, supporting partial-order structures or restricted hierarchies. A further enhancement, *Constrained RBAC*, enforces separation of duty constraints, static or dynamic, to prevent incompatible roles from being assigned or activated simultaneously. The *Symmetric RBAC* (or RBAC3), retains all the functionality of Constrained RBAC and adds the crucial capability of permission-role review. In effect, the four RBAC levels—Flat, Hierarchical, Constrained, and Symmetric—form a cumulative hierarchy in which each level builds upon the previous one by introducing a single new dimension of control: role hierarchy, separation of duties, and role–permission auditability, respectively [17].

In this work, we present a technique to incorporate the principles of constrained role-based access Control within the PPVKG framework. This is achieved by enforcing access control through the concept of sessions, where each session represents a specific user operating under a defined access role.

We formalize the Flat RBAC model through the following notion.

Definition 4 (RBAC specification). A *role-based access control (RBAC) specification* is a tuple $\langle \mathcal{O}, \mathcal{A}, \mathcal{U}, \mathcal{R}, \mathsf{UR}, \mathsf{P}, \mathsf{perms} \rangle$, where:

- $\mathcal{O}$ is a finite set of *objects*, representing protected resources;
- $\mathcal{A}$ is a finite set of *actions*, representing operations that can be performed on the objects;
- $\mathcal{U}$ is a finite set of *users*;
- $\mathcal{R}$ is a finite set of *roles* (e.g., *doctor*, *nurse*);
- $\mathsf{UR} \subseteq \mathcal{U} \times \mathcal{R}$ is a *user-role assignment* relation;
- $\mathsf{P} \subseteq \mathcal{O} \times \mathcal{A}$ represents *permissions*, where each pair $\langle o, a \rangle \in \mathsf{P}$ represents a permission to perform action a on object o; and
- $\mathsf{perms} : \mathcal{R} \to 2^{\mathsf{P}}$ is a function mapping roles to sets of their permissions. ◁

Modern RBAC implementations increasingly incorporate dynamic policy-driven mechanisms where role assignments become context-sensitive rather than static. Such user to role mappings may dynamically activate based on temporal constraints (e.g., emergency room physicians gaining after-hours access), environmental conditions (e.g., location-based restrictions for mobile clinicians) or runtime attributes (e.g., current patient treatment relationships). This dynamic capability transforms RBAC from a rigid structure into an adaptive policy enforcement engine, where roles serve as policy anchors that respond to real-world operational contexts. We will take into account the dynamicity of actual role assignments by allowing users to log into a PPOBDA system with a specific role, and it is such role that determines the access rights of the user, and hence the PPVKG policies that should be applied.

The GDPR compliance [8] advantages of RBAC are particularly pronounced in policy-driven implementations. *Article 5(1)(c)*'s data minimization principle is reinforced through just-in-time permission activation, ensuring users access only currently necessary data. Similarly, *Article 32(1)(b)*'s confidentiality requirements benefit from dynamic permission revocation when contexts change (e.g., automatically downgrading access when clinicians rotate between departments). By enabling fine-grained, context-aware authorization that restricts data visibility to the minimal necessary set, policy-driven RBAC provides a robust technical foundation for fulfilling legal obligations while enforcing least-privilege access across evolving operational scenarios.

4 PPVKG with RBAC Functionalities

In this section, we extend the PPVKG framework by integrating RBAC to overcome the limitations of traditional policy-protected VKGs that expose the same data to all users regardless of their roles, thus enforcing static policies that lack adaptability and failing to align with real-world organizational structures. We do so by *(i)* suitably duplicating mappings according to user roles and embedding policies associated to roles where appropriate, *(ii)* enabling for each user role only the relevant mapping copies during query unfolding, and *(iii)* ensuring that the proper mappings are taken into account when, for query answering, a user logs into the system with a specific roles.

4.1 Formalizing PPVKG with RBAC Functionalities

We start by providing the formalization of the proposed extension of PPVKG with RBAC functionalities.

Definition 5 (RBAC-PPVKG specification). An *RBAC policy-protected VKG* (RBAC-PPVKG) is a tuple $\mathcal{H} = \langle \mathcal{E}, \mathcal{R}, \mathsf{RP}, \mathcal{U}, \mathsf{UR} \rangle$, where:

- $\mathcal{E} = \langle \mathcal{T}, \mathcal{S}, \mathcal{M}, \mathcal{P} \rangle$ is a PPVKG,
- $\mathcal{R}$ is a finite set of roles,
- $\mathsf{RP} : \mathcal{R} \rightarrow 2^{\mathcal{P}}$ is a function assigning to each role a set of policies,
- $\mathcal{U}$ is a finite set of users, and
- $\mathsf{UR} \subseteq \mathcal{U} \times \mathcal{R}$ is a user-role assignment. ◁

To relate an RBAC-PPVKG specification to an RBAC specification according to Definition 4, we observe that we are interested in answering queries posed by a user over a PPVKG, and the information the user is allowed to access, according to their role, depends on the policy-protected mappings that they can access while unfolding the query. Hence, $\mathcal{A}$ consists of the single action of accessing a mapping during query unfolding. Instead, the set $\mathcal{O}$ of objects (to be accessed) consists of the union of suitable copies of mapping assertions, with one copy $\mathcal{M}_r$ for each role $r \in \mathcal{R}$, obtained from $\mathcal{M}$ by applying to all its mapping assertions the policies $\mathsf{RP}(r)$ assigned to r.

With this correspondence in mind, we want now to transform an RBAC-PPVKG specification into an ordinary VKG specification, which can be processed by a VKG engine like Ontop, and that automatically enforces the appropriate policies for the role with which a user logs into the systems for query answering. We perform this transformation in two steps:

1. For each role $r \in \mathcal{R}$, we create from $\mathcal{M}$ the set $\mathcal{M}_r$ of mapping assertions that embed in $\mathcal{M}$ the policies relevant for role r.
2. We ensure that the mapping assertions in $\mathcal{M}_r$ are activated only for users that log into the system with role r.

Step 2 of the transformation is achieved by making use of a special boolean SQL function `ontop_contains_role`(r), which takes a role r as argument, and evaluates to true only if it is executed while the user is logged in the system with role r. Such function is treated by Ontop in a special way during query processing. Specifically, Ontop evaluates the function at query unfolding time (rather than leaving it in the generated SQL query), and substitutes the function call with the boolean constant `true` or `false` resulting from this evaluation. Then, the query optimization algorithm of Ontop can take into account the value of such constant and simplify the generated SQL query accordingly. We exploit `ontop_contains_role` by embedding in the source query of a mapping assertion m the condition `ontop_contains_role`(r), in conjunction (at the top level of the `WHERE` clause) with the existing condition. Hence: *(i)* If the user is logged with role r, at query unfolding time the condition evaluates to `true` and the mapping assertion m is actually used for the unfolding. *(ii)* Instead, if the user is logged with a role different from r, the condition evaluates to `false` and the overall source query of m is considered by Ontop as equivalent to a query that returns the empty answer. As a result, m is effectively ignored in query unfolding.

To concretely realize the described mapping transformation approach, we make use of two functions that manipulate mapping assertions:

- The function EncodeP takes as arguments a set $\mathcal{M}$ of mapping assertions and a set $\mathcal{P}$ of policies, and returns the set $\text{EncodeP}(\mathcal{M}, \mathcal{P})$ of mapping assertions in which all policies in $\mathcal{P}$ have been encoded into the mapping assertions of $\mathcal{M}$, according to the technique described in [2, 6].[2]
- The function EncodeR takes as arguments a mapping assertion m and a role $r \in \mathcal{R}$ and returns a new mapping assertion $\text{EncodeR}(m, r)$ in which the source query of m is modified by conjoining the condition it its `WHERE` clause with an additional filtering condition of the form `ontop_contains_role`(r). We then extend EncodeR to a set $\mathcal{M}$ of mapping assertions by defining $\text{EncodeR}(\mathcal{M}, r) = \bigcup_{m \in \mathcal{M}} \{\text{EncodeR}(m, r)\}$.

We are now ready to describe how to construct, given an RBAC-PPVKG specification $\mathcal{H} = \langle \mathcal{E}, \mathcal{R}, \mathsf{RP}, \mathcal{U}, \mathsf{UR} \rangle$, where $\mathcal{E} = \langle \mathcal{T}, \mathcal{S}, \mathcal{M}, \mathcal{P} \rangle$, a VKG specification $\mathcal{V}_{\mathcal{H}} = \langle \mathcal{T}, \mathcal{S}, \mathcal{M}_{\mathcal{H}} \rangle$ with an enriched set $\mathcal{M}_{\mathcal{H}}$ of mapping assertions that encode both the appropriate role-specific policies according to $\mathcal{P}$ and RP, and the role-based activation conditions for each role in $\mathcal{R}$. Specifically, for a role $r \in \mathcal{R}$, we define $\text{EncodeRP}(\mathcal{M}, r) = \text{EncodeR}(\text{EncodeP}(\mathcal{M}, \mathsf{RP}(r)), r)$. Then, $\mathcal{M}_{\mathcal{H}} = \bigcup_{r \in \mathcal{R}} \text{EncodeRP}(\mathcal{M}, r)$. Algorithm 1 (EncodeRolePolicies) computes the function $\mathcal{M}_{\mathcal{H}}$ according to this definition.

Example 6 (Example 3 cont'd). Consider a healthcare domain with users $\mathcal{U} = \{\mathit{alice}, \mathit{bob}\}$ and roles $\mathcal{R} = \{\mathit{pharmacist}, \mathit{receptionist}\}$, and assume that $\mathsf{RP}(\mathit{pharmacist}) = \{p_1\}$ and $\mathsf{RP}(\mathit{receptionist}) = \{p_2\}$. Then, $\text{EncodeP}(\{m_1, m_2\}, \mathsf{RP}(\mathit{pharmacist})) = \{mp_1, m_2\}$ and $\text{EncodeP}(\{m_1, m_2\}, \mathsf{RP}(\mathit{receptionist})) = \{m_1, mp_2\}$. Moreover, by embedding also the role-based activation conditions, we obtain the following mapping assertions:

```
mappingId mrp1
target :Patient/{person_id} a :Patient ;
        :Patient.gender {gender}^^xsd:string .
source SELECT person_id, gender FROM Person
       WHERE location_id IS NULL AND
             ontop_contains_role ('pharmacist')

mappingId mr2
target :Prescription/{drug_exposure_id} a :Prescription ;
        :Prescription.subject :Patient/{person_id} .
```

[2] The function EncodeP was called EncodeMapping in [2].

Algorithm 1: ENCODEROLEPOLICIES

Input: A PPVKG specification $\langle \mathcal{T}, \mathcal{S}, \mathcal{M}, \mathcal{P} \rangle$, a finite set $\mathcal{R}$ of roles, a role-policy assignment RP

Output: A role-based permission assigned mapping $\mathcal{M}_{\mathcal{H}}$

$\mathcal{M}_{\mathcal{H}} \leftarrow \emptyset$;
for each role $r \in \mathcal{R}$ **do**
    $\mathcal{M}_0 \leftarrow \text{ENCODEP}(\mathcal{M}, \mathsf{RP}(r))$;
    $\mathcal{M}_r \leftarrow \emptyset$;
    for each mapping assertion $m \in \mathcal{M}_0$ **do**
        $\mathcal{M}_r \leftarrow \mathcal{M}_r \cup \{\text{ENCODER}(m, r)\}$;
    $\mathcal{M}_{\mathcal{H}} \leftarrow \mathcal{M}_{\mathcal{H}} \cup \mathcal{M}_r$;
return $\mathcal{M}_{\mathcal{H}}$;

```
source SELECT drug_exposure_id, person_id FROM DrugExposure
       WHERE ontop_contains_role ('pharmacist')

mappingId mr1
target :Patient/{person_id} a :Patient ;
         :Patient.gender {gender}^^xsd:string .
source SELECT person_id, gender FROM Person P
       WHERE ontop_contains_role ('receptionist')

mappingId mrp2
target :Prescription/{drug_exposure_id} a :Prescription ;
         :Prescription.subject :Patient/{person_id} .
source SELECT drug_exposure_id, person_id FROM DrugExposure
       WHERE person_id IS NULL AND
             ontop_contains_role ('receptionist')
```

Specifically, $\text{ENCODERP}(\{m_1, m_2\}, \textit{pharmacist}) = \{mrp_1, mr_2\}$ and $\text{ENCODERP}(\{m_1, m_2\}, \textit{receptionist}) = \{mr_1, mrp_2\}$. $\triangleleft$

These role-conditioned mappings ensure that only users with the appropriate role can access the corresponding data, thereby enforcing both confidentiality and RBAC-based access control. The approach of extending PPVKG with RBAC enables for the policy embedded mappings activation of particular mapping assertion based on the roles of the user querying the data . Hence, the permission now represents the activation or inactivation of such mappings in accordance of user's roles.

4.2 RBAC-PPVKG Query Answering Sessions

In an RBAC specification, a user may be assigned multiple roles by the user-role assignment UR. However, when a user interested in query answering logs into an RBAC-PPVKG system, they will do so by assuming a specific role, and such role should be compatible with the roles assigned to that user. We capture this through the following definition.

Definition 7 (RBAC-PPVKG session). Let $\mathcal{H} = \langle \mathcal{E}, \mathcal{R}, \mathsf{RP}, \mathcal{U}, \mathsf{UR} \rangle$ be an RBAC-PPVKG specification, $r \in \mathcal{R}$ a role, and $u \in \mathcal{U}$ a user. An *RBAC-PPVKG session* is a triple $\langle \mathcal{H}, r, u \rangle$ such that $(u, r) \in \mathsf{UR}$. $\triangleleft$

Intuitively, an RBAC-PPVKG session $\langle\mathcal{H}, r, u\rangle$ represents a user u logged into an RBAC-PPVKG system formalized by $\mathcal{H}$ with a role r that is one of the roles assigned to u in $\mathcal{H}$.

In its setup phase, an RBAC-PPVKG system with specification $\mathcal{H} = \langle\mathcal{E}, \mathcal{R}, \mathsf{RP}, \mathcal{U}, \mathsf{UR}\rangle$, where $\mathcal{E} = \langle\mathcal{T}, \mathcal{S}, \mathcal{M}, \mathcal{P}\rangle$, uses algorithm EncodeRoles to construct the mapping $\mathcal{M}_{\mathcal{H}}$, and exposes the VKG $\mathcal{V}_{\mathcal{H}} = \langle\mathcal{T}, \mathcal{S}, \mathcal{M}_{\mathcal{H}}\rangle$ as a SPARQL endpoint. Consider now a session $\langle\mathcal{H}, r, u\rangle$, where a user u, who is logged into $\mathcal{H}$ with role r, issues queries to the SPARQL endpoint. Then, a given query is processed by the system according to $\mathcal{V}_{\mathcal{H}}$. However, of the overall set $\mathcal{M}_{\mathcal{H}}$ of mapping assertions, the ones actually used to unfold the query are only those in $\mathcal{M}_r \subseteq \mathcal{M}_{\mathcal{H}}$, i.e., those policy-protected according to the policies relevant for role r and activated for r. This ensures that while processing the query, the role-specific RBAC permissions are enforced on top of the role-specific PPVKG confidentiality policies.

5 Use Case in the Healthcare Domain

We now describe the main elements of a use case in the healthcare domain on which we carried out our experimental evaluation. We detail the VKG-related aspects of the use case and discuss a categorization of relevant roles.

VKG Specification and Dataset. For our experimentation, we make use of the *Medical Information Mart for Intensive Care* (MIMIC)-III clinical dataset [13], which aggregates de-identified healthcare records spanning 46,000 unique patients and over 60,000 intensive-care unit admissions across two ICU systems at Boston's Beth Israel Deaconess Medical Center (2001–2012). To standardize this heterogeneous data for semantic processing, we employed the open-source MIMIC-OMOP ETL Tool[3] for conversion into the *Observational Medical Outcomes Partnership Common Data Model* (OMOP-CDM)[4]. This community-developed standard normalizes observational health data structures to facilitate reproducible evidence generation through systematic analysis [1]. To represent the information of OMOP-CDM at the semantic level, me make use of the *FHIR Model Ontology* specified in the OWL ontology language[5].

For relating the FHIR Ontology to OMOP-CDM, we adopted the R2RML-compliant mappings developed by Xiao et al. [22], which establish precise correspondences between OMOP-CDM schemas and *Fast Healthcare Interoperability Resources* (FHIR) RDF representations. These mappings, executable within the Ontop platform, maintain a predominantly direct alignment where OMOP elements map to singular FHIR components (e.g., `person_id` maps to `Resource.id`). We observe, however, that certain temporal constructs necessitate intermediate blank node generation and therefore have a more complex

[3] https://github.com/MIT-LCP/mimic-omop.
[4] https://ohdsi.github.io/CommonDataModel.
[5] http://build.fhir.org/fhir.ttl.

Table 1. Mapping of roles to accessible FHIR classes

RBAC roles	Role name	Accessible FHIR class
Physician/Nurse	r_1	all classes
Receptionist	r_2	*Encounter*, *Practitioner*
Researcher	r_3	all classes (since data are de-identified)
Data Administrator	r_4	*Encounter*
Pharmacist	r_5	*Prescription*, *Observation*
Patient	r_6	no class

structure, and that the mappings intentionally exclude redundant OMOP elements to optimize semantic representation, such as omitting duplicate temporal fields.

Role Definitions for the Hospital Domain (MIMIC-III Dataset). Using the MIMIC-III clinical care dataset, we define representative hospital roles and their access scopes for the RBAC framework (See Table 1):

- **Clinician (Physician/Nurse):** Full access to patient-level clinical data including demographics, lab results, medication orders, and clinical notes.
- **Receptionist:** Access limited to non-clinical patient demographics (e.g., contact information, appointment records) with exclusion of medical data.
- **Researcher:** Access to data of any type, provided they are de-identified.
- **Data Administrator:** Access to current medical procedures, and data for system maintenance, excluding patient-level clinical data.
- **Pharmacist:** Access to medication-focused data (e.g., medication order, dosage, and allergy) without direct patient identifiers.
- **Patient:** Access to their own records, but not to those of other patients.

5.1 Implementation

To implement the RBAC-PPVKG framework, we utilize the open-source implementation of PPVKG described in [2], based on Ontop. Our system operates over an RBAC-PPVKG specification $\mathcal{H} = \langle \mathcal{E}, \mathcal{R}, \mathsf{RP}, \mathcal{U}, \mathsf{UR} \rangle$, where $\mathcal{E} = \langle \mathcal{T}, \mathcal{S}, \mathcal{M}, \mathcal{P} \rangle$, and creates, for each role $r \in \mathcal{R}$, a separate copy $\textsc{EncodeP}(\mathcal{M}, \mathsf{RP}(r))$ of the mapping $\mathcal{M}$ in which all role-specific policies $\mathsf{RP}(r)$ are embedded.

To merge these into a unified, role-aware mapping $\mathcal{M}_{\mathcal{H}}$, we developed a Java utility class called `MappingCombinerRBAC`, which takes as input a file with multiple paths to a mapping file and an associated role. It then implements $\textsc{EncodeR}$ by parsing the mappings and injecting directly into the SQL `WHERE` clause of the source part of each mapping assertion conditions that make use of Ontop's built-in Boolean function `ontop_contains_role(`*`roleName`*`)`. When Ontop is configured with `ontop.authorization=true`, this function checks if the user role specified in the `x-roles` HTTP header equals *roleName*. As a result, such conditions encode RBAC that activates the mapping assertion according to the user role.

To execute queries against this RBAC-enforced mapping, we developed another Java class called `QueryExecutor`, which allows users to authenticate with their role and issue SPARQL queries. It issues an HTTP POST request to the Ontop endpoint using `HttpURLConnection`, setting headers such as `Content-Type`, `Accept`, and most importantly, `x-roles` to specify the user's role. One can also use a reverse proxy to let the user authenticate, inject these HTTP headers and send the modified HTTP request to the Ontop WebAPI.

The Ontop engine evaluates the query while applying the embedded access control logic from the mappings. Results are returned in SPARQL-JSON format. For our evaluation, we executed queries across different roles over multiple iterations to compute and report the average query execution time.

5.2 Experimental Evaluation

For our evaluation, we examine the impact of six roles and their associated policy sets on query execution, specifically analyzing how they affect the number of query results returned as well as the overall query evaluation time using the setup described above.

We employ the following seven denial policies (expressed over FHIR)[6]:

p_1: $\forall x.\forall y.\forall z.\mathit{Patient.gender}(x, y) \wedge \mathit{Patient.address}(x, z) \rightarrow \bot$
p_2: $\forall x.\forall y.\mathit{Prescription}(x) \wedge \mathit{Prescription.subject}(x, y) \rightarrow \bot$
p_3: $\forall x.\forall y.\forall z.\mathit{Patient.condition}(x, y) \wedge \mathit{Condition.code}(y, z) \rightarrow \bot$
p_4: $\forall x.\forall y.\forall z.\mathit{Practitioner.qualification}(x, y) \wedge \mathit{Practitioner.qualification.code}(y, z) \rightarrow \bot$
p_5: $\forall x.\forall y.\forall z.\mathit{Procedure.code}(x, y) \wedge \mathit{Procedure.performedDateTime}(y, z) \rightarrow \bot$
p_6: $\forall x.\forall y.\forall z.\mathit{Observation.valueQuantity}(x, y) \wedge \mathit{Quantity.value}(y, z) \rightarrow \bot$
p_7: $\forall x.\forall y.\forall z.\mathit{Encounter.location}(x, y) \wedge \mathit{Location.name}(y, z) \rightarrow \bot$

They are assigned to roles r_1–r_6 as specified below, coherently with Table 1:

r_1: All data are accessible. Hence, no policy applies.
r_2: p_2, p_3, p_5, and p_6.
r_3: All data are de-identified. Hence, no policy applies.
r_4: p_1, p_2, p_3, p_4, p_5 and p_6.
r_5: p_1, p_3, p_4, p_5 and p_7.
r_6: p_1, p_2, p_3, p_4, p_5, p_6, and p_7.

Notice that, since patients can access only information about themselves, and in RBAC-PPVKG policies are enforced based on roles (and not individual users), in order to avoid that a user accesses the data of other patients, we also have to forbid that they access their own data. Hence, for role Patient (r_6) we have to activate all policies. This represents a limitation of utility (in line with the tradeoff between utility and privacy).

[6] Additional policies are available in our Git repository.

Table 2. Impact of privacy policies with roles on execution times (in ms) and number of query results

Pol.	q_1		q_2		q_3		q_4		q_5		q_6	
	time	#res.	time	#res.	time	#res.	time	#res.	time	#res.	time	#res.
None	815	4730	1655	34	15282	4547	1042	944	18136	5155	1253	93
r_1	820	4730	1668	34	15376	4547	1186	944	18198	5155	1498	93
r_2	823	4730	490	0	1709	0	298	0	18193	5155	1499	93
r_3	828	4730	1664	34	15372	4547	1184	944	18194	5155	1496	93
r_4	212	0	491	0	1702	0	285	0	1191	0	1493	93
r_5	227	0	496	0	15371	4547	1181	944	1189	0	158	0
r_6	223	0	493	0	1692	0	286	0	1159	0	154	0

We have chosen six representative SPARQL queries, some of which are adapted from [22], covering a variety of clinical use cases. For each query, we have specified the roles for which it is answered, according to the policies:

q_1: Male patients with inpatient admissions lasting $\geq$ 5 days. – r_1, r_2, and r_3
q_2: Patients who delivered a baby. – r_1 and r_3
q_3: Patients taking Trazodone. – r_1, r_3 and r_5
q_4: Patients with an HbA1c $\geq$ 10%. – r_1, r_3 and r_5
q_5: Specialty of a Practitioner. – r_1, r_2 and r_3
q_6: Patient's treatment location name. – r_1, r_2, r_3 and r_4

Our experiments demonstrate that data can be effectively hidden based on user roles, validating the correctness of role-based access control within the PPVKG framework (Table 2). Furthermore, the results show that enforcing role-based privacy policies introduces only a modest overhead in query execution time, thereby confirming the practical feasibility of such mechanisms for privacy-preserving query answering. As expected, execution times are notably reduced in cases where the applied policies yield empty query results, due to early pruning of the result set during query evaluation.

6 Conclusions

In this paper, we presented an extension of the PPVKG framework to support Role-Based Access Control (RBAC), enabling privacy-preserving query answering that adapts to the access privileges of different user roles. Our approach enforces the principle of least privilege by injecting role-specific policy constraints directly into the VKG mappings, ensuring that users can only access ontology classes and properties permitted by their assigned roles. Our experiments confirm that this mechanism effectively restricts access according to roles, while introducing only minimal query performance overhead. These findings validate the practical applicability of role-based privacy enforcement in real-world knowledge graph settings. For future work, we aim to extend the RBAC-PPVKG framework by incorporating *Hierarchical Role-Based Access Control*, enabling structured role inheritance and more flexible policy management.

Acknowledgments. This research has been partially supported by the Wallenberg AI, Autonomous Systems and Software Program (WASP), funded by the Knut and Alice Wallenberg Foundation, by the HEU project CyclOps (GA n. 101135513), by the Province of Bolzano and FWF through project OnTeGra (DOI 10.55776/PIN8884924). by the Province of Bolzano and EU through projects ERDF-FESR 1078 CRIMA, and ERDF-FESR 1047 AI-Lab, by MUR through the PRIN project 2022XERWK9 S-PIC4CHU, and by the EU and MUR through the PNRR project PE0000013-FAIR.

References

1. Baura, D., Calvanese, D.: Assessing privacy requirements for controlled query evaluation in OBDA. In: Proceedings of the 22nd International Conference on Modeling Decisions for Artificial Intelligence (MDAI). Lecture Notes in Computer Science. Springer, Cham (2025) Decisions for Artificial Intelligence (MDAI). Lecture Notes in Computer Science, Springer (2025)
2. Baura, D., Calvanese, D., Marconi, L.: Implementing controlled query evaluation in OBDA. In: Proceedings of the Joint Ontology Workshops Episode 10: The Tukker Zomer of Ontology (JOWO). CEUR-WS.org, vol. 3882. CEUR Workshop Proceedings (2024). https://ceur-ws.org/Vol-3882/st4dm-1.pdf
3. Bonatti, P.A., Sauro, L.: A confidentiality model for ontologies. In: Alani, H., et al. (eds.) ISWC 2013. LNCS, vol. 8218, pp. 17–32. Springer, Heidelberg (2013). https://doi.org/10.1007/978-3-642-41335-3_2
4. Calvanese, D., et al.: Ontop: answering SPARQL queries over relational databases. Semantic Web J. **8**(3), 471–487 (2017). https://doi.org/10.3233/SW-160217
5. Calvanese, D., De Giacomo, G., Lembo, D., Lenzerini, M., Rosati, R.: Tractable reasoning and efficient query answering in description logics: the DL-Lite family. J. Autom. Reason. **39**(3), 385–429 (2007). https://doi.org/10.1007/s10817-007-9078-x
6. Cima, G., Lembo, D., Marconi, L., Rosati, R., Savo, D.F.: Controlled query evaluation in ontology-based data access. In: Pan, J.Z., et al. (eds.) ISWC 2020. LNCS, vol. 12506, pp. 128–146. Springer, Cham (2020). https://doi.org/10.1007/978-3-030-62419-4_8
7. Das, S., Sundara, S., Cyganiak, R.: R2RML: RDB to RDF mapping language. W3C Recommendation, World Wide Web Consortium (2012). http://www.w3.org/TR/r2rml/
8. European Parliament and Council of the European Union: General data protection regulation (GDPR). regulation (EU) 2016/679 of the European Parliament and of the Council (2016), articles 5(1)(c) and 32(1)(b). https://eur-lex.europa.eu/eli/reg/2016/679/oj
9. Ferrari, E.: Role-based access control. In: Access Control in Data Management Systems, pp. 61–75. Springer (1992)
10. Glimm, B., Ogbuji, C.: SPARQL 1.1 entailment regimes. W3C Recommendation, World Wide Web Consortium (2013). http://www.w3.org/TR/sparql11-entailment/

11. Grau, B.C., Kharlamov, E., Kostylev, E.V., Zheleznyakov, D.: Controlled query evaluation for Datalog and OWL 2 Profile ontologies. In: Proceedings of the 24th International Joint Conference on Artificial Intelligence (IJCAI), pp. 2883–2889. AAAI Press (2015). https://ijcai.org/Abstract/15/408
12. Harris, S., Seaborne, A.: SPARQL 1.1 query language. W3C Recommendation, World Wide Web Consortium (2013). http://www.w3.org/TR/sparql11-query
13. Johnson, A.E., et al.: MIMIC-III, a freely accessible critical care database. Sci. Data **3**(1), 1–9 (2016)
14. Lembo, D., Rosati, R., Savo, D.F.: Revisiting controlled query evaluation in description logics. In: Proceedings of the 28th International Joint Conference on Artificial Intelligence (IJCAI), pp. 1786–1792. ijcai.org (2019). https://doi.org/10.24963/IJCAI.2019/247
15. Motik, B., Cuenca Grau, B., Horrocks, I., Wu, Z., Fokoue, A., Lutz, C.: OWL 2 Web Ontology Language profiles (second edition). W3C Recommendation, World Wide Web Consortium (2012). http://www.w3.org/TR/owl2-profiles/
16. Poggi, A., Lembo, D., Calvanese, D., De Giacomo, G., Lenzerini, M., Rosati, R.: Linking data to ontologies. J. Data Semant. **10**, 133–173 (2008). https://doi.org/10.1007/978-3-540-77688-8_5
17. Sandhu, R., Ferraiolo, D., Kuhn, R.: The NIST model for role-based access control: towards a unified standard. Technical report, NIST (2000), also known as ANSI/INCITS 359-2004
18. Sandhu, R.S.: Role-based access control. In: Advances in Computers, vol. 46, pp. 237–286. Elsevier (1998)
19. Voigt, P., von dem Bussche, A.: The EU General Data Protection Regulation (GDPR) – A Practical Guide. Springer (2017)
20. Xiao, G., et al.: Ontology-based data access: a survey. In: Proceedings of the 27th International Joint Conference on Artificial Intelligence (IJCAI), pp. 5511–5519. IJCAI Org (2018). https://doi.org/10.24963/ijcai.2018/777
21. Xiao, G., et al.: The virtual knowledge graph system ontop. In: Pan, J.Z. (ed.) ISWC 2020. LNCS, vol. 12507, pp. 259–277. Springer, Cham (2020). https://doi.org/10.1007/978-3-030-62466-8_17
22. Xiao, G., et al.: FHIR-Ontop-OMOP: building clinical knowledge graphs in FHIR RDF with the OMOP Common Data Model. J. Biomed. Inform. **134**, 104201 (2022). https://doi.org/10.1016/J.JBI.2022.104201

Realizing Bidirectional Virtual Knowledge Graphs Using Ontop

Romuald Esdras Wandji[1(✉)] and Diego Calvanese[2]

[1] Department of Computing Science, Umeå Universitet, Umeå, Sweden
romuald.esdras.wandji@umu.se
[2] Faculty of Engineering, Free University of Bozen-Bolzano, Bolzano, Italy
diego.calvanese@unibz.it

Abstract. The Virtual Knowledge Graph (VKG) paradigm enables querying heterogeneous relational databases through a unified semantic layer comprising an ontology and declarative mappings (typically in R2RML). While querying is well-supported, propagating updates from the virtual RDF graph (ABox) back to the source databases remains a challenge. This paper addresses the problem of translating updates (expressed in SPARQL Update) applied over the ABox into equivalent SQL updates over the underlying databases, specifically within the Ontop VKG system. Key difficulties arise from the non-injectivity inherent in R2RML mappings, where a single ABox update can correspond to multiple source update possibilities, and the potential for these source updates to cause unintended side effects—additional insertions or deletions in the ABox beyond the user's original intent. While relying on *Ontop*'s query rewriting engine, our method employs lineage computation to identify source tuples for deletion and a strategy for handling existential variables during insertion. It generates candidate SQL translations, analyzes their potential side effects on the virtual ABox, and selects the ones that minimize these unintended consequences. This work represents a step toward closing the Linked Data Life Cycle loop, allowing changes in the knowledge graph to be reflected in the corresponding source.

Keywords: Knowledge Representation · Virtual Knowledge Graph (VKG) · Ontology-based Data Access · View Updates

1 Introduction

In the evolving landscape of data integration, Virtual Knowledge Graphs (VKGs), previously known as Ontology-Based Data Access (OBDA), have emerged as a powerful paradigm for querying relational data sources through an ontological lens. VKGs enable users to pose high-level semantic queries using standards like SPARQL over an ontology, which are automatically rewritten and unfolded into executable low-level queries (e.g., SQL) via declarative schema mappings, typically expressed in languages like R2RML [10]. These mappings link a lightweight ontology, often specified in a description logic such as DL-Lite [5], to the underlying data source, facilitating efficient query answering through query rewriting and unfolding [38]. Current approaches, such as

H. Takeda et al. (Eds.): IJCKG 2025, LNAI 16297, pp. 220–235, 2026.
https://doi.org/10.1007/978-981-95-5009-8_15

Ontop [4], Morph [29], SparqlMap [32], and Mastro [6] primarily focus on exposing structured data as virtual RDF for integration into Linked Data workflows without the need for costly data extraction, transformation, and loading (ETL) processes. However, these systems lack mechanisms for bidirectional integration, where updates applied to the virtual layer are propagated back to the source.

The task of updating VKGs, particularly translating *instance-level* (ABox) updates into equivalent updates over relational data sources, is a relatively underexplored area that intersects OBDA [36], view updates in databases [1], and SPARQL-to-SQL translation [31]. Updating the extensional data (ABox) of a knowledge base (KB) while preserving the ontology is a well-studied problem in description logics (DLs) and knowledge representation. The complexity arises from logical inference in the ontology, which can lead to inconsistencies and therefore to multiple update outcomes, aiming at (minimally) removing the inconsistencies [16,22]. Existing semantics for ABox updates are broadly classified into formula-based and model-based approaches. Formula-based semantics represent updates as logical formulas, achieving consistency by minimally removing conflicting ABox facts [7,14]. For instance, [7] proposed a formula-based semantics for DL-Lite, leveraging its first-order rewritability to ensure a unique repaired ABox after an update. Model-based semantics, conversely, update all models of the KB [13,35]. The ComputeUpdate algorithm proposed in [13] extends [35] to compute ABox updates for DL-Lite$_{\mathcal{F}}$ KBs. More expressive DLs, however, face expressibility challenges for updates [26].

Translating KB updates into source data updates is closely related to the view update problem in relational databases, where updates to a view must be reflected in the base tables [1,11,12]. The ambiguity introduced by mappings, where multiple source updates may correspond to a single view update, poses significant challenges [24,25]. SPARQL-to-SQL translation is a cornerstone of OBDA systems, enabling relational data to be queried as virtual RDF. State-of-the-art systems like *Ontop* [4], Morph-RDB [29], and SparqlMap [32] use R2RML mappings [10] to rewrite SPARQL queries into SQL. However, these systems primarily focus on read-only access, with no support for updates. SPARQL Update [19] provides a standardized mechanism for modifying RDF data, but translating such updates into SQL DML statements remains challenging due to schema constraints and mapping ambiguity.

D2RQ/update [15] introduced a schema-aware SPARQL Update processor that optimizes SQL statement generation and prevents constraint violations by grouping related triples. D2RQ++ [30] proposed integrating a native triple store to handle unmapped triples or those violating constraints, later propagating them to the database when feasible. OntoAccess [21] developed a custom mapping language (R3M) for bidirectional mappings, addressing insert and delete operations considering schema constraints. Additionally, [18] systematically explored SPARQL-to-SQL translations for both read and write scenarios, building on [8]. However, this work is conceptual and lacks detailed handling of complex R2RML templates or empirical evaluation.

The literature reveals several gaps: existing KB update semantics do not account for source data translation via mappings, which limit their application to VKG, *(1)* view update solutions are not tailored to the complex, ontology-driven mappings in VKGs, and *(2)* SPARQL-to-SQL update mechanisms lack robust strategies for minimizing side effects. In this paper, we address these gaps by proposing a comprehensive framework for *bidirectional VKGs* within *Ontop* that translates ABox updates into source updates using the mappings, allowing combined insertion and deletion operations to achieve minimal or no side effects. Thus we improve on our prior work [33], which restricted the translation of ABox insertions and deletions to corresponding source updates of the same type. Building upon the *Ontop* system, we propose an implementation of the translation algorithms proposed in [33] that prioritizes translations with minimal side effects, with R2RML mappings to ensure precise propagation of updates.

The rest of the paper is organized as follows. In Sect. 2 we provide some technical preliminaries on VKGs, the *Ontop* system, and updates in VKGs. In Sect. 3 we discuss how maximum recoveries are computed in *Ontop*, and in Sect. 4 how *Ontop* performs the translation of updates. In Sect. 5 we present some experimental results, and in Sect. 6 we conclude the paper.

2 Preliminaries

Formerly referred to as Ontology-Based Data Access (OBDA), the VKG paradigm is designed to facilitate access to heterogeneous data sources, in particular relational databases, through a high-level conceptual schema called *ontology* [4]. Formally, a VKG *specification* is a triple $\mathcal{P} = \langle \mathcal{T}, \mathcal{M}, \mathcal{S} \rangle$ consisting of *(i)* an *ontology*, which provides to the user a convenient and familiar vocabulary (of classes and properties), and extends the data with background knowledge, *(ii)* a relational *source schema* $\mathcal{S}$ for the underlying repositories, and *(iii)* a *schema mapping* $\mathcal{M}$, which describes through queries over the signature of $\mathcal{S}$ how to populate the classes and properties in $\mathcal{T}$. A VKG instance is given as a pair $\langle \mathcal{P}, D \rangle$ where $\mathcal{P}$ is a VKG specification and D is a source instance compliant with $\mathcal{S}$. While component *(ii)* is specified in standard SQL, W3C published recommended languages for specifying components *(i)* and *(iii)*, namely OWL 2 QL [27] and R2RML [10], respectively.

In practice, each assertion in the provided mapping $\mathcal{M}$ of a traditional VKG has one of the forms

$$\begin{array}{lll} \texttt{sql}(\boldsymbol{y}) & \rightsquigarrow & f(\boldsymbol{x})\ \texttt{:rdfType C}, \\ \texttt{sql}(\boldsymbol{y}) & \rightsquigarrow & f_1(\boldsymbol{x}_1)\ \texttt{:P}\ f_2(\boldsymbol{x}_2), \end{array}$$

where $\boldsymbol{x} \subseteq \boldsymbol{y}$, $\boldsymbol{x}_1 \subseteq \boldsymbol{y}$, $\boldsymbol{x}_2 \subseteq \boldsymbol{y}$, $\texttt{sql}$ is an SQL query over the signature of $\mathcal{S}$ projecting the columns $\boldsymbol{y}$, and $f(\boldsymbol{x})$, $f_1(\boldsymbol{x}_1)$, and $f_2(\boldsymbol{x}_2)$ are *IRI-templates*[1], obtained by embedding the answer variables of the source query (enclosed in curly brackets) into a fixed string. For instance, if we have the IRI-template `:univ/student/{id}`, where `id` is a placeholder for a student ID, then we can obtain an IRI `:univ/student/7` by instantiating the variable `id` with the value "7". In this paper, we make the common assumption that IRIs in VKGs are generated in a unique way by IRI-templates, i.e., IRI-templates are *injective*. Moreover we assume that for each IRI-template $f(x_1, \ldots, x_n)$, we have also n *inverse templates* $f^1, \ldots, f^n$ such that $f^i(f(v_1, \ldots, v_n)) = v_i$, for each $i \in [1..n]$ and for all possible values $v_1, \ldots, v_n$ instantiating the variables $x_1, \ldots, x_n$. When a mapping $\mathcal{M}$ is applied over a source instance D, we obtain a set $\mathcal{M}(D)$ of (*virtual*) RDF triples that represent a knowledge graph (also called ABox).

[1] IRI-templates correspond to the R2RML *string templates*.

In VKGs, queries are expressed over the terms in the ontology $\mathcal{T}$ and are formulated in SPARQL [27]. They are evaluated over the knowledge base $\langle \mathcal{T}, \mathcal{M}(D) \rangle$. Notice that the knowledge graph is not materialized, but is kept virtual, which ensures that the VKG system always exposes the actual up-to-date information.

2.1 The Ontop System

The *Ontop* framework is an open-source[2] VKG system designed to expose relational databases as virtual RDF graphs, allowing SPARQL queries to be answered over a semantic layer defined by an ontology [4]. Its virtual approach avoids materializing RDF triples by rewriting SPARQL queries into optimized SQL queries, which are then executed by the underlying database engine.

Ontop supports all W3C standards relevant to VKG, including SPARQL 1.0 (with partial SPARQL 1.1 support), OWL 2 QL [28], RDFS [3] and is compatible with major relational databases (e.g., PostgreSQL, MySQL, Oracle) and data federation systems, such as Teiid[3], Dremio[4], and Denodo[5], enabling data integration across heterogeneous sources. A core feature of *Ontop* is its query rewriting mechanism, which translates SPARQL queries over the ontology into SQL queries executable on the underlying database. Since version 3, *Ontop* has adopted an algebra-based data structure called Intermediate Query (IQ) to facilitate this translation. While Ontop focuses on query answering, understanding its query rewriting process is crucial for exploring update translation in VKGs, where similar techniques may be adapted to propagate ontology-level updates to the database.

IQ is a variant of relational algebra designed to represent both SPARQL queries and SQL queries from the mapping, bridging the semantic gap between the two [37]. Based on that notion, when the user poses a SPARQL query over a VKG instance, *Ontop* proceeds as follows: *(i)* the query is translated into IQ; *(ii)* the translated IQ is expanded taking into account the ontology (e.g., subclass and subproperty axioms, and domain and range assertions); *(iii)* the expanded IQ is *unfolded* using the mapping; and *(iv)* after a series of optimizations carried out by *Ontop*, the optimized IQ is converted into an SQL query for the underlying data source.

Example 1. As an example, assume that our mapping $\mathcal{M}$ has the following assertions:

```
T_1(x,y) ⇝ :b/{x} :p y ,
T_2(x,y) ⇝ :b/{x} :p y .
```

where `T_1` and `T_2` are source relations with `x` as primary key of type `TEXT`, and `y` of type `INTEGER`. Each assertion in $\mathcal{M}$ translates the tuples in the source relation into RDF triples with IRI subjects `:b/{x}`, predicate `:p` and object `y`. Assume that the user poses the following SPARQL query:

```
SELECT ?x ?y WHERE { ?x :p ?y. }
```

Then, in Step *(i)*, the query is translated into the following IQ:

[2] https://github.com/ontop/ontop.

[3] https://teiid.io/.

[4] https://www.dremio.com/.

[5] https://www.denodo.com/.

```
ans(x,y) INTENSIONAL triple(x,:p,y)
```

For simplicity, we assume that the ontology does not contain any axiom, hence Step *(ii)* leaves the IQ unchanged. In Step *(iii)*, the triple atom obtained in the previous step is replaced with the mapping assertions for `:p`, which in this case involve `T_1` and `T_2`. The unfolded IQ is:

```
ans(x,y)
CONSTRUCT [x,y] [x/RDF(:b{}(z),IRI), y/RDF(i2t(z1), xsd:integer)]
    UNION [z,z1]
        EXTENSIONAL T_1(z,z1)
        EXTENSIONAL T_2(z,z1)
```

In Step *(iv)*, the final one, the IQ is translated into SQL, tailored to the DBMS dialect. For example, in PostgreSQL, it might become:

```
SELECT z AS x, CAST(z1 AS TEXT) AS y FROM T_1
UNION ALL
SELECT z AS x, CAST(z2 AS TEXT) AS y FROM T_2
```

◁

In the above example, we observe that using IQ facilitates handling complex SPARQL queries and aligns with SQL semantics, which is a potential foundation for update translation in VKGs. Extending this framework to support ontology-to-database updates involves defining update operations in IQ and ensuring consistency, which is what we are investigating in this paper.

2.2 Updates in VKGs and Instance Recoveries

In this paper, we focus on two types of instance-level updates in VKGs: either a set of *deletions* or a set of *insertions* over the *retrieved* ABox $\mathcal{M}(D)$. We are interested in translating such ABox update requests into suitable source update requests. For our techniques to be applicable, We need to make the simplifying assumption that the source of each mapping assertion in $\mathcal{M}$ is a union of conjunctive queries (UCQs). Considering more general forms of mappings is left for future work.

We now present the notion of update translation in VKGs as introduced in [33]. We start with the definition of translation for ABox deletions $\mathcal{U}_A^-$ in a VKG instance $\mathcal{J}$. The goal is to find a set of source deletions $\mathcal{U}_D^-$ that can be applied over the database D to generate the desired (*virtual*) ABox through the mapping $\mathcal{M}$. We consider here *direct translations*, which are translations that are of the same nature as the original ABox update, i.e., ABox deletions translate to source deletions, and similarly, ABox insertions translate to source insertions.[6]

Definition 1 (Direct deletion translation). Let $\mathcal{U}_A^-$ be a set of ABox deletions. A source deletion $\mathcal{U}_D^-$ is a *direct translation of* $\mathcal{U}_A^-$ *in* $\mathcal{J}$ if $\mathcal{U}_A^- \cap \mathcal{M}(D \setminus \mathcal{U}_D^-) = \emptyset$. We say that $\mathcal{U}_D^-$ is an *exact direct translation of* $\mathcal{U}_A^-$ *in* $\mathcal{J}$ if $\mathcal{M}(D \setminus \mathcal{U}_D^-) = \mathcal{M}(D) \setminus \mathcal{U}_A^-$. ◁

Note that in our definition, we use the term *exact translation* for translations whose execution over the data source would create a (*virtual*) ABox that perfectly reflects the

[6] More general forms of translations, where ABox deletions (or insertions) are translated using both deletions and insertions are considered in [34].

requested deletion. However, translations that are not exact will lead to extra deletions in the ABox, and these unintendedly deleted ABox assertions are referred to as *side effects*. We provide a similar definition of translation for ABox insertions.[7]

Definition 2 (Direct insertion translation). Let $\mathcal{U}_A^+$ be a set of ABox insertions. A source insertion $\mathcal{U}_D^+$ is a *direct translation of* $\mathcal{U}_A^+$ *in* $\mathcal{J}$ if $\mathcal{U}_A^+ \subseteq \mathcal{M}(\mathcal{U}_D^+)$ and $\langle \mathcal{T}^-, \mathcal{M}(D \cup \mathcal{U}_D^+)\rangle$ is consistent. $\mathcal{U}_D^+$ is an *exact direct translation of* $\mathcal{U}_A^+$ *in* $\mathcal{J}$ if $\mathcal{M}(D \cup \mathcal{U}_D^+) = \mathcal{M}(D) \cup \mathcal{U}_A^+$. ◁

We also require that translations comply with the minimal change principle, i.e., the state of the new VKG system should remain as close as possible to the previous one [16, 17].

Definition 3 (Minimal direct translation). Let $\mathcal{U}_A^-$ be a set of ABox deletions. A direct translation $\mathcal{U}_D^-$ of $\mathcal{U}_A^-$ in $\mathcal{J}$ is *minimal* if every proper subset of $\mathcal{U}_D^-$ is not a translation of $\mathcal{U}_A^-$ in $\mathcal{J}$. Similarly, let $\mathcal{U}_A^+$ be a set of ABox insertions. A direct translation $\mathcal{U}_D^+$ of $\mathcal{U}_A^+$ in $\mathcal{J}$ is *minimal* if every proper subset of $\mathcal{U}_D^+$ is not a translation of $\mathcal{U}_A^+$ in $\mathcal{J}$. ◁

The main challenge related to translating instance-level updates in VKGs (which is also strongly related to the *view update problem*), is to compute the set of source tuples from which a given set of ABox assertions (to be deleted or inserted) can be generated through the provided schema mapping. In other words, given an ABox deletion $\mathcal{U}_A^-$, how can we compute its *lineage* in the source? Similarly, given an ABox insertion $\mathcal{U}_A^+$, how can we compute the minimal sets of source tuples that generate $\mathcal{U}_A^+$? Given the *non-injective* nature of VKG mappings, there is no unique way to map ABox assertions to their corresponding tuples in the source. To address this, [33] has proposed to compute a reverse mapping $\widehat{\mathcal{M}}$ that approximates the inverse of the original VKG mapping $\mathcal{M}$. This reverse mapping, called the *MR-os-mapping*[8] of $\mathcal{M}$, is constructed by associating each target atom in $\mathcal{T}$ with its rewriting over $\mathcal{S}$ via $\mathcal{M}$. Conceptually, applying $\widehat{\mathcal{M}}$ to an ABox assertion f computes its logical rewriting over the source, thereby capturing all possible source tuples that could produce f.

Example 2. Assume that our mapping $\mathcal{M}$ contains the assertion `T1(x,y,z)` $\rightsquigarrow$ `:b/{x} :p y`. Then, when the MR-os-mapping $\widehat{\mathcal{M}}$ of $\mathcal{M}$ is applied to the ABox assertion $f =$ `(:b/{a} :p c)`, it generates $\widehat{\mathcal{M}}(\{f\}) = \exists z.\texttt{T1}(\texttt{a},\texttt{c},z)$. ◁

Our goal is to compute the source tuples from which a set of ABox assertions can be derived. To achieve this, we leverage the MR-os-mapping recovery of each assertion, producing a logical condition over $\mathcal{S}$ that describes all generating source sets. We formalize this with the notion of *instance recovery*, defined as follows, where we assume that databases are expressed over a predefined domain Δ of values, and we use $\mathcal{S}(\Delta)$ to express the set of all facts obtained by applying the predicates in $\mathcal{S}$ to the values in Δ.

[7] Notice that we adopt here the definition of direct translation of ABox insertions proposed in [34], where the translation contains *all* source tuples necessary to generate the inserted ABox facts, including tuples already present in the data source, as opposed to the definition of [33], where the translation contains only the new tuples to be added to the data source to accomplish the ABox insertion.

[8] MR-os-mapping stands for *maximum recovery* from the ontology to the source schema.

Definition 4 **($\widehat{\mathcal{M}}$-instance recovery** [33]**).** Let $\mathcal{P} = \langle \mathcal{T}, \mathcal{M}, \mathcal{S} \rangle$ be a VKG instance and $\widehat{\mathcal{M}}$ the MR-os-mapping of $\mathcal{M}$. The *$\widehat{\mathcal{M}}$-instance recovery* is the function $\widehat{\mathcal{M}}_{\mathcal{S}}$ from ABox facts over $\mathcal{T}$ to $2^{2^{\mathcal{S}(\Delta)}}$ such that

$$\widehat{\mathcal{M}}_{\mathcal{S}}(f) = \{B \subseteq \mathcal{S}(\Delta) \mid B \text{ is true in } \widehat{\mathcal{M}}(\{f\}), \text{ and } \forall B' \subsetneq B, B' \text{ is not true in } \widehat{\mathcal{M}}(\{f\})\}.$$

◁

Notice that the above definition of instance recovery differs from the one provided in [33], which relied on a specific source instance D. Here instead, we abstract from such a source instance and consider instead the minimal source instances B contained in the (infinite) set $\mathcal{S}(\Delta)$ of facts from which an ABox fact f can be generated through $\mathcal{M}$. Notice that, while each B in $\widehat{\mathcal{M}}_{\mathcal{S}}(f)$ is finite (since minimal), in general $\widehat{\mathcal{M}}_{\mathcal{S}}(f)$ will contain an infinite number of such source instances. For ABox deletions, we will restrict the set of such source instances to those contained in the given source instance D (for $\mathcal{S}$). Instead, for ABox insertions we will have to find suitable representatives in $\widehat{\mathcal{M}}_{\mathcal{S}}(f)$ that lead to minimal insertion side-effects through $\mathcal{M}$.

3 Maximum Recovery in *Ontop*

As we mentioned earlier, *Ontop* excels at exposing relational databases as virtual RDF graphs through the ontologies and the mappings, which allows SPARQL queries to be answered without materializing the triples. *Ontop*'s query answering process relies on rewriting SPARQL queries into SQL queries using mappings and ontology inference [4]. However, translating ABox assertions (instance-level facts) into database tuples for updates is not supported in *Ontop*. In this section, we show how we can exploit the rewriting capabilities of *Ontop* to efficiently compute the rewriting of ABox assertions, which then allows us to compute source updates for ABox updates.

In *Ontop*, given a VKG specification $\langle \mathcal{T}, \mathcal{M}, \mathcal{S} \rangle$, a SPARQL query $\mathcal{Q}_{\mathcal{T}}$ over $\mathcal{T}$ is translated into an SQL query $\mathcal{Q}_{\mathcal{S}}$ over $\mathcal{S}$ by using Intermediate Query (IQ), which is a language combining basic algebraic operations like `UNION`, `JOIN`, `PROJECT`, and `SLICE`. In our proposed method, we leverage that translation by deriving from the target triple (e.g., f = `(:b/{a} :p b)`) a ground `ASK` query (e.g., `ASK { ` f ` .}`), and *Ontop*'s query engine then translates this SPARQL query into an IQ. For `ASK` queries, the unfolded IQ typically includes a `SLICE` (or `LIMIT` 1) construct which checks for the existence of any set of source tuples satisfying the IQ. By programmatically removing this `SLICE` construct from the unfolded IQ, the resulting IQ transforms the existential check (with the `SLICE` constructor) into a comprehensive retrieval mechanism.

Example 3 (Example 1 *cont'd).* If we have the RDF triple f = `(:b/{a} :p b)`, from which we form the `ASK` query `ASK { ` f ` . }`, the unfolded IQ will be:

```
ans1() SLICE limit=1
UNION []
    SLICE limit=1 EXTENSIONAL T_1(0:a^^TEXT,1:b^^INTEGER)
    SLICE limit=1 EXTENSIONAL T_2(0:a^^TEXT,1:b^^INTEGER)
```

◁

Another challenge in computing the MR-os-mapping is matching the assertions' IRIs to mapping assertions, which requires handling IRI templates and their inverses. For an ABox assertion f = `(:b/{a} :p b)`, *Ontop* identifies mappings whose target triples unify with f. This involves: *(i)* matching `p` to the mapping's predicate, *(ii)* applying the inverse templates to `:b/{a}` and `b` to extract database values, and *(iii)* using the extracted values to construct the unfolded IQ. In the next sections, we will show how the notion of MR-os-mapping can be used in the translation of updates within the *Ontop* framework.

4 Update Translation in *Ontop*

SPARQL [20] is known as the standard query language for RDF graphs, however, it is limited to read-only access to a knowledge graph. In order to overcome the limitation, W3C introduced SPARQL Update [19] as a standardized language for modifying RDF graphs, supporting operations such as insertions and deletions of RDF triples. Insertions are specified using the `INSERT DATA` construct, which adds new triples to the graph. For example, `INSERT DATA { :student/1 :hasName"Alice" }` adds a triple stating that a student has a specific name. Similarly, deletions are specified using `DELETE DATA` to remove specific triples from the graph. In VKGs, these update operations must be translated into equivalent SQL update statements that modify the underlying database through mappings. In this section, we first provide a running example that we use throughout the rest of the paper to illustrate the update problems in VKGs and our solution in *Ontop*.

4.1 Running Example

We present a running example involving a university database with two relational tables: `Student(id,name,faculty)` and `Faculty(id,course)`. The `Student` table stores student records with a unique identifier (`id`), name (`name`), and their enrolled faculty (`faculty`). The `Faculty` table specifies the mandatory courses (`course`, e.g., "Math") for each faculty. Based on that source schema, we define the following mappings:

- Mapping m_1 maps `Student` records to the `:hasName` data property:

```
SELECT id, name FROM Student
  ⇝ :uni/student/{id} :hasName {name}.
```

- Mapping m_2 maps `Student` and `Faculty` records to the `:isTaking` data property:

```
SELECT s.id AS id, f.course AS course
FROM Student s JOIN Faculty f ON s.faculty = f.id
  ⇝ :uni/student/{id} :isTaking {course}^^xsd:string .
```

The above mapping $\mathcal{M} = \{m_1, m_2\}$ is specified in *Ontop*'s native mapping language[9], where each mapping assertion consist of *(i)* a source, which is an SQL query,

[9] See https://github.com/ontop/ontop/wiki/ontopOBDAModel for a full description of *Ontop*'s mapping language.

and *(ii)* a target, which is an RDF triple pattern with placeholders (enclosed in '{' and '}') for the values of answer variables from the source query. When applied over a source instance, these mappings create a virtual RDF graph where students are linked to their names and the courses they are taking. Consider the database D shown below on the left-hand side:

Student		
id	*name*	*faculty*
s1	john	f1
s1	john	f2
s2	paul	f2

Faculty	
id	*course*
f1	ethics
f2	ethics
f1	law

$\mathcal{M}(D)$:

```
:uni/student/s1 :hasName "john"
:uni/student/s2 :hasName "paul"
:uni/student/s1 :isTaking "ethics"
:uni/student/s1 :isTaking "law"
:uni/student/s2 :isTaking "ethics"
```

By applying the mapping $\mathcal{M}$ over D, we obtain the (virtual) ABox $\mathcal{M}(D)$ shown above on the right-hand side. Assume now that we have also an OWL 2 QL ontology containing the following inclusion assertion:

```
:isTaking rdfs:subPropertyOf :isStudying
```

By using T-mappings [23] w.r.t. the above ontology (which essentially encode the TBox axioms into the mapping), the ABox is enriched with the assertions: `(:uni/student/s1 :isStudying"ethics")`, `(:uni/student/s1 :isStudying "law")`, and `(:uni/student/s2 :isStudying "ethics")`.

4.2 Translation of SPARQL `DELETE`

To translate a SPARQL `DELETE` operation into an update over the underlying data source in *Ontop*, we propose an approach grounded in the notion of data *lineage* [9,12] and MR-os-mapping for VKG mappings (see Sect. 3). When a user requests the deletion of an ABox assertion f, our goal is to compute the minimal set of deletions over the source instance that ensures f is no longer entailed in the virtual ABox generated by the VKG. Central to our approach is the observation that if f currently holds in the virtual ABox, it must have been generated by at least one mapping assertion applied to a subset of the database. We define the *lineage* of f as the set of all minimal subsets of the source instance (called *lineage branches*) that individually suffice to produce f. This is captured by the following definition:

Definition 5 (Lineage [33]**).** Let $\mathcal{P} = \langle \mathcal{T}, \mathcal{M}, \mathcal{S} \rangle$ be a VKG specification, $\langle \mathcal{P}, D \rangle$ a VKG instance, and $f \in \mathcal{M}(D)$ an ABox assertion. A subset $B \subseteq D$ is a *lineage branch of* f if $f \in \mathcal{M}(B)$, and for every $B' \subsetneq B$, $f \notin \mathcal{M}(B')$. The *lineage of* f, denoted $\mathsf{lineage}(f, \mathcal{P}, D)$, is the set of all lineage branches of f. ◁

Each lineage branch represents a distinct witness of f in the data source through $\mathcal{M}$. It was proven in [33] that for the MR-os-mapping $\widehat{\mathcal{M}}$, every set B in $\widehat{\mathcal{M}}_{\mathcal{S}}(f)$ such that $B \subseteq D$ is a lineage branch of f. Hence, to formally capture these branches in a source instance D, we select from its instance recovery $\widehat{\mathcal{M}}_{\mathcal{S}}(f)$ those sets that are included in D. In practice, these branches are obtained by searching sets of tuples in D that are true in $\widehat{\mathcal{M}}(\{f\})$ (which is the source IQ).

Example 4 (Continued from Sect. 4.1*).* The lineage in D of the ABox assertion f = `(:uni/student/s1 :isTaking "ethics")` is $\mathsf{lineage}(f, \mathcal{P}, D) = \{B_1, B_2\}$, where

$$\begin{aligned} B_1 &= \{\texttt{Student(s1,john,f1), Faculty(f1,ethics)}\}, \\ B_2 &= \{\texttt{Student(s1,john,f2), Faculty(f2,ethics)}\}. \end{aligned} \qquad \lhd$$

Since the lineage branches are minimal (according to their definition), removing at least one tuple from each of them is enough to delete f from $\mathcal{M}(D)$. However, to minimize side effects on other ABox assertions, we select combinations of deletions that include exactly one tuple per branch, and that intersect minimally with the lineage of other assertions. This corresponds to computing a *hitting set* over the lineage branches, constrained to avoid tuples that also contribute to other assertions in the ABox.

Example 5 (Continued from Example 4*).* The *hitting set* over the lineage branches of f is $\mathbf{T} = \{T_1, T_2, T_3, T_4\}$ where

$$\begin{aligned} T_1 &= \{\texttt{Student(s1,john,f1), Student(s1,john,f2)}\}, \\ T_2 &= \{\texttt{Student(s1,john,f1), Faculty(f2,ethics)}\}, \\ T_3 &= \{\texttt{Faculty(f1,ethics), Student(s1,john,f2)}\}, \\ T_4 &= \{\texttt{Faculty(f1,ethics), Faculty(f2,ethics)}\}. \end{aligned} \qquad \lhd$$

We observe that each *hitting set* over the lineage branches of an ABox assertion represents a minimal set of source tuples to delete in D to realize the deletion of that assertion. Algorithm 1 formalizes the translation of a SPARQL `DELETE` request into a minimal SQL `DELETE` request over the source instance of $\mathcal{J}$. Our algorithm takes as input $\mathcal{J}$, a SPARQL `DELETE` request $\mathcal{Q}_{\mathcal{T}}^{-}$, and a distance metric $\Delta_{\mathcal{J}}^{-}$ to guide side-effect minimization[10]. The algorithm begins by extracting from $\mathcal{Q}_{\mathcal{T}}^{-}$ the set F of ABox assertions targeted for deletion (at Line 1), and the lineage $\mathbf{B}$ of F in D (which is the union of the lineage branches of all assertions in F). Since deleting the set F amounts to deleting at least one tuple in each lineage branch of its elements, the algorithm computes the *hitting set* of $\mathbf{B}$ (at Line 3) using the ComputeHittingSets algorithm. Then the algorithm selects the subset $\mathbf{U}_D^{-} \subseteq \mathbf{T}$ of minimal translations according to $\Delta_{\mathcal{J}}^{-}$. Finally, each selected translation is converted into a concrete SQL `DELETE` over $\mathcal{S}$ (using ConvertToDelSQL) and returned to the user as a possible translation.

Example 6 (Cont.'d from Example 5*).* Consider the request $\mathcal{Q}_{\mathcal{J}}^{-}$: `DELETE DATA { f }`. For any comparison metric $\Delta_{\mathcal{J}}^{-}$, the only set of tuples to delete is T_3, since its deletion leads to an empty side-effect. Therefore, Algorithm 1 returns the SQL `DELETE`:

```
DELETE FROM Student WHERE id="s1" AND name="john" AND faculty="f2";
DELETE FROM Faculty WHERE id="f1" AND course="ethics".
```
$\lhd$

[10] $\Delta_{\mathcal{J}}^{-}$ can also encode heuristics such as tuple preservation or data provenance cost.

Algorithm 1: ImpTransDel

input : A VKG instance $\mathcal{J} = \langle \mathcal{P}, D \rangle$ with $\mathcal{P} = \langle \mathcal{T}, \mathcal{M}, \mathcal{S} \rangle$ where $\mathcal{T}$ is specified in OWL 2 QL and $\mathcal{M}$ in R2RML.
A SPARQL `DELETE` request $\mathcal{Q}_{\mathcal{T}}^{-}$ in $\mathcal{T}$ and distance metric $\Delta_{\mathcal{J}}^{-}$.

output: A set of SQL `DELETE` request expressed in $\mathcal{S}$.

$F \leftarrow$ Set of ABox assertions contained in $\mathcal{Q}_{\mathcal{T}}^{-}$;
$\mathbf{B} \leftarrow \bigcup_{f \in F} \mathsf{lineage}(f, \mathcal{P}, D)$;
$\mathbf{T} \leftarrow \textsc{ComputeHittingSets}(\mathbf{B})$;
$\mathbf{U}_{D}^{-} \leftarrow$ Translations in $\mathbf{T}$ that are minimal according to $\Delta_{\mathcal{J}}^{-}$;
return $\{\textsc{ConvertToDelSQL}(T) \mid T \in \mathbf{U}_{D}^{-}\}$

4.3 Translation of SPARQL INSERT

Unlike deletions, the translation of a SPARQL `INSERT` request for a set F of ABox assertions in VKGs presents a fundamentally different challenge: the ABox assertions targeted for insertion do not yet exist in the database. Therefore, we cannot simply manipulate the database to realize the intended insertion. Instead, we must construct new source tuples such that, when the VKG mapping is applied, the desired assertion appears in the virtual ABox. To address this, we apply the MR-os-mapping over the set of ABox assertions to be inserted. This leads to an unfolded IQ, which serves as a logical specification of the source data needed to generate F. The problem then becomes one of finding a suitable set of source tuples that satisfy the unfolded IQ.

To accomplish this, we propose to rewrite the set F as a whole in the source schema. A key advantage of using *Ontop* to achieve this, lies in the fact that the resulting rewritten IQ reflects a unified logical condition that captures all possible ways to jointly generate F through the original VKG mapping $\mathcal{M}$. *Ontop* leverages a unification mechanism at the level of the source schema $\mathcal{S}$, which merges repeated variables and shared join paths across the different assertions in F, and also correctly deals with IRI templates. This avoids redundancy and reveals shared existential dependencies.

Example 7 (Cont.'d from Sect. 4.1*).* Consider the set F of ABox assertions:

```
{ (:uni/student/s3 :hasName "smith");
  (:uni/student/s3 :isTaking "art")    }
```

Rewriting each assertion separately yields two IQs that are respectively equivalent to the CQs $\exists id.\mathtt{Student}(\mathtt{"s3"}, \mathtt{"smith"}, id)$ and $\exists name, id.\mathtt{Student}(\mathtt{"s3"}, name, id) \wedge \mathtt{Faculty}(id, \mathtt{"art"})$. Instead, by rewriting the entire set F as a single query, *Ontop* produces the unified IQ $\exists id.\mathtt{Student}(\mathtt{"s3"}, \mathtt{"smith"}, id) \wedge \mathtt{Faculty}(id, \mathtt{"art"})$, where the only existential variable id is shared across the two atoms. ◁

The first challenge when translating SPARQL `INSERT` operations arises when the set of RDF triples includes multiple distinct subjects. Since each subject typically corresponds to a distinct entity in the domain, a set of triples with different subjects will generate several distinct IQs in the source. Therefore, to ensure an efficient translation, we group the ABox assertions by subject before applying the rewriting. Within each

Algorithm 2: IMPTRANSINS

input : A VKG instance $\mathcal{J} = \langle \mathcal{P}, D \rangle$ with $\mathcal{P} = \langle \mathcal{T}, \mathcal{M}, \mathcal{S} \rangle$, where $\mathcal{T}$ is specified in OWL 2 QL and $\mathcal{M}$ in R2RML.
A SPARQL `INSERT` request $\mathcal{Q}^+_{\mathcal{T}}$ in $\mathcal{T}$ and distance metric $\Delta^+_{\mathcal{J}}$.
output: A set of SQL `INSERT` request expressed in $\mathcal{S}$.

1 $\mathbf{T} \leftarrow \{\}$; $F \leftarrow$ set of ABox assertions contained in $\mathcal{Q}^+_{\mathcal{T}}$;
2 **for each** group $G \in$ GROUPBYSUBJECT(F) **do**
3   **for each** $q \in \widehat{\mathcal{M}}(G)$ **do**
4         $\Sigma \leftarrow$ SATSUBSET$(q, D) \cup \{(\emptyset, \emptyset)\}$;
5   **for each** $(B_\sigma, \sigma) \in \Sigma$ **do**
6             $v_q \leftarrow \mathsf{vars}(q \setminus B_\sigma)$;
7             $\Delta_q = \{b_1, \ldots, b_{|v_q|}\} \cup \Delta_A$, where Δ_A are the constants in G and each b_i is a fresh value not in Δ_A and not appearing in D;
8             $\Gamma \leftarrow$ ASSIGN$(q \setminus B_\sigma, \Delta_q, D)$;
9   **for each** $(B_\gamma, \gamma) \in \Gamma$ **do**
10                 $\mathbf{T} \leftarrow \mathbf{T} \cup \{B_\sigma[\sigma] \cup B_\gamma[\gamma]\}$;
11     $\mathbf{U}^+_D \leftarrow$ translations in $\mathbf{T}$ that are minimal according to $\Delta^+_{\mathcal{J}}$
12 **return** $\{$CONVERTTOINSSQL$(T) \mid T \in \mathbf{U}^+_D\}$

group, the triples collectively represent attributes or relationships of the same entity and are therefore more likely to share a common source predicate or join path.

Once an MR-os-mapping $\widehat{\mathcal{M}}$ is applied to each group of ABox assertions in the provided SPARQL `INSERT` request, we obtain an unfolded IQ $\widehat{\mathcal{M}}(F)$ over the source, from which we have to search for sets B of source tuples such that B is true in $\widehat{\mathcal{M}}(F)$ with minimum side-effect. Since the sets B are not restricted to $B \subseteq D$ (as is the case for deletion), the search space of such B can be infinite, due to the presence of existential variables in $\widehat{\mathcal{M}}(F)$. To narrow this space, we decompose each CQ q in $\widehat{\mathcal{M}}(F)$ and compute its *maximum satisfiable subsets* [2]—the subqueries that evaluate to true over D, with the additional condition that all variables that occur in such a maximum satisfiable subset $s \subseteq q$ do not occur in any atom of q not in s. These satisfiable fragments allow for partial grounding of variables, from which we infer the values of the remaining existential variables. Notice that the additional condition that no variable x occurring in s occurs also outside of s is necessary in order to ensure that we do not lose solutions with minimal side-effect. Indeed, when considering a maximum satisfiable subset s, we enforce the instantiation of all variables in s via the facts already in D, and we want to avoid that such instantiation constains also atoms (not already in D) that need to be inserted to accomplish the ABox insertion according to q.

Algorithm 2 outlines the process of translating a SPARQL `INSERT` request into SQL `INSERT` statements over a VKG instance $\mathcal{J}$. The algorithm begins by grouping the target ABox assertions F by subject to ensure that related triples are rewritten together (using GROUPBYSUBJECT), capturing shared existential dependencies. For each group G, the algorithm applies the MR-os-mapping to get $\widehat{\mathcal{M}}(G)$ (at Line 3), returning a set of CQs in $\mathcal{S}$. Each CQ is then analyzed to identify its maximal satisfiable subsets with respect to the current database D (with the additional condition on variables described above). This is achieved by using the SATSUBSET algorithm, which takes as input a CQ q and a source instance D, and returns a set of pairs (B_σ, σ), where σ is a partial binding of variables in q with values in D, and $B_\sigma \subseteq q$ is the set of atoms corresponding to σ. For a variable binding σ and a set B_σ of atoms, we use $B_\sigma[\sigma]$ to denote the set of atoms obtained from q by binding the variables in q according to σ.

The remaining variables in $q \setminus B_\sigma$ (from the unsatisfiable subset) are instantiated either from known values in D or by introducing fresh values (at Line 8). This is achieved with the ASSIGN function, which takes as input $q \setminus B_\sigma$ and by using its existential variables as answer variables, queries from D the set of possible assignments, on top of which the fresh values from Δ_q are added. For each assignment returned by the ASSIGN function, we create an insertion candidate, which we add to $\mathbf{T}$ (at Lines 10) from which a minimal subset $\mathbf{U}_D^+$ is selected using a comparison metric $\Delta_{\mathcal{J}}^+$. The selected instantiations are then translated into SQL `INSERT` statements (using CONVERTTOINSSQL), ensuring that the intended ABox assertions are derivable from the updated source.

Example 8 (Cont.'d from Example 7). Consider the request $\mathcal{Q}_{\mathcal{T}}^+$: `INSERT DATA {` F `}`. From its unified source IQ (which in this case has no satisfiable subset), we observe that a fresh value assigned to the variable id will lead to an empty side-effect (assuming $\Delta_{\mathcal{J}}^+$ is based on cardinality). Therefore, a possible source translation is the following:

```
INSERT INTO student VALUES (s3, "smith", f8)
INSERT INTO faculty VALUES (f8, "art")
```
◁

5 Empirical Results

To evaluate the practical applicability and scalability of our proposed translation algorithms for SPARQL Update, we conducted two complementary experiments based on our running example. In the first experiment, we varied the database size from 1 to 2,000 students and from 1 to 500 faculties (with each faculty having two students). For insertion, we tested adding a fresh grouped triple that requires creating new source tuples, since no satisfiable rewriting exists. For deletion, we focused on removing an existing `foaf:isTaking` triple, which consists of two relations in the source schema. The results show that insertion time grows significantly with database size due to the need to search for translations that minimize side effects, while deletion time remains relatively stable since it depends only on the fixed lineage branch of the target assertion (Fig. 1, left).

In the second experiment, we fixed the database size but progressively increased the number of lineage branches per assertion from 1 to 15. We observed that deletion time can grow exponentially with the number of branches, reaching up to 23 min for 15 branches. However, by employing early stopping strategies (terminating as soon as an exact translation is found or one below a side-effect threshold), runtime was drastically reduced. Together, these experiments show that the insertion cost scales primarily with the database size, while the deletion cost scales with lineage complexity, and both can be significantly optimized using practical heuristics such as early stopping, prioritizing fresh value insertions, or even allowing null values in the data source (Fig. 1, right).

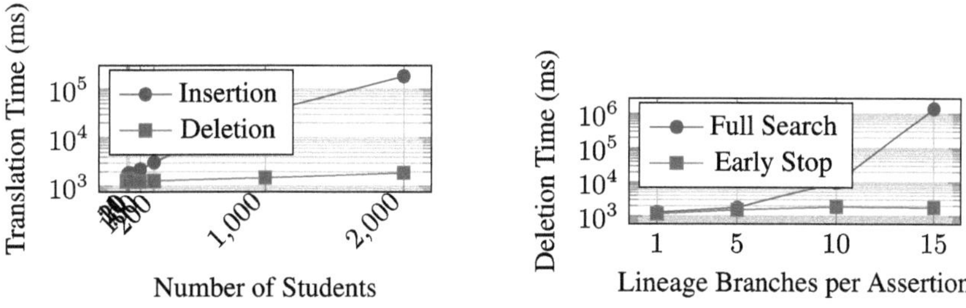

Fig. 1. Translation time scaling: (a) with database size; (b) with lineage branch complexity.

6 Conclusions

In this paper, we addressed the challenge of translating instance-level update requests in Virtual Knowledge Graphs into executable source-level operations within the *Ontop* framework. We introduced a novel approach based on maximum recovery, which enables us to systematically rewrite ABox assertions into source queries and thus compute the minimal source updates. We have demonstrated how such a mapping can be obtained by leveraging the *Ontop* rewriting engine, and we proposed translation algorithms for both insertion and deletion of ABox assertions. Through extensive experiments, we demonstrated how the insertion cost scales with the database size, while the deletion cost scales with lineage complexity, and we showed that practical heuristics such as early stopping can significantly improve performance. Our results highlight the feasibility of supporting precise and side-effect-aware updates in VKG systems like *Ontop*, paving the way for more expressive and update-capable virtualized semantic data management.

Acknowledgments. This research has been partially supported by the Wallenberg AI, Autonomous Systems and Software Program (WASP), funded by the Knut and Alice Wallenberg Foundation, by the HEU project CyclOps (GA n. 101135513), by the Province of Bolzano and FWF through project OnTeGra (DOI 10.55776/PIN8884924). by the Province of Bolzano and EU through projects ERDF-FESR 1078 CRIMA, and ERDF-FESR 1047 AI-Lab, by MUR through the PRIN project 2022XERWK9 S-PIC4CHU, and by the EU and MUR through the PNRR project PE0000013-FAIR.

References

1. Bancilhon, F., Spyratos, N.: Update semantics of relational views. ACM Trans. Database Syst. **6**(4), 557–575 (1981). https://doi.org/10.1145/319628.319634
2. Bendík, J.: On decomposition of maximal satisfiable subsets. In: Proceedings of Formal Methods in Computer Aided Design (FMCAD), pp. 212–221. IEEE (2021). https://doi.org/10.34727/2021/ISBN.978-3-85448-046-4_30
3. Brickley, D., Guha, R.V.: RDF vocabulary description language 1.0: RDF Schema. W3C Recommendation, World Wide Web Consortium (2004). https://www.w3.org/TR/rdf-schema/

4. Calvanese, D., et al.: Ontop: answering SPARQL queries over relational databases. Semantic Web J. **8**(3), 471–487 (2017). https://doi.org/10.3233/SW-160217
5. Calvanese, D., et al.: Ontologies and databases: the *DL-Lite* approach. In: Tessaris, S., et al. (eds.) Reasoning Web 2009. LNCS, vol. 5689, pp. 255–356. Springer, Heidelberg (2009). https://doi.org/10.1007/978-3-642-03754-2_7
6. Calvanese, D., et al.: The Mastro system for ontology-based data access. Semantic Web J. **2**(1), 43–53 (2011). https://doi.org/10.3233/SW-2011-0029
7. Calvanese, D., Kharlamov, E., Nutt, W., Zheleznyakov, D.: Updating ABoxes in *DL-Lite*. In: Proceedings of the 4th Alberto Mendelzon International Workshop on Foundations of Data Management (AMW). CEUR Workshop Proceedings, vol. 619, pp. 3.1–3.12. CEUR-WS.org (2010)
8. Chebotko, A., Lu, S., Fotouhi, F.: Semantics preserving SPARQL-to-SQL translation. Data Knowl. Eng. **68**(10), 973–1000 (2009). https://doi.org/10.1016/J.DATAK.2009.04.001
9. Cui, Y., Widom, J.: Lineage tracing for general data warehouse transformations. VLDB J. **12**(1), 41–58 (2003). https://doi.org/10.1007/s00778-002-0083-8
10. Das, S., Sundara, S., Cyganiak, R.: R2RML: RDB to RDF mapping language. W3C Recommendation, World Wide Web Consortium (2012). https://www.w3.org/TR/r2rml/
11. Dayal, U., Bernstein, P.A.: On the updatability of relational views. In: Proceedings of the 4th International Conference on Very Large Data Bases (VLDB), vol. 78, pp. 368–377 (1978). https://doi.org/10.14778/3415478.3415503
12. Dayal, U., Bernstein, P.A.: On the correct translation of update operations on relational views. ACM Trans. Database Syst. **7**(3), 381–416 (1982). https://doi.org/10.1145/319732.319740
13. De Giacomo, G., Lenzerini, M., Poggi, A., Rosati, R.: On instance-level update and erasure in description logic ontologies. J. Logic Comput. **19**(5) (2009). https://doi.org/10.1093/logcom/exn051
14. De Giacomo, G., Oriol, X., Rosati, R., Savo, D.F.: Instance-level update in DL-Lite ontologies through first-order rewriting. J. Artif. Intell. Res. **70**, 1335–1371 (2021). https://doi.org/10.1613/jair.1.12414
15. Eisenberg, V., Kanza, Y.: D2RQ/update: updating relational data via virtual RDF. In: Proceedings of the 21st International World Wide Web Conference (WWW), pp. 497–498 (2012). https://doi.org/10.1145/2187980.2188095
16. Eiter, T., Gottlob, G.: On the complexity of propositional knowledge base revision, updates and counterfactuals. Artif. Intell. **57**, 227–270 (1992). https://doi.org/10.1016/0004-3702(92)90018-S
17. Flouris, G., Manakanatas, D., Kondylakis, H., Plexousakis, D., Antoniou, G.: Ontology change: classification and survey. Knowl. Eng. Rev. **23**(2), 117–152 (2008). https://doi.org/10.1017/S0269888908001367
18. Garrote Hernandez, A., García, M.N.M.: RESTful writable APIs for the web of linked data using relational storage solutions. In: Proceedings of WWW Workshop on Linked Data on the Web (LDOW). CEUR Workshop Proceedings, vol. 813. CEUR-WS.org (2011). https://ceur-ws.org/Vol-813/ldow2011-paper04.pdf
19. Gearon, P., Passant, A., Polleres, A.: SPARQL 1.1 update. W3C Recommendation, World Wide Web Consortium (2013). https://www.w3.org/TR/sparql11-update/
20. Harris, S., Seaborne, A.: SPARQL 1.1 query language. W3C Recommendation, World Wide Web Consortium (2013). https://www.w3.org/TR/sparql11-query
21. Hert, M., Reif, G., Gall, H.C.: Updating relational data via SPARQL/update. In: Proceedings of the EDBT/ICDT Workshops, pp. 1–8 (2010). https://doi.org/10.1145/1754239.1754266
22. Katsuno, H., Mendelzon, A.: On the difference between updating a knowledge base and revising it. In: Proceedings of the 2nd International Conference on Principles of Knowledge Representation and Reasoning (KR), pp. 387–394 (1991)

23. Kontchakov, R., Rezk, M., Rodríguez-Muro, M., Xiao, G., Zakharyaschev, M.: Answering SPARQL queries over databases under OWL 2 QL entailment regime. In: Mika, P., et al. (eds.) ISWC 2014. LNCS, vol. 8796, pp. 552–567. Springer, Cham (2014). https://doi.org/10.1007/978-3-319-11964-9_35
24. Kotidis, Y., Srivastava, D., Velegrakis, Y.: Updates through views: a new hope. In: Proceedings of the 22nd IEEE International Conference on Data Engineering (ICDE), p. 2 (2006). https://doi.org/10.1109/ICDE.2006.167
25. Langerak, R.: View updates in relational databases with an independent scheme. ACM Trans. Database Syst. **15**(1), 40–66 (1990). https://doi.org/10.1145/77643.77645
26. Liu, H., Lutz, C., Milicic, M., Wolter, F.: Foundations of instance level updates in expressive description logics. Artif. Intell. **175**(18), 2170–2197 (2011). https://doi.org/10.1016/J.ARTINT.2011.08.003
27. Motik, B., Cuenca Grau, B., Horrocks, I., Wu, Z., Fokoue, A., Lutz, C.: OWL 2 Web Ontology Language profiles (second edition). W3C Recommendation, World Wide Web Consortium (2012). https://www.w3.org/TR/owl2-profiles/
28. Motik, B., Cuenca Grau, B., Horrocks, I., Wu, Z., Fokoue, A., Lutz, C.: OWL 2 Web Ontology Language profiles (second edition). W3C Recommendation, World Wide Web Consortium (2012). http://www.w3.org/TR/owl2-profiles/
29. Priyatna, F., Corcho, O., Sequeda, J.F.: Formalisation and experiences of R2RML-based SPARQL to SQL query translation using morph. In: Proceedings of the 23rd International World Wide Web Conference (WWW), pp. 479–490 (2014). https://doi.org/10.1145/2566486.2567981
30. Ramanujam, S., Khadilkar, V., Khan, L., Seida, S., Kantarcioglu, M., Thuraisingham, B.: Bi-directional translation of relational data into virtual RDF stores. In: Proceedings of the 4th IEEE International Conference on Semantic Computing (ICSC), pp. 268–276. IEEE Computer Society (2010). https://doi.org/10.1109/ICSC.2010.61
31. Rodriguez-Muro, M., Rezk, M.: Efficient SPARQL-to-SQL with R2RML mappings. J. Web Seman. **33**, 141–169 (2015). https://doi.org/10.1016/j.websem.2015.03.001
32. Unbehauen, J., Martin, M.: Executing SPARQL queries over mapped document store with SparqlMap-M. In: Proceedings of the 12th International Conference on Semantic Systems (SEMANTICS), pp. 137–144. ACM (2016). https://doi.org/10.1145/2993318.2993326
33. Wandji, R.E., Calvanese, D.: Ontology-based update in virtual knowledge graphs via schema mapping recovery. In: Proceedings of the 8th International Joint Conference on Rules and Reasoning (RuleML+RR). Lecture Notes in Computer Science, vol. 15183, pp. 59–74. Springer, Cham (2024). https://doi.org/10.1007/978-3-031-72407-7_6
34. Wandji, R.E., Calvanese, D.: Minimizing side-effects in virtual knowledge graph updates. In: Proceedings of the 9th International Joint Conference on Rules and Reasoning (RuleML+RR). Lecture Notes in Computer Science. Springer, Cham (2025)
35. Winslett, M.: A model-based approach to updating databases with incomplete information. ACM Trans. Database Syst. **13**(2), 167–196 (1988). https://doi.org/10.1145/42338.42386
36. Xiao, G., et al.: Ontology-based data access: a survey. In: Proceedings of the 27th International Joint Conference on Artificial Intelligence (IJCAI), pp. 5511–5519. IJCAI Org (2018). https://doi.org/10.24963/ijcai.2018/777
37. Xiao, G., et al.: The virtual knowledge graph system ontop. In: Pan, J.Z., et al. (eds.) ISWC 2020. LNCS, vol. 12507, pp. 259–277. Springer, Cham (2020). https://doi.org/10.1007/978-3-030-62466-8_17
38. Zheleznyakov, D., Kharlamov, E., Nutt, W., Calvanese, D.: On expansion and contraction of DL-Lite knowledge bases. J. Web Seman. **57**, 100484 (2019). https://doi.org/10.1016/j.websem.2018.12.002

Trajectory Encoding Temporal Graph Networks

Jiafeng Xiong(✉) and Rizos Sakellariou

Department of Computer Science, University of Manchester, Manchester, UK
{jiafeng.xiong,rizos}@manchester.ac.uk

Abstract. Temporal Graph Networks (TGNs) have demonstrated significant success in dynamic graph tasks such as link prediction and node classification. Such tasks comprise transductive settings, where the temporal graph learning model predicts links among known nodes, and inductive settings, where it generalises learned patterns to previously unseen nodes. Existing TGN designs face a dilemma under these dual scenarios. Anonymous TGNs, which rely solely on temporal and structural information, offer strong inductive generalisation but struggle to distinguish known nodes. In contrast, non-anonymous TGNs leverage node features to excel in transductive tasks yet fail to adapt to new nodes. To address these challenges, this paper proposes *Trajectory Encoding TGN* (TETGN), which introduces automatically expandable node identifiers (IDs) as learnable temporal positional features and performs message passing over these IDs to capture each node's historical context. By integrating this trajectory-aware module with a standard TGN using multi-head attention, TETGN effectively balances transductive accuracy with inductive generalisation. TETGN uses the trajectory encoding method to support anonymous inductive generalisation, and augments non-anonymous transductive prediction with extendable learnable IDs. Experimental results on three real-world datasets show that TETGN significantly outperforms strong baselines on both link prediction and node classification tasks, demonstrating its ability to unify the advantages of anonymous and non-anonymous models for dynamic graph learning (Code is available at: https://github.com/xjiaf/TETGN.git).

Keywords: Temporal Networks · Link Prediction · Graph Neural Networks · Graph Representation Learning · Machine Learning

1 Introduction

Graph representation learning and Graph Neural Networks (GNNs) [13,28] have achieved remarkable success in diverse applications by utilising message-passing mechanisms over static graphs [1]. Nonetheless, many real-world systems such as social networks, communication networks and recommendation systems [9,18,27] are naturally dynamic, in that relationships and entities change over time. To capture the dynamics, temporal graph representation learning and Temporal

H. Takeda et al. (Eds.): IJCKG 2025, LNAI 16297, pp. 236–251, 2026.
https://doi.org/10.1007/978-981-95-5009-8_16

Graph Networks (TGNs) [7,11,21] have been developed, integrating both temporal and structural information into node embeddings. By incorporating time-aware message passing, TGNs adapt to shifting node interactions to predict tasks such as node classification [32] and temporal link prediction (TLP) [5,30].

In temporal graph representation learning, models typically predict under two settings: transduction and induction. Transductive prediction is for nodes already observed, whereas inductive prediction aims to handle previously unseen nodes. Despite notable advancements, developing a method that effectively balances transductive and inductive requirements remains challenging. This fundamental challenge is reflected in current approaches, where existing methods typically excel in one setting whilst compromising performance in another. Specifically, temporal graph representation learning can be classified into anonymous and non-anonymous strategies, each demonstrating this inherent trade-off.

Anonymous methods, such as causal anonymous walks (CAWs) [24], rely solely on relatively temporal and positional relationships without leveraging explicit node features. This design choice promotes strong inductive performance, as it does not depend on pre-obtained node features or identifiers (IDs). However, anonymity leads to a shortage in transductive tasks involving complex topologies (e.g. symmetrical or cyclic structures), where the absence of unique node information limits the model's ability to distinguish structurally similar nodes. In contrast, non-anonymous methods incorporate node features to generate node embeddings and make predictions. This inclusion enhances the expressiveness required to solve transductive tasks since node-specific information is unique. Yet, relying on pre-existing features can hinder inductive generalisation for unseen nodes. Moreover, purely feature-based approaches may capture a network's temporal and positional dynamics rather poorly.

To address these issues, we propose **T**rajectory **E**ncoding **TGN**s (**TETGN**), which integrates the strengths of both anonymous and non-anonymous strategies. Our model introduces two key innovations: *automatically extendable learnable node IDs* and a *trajectory encoding* module. The learnable node IDs serve as initial position features whilst being assignable to unseen nodes, thereby overcoming the inductive limitations of traditional non-anonymous methods. The trajectory encoding module leverages message passing to capture nodes' historical temporal-positional relationships based on these learnable IDs, preserving the generalisation benefits of anonymous approaches whilst maintaining distinctive node-level information, crucial for complex transductive tasks.

Architecturally, TETGN extends the MP-TGN framework [16] by incorporating a parallel *trajectory message passing* module alongside the original encoder. This dual-stream design enables separate processing of node features and trajectory information stored in memory. Subsequently, multi-head attention [22] is used to fuse these representations, node states and trajectory encodings, into final node embeddings. Finally, TETGN achieves better expressiveness and performance in both transductive and inductive tasks.

The **contributions** of this paper are:

- **(i)** We propose exploiting *automatically extendable learnable node IDs* as temporal positional features to handle unseen nodes and resolve the transductive indistinguishability issues inherent in anonymous approaches.
- **(ii)** We introduce a *trajectory encoding* component via message passing to capture each node's historical temporal positional relationships, helping non-anonymous methods in inductive generalisation.
- **(iii)** We evaluate TETGN through extensive experiments on TLP and node classification across three real-world benchmark datasets and two masked datasets. Our results demonstrate that TETGN consistently and significantly outperforms strong baselines in transductive and inductive settings.

In the remainder of the paper, Sect. 2 discusses related work; Sect. 3 outlines the problem definition and some background; Sect. 4 introduces the methodology; Sect. 5 presents experimental results; and Sect. 6 concludes the paper and discusses future work.

2 Related Work

2.1 Temporal Graph Neural Networks

Temporal networks can be modelled as *discrete-time dynamic graphs* (DTDGs) or *continuous-time dynamic graphs* (CTDGs). In the DTDG setting, a sequence of graph snapshots is recorded at uniform time intervals. Earlier studies [2,6,12] apply traditional GNNs [17] such as GCN [10] or GAT [23] to these snapshots and aggregate the resulting embeddings for downstream tasks [24,31]. However, the DTDG model requires regular time intervals, limiting its expressiveness.

CTDGs treat temporal networks as a series of timestamped events with irregular time intervals. For instance, DyRep [21] and JODIE [11] employ RNNs to refresh node embeddings after each event, while TGAT [31], TGN [16] and PINT [19] adopt message-passing to aggregate temporal neighbourhood. Meanwhile, random-walk-based methods such as CTDNE [15] and CAW [24] learn representations by exploring local structures. Furthermore, recent works have explored diverse approaches to enhance dynamic graph representation learning: GraphMixer [4] utilises efficient neighbour pooling mechanisms, DyGFormer [34] adopts specialised transformer architectures for temporal dynamics, FreeDyG [20] introduces innovative node frequency-based modelling, and GTGIB [29] leverages graph structure learning and the information bottleneck principle to enhance graph representations.

2.2 Position Encoding in Temporal Graph Learning

Position encoding in temporal graph learning aims to capture structural or temporal information to enhance node embeddings. TGAT [31] extends graph attention networks with sinusoidal time encoding for temporal neighbours, whilst CAW [24] and PINT [19] use temporal anonymous walks to count motifs and encode the relative positional information. Despite these advances, methods without node features struggle to discriminate between nodes in complex topological structures, presenting a critical challenge in effective temporal graph learning.

3 Preliminaries

3.1 Notations and Problem Definition

In a CTDG temporal graph $G(t) = (V(t), E(t), \mathcal{X}, \mathcal{E})$ defined over $T = \{t_1, t_2, \ldots \mid t_1 < t_2 < \ldots\}$, the set $V(t) \subseteq V$ contains the nodes present at time t, where $V = \{1, 2, \ldots, n\}$. The set $E(t) \subseteq \{(i, j, t') \mid i, j \in V(t),\, t' \leq t\}$ comprises timestamped edge events between nodes i and j at time t'. Each event may carry an edge feature $e_{ij}(t) \in \mathcal{E}$ with $\mathcal{E} \in \mathbb{R}^k$ and every node i has an initial feature $x_i \in \mathcal{X}$ where $\mathcal{X} \in \mathbb{R}^d$. Finally, the temporal neighbourhood of node i is defined as $\mathcal{N}_i(t) = \{(j, t') \mid \exists (j, i, t') \in G(t) \text{ with } t' < t\}$.

Both TLP and node classification are both transductive and inductive [30]. In the transductive setting, the model predicts tasks among the nodes accessed during training. The inductive setting requires the model to generalise its learned temporal patterns to predict tasks involving nodes unseen during training.

3.2 Message-Passing Temporal Graph Networks

Message-passing Temporal Graph Networks (MP-TGNs) [16] form a framework for temporal graph representation learning comprising three key components: aggregation, update, and memory. The memory comprises vectors that summarise each node's history, updated as events occur. The update process uses the MEM_AGG module to gather information from the temporal neighbours of node i, followed by the MEM_UPDATE module, which typically employs a gated recurrent unit (GRU) [3]. When events involving node i occur at timestamp t, the event message $m_i(t)$ and the node's memory state are updated as:

$$m_i(t) = \text{MEM\_AGG}\left(s_i(t), s_j(t), \Delta t, e_{ij}(t)\right), \tag{1}$$

$$s_i(t) = \text{MEM\_UPDATE}\left(s_i(t), m_i(t)\right), \tag{2}$$

where the initial state $s_i(0)$ is the initial node feature x_i; $s_i(t)$ denotes the updated state of i at timestamp t; and Δt is the time difference between the current timestamp and the last update timestamp. MP-TGNs compute the node embedding $z_i^{(l)}(t)$ of the node i at the layer l by recursively applying:

$$\tilde{z}_i^{(l)}(t) = \text{AGG}^{(l)}\left(z_j^{(l-1)}(t), \Delta t', e_{ij}(t')\right), \tag{3}$$

$$z_i^{(l)}(t) = \text{UPDATE}^{(l)}\left(z_i^{(l-1)}(t), \tilde{z}_i^{(l)}(t)\right), \tag{4}$$

where edge features $e_{ij}(t) \in \mathcal{E}$; $z_i^{(0)}(t) = s_i(t)$ represent the state of i at timestamp t; $\Delta t'$ is the time difference between the current timestamp and the edge's timestamp t'; (j, t') belongs to the temporal neighbourhood $\mathcal{N}_i(t)$; $\text{AGG}^{(l)}$ and $\text{UPDATE}^{(l)}$ are parameterised functions.

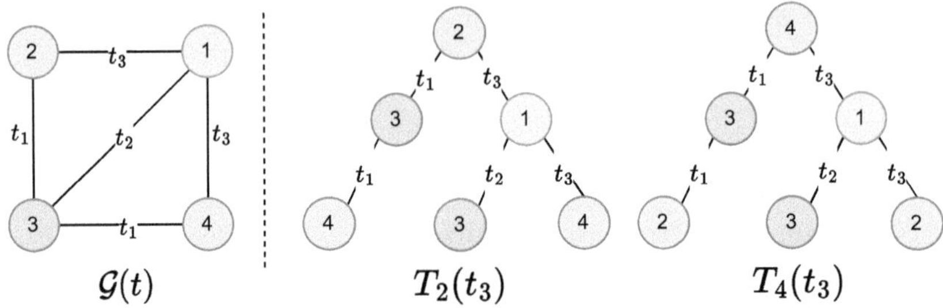

Fig. 1. An illustrative example of symmetric and cyclic structures in temporal graphs. The left-hand side graph shows a temporal graph where nodes 2 and 4 exhibit structural symmetry and are part of a cycle. The right panels display their corresponding temporal computation trees (TCTs), which are isomorphic.

3.3 Expressiveness and Limitations of TGNs

The Weisfeiler-Lehman (WL) test [25] serves as a fundamental benchmark for assessing GNN expressiveness, establishing that standard message-passing architectures cannot exceed its discriminative power [14,33]. For temporal graphs, this concept extends to the T-WL test [19], which evaluates a TGN's ability to distinguish nodes in dynamic settings.

Central to this analysis is the *temporal computation tree* (TCT) [24], denoted as $T_i(t)$, which captures L layers of message passing around node i up to timestamp t. Two nodes j and j' cannot be distinguished by an L-layer TGN if their TCTs, $T_j(t)$ and $T_{j'}(t)$, are isomorphic, meaning they share identical structure, timestamps, and embedding.

As illustrated in Fig. 1, nodes 2 and 4 generate isomorphic TCTs in a graph containing both symmetric and cyclic structures. Anonymous TGNs, which rely solely on relative temporal and positional information, fail the T-WL test in such scenarios as they cannot break this structural symmetry without node features.

4 Methodology

4.1 Overview

TETGN comprises two main phases, encoding and decoding, as illustrated in Fig. 2. The framework extends the MP-TGN architecture [16] by incorporating our trajectory encoding (TE) method.

In the encoding phase (top part of Fig. 2), incoming interactions (i, j, t_k) are processed through two parallel streams, the trajectory message encoder and the TGN message module, producing trajectory encoding $v_i(t_k)$ and message $m_i(t_k)$, respectively. These outputs are concatenated ($\|$) and fed into both the trajectory and TGN aggregators as per Equation (1), resulting in updated representations $v'_i(t_k)$ and $m'_i(t_k)$, which are then stored in the memory module [16].

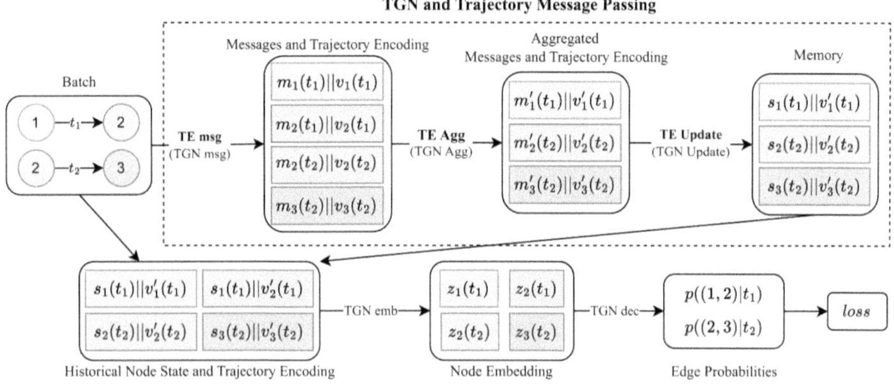

Fig. 2. Overview of TETGN architecture. The framework consists of two main phases: (a) An encoding phase (top) where temporal interactions (i.e., edges $(1 \rightarrow 2, 2 \rightarrow 3)$) are processed through parallel streams, a trajectory message encoding (TE msg) for node trajectories and a TGN message encoder for standard message passing, with outputs aggregated (TE Agg and TGN Agg) and stored in memory (TE Update and TGN Update); (b) A decoding phase (bottom) where historical node states and trajectory encodings from memory are fused via multi-head attention embedding (TGN emb) to produce the final node embeddings for downstream tasks. Here, $s_i(t_k)$ represents node states in memory, $v_i(t_k)$ and $v_i'(t_k)$ denote raw and aggregated trajectory encodings respectively, $m_i(t_k)$ and $m_i'(t_k)$ represent raw and aggregated messages, and $z_i(t_k)$ denotes the final node embeddings used for downstream tasks.

The decoding phase (bottom part of Fig. 2) focuses on generating node embeddings for downstream tasks. Historical node states and trajectory encodings are retrieved based on previous interactions and concatenations. These combined representations are processed through a multi-head attention module to produce the final node embeddings, which are subsequently used for TLP. The detailed embedding generation process is described in Sect. 4.3.

4.2 Trajectory in Message Passing

Concept of Trajectory. We define the trajectory in a temporal graph to represent the evolving history of a node within the network. Specifically, the trajectory describes the sequence of interacting node IDs that a node experiences over time. Through a message-passing mechanism, nodes exchange trajectory messages that carry their temporal positional information; each node then updates its trajectory by integrating these incoming messages.

Automatically Extendable Learnable Node ID. To leverage the non-anonymous settings, each node is assigned an automatically extendable learnable unique ID vector via a multi-layer perceptron (MLP). This approach helps resolve issues related to symmetric and cyclic structures in temporal networks. The initial ID lookup matrix $\mathbf{M} \in \mathbb{R}^{N \times d}$ maps N nodes to dense vectors of dimension d. This node ID serves as the node's initial temporal positional feature $\mathrm{TP}_i(0)$. For a given node i, its initial ID vector $\mathrm{ID}_i(0)$ at the beginning is:

$$\mathrm{TP}_i(0) = \mathrm{ID}_i(0) = \mathrm{MLP}(\mathbf{M}[i]) \tag{5}$$

where $\mathbf{M}[i]$ denotes the i-th row of the lookup matrix $\mathbf{M}$. These temporal positional features will be encoded into trajectory messages through the trajectory message encoding method $\mathrm{TE}(\cdot)$ below, which will be used for message passing.

Trajectory Messages Encoding. Each trajectory message encoding $\mathrm{TE}(\cdot)$ may undergo multiple operations in a nested manner (e.g. $\mathrm{TE}(\mathrm{TE}(\cdot))$) as it propagates along the TCT based on the message-passing mechanism. To accurately capture both structural (TCT layer-depth) and temporal changes, this encoding must satisfy two key properties:

1. As a message traverses deeper TCT layers, successive applications of $\mathrm{TE}(\cdot)$ yield cumulative positional shifts.
2. For sequential temporal gaps (e.g. Δt_1 then Δt_2), nested encodings should equate to a single encoding of $\Delta t_1 + \Delta t_2$.

Concretely, let $\mathrm{TE}(x_0, x)$ represent updating a *state* x_0 with two incremental changes x and y. Formally:

$$\mathrm{TE}\big(\mathrm{TE}(x_0, x), y\big) = \mathrm{TE}\big(x_0, x + y\big). \tag{6}$$

Define $\mathrm{TE}(x_0, x) = x_0 \cdot f(x)$, where f encodes the partial update. Then,

$$\mathrm{TE}\big(\mathrm{TE}(x_0, x), y\big) = x_0 \cdot f(x) \cdot f(y), \tag{7}$$

$$\mathrm{TE}\big(x_0, x + y\big) = x_0 \cdot f\big(x + y\big). \tag{8}$$

Hence, according to the time consistency required for the second property:

$$f(x) \cdot f(y) = f\big(x + y\big), \tag{9}$$

which implies that $f(\cdot)$ is a *Cauchy function.* Thus, we adopt a generic exponential function from the Cauchy function family to represent time decay [26]:

$$\mathrm{TE}_{\exp}(x, \Delta t) = \alpha x \exp(-\beta \Delta t), \tag{10}$$

where x is the node temporal positional feature, α is a step-count hyperparameter, β the time decay factor, and Δt the time interval between updates. The power of α naturally increases with the layer depth as the trajectory message passes through the TCT, reflecting the positional information of the node trajectory, whilst $\exp(-\beta \Delta t)$ ensures coherent temporal accumulation. For instance, if a trajectory message $\alpha x_0 \exp(-\beta \Delta t_1)$ is encoded again at a subsequent layer with an additional interval Δt_2, then:

$$\alpha\big(\alpha x_0 \exp(-\beta \Delta t_1)\big) \exp(-\beta \Delta t_2) = \alpha^2 x_0 \exp(-\beta(\Delta t_1 + \Delta t_2)), \tag{11}$$

where the power of α increases by 1 to represent advancing one layer in the TCT, and $\Delta t_1 + \Delta t_2$ tracks the total time elapsed. Thus, this design preserves a coherent representation of both layer-depth progression and accumulated time.

Trajectory Aggregation. As with conventional message passing, node i at time t aggregates trajectory messages from its temporal neighbours $\mathcal{N}_i(t)$. Each neighbour j supplies its latest trajectory encoding $v_j(t')$, which is then combined with the node's temporal positional feature $v_i(t)$ via a summation aggregator:

$$v_i'(t) = \text{TE\_AGG}(v_i(t), v_j(t')) = v_i(t) + \sum_{(j,t') \in \mathcal{N}_i(t)} v_j(t'), \tag{12}$$

where TE_AGG is the trajectory aggregation function and node j is the temporal neighbours of node i. This approach implicitly tracks higher-order connections within the TCT without explicitly enumerating them.

Temporal Positional Feature Update. The Trajectory Message Passing module updates and derives a new temporal positional feature for node i based on its learned node ID and the trajectory messages from neighbouring nodes whenever node i interacts. We define the updated temporal positional feature, $\text{TP}_i(t)$, at time t for node i as a function of its current learned ID, $\text{ID}_i(t)$, and the aggregated trajectory message, $v_i'(t)$, at timestamp t:

$$\text{TP}_i(t) = \text{TE\_UPDATE}(\text{ID}_i(t), v_i'(t)) = \text{ID}_i(t) + v_i'(t), \tag{13}$$

where TE_UPDATE is the update function of trajectory encoding, $\text{ID}_i(t)$ is the newly learned ID at current timestamp t. The updated new temporal positional feature $\text{TP}_i(t)$ is stored in memory for future message passing, similar to the memory module in MP-TGN [16].

Complexity Implications. The TGN has time complexity $\mathcal{O}(L(\bar{n}+d)d|E|)$, where $\bar{n}$ is the average node degree, L is the number of layers, $|E|$ is the number of edges and d the embedding dimension. Since TETGN has an extra trajectory message-passing module, time complexity becomes $\mathcal{O}(2L(\bar{n}+d)d|E|)$. Besides, the storage complexity remains $\mathcal{O}(2|V|d)$. Despite the extra trajectory message-passing module, TETGN's computational overhead is merely a constant factor which is asymptotically negligible, making it highly efficient for TGN.

4.3 Node Embedding and Prediction

Following MP-TGN [16], TETGN maintains historical states $s_i(t_k)$ alongside trajectory encodings $v_i'(t_k)$. These are concatenated and passed through MLPs, then further fused by multi-head attention layers [22] to get final embedding $z_i(t_k)$ for downstream tasks:

$$z_i(t_k) = \text{Multi-head\_Attn}(\text{MLP}([s_i(t_k) \parallel \text{MLP}'(v_i'(t_k))])). \tag{14}$$

5 Experimental Evaluation

5.1 Experiment Setup

Datasets. We apply TLP on three real-world datasets, Wikipedia, Reddit, LastFM and two masked versions Wikipedia-fm, and Reddit-fm [11], following the protocols established by [16,31]. For Wikipedia and Reddit, edge feature vectors are intact, while for Wikipedia-fm and Reddit-fm, these edge features are masked. LastFM, however, is a natural non-attributed network. For consistency with earlier studies, all datasets employ zero vectors for node features.

Baselines. We compare TETGN against both static and temporal GNN baselines. GAT [23] and GraphSAGE [8] are popular static GNNs. Jodie [11], DyRep [21], TGAT [31] and MP-TGN [16] are non-anonymous temporal GNNs, while CAW-N [24] and PINT [19] are anonymous models. The results adhere to the implementations and guidelines from the official repositories. Baseline figures for GAT, GraphSAGE, Jodie, DyRep, TGAT and MP-TGN on Wikipedia, Reddit and LastFM are taken from [16,19]. The remaining results were obtained utilising the official implementations. For the masked datasets Wikipedia-fm and Reddit-fm, the settings of Wikipedia and Reddit have been adopted.

Implementation Details. The experiments use a 70%-15%-15% (train-val-test) temporal split for all datasets, with average precision (AP) as the performance metric. Results are analysed for transductive (seen nodes) and inductive (unseen nodes) predictions, reporting the mean and standard deviation of AP over 10 runs. All methods are implemented using PyTorch, accelerated by an NVIDIA V100 GPU. The batch size is 200; negative sample size is 5; β is 0.1; α is 2. The trajectory encoding dimension d is tuned in $\{4, 12, 20, 28\}$ and the learning rate in $[1e-6, 1e-4]$. Other settings are the same as MP-TGNs [16].

5.2 Performance Comparison

Table 1 shows the average precision results. The following observations are made:

(1) Comparing the baselines and TETGN on the Wikipedia, Wikipedia-fm, Reddit and Reddit-fm datasets reveals TETGN suffers the smallest performance drop among all methods. This suggests it has low dependency on edge features and is robust in capturing the temporal dynamics of the graph structure.
(2) On the non-attributed LastFM dataset, TETGN significantly outperforms the baselines, demonstrating its effectiveness when node and edge features are absent. This underscores TETGN's superior ability to capture and express temporal graph structural information, with improvements of 2.4% and 2.8% in the transductive and inductive settings, respectively.
(3) When comparing baselines in non-anonymous settings, such as MP-TGN, with those in anonymous settings like CAW-N, CAW-N generally outperforms the non-anonymous baselines, except on the Reddit-fm dataset. This highlights the importance of capturing relative positional relationships. Moreover,

Table 1. Temporal link prediction performance (AP, mean ± std over 10 runs) for various models. The best result is in bold and the second best is underlined.

Setting	Model	Reddit	Wikipedia	Reddit-fm	Wikipedia-fm	LastFM
Transductive	GAT	97.33 ± 0.2	94.73 ± 0.2	-	-	-
	GraphSAGE	97.65 ± 0.2	93.56 ± 0.3	-	-	-
	Jodie	97.11 ± 0.3	94.62 ± 0.5	96.94 ± 0.8	95.19 ± 0.5	69.32 ± 1.0
	DyRep	97.98 ± 0.1	94.59 ± 0.2	96.95 ± 1.0	94.58 ± 0.2	69.24 ± 1.4
	TGAT	98.12 ± 0.2	95.34 ± 0.1	75.11 ± 0.1	77.91 ± 0.5	54.77 ± 0.4
	MP-TGN	98.70 ± 0.1	98.46 ± 0.1	<u>98.54 ± 0.1</u>	98.54 ± 0.1	80.69 ± 0.2
	CAW-N	98.39 ± 0.1	<u>98.63 ± 0.1</u>	98.41 ± 0.1	<u>98.58 ± 0.1</u>	<u>81.29 ± 0.1</u>
	PINT	<u>98.79 ± 0.1</u>	98.40 ± 0.1	97.35 ± 0.1	98.55 ± 0.1	77.06 ± 1.5
	TETGN (ours)	**98.90 ± 0.1**	**98.74 ± 0.1**	**98.79 ± 0.1**	**98.73 ± 0.1**	**83.26 ± 0.1**
Inductive	GAT	95.37 ± 0.3	91.27 ± 0.4	-	-	-
	GraphSAGE	96.27 ± 0.2	91.09 ± 0.3	-	-	-
	Jodie	94.36 ± 1.1	93.11 ± 0.4	93.77 ± 1.7	93.77 ± 0.3	80.32 ± 1.4
	DyRep	95.68 ± 0.2	92.05 ± 0.3	93.33 ± 0.2	92.64 ± 0.5	82.03 ± 0.6
	TGAT	96.62 ± 0.3	93.99 ± 0.3	69.72 ± 0.1	78.45 ± 0.5	56.76 ± 0.9
	MP-TGN	97.55 ± 0.1	97.81 ± 0.1	97.26 ± 0.1	97.80 ± 0.2	84.66 ± 0.1
	CAW-N	97.81 ± 0.1	<u>98.22 ± 0.1</u>	<u>97.45 ± 0.1</u>	<u>98.15 ± 0.1</u>	<u>85.67 ± 0.5</u>
	PINT	<u>98.25 ± 0.1</u>	97.14 ± 0.1	94.72 ± 0.1	97.42 ± 0.2	81.57 ± 0.1
	TETGN (ours)	**98.29 ± 0.1**	**98.28 ± 0.1**	**98.26 ± 0.2**	**98.23 ± 0.2**	**88.11 ± 0.1**

Table 2. Ablation study (AP, mean ± std over 10 runs). The best result is in bold and the second best is underlined. (rand feat.) denotes randomly initialised node features, while "(*w/o* exp.)" indicates using raw IDs directly.

Setting	Model	Reddit	Wikipedia	Reddit-fm	Wikipedia-fm	LastFM
Transductive	MP-TGN	98.70 ± 0.1	98.46 ± 0.1	98.54 ± 0.1	98.40 ± 1.4	80.69 ± 0.2
	MP-TGN (rand feat.)	98.71 ± 0.1	98.49 ± 0.2	98.50 ± 0.2	98.39 ± 0.3	74.71 ± 1.5
	TETGN (*w/o* exp.)	98.83 ± 0.3	98.68 ± 0.2	98.56 ± 0.1	98.59 ± 0.2	82.52 ± 0.2
	TETGN	**98.90 ± 0.2**	**98.74 ± 0.1**	**98.79 ± 0.1**	**98.73 ± 0.1**	**83.26 ± 0.1**
Inductive	MP-TGN	97.55 ± 0.1	97.81 ± 0.1	97.26 ± 0.1	97.80 ± 0.2	84.66 ± 0.1
	MP-TGN (rand feat.)	97.41 ± 0.4	97.77 ± 0.3	97.16 ± 0.2	97.67 ± 0.1	79.20 ± 3.9
	TETGN (*w/o* exp.)	97.93 ± 0.2	98.15 ± 0.1	97.31 ± 0.2	98.12 ± 0.1	87.51 ± 0.1
	TETGN	**98.29 ± 0.1**	**98.28 ± 0.1**	**98.26 ± 0.1**	**98.23 ± 0.2**	**88.11 ± 0.1**

TETGN leverages the non-anonymous setting to learn network attributes while effectively capturing these relationships.

Overall, TETGN consistently outperforms all baselines across the five datasets in both inductive and transductive AP.

5.3 Ablation Study

Ablation Study on Non-anonymous Setting. To validate the effect of the learnable ID, experiments were conducted using a version of MP-TGN with randomly initialised node features, referred to as MP-TGN (rand feat.), and a variant of TETGN where the exponential trajectory message encoder was removed, retaining only the learnable ID module, referred to as TETGN (*w/o* exp.). These models were trained using the same settings that were used for the results in Table 1. Table 2 summarises the results; we observe:

(1) MP-TGN (rand feat.) and MP-TGN perform similarly on the attributed datasets (Wikipedia and Reddit) but drop on the three non-attributed datasets. This indicates that utilising node IDs for a lookup table offers little benefit to TGNs and may even hinder performance when capturing the graph structure is essential in the inductive setting.
(2) The performance of TETGN (*w/o* exp.) is higher than MP-TGN (rand feat.), indicating the learnable ID, even without the exponential encoding, provides a significant enhancement by effectively capturing node-specific information. This enhancement suggests that the non-anonymous setting plays a crucial role in representing nodes better than the anonymous setting.
(3) When comparing the results of the MP-TGN and TETGN (*w/o* exp.) experiments with those of the MP-TGN (rand feat.) and TETGN (*w/o* exp.) experiments, it is evident that the independent trajectory encoding architecture (which captures temporal and relative positional relationships) outperforms the merely message-passing approach for representing node features.

Ablation Study on Exponential Trajectory Message Encoder. Further investigation into how the trajectory message encoder contributes to the overall performance of TETGN reveals that TETGN consistently outperforms TETGN (*w/o* exp.). This suggests that the exponential trajectory message encoder is critical for the model. The consistent performance improvement highlights the importance of this component in leveraging the temporal and structural information, thereby enhancing the overall effectiveness of TETGN.

5.4 Hyper-Parameter Analysis on α and β

Figure 3 illustrates the impact of the step-count parameter α and time decay factor β on performance. Two experiments were conducted: one with α fixed at 2 while β was varied, and another with β fixed at 0.1 while α was varied. The mean AP for each setting, averaged over 10 runs, is reported.

The performance exhibits a hump-shaped curve. Optimal performance is achieved with α around 2 and β near 0.1. Values deviating significantly from these optima result in lower accuracy, highlighting the sensitivity of the system to the choice of these parameters. Notably, setting either α or β to 1 leads to a marked performance decline; this is due to the value of 1 making the corresponding parameter inactivate, thereby diminishing its contribution.

5.5 Further Hyper-Parameter Analysis on α, β and d

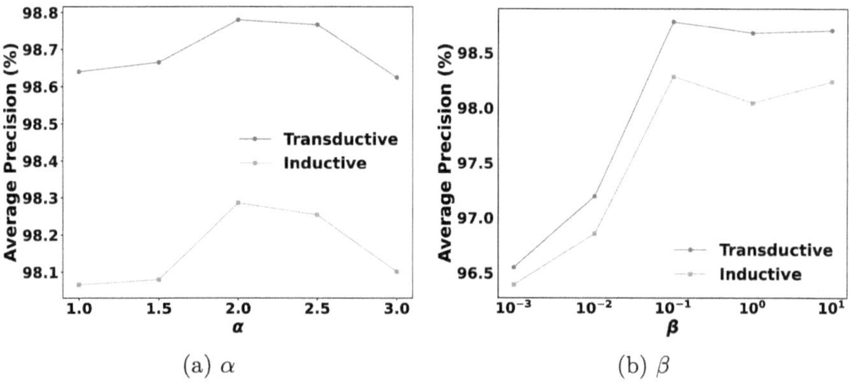

Fig. 3. AP scores of α and β analysis.

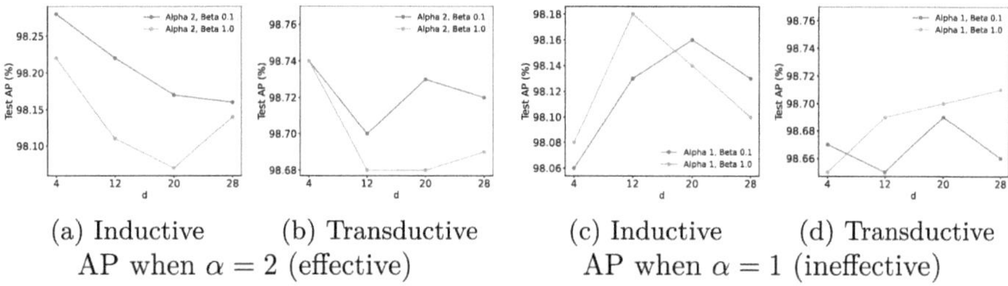

Fig. 4. Hyper-parameter analysis for different settings of α on Wikipedia.

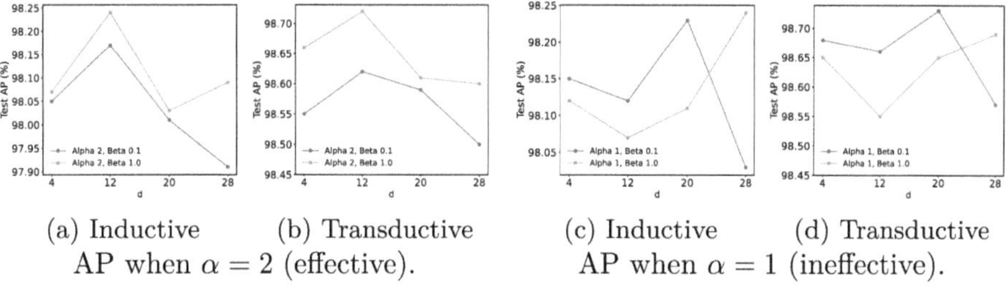

Fig. 5. Hyper-parameter analysis for different settings of α on Wikipedia-fm.

In this section, we specify the "effective" and "ineffective" states of parameters α and β: when $\alpha = 1$ or $\beta = 1$, step scaling or time decay yields negligible improvement (*ineffective*); when $\alpha = 2$ or $\beta = 0.1$, they produce a significant positive impact on performance (*effective*).

(1) Impact of Dimensionality d. We distinguish between two cases.

- Attributed Wikipedia: In both inductive and transductive tasks, the highest mean AP is achieved at $d = 4$ in Fig. 4. As d increases, performance grad-

Table 3. Node classification (AUC, mean ± std over 5 runs).

Model	Wikipedia	Reddit
JODIE	83.91 ± 0.3	61.93 ± 0.3
TGAT	83.71 ± 0.2	65.56 ± 0.5
DyRep	83.29 ± 0.3	62.61 ± 0.5
TGN-Att	87.81 ± 0.6	67.26 ± 0.6
PINT	87.04 ± 0.3	67.34 ± 0.5
TETGN	**89.21 ± 0.4**	**68.31 ± 0.3**

ually decreases, indicating that the model relies primarily on inherent node attributes with trajectory encoding playing a secondary role.

- Non-attributed Wikipedia-fm: Under the same settings, performance consistently improves as d increases from 4 to 16, with the optimum at higher dimensions ($d = 16$) in Fig. 5. This suggests that the model depends more heavily on trajectory encoding in the absence of node and edge features and requires higher dimensionality to capture temporal and structural information.

(2) Impact of α. We distinguish between two cases:

- $\alpha = 1$ (*Ineffective*): On both datasets and in both task settings, performance fluctuates without a consistent optimal dimension and the differences across d are not significant.
- $\alpha = 2$ (*Effective*): Regardless of attribute availability, setting $\alpha = 2$ yields a significant improvement in mean AP than $\alpha = 1$, highlighting α's critical role in capturing temporal dependencies and enhancing model robustness.

(3) Impact of β (under effective α). When $\alpha = 2$ and $\beta = 0.1$, performance consistently surpasses that with $\beta = 1$ on both datasets and across both tasks, with the optimal dimension remaining $d = 4$. This demonstrates that activating time decay enables the model to learn temporal information efficiently even at lower dimensionalities.

5.6 Node Classification

In Table 3, we compare the performance of several models on node classification tasks over the Wikipedia and Reddit to show the effectiveness and versatility for different downstream tasks. TETGN achieves the highest AUC scores on both Wikipedia and Reddit, outperforming the strongest baselines by approximately 1.6% over TGN-Att on Wikipedia and 1.5% over PINT on Reddit. The relatively low standard deviations highlight that TETGN improves average performance and maintains consistent results across multiple runs. This stability suggests that our trajectory encoding mechanism yields better node embeddings in the face of temporal dynamics.

6 Conclusion, Limitations and Future Work

This paper introduced TETGN to address longstanding limitations in both anonymous and non-anonymous Temporal Graph Networks (TGNs). By encoding node trajectories within the message-passing process, TETGN effectively captures temporal and positional relationships without requiring complex preliminary walks. As a result, it can distinguish nodes in symmetric or cyclic structures where previous methods fall short. Experimental results on three real-world datasets confirm its superior performance in modelling dynamic graphs.

Whilst TETGN demonstrates strong performance, we acknowledge two limitations: (1) the dual message-passing streams introduce additional computational overhead; (2) the exponential trajectory encoding may miss periodic patterns in temporal interactions. Future work could explore more efficient encoding mechanisms and extend to large-scale complex temporal networks.

Overall, TETGN takes a step forward in unifying anonymous and non-anonymous approaches, and we anticipate it will inspire further innovations in temporal graph representation learning.

References

1. Battaglia, P.W., et al.: Relational inductive biases, deep learning, and graph networks 10.48550/arXiv, 1806.01261 (2018)
2. Chen, J., et al.: E-LSTM-D: a deep learning framework for dynamic network link prediction. IEEE Trans. Syst. Man Cybern. Syst. **51**(6), 3699–3712 (2021)
3. Cho, K., et al.: Learning phrase representations using RNN encoder–decoder for statistical machine translation. In: EMNLP, pp. 1724–1734. Association for Computational Linguistics, Doha, Qatar (2014). https://doi.org/10.3115/v1/D14-1179
4. Cong, W., et al.: Do We Really Need Complicated Model Architectures For Temporal Networks? (2023). https://doi.org/10.48550/arXiv.2302.11636
5. Ghorbanzadeh, H., Sheikhahmadi, A., Jalili, M., Sulaimany, S.: A hybrid method of link prediction in directed graphs. Expert Syst. Appl. **165**, 113896 (2021). https://doi.org/10.1016/j.eswa.2020.113896
6. Goyal, P., Kamra, N., He, X., Liu, Y.: DynGEM: Deep Embedding Method for Dynamic Graphs (2018). https://arxiv.org/abs/1805.11273
7. Gravina, A., Lovisotto, G., Gallicchio, C., Bacciu, D., Grohnfeldt, C.: Long range propagation on continuous-time dynamic graphs. In: Salakhutdinov, R., et al. (eds.) Proceedings of the 41st International Conference on Machine Learning, vol. 235, pp. 16206–16225. PMLR (2024)
8. Hamilton, W.L., Ying, R., Leskovec, J.: Inductive representation learning on large graphs. In: Proceedings of the 31st International Conference on Neural Information Processing Systems, NIPS 2017, pp. 1025–1035. Curran Associates Inc., Red Hook (2017)
9. Jiang, W.: Graph-based deep learning for communication networks: a survey. Comput. Commun. **185**, 40–54 (2022). https://doi.org/10.1016/j.comcom.2021.12.015
10. Kipf, T.N., Welling, M.: Semi-Supervised Classification with Graph Convolutional Networks (2017). https://doi.org/10.48550/arXiv.1609.02907

11. Kumar, S., Zhang, X., Leskovec, J.: Predicting dynamic embedding trajectory in temporal interaction networks. In: Proceedings of the 25th ACM SIGKDD International Conference on Knowledge Discovery & Data Mining, KDD 2019, pp. 1269–1278. Association for Computing Machinery, New York (2019). https://doi.org/10.1145/3292500.3330895
12. Li, T., Zhang, J., Yu, P.S., Zhang, Y., Yan, Y.: Deep dynamic network embedding for link prediction. IEEE Access **6**, 29219–29230 (2018). https://doi.org/10.1109/ACCESS.2018.2839770
13. Liang, F., Qian, C., Yu, W., Griffith, D., Golmie, N.: Survey of graph neural networks and applications. Wirel. Commun. Mob. Comput. **2022**, 1–18 (2022). https://doi.org/10.1155/2022/9261537
14. Morris, C., et al.: Weisfeiler and leman go neural: higher-order graph neural networks. In: Proceedings of the AAAI Conference on Artificial Intelligence, vol. 33, no. 01, pp. 4602–4609 (2019). https://doi.org/10.1609/aaai.v33i01.33014602
15. Nguyen, G.H., Lee, J.B., Rossi, R.A., Ahmed, N.K., Koh, E., Kim, S.: Continuous-time dynamic network embeddings. In: Companion Proceedings of the Web Conference 2018, pp. 969–976. Republic and Canton of Geneva, CHE (2018). https://doi.org/10.1145/3184558.3191526
16. Rossi, E., Chamberlain, B., Frasca, F., Eynard, D., Monti, F., Bronstein, M.: Temporal Graph Networks for Deep Learning on Dynamic Graphs (2020). http://arxiv.org/abs/2006.10637
17. Scarselli, F., Gori, M., Tsoi, A.C., Hagenbuchner, M., Monfardini, G.: The graph neural network model. IEEE Trans. Neural Networks **20**(1), 61–80 (2009). https://doi.org/10.1109/TNN.2008.2005605
18. Skarding, J., Gabrys, B., Musial, K.: Foundations and modeling of dynamic networks using dynamic graph neural networks: a survey. IEEE Access **9**, 79143–79168 (2021). https://doi.org/10.1109/ACCESS.2021.3082932
19. Souza, A.H., Mesquita, D., Kaski, S., Garg, V.: Provably expressive temporal graph networks (2022). https://arxiv.org/abs/2209.15059
20. Tian, Y., Qi, Y., Guo, F.: FreeDyG: frequency enhanced continuous-time dynamic graph model for link prediction. In: The Twelfth International Conference on Learning Representations (2024)
21. Trivedi, R., Farajtabar, M., Biswal, P., Zha, H.: Dyrep: learning representations over dynamic graphs. In: ICLR (2019)
22. Vaswani, A., et al.: Attention is all you need. Adv. Neural. Inf. Process. Syst. **30**, 5998–6008 (2017)
23. Veličković, P., Cucurull, G., Casanova, A., Romero, A., Liò, P., Bengio, Y.: Graph Attention Networks (2018). https://doi.org/10.48550/arXiv.1710.10903
24. Wang, Y., Chang, Y.Y., Liu, Y., Leskovec, J., Li, P.: Inductive Representation Learning in Temporal Networks via Causal Anonymous Walks (2022). https://doi.org/10.48550/arXiv.2101.05974
25. Weisfeiler, B., Leman, A.: The reduction of a graph to canonical form and the algebra which appears therein. NTI Series **2**(9), 12–16 (1968). (translation)
26. Wen, Z., Fang, Y.: TREND: temporal event and node dynamics for graph representation learning. In: Proceedings of the ACM Web Conference 2022, WWW 2022, pp. 1159–1169. Association for Computing Machinery, New York (2022). https://doi.org/10.1145/3485447.3512164
27. Wu, S., Sun, F., Zhang, W., Xie, X., Cui, B.: Graph neural networks in recommender systems: a survey. ACM Comput. Surv. **55(5), 97: 1 –97**, 37 (2022). https://doi.org/10.1145/3535101

28. Wu, Z., Pan, S., Chen, F., Long, G., Zhang, C., Yu, P.S.: A comprehensive survey on graph neural networks. IEEE Trans. Neural Netw. Learn. Syst. **32**(1), 4–24 (2020). https://doi.org/10.1109/TNNLS.2020.2978386
29. Xiong, J., Sakellariou, R.: Graph structure learning with temporal graph information bottleneck for inductive representation learning (2025). https://arxiv.org/abs/2508.14859
30. Xiong, J., Zareie, A., Sakellariou, R.: A Survey of Link Prediction in Temporal Networks (2025). https://doi.org/10.48550/arXiv.2502.21185
31. Xu, D., Ruan, C., Korpeoglu, E., Kumar, S., Achan, K.: Inductive representation learning on temporal graphs (2020). https://arxiv.org/abs/2002.07962
32. Xu, D., et al.: Adaptive neural network for node classification in dynamic networks. In: 2019 IEEE International Conference on Data Mining (ICDM), pp. 1402–1407. IEEE, Beijing, China (2019). https://doi.org/10.1109/ICDM.2019.00181
33. Xu, K., Hu, W., Leskovec, J., Jegelka, S.: How Powerful are Graph Neural Networks? (2019). https://doi.org/10.48550/arXiv.1810.00826
34. Yu, L., Sun, L., Du, B., Lv, W.: Towards better dynamic graph learning: new architecture and unified library. In: NeurIPS. Curran Associates Inc., Red Hook (2023)

Evaluation and Comparison Semantics for ODRL

Jaime Osvaldo Salas, Paolo Pareti(✉), Semih Yumuşak, Soulmaz Gheisari, Luis-Daniel Ibáñez, and George Konstantinidis

University of Southampton, University Road, Southampton SO17 1BJ, UK
p.pareti@soton.ac.uk

Abstract. We consider the problem of evaluating, and comparing computational policies in the Open Digital Rights Language (ODRL), which has become the de facto standard for governing the access and usage of digital resources. Although preliminary progress has been made on the formal specification of the language's features, a comprehensive formal semantics of ODRL is still missing. In this paper, we provide a simple and intuitive formal semantics for ODRL that is based on query answering. Our semantics refines previous formalisations, and is aligned with the latest published specification of the language (2.2). Building on our evaluation semantics, and motivated by data sharing scenarios, we also define and study the problem of comparing two policies, detecting equivalent, more restrictive or more permissive policies.

Keywords: ODRL · Data Sharing · Policy Evaluation

1 Introduction

The Open Digital Rights Language (ODRL) is a W3C Recommendation that aims at providing a flexible and interoperable way to express rights and obligations related to digital content. Notably, ODRL has been adopted by developers of Data Spaces[1] to regulate access to datasets, and is set to play a key role in regulating data usage within AI pipelines [7]. However, to be useful in practice, policy specifications need to be implementable, enforceable, and universally understandable within the Data Spaces ecosystem. An essential step towards ODRL implementation is defining the semantics of the *Policy Evaluation Problem*: given a usage policy over certain resources, e.g., a dataset, and a *state of the world* capturing the actions of actors, evaluate whether the policy is violated or satisfied in this state.

At the same time, a major function of Data Spaces is data exchange in scenarios where a data consumer wants to use data accompanied by a usage policy (written by the data provider). In this paper, we model the request of the consumer as another ODRL policy and define the problem of policy comparison: given a provider policy defining usage constraints over a resource, a

[1] https://digital-strategy.ec.europa.eu/en/policies/strategy-data.

H. Takeda et al. (Eds.): IJCKG 2025, LNAI 16297, pp. 252–267, 2026.
https://doi.org/10.1007/978-981-95-5009-8_17

usage request policy describing intended actions over that resource, decide if the request conflicts with provider policy, before the exchange takes place.

These scenarios require a precise formal semantics for ODRL evaluation and comparison. ODRL's origins come from the area of Deontic Logic [5], which provides the formal tools to define such a language. However, as of August 2025, the formal semantics of ODRL is still in a draft state[2], and most importantly, it is in natural language. Only a few academic works have attempted to define semantics for ODRL [1,4,15,18], that either consider older versions of ODRL, or are at a preliminary and partial state of formalisation. A more detailed comparison with previous work is provided in Sect. 5.

The problem of evaluation is arguably the most fundamental problem of computational policies, but the fact that it has not been treated in a uniform way is not surprising: ODRL has been used in a large variety of very different applications and each has its own description of operations, systems and functions that the policies should be checked against. Each such system has implemented its own ad-hoc policy evaluation, having to closely synchronise with other systems to achieve interoperability. In this paper, and in line with the recently published draft semantics, we model a state of the world as a set of *events* and implement it as a database relation. Although implementation is orthogonal (e.g., one could use an event recognition system within an organisation, or simply a triplestore with a log of events) our mathematical representation of this is a simple relation that we can query to decide our evaluation and comparison problems.

Our main contributions are the following: (1) a novel, intuitive and simple formal semantics for ODRL, and for a subset of this language that we call ODRL Lite, based on translating policies to queries on a state of the world and (2) a semantics for policy comparison grounded on the notion of query containment and equivalence - giving rise to different comparison semantics (symmetric and asymmetric) based on the application.

Another limitation of the current ODRL Semantics is the lack of strict definition of how to handle reasoning when evaluating policies. In this paper, instead of committing our semantics to a specific approach, we chose to decouple the reasoning dimension and show how our semantics can easily be extended with different types of inferences through a separate materialisation step. This allows us to provide a core semantics for the basic case without inference, and the flexibility to adapt to different reasoning algorithms.

2 Background

ODRL is a policy expression language that provides a flexible and interoperable information model, vocabulary, and encoding mechanism to represent *rules* about the usage of content and services. ODRL policies have four *core components*: (1) *Asset*: a resource that is subject of a policy; (2) *Party*: an entity that undertakes functional roles as the *assigner* or *assignee* of a policy; (3) *Action*:

[2] https://w3c.github.io/odrl/formal-semantics/.

an operation (for example Read, Write, Copy, Anonymise) that can be exercised on a target asset and (4) *Rule*: defines the *permission*, *prohibition*, or *obligation* of the assignees of the policy to execute actions over assets. A single policy may define multiple rules on different core components.

Components of an ODRL Policy can be constrained. Constraints are boolean expressions that either dictate conditions applicable to a Rule, or refine the semantics of actions, parties or assets. In ODRL, the former usage is called *Constraints* and the latter *Refinements.* For example, a permission rule can be constrained to a particular datetime, or to a particular location (e.g., the European Union); the Action *Print* can be refined with the boolean expression *with resolution less than 500dpi*; and the asset *Document D* can be refined with *subset of pages 1 to 10.*

Complex ODRL policies can also make use of additional types of rules, which build on the three basic types of rules previously mentioned. Permissions can have *duties*, which are types of obligations that need to be fulfilled before the assignee exercises the permission. The *remedy* and *consequence* rules specify remedial actions that can restore validity in a state of the world in the case of prohibition violations, or in the failure to fulfill an obligation or a duty.

ODRL policies are usually read as *Rule from Assigner for Assignee to Action Asset subject to Constraints and Refinements, e.g. Permission from Bob for Alice to Read Document D subject to the constraint the day being Monday and Document D to be refined to its Subset of Pages 1–10.* While all rules must define the Action component, all other elements are optional. We assume all instances of components are drawn from a known vocabulary or Knowledge Graph.

Policies act upon a *world*, e.g., a Socio-Technical System or a Data Space. A *state of the world* is a formal representation of the state of the assets, parties and any other environmental information relevant for the constraints. For example, *Bob Read document D, Alice Read Document D, The day is Tuesday.* An ODRL Evaluator receives as input an ODRL policy and a state of the world and decides which rules of the policy are violated or fulfilled in the state of the world. An ODRL Evaluator may work in two different scenarios or modalities: *Access Control*, adds to the input an action that is *requested*, the evaluator must decide if the execution of the action would create a state of the world that violates any policy; and *Monitoring*, where the state of the world includes actions already performed and the evaluator decides violation/fulfillment. In this paper, we focus on the problem of violation and fulfillment of policies given a state of the world, conforming to the Monitoring scenario. Note however, that the Access Control modality can be reduced to the Monitoring one if the state of the world that would result after performing the requested action can be predicted.

3 First-Order ODRL Semantics

An ODRL policy is evaluated against a *state of the world*, which we model as a set of *events*. For a state of the world ω and an event e we write $\omega(e)$ when $e \in \omega$. An event is described by a finite number of features $i \in [0, n]$, corresponding to core components, constraints and refinements of the language, e.g., time, actor, action, etc. These features can also describe relationships with other events. For example, a feature can represent the number of times the same action has been already performed, as defined by the ODRL *count* constraint. Thus an event e can be defined as an n-tuple $< e_0, e_1, e_2, \ldots, e_n >$, where the value of a feature e_i is either null (unspecified), a constant, or a set of constants. For simplicity, we assume that the first value e_0 of event e is always the timestamp of when the action has been performed, and e_1 is the action. All other values correspond to the rest of the ODRL components, refinements and constraints an ODRL rule can have. Each feature is unique, and its relationship with a specific ODRL component known. Two features i and j might both represent the "is located in" relation, but the first one might represent this relation as a property (refinement) of the assignee component, while the second one as a refinement of the asset. For any feature i, let γ_i be the ODRL component of i, that is, if i represents a refinement of the value e_k, then $\gamma_i = k$. For example, if i is "has the creation date" of the asset component, then $\gamma_i = Asset$. When a feature captures a rule-wide property (corresponding to a constraint and not a refinement), or the timestamp e_0, then γ_i equals a special constant ρ. In all other cases, that is, when a feature i is a component itself, then simply $\gamma_i = i$.

Table 1. A state of the world corresponding to Example 1

Datetime	Action	Actor	Asset	Print.Resolution	Book.Pages
1	Print	Alice	Picture	500 dpi	null
2	Read	Bob	Book	null	450
3	Print	Alice	Book	600 dpi	300

Example 1. Consider a world with actors Alice and Bob, assets Book and Picture, and actions Read and Print. Book has the property "number of pages", and action Print has the property "resolution" that can take an integer value in the range 100-5000dpi. Table 1 shows the model of the events "Actor Alice executes Action Print at 500dpi on Asset Picture at datetime 1", "Actor Bob executes Action Read on Asset Book with 450 pages at datetime 2" and "Actor Alice executes Action Print at 600 dpi on Asset Book with 300 pages at datetime 3".

Rules in an ODRL policy apply only to specific events, based on restrictions on the values of the features of an event. For example, a prohibition to "print" can be restricted to apply only to print actions over certain resolutions. To restrict the value of a feature i we write triples of the form $< i, op, v >$, where i is one of the n features of an event, v is a constant or set of constants, and op is one of the following binary comparison operators: `odrl:equals` $(=)$, `odrl:gt` $(>)$, `odrl:gteq` $(\geq)$, `odrl:lt` $(<)$, `odrl:lteq` $(\leq)$, `odrl:neq` $(\neq)$, `odrl:isA` (or `rdf:type`), `odrl:hasPart` $(\supset)$, `odrl:isPartOf` $(\subset)$, `odrl:isAllOf` $(\equiv)$, `odrl:isAnyOf` ($\in$), and `odrl:isNoneOf` ($\notin$). A *condition* C is either a single triple $< i, op, v >$, known as a *simple* condition, or a *complex* condition which is a boolean combination of simple conditions, combined with the $\wedge$, $\vee$ and $\neg$ operators.[3] We denote with $C(e)$ the boolean result of the evaluation of condition C on event e. Note that, ODRL does not provide negation of conditions directly but provides the exclusive-or operator $\nleftrightarrow$ that we can use to obtain negation when features are not null.[4] A complex condition is evaluated in the natural way as a boolean combination of simple conditions. A simple condition $< i, op, v >$ is evaluated to true on an event e if the value comparison of e_i and v using operator op is true, and false otherwise. We ground our condition evaluation on SPARQL Filter Evaluation semantics, by defining the value comparison as true if the equivalent SPARQL filter expression would keep the bound solution. Thus evaluation errors, for example due to incomparable types, behave as evaluations to false. Condition $< i, \texttt{odrl:isA}, v >$ is evaluated as $< i_c, \texttt{odrl:hasPart}, \{v\} >$, where i_c is the set of classes of i. The remaining set operators are evaluated in their natural way. When the value of feature i is unspecified (i.e., null) in the event, then any simple condition C for that feature evaluates to false.

Corresponding to ODRL rules, an *event rule* τ is defined as a conjunction of conditions. In the next subsection we will see how event rules are assigned a specific type, such as permissions or prohibitions, when used within a policy.

Example 2. Consider a policy with the following rules (1) "Permission for Alice to Print Picture", is translated as permission $< Actor, =, Alice > \wedge < Action, = , Print > \wedge < Asset, =, Picture >$; (2) "Prohibition for Bob to Read Book between datetimes 3 and 5 and number of pages is greater than 250" translates as prohibition $< Actor, =, Bob > \wedge < Action, =, Read > \wedge < Asset, =, Book > \wedge (< Datetime, \leq, 5 > \wedge < Datetime, \geq, 3 >) \wedge < Book.Pages, >, 250 >$ (demonstrating a complex condition on the Datetime feature); and (3) "Obligation for Bob to Read Book before datetime 3" is translated as obligation $< Actor, =, Bob > \wedge < Action, =, Read > \wedge < Asset, =, Book > \wedge < Datetime, <, 3 >$.

Given a condition C, let $\bar{C}$ be the set of simple conditions in C, and I_C be the set of features that appear in it, that is, the set $\{i \mid \exists\ op,\ v\ .\ < i, op, v > \in \bar{C}\}$. Correspondingly, given an event rule I_τ, let $I_\tau = \bigcup_{C \in \tau} I_C$. To align with the ODRL specification we require all our rules to be well-formed, as follows. An event rule τ is said to be *well-formed* if:

1. there is an action a such that $< Action, =, a > \in \tau$; that is, each ODRL rule must specify the action a it applies to,
2. for all $i \in I_\tau$, if $\gamma_i \in \{Action, Asset, Party\}$, then $< \gamma_i, =, v > \in \tau$, and there exists no other condition $C \in \tau$ such that $\gamma_i \in I_C$ (in other words, each

[3] We omit the `odrl:andSequence` logic operator due to its lack of a clear semantic interpretation.

[4] For a condition c, holds $\neg c \equiv c \nleftrightarrow \top$ where $\top$ is always true; $\top$ can be expressed with a disjunction of complementary conditions, such as $< i, =, x > \vee < i, \neq, x >$.

ODRL core component used in a rule must be specified once, and not used in any other condition),
3. for all complex conditions $C \in \tau$ there do not exist i and i' in I_C with $\gamma_i \neq \gamma_{i'}$; in other words, each boolean combination of constraints must use features of a single ODRL component.

From now on, we will assume that event rules are well-formed. We say that an event rule τ *matches* an event e, denoted by $\mathsf{match}(\tau, e)$ if all its conditions evaluate to true on e, that is when $\bigwedge_{C\in\tau} C(e)$ is true. Note that $\mathsf{match}(\tau, e)$ has a natural correspondence to the evaluation of a boolean query on tuple e, and we can easily implement it with SQL or SPARQL.

Example 3. All three rules from Example 2 are well-formed. The rule permitting Alice to Print Picture matches the event with datetime 1 from Table 1. The same rule does not match the event with datetime 2, because, among other things, Action and Actor don't match, nor with the event with datetime 3 because the Assets don't match. Intuitively, the permission did not specify a refinement on the Print action, hence Alice is allowed to print at any resolution. The rule prohibiting Bob to Read the Book does not match the event with datetime 1 or 3 because either the Action and Actor don't match. The rule does not match with the event with datetime 2 either: though the event specifies *Book.Pages* must be greater than 250, its datetime falls outside the specified range. Finally, the rule obligating Bob to Read the Book does not match the first and third event, but it does match the second.

3.1 ODRL Lite

An *ODRL Lite policy* is a tuple $< P, F, O >$, where P, F and O are the sets of permission, prohibition and obligation event rules, respectively. Faithful to the deontic semantics of these concepts permissions model what is allowed in a state of the world, but not compulsory, prohibitions state what is not allowed, and obligations require certain actions to be performed to be fulfilled.

Definition 1. *We say that a state of the world ω violates a policy $< P, F, O >$ if any of the following three cases occurs:*

- ***(permissions)*** *there exists an event $e \in \omega$ and for all permission rules $\tau \in P$, $match(\tau, e)$ is false; equivalently if the following is true:*

$$\exists e\ \omega(e) \wedge \bigwedge_{\tau\in P} \neg match(\tau, e) \tag{1}$$

- ***(prohibitions)*** *there exists an event $e \in \omega$ and a prohibition rule $\tau \in F$, such that $match(\tau, e)$ is true; equivalently if the following is true:*

$$\exists e\ \omega(e) \wedge \bigvee_{\tau\in F} match(\tau, e) \tag{2}$$

– ***(obligations)*** *for any of the obligation rules $\tau \in O$, there does not exist an event $e \in \omega$, such that $match(\tau, e)$ is true; equivalently:*

$$\bigvee_{\tau \in O} \nexists e \ \omega(e) \wedge match(\tau, e) \tag{3}$$

Conversely, a state of the world ω is said to be *valid* with respect to a policy $p =< P, F, O >$ if it does not violate it, or equivalently, the policy is *satisfied* or *true on* ω, writing $p(\omega) = T$. Following from Definition 1, to decide violations one should evaluate the following logical expression:

$$\left(\exists e \ \omega(e) \wedge (\bigwedge_{\tau \in P} \neg match(\tau, e)) \vee (\bigvee_{\tau \in F} match(\tau, e)) \right) \vee \left(\bigvee_{\tau \in O} \nexists e \ \omega(e) \wedge match(\tau, e) \right) \tag{4}$$

Since the predicate *match* is just a shorthand notation for the satisfaction of the complex conditions of a rule on an event tuple, as discussed above, expression 4 is just a first-order query on relation ω (the state of the world) and can be easily implemented via SQL or SPARQL.

Example 4. Consider the policy $p =< P, F, O >$ where each of P, F, O contain the single event rule p_1, f_1 and o_1, respectively, as defined in Example 2. Permission p_1 allows Alice to Print Picture, f_1 prohibits Bob to Read Book at a certain time and over a number of pages, and o_1 obliges Bob to Read Book before a timestamp. The evaluation of p over the state of the world ω in Table 1 is as follows: for p_1, the matches with events at datetime 2 and 3 return false, meaning ω violates the permissions of the policy (e.g. the policy does not explicitly grant permission for Bob to Read Book); for f_1, none of the events match as explained in Example 3, hence, ω is valid wrt the prohibitions of the policy; for O_1, there is an event where Bob Read[s] Book before datetime 3, so ω is valid wrt the obligations of the policy. However, as permissions are violated, ω violates the whole policy.

One can observe that our semantics adopts the *prohibited-by-default* approach, that is, everything not explicitly permitted is implicitly prohibited. Prohibitions are then primarily useful to "carve out" prohibited subsets of a permission. We believe this is a practical choice in several use cases, such as data sharing, which require a cautious approach to permissions. However, one can easily simulate the alternative semantics, *permitted-by-default*, by adding to all policies a universal permission that matches every event. This can be easily done with inferences over actions, by defining a single permission for a generic top-level action that all other actions are specialisations of. The only limitation of simulating this alternative semantics is that it does not allow the "carving out" of permitted subsets of prohibitions.

It should also be noted that our semantics is *closed-world*; any event that does not appear in the state of the world is false, that is, has not happened. Closed-world semantics are more natural for a usage policy, which by nature intends to govern the access and use of resources.

3.2 Full ODRL

In this section we extend ODRL Lite to cover additional ODRL features such as duties, remedies and consequences to express the core of the ODRL language.

Definition 2. *An ODRL Lite policy* $< P, F, O >$ *can be extended into an ODRL Policy* $< P, F, O, DP, DPC, FR, OC >$, *where:*

- DP *is a set of pairs of event rules* $< \tau, \tau' >$, *with* $\tau, \tau' \in P$, *specifying that* τ' *is a duty of permission* τ.
- DPC *is a set of triples of event rules* $< \tau, \tau', \tau'' >$, *with* $\tau, \tau', \tau'' \in P$ *specifying that* τ' *is a duty of permission* τ *with consequence* τ''.
- FR *is a set of pairs of event rules* $< \tau, \tau' >$, *with* $\tau \notin F$ *and* $\tau' \in P$ *specifying that* τ' *is a remedy of prohibition* τ.
- OC *is a set of pairs of event rules* $< \tau, \tau' >$, *with* $< Datetime, \leq, t > \in \tau$, *where* t *is a timestamp,* $\tau \notin O$, *and* $\tau' \in P$, *specifying that* τ' *is a consequence of not fulfilling the obligation* τ *by time* t.

Note that per the specification "the Party obligated to perform the duty MUST have the ability to exercise the Duty Action" [10]. Thus, every duty rule τ' in a pair $< (\tau, \tau') >$ in DP must explicitly be in P (and this sentence is guaranteed for consistent policies). We have made similar choices for the rest of these complex rules as per the specification.

Unlike ODRL Lite, this extended definition makes direct use of event timestamps, to establish the temporal ordering of events. Intuitively, DP describes duties that must have been implemented before a permission was exercised; DPC extends this with consequences that must come, alongside the original duty if a permission was exercised without fulfilling its duty; FR contains prohibitions and remedies which must follow the violation of the former; lastly, OC dictates consequences that should occur, in addition to the obliged event, if the latter is not performed in time. Given a rule τ and an event e, a time-independent match is denoted $softmatch(\tau, e)$ and is true if $match(\tau \setminus \{ \langle Datetime, \leq, t \rangle \mid t \in T \}, e)$, where T is the set of all time points.

Definition 3. *An ODRL Policy* $< P, F, O, DP, DPC, FR, OC >$ *is violated on a state of the world* ω *if the ODRL Lite Policy* $< P, F, O >$ *is violated on* ω, *or one of the following holds:*

- ***(permission duties)*** *there exists an event* $e \in \omega$ *and a pair of event rules* $<\tau, \tau'>$ *in* DP *such that* $match(\tau, e)$ *is true and there is no event* $e' \in \omega$ *with* $e'_0 \leq e_0$ *and* $match(\tau', e')$ *true; equivalently:*

$$\exists e \; \omega(e) \wedge \bigvee_{<\tau,\tau'>\in DP} match(\tau, e) \wedge \nexists e' \; \omega(e') \wedge match(\tau', e') \wedge (e'_0 \leq e_0) \quad (5)$$

- ***(permission duties with consequences)*** *there exists an event* $e \in \omega$ *and a triple of rules* $<\tau, \tau', \tau''>$ *in DPC such that* $match(\tau, e)$ *is true (permission exercised), and (a) there is no prior (duty) event* $e' \in \omega$ *with* $match(\tau', e')$ *and* $e'_0 \leq e_0$ *and (b) there is no subsequent duty event* $e'' \in \omega$ *with* $match(\tau', e'')$ *and* $e_0 \leq e''_0$*, or, there is no subsequent consequence event* $e''' \in \omega$ *with* $match(\tau'', e''')$ *and* $e_0 \leq e'''_0$*; equivalently:*

$$\exists e\ \omega(e) \wedge \bigvee_{<\tau,\tau',\tau''>\in DPC} match(\tau, e) \wedge \left(\begin{array}{c} \nexists e'\ \omega(e') \wedge match(\tau', e') \wedge (e'_0 \leq e_0) \wedge \\ \left(\begin{array}{c} \nexists e''\ \omega(e'') \wedge match(\tau', e'') \wedge (e_0 \leq e''_0) \\ \vee \\ \nexists e'''\ \omega(e''') \wedge match(\tau'', e''') \wedge (e_0 \leq e'''_0) \end{array} \right) \end{array} \right) \quad (6)$$

- ***(prohibition remedies)*** *there exists an event* $e \in \omega$ *and a pair of event rules* $<\tau, \tau'>$ *in FR with* $match(\tau, e)$ *and there is no event* $e' \in \omega$ *with* $match(\tau', e')$ *and* $e'_0 \geq e_0$*; equivalently:*

$$\exists e\ \omega(e) \wedge \bigvee_{<\tau,\tau'>\in FR} match(\tau, e) \wedge \nexists e'\ \omega(e') \wedge match(\tau', e') \wedge (e'_0 \geq e_0) \quad (7)$$

- ***(obligation consequences)*** *for any pair of event rules* $< \tau, \tau' >$ *in OC, such that* $< Datetime, \leq, t > \in \tau$ *there is no event* $e \in \omega$ *such that* $match(\tau, e)$ *(obligation not fulfilled) and there is no pair of events* $e', e'' \in \omega$ *such that* $softmatch(\tau, e')$ *(obligation implemented late),* $match(\tau', e'')$ *and* $e''_0 \geq t$ *(consequence implemented); equivalently:*

$$\bigvee_{\substack{<\tau,\tau'>\in OC, \\ <0,\leq,t>\in\tau}} \left(\begin{array}{c} (\nexists e.\ \omega(e) \wedge match(\tau, e)) \wedge (\nexists e', e''.\ \omega(e') \wedge \omega(e'') \\ \wedge\ softmatch(\tau, e') \wedge match(\tau', e'') \wedge (e''_0 \geq t)) \end{array} \right) \quad (8)$$

Again, all elements of Definition 3 are first-order queries and thus detecting violations of ODRL policies can be done via query answering.

Example 5. Consider the following Policy: "Permission for Alice to Print Book provided Bob has Read Book". We translate this as a pair of event rules $< \tau, \tau' >$ where the permission component is $< Actor, =, Alice > \wedge < Action, =, Print > \wedge < Asset, =, Book >$ and the duty component is translated as $< Actor, =, Bob > \wedge < Action, =, Read > \wedge < Asset, =, Book >$. In this case, each individual component matches an event in Table 1– the event with datetime 1 for the permission component, and the event with datetime 2 for the duty component. However the datetime feature of the duty component is greater than that of the permission component, indicating that the permitted event occurred before the duty was carried out. Thus, the state of the world violates the policy.

Note that Eq. 5 does not examine if there is a permission for all events, as this is already performed in Eq. 1. This is why any event detected here (and equally for should also be explicitly permitted by Eq. 1; this is required by Definition 2 which makes sure that for all pairs $< (\tau, \tau') >$ in DP, rule τ also is in P. The same is true for permission rules in DPC. This is not true however for prohibitions in FR, i.e., if a for a pair $< (\tau, \tau') >$ a rule τ is also in F then Eq. 2 will capture a violation independent of remedies, a case we want to avoid as it would make the remedy meaningless. Similarly, an obligation rule τ that exists in a pair $< (\tau, \tau') >$ in OC should not exist in O since the absence of an obliged event in O, inside the "deadline", would trigger a violation by Eq. 3, regardless of whether the obligation and its consequence were later fulfilled.

3.3 Incorporating Reasoning

Instances of ODRL components may come from an ontology that defines relationships between them that are relevant to the problem of policy evaluation. The ODRL Vocabulary[5] defines the Actions *Use* and *Transfer* as the top level types, with more specific instances that encompass their operational semantics specified with the *included in* relation. For example, the operational semantics of action *Display* (e.g. an image) is included in the operational semantics of action *Play* (e.g. a video). A permission to perform an action, such as *Play*, naturally also implies the permission to perform its sub-actions, such as *Display*. Our semantics can easily be extended to capture this type of reasoning, by considering extended policies where all such implied permissions are materialized. Other types of reasoning can equally be dealt with similar materialisation steps, or through extensions of our definitions of query evaluation, containment and equivalence to include reasoning dependencies. For example, conjunctive query containment under dependencies [11] can be answered either by (a) materialising new event rules, by the use of forward-chaining, algorithms that use the dependencies, resulting in saturated policies which can then be checked for containment; or (b) directly rewriting the query form of the initial policy into a new policy and then checking for containment [2].

We do not discuss other rules of inference in this paper, as no obvious interpretation of them can be extracted from the ODRL specification. For example, it is debatable whether a prohibition to perform the *Play* action should also imply a prohibition to perform the *Display* action. Conversely, a prohibition to *Display* might also imply a prohibition to *Play*. How reasoning should be dealt with is also unclear for obligations or other ODRL components. Once a suitable inference strategy is chosen, it can be applied in a similar way as for permissions; for example, by materialising the more general or more specific forms of each prohibition.

[5] https://www.w3.org/TR/odrl-vocab.

4 Policy Comparison Semantics

In this section we model policy comparison using query containment and equivalence. In particular, query containment models policy implication, intuitively stating when a policy is broader or stricter than another policy. On the other hand, query equivalence models policies that are fulfilled on exactly the same states of the world, that is, they are not conflicting. A query q is *contained* in a query q' ($q \sqsubseteq q'$) if for all database instances D, the result of q on D, denoted $q(D)$ is included in the result of q' on the same D, that is $q(D) \subseteq q'(D)$. When $q \sqsubseteq q'$ and $q' \sqsubseteq q$ we say that the two queries are equivalent and write $q \equiv q'$.

Given ODRL policies p and p', p is contained in p', denoted $p \sqsubseteq p'$, if: for any state of the world ω, whenever $p(\omega)$ is true then $p'(\omega)$ is true. Thus by translating the policies p and p' to queries, q_p, $q_{p'}$ correspondingly, as dictated by Def 3, it holds that $p \sqsubseteq p'$ iff $q_p \sqsubseteq q_{p'}$. If $p \sqsubseteq p'$ and $p' \sqsubseteq p$ then we say that p is equivalent to p', denoted $p \equiv p'$.

In the general case, any semantic difference between two policies could be considered a cause for conflict, such as in a policy negotiation scenario when each party is not willing to compromise on their chosen set of policy rules. We call this a *symmetric conflict.* When a symmetric conflict exists between policies p and p', then it also exists between p' and p.

Definition 4. *Given policies p and p', we say that there is a symmetric conflict between p and p' if $p \not\equiv p'$.*

Often, it is useful to determine whether a policy is strictly more/less permissive than another one. For example, in a data sharing scenario, no conflict arises if the policy proposed by a data requester is stricter than the policy offered by the data provider. This relaxation leads to the notion of asymmetric conflicts.

Definition 5. *Given a requester policy p^r and a provider policy p^o, we say that there is an asymmetric conflict between them, if $p^r \not\sqsubseteq p^o$.*

4.1 Comparison of ODRL Lite Policies

We now provide an approach, alternative to query containment, to detect asymmetric conflicts (and thus also symmetric ones) among ODRL Lite policies that is less computationally complex. As evidence of this, we developed an implementation [14] which is currently used in the UPCAST project [19] to facilitate policy negotiation in Data Marketplaces. This approach works by directly comparing the rules of two policies against each other. For this, we are going to use the notions of *event rule containment* and *event rule overlap.*

Definition 6. *For all event rules τ, τ', we say that τ is contained in τ', denoted $\tau \sqsubseteq \tau'$ if for all events e, $match(\tau, e)$ implies $match(\tau', e)$.*

Definition 7. *For all event rules τ, τ', we say that τ overlaps with τ', denoted $\tau \sqcap \tau'$ if there exists an event e such that $match(\tau, e) \wedge match(\tau', e)$.*

These definitions are trivially lifted to sets of event rules on either side of the containment and overlap operator. This approach also assumes that the policies are in a *consistent* normal form, that is, when there is no overlap between prohibitions and permissions/obligations, and every event that matches an obligation also matches a permission. While the existing implementation assumes that policies are already in this format, any ODRL Lite policy can be made consistent by removing from a policy the following redundant elements: (1) every part of a permission and obligation that is explicitly forbidden, and (2) every part of an obligation that is not permitted. Note how these elements are redundant, and thus their removal does not change the meaning of the policy. Also note how prohibitions can be removed too, as they are redundant in a consistent policy.

Definition 8. *An ODRL Lite policy* $p =< P, F, O >$ *is consistent if* $P \not\sqcap F$, $O \not\sqcap F$, *and* $O \sqsubseteq P$.

Theorem 1. *Given a consistent ODRL Lite requester policy* $p =< P, F, O >$, *and a consistent ODRL Lite provider policy* $p' =< P', F', O' >$, $p \not\sqsubseteq p'$ *iff:*

1. *The requester's permissions are not contained in the provider's permissions, that is,* $P \not\sqsubseteq P'$ *is true, or*
2. *The requester is not explicitly agreeing to every obligation set by the provider, that is,* $\exists \tau' \in O', \nexists \tau \in O \; (\tau \sqsubseteq \tau')$

Proof. $\Leftarrow$ If $p \not\sqsubseteq p'$, then there must exist a state of the world ω where p is valid but p' is not. Since prohibitions are redundant in consistent policies, there must exist a violation for p' in ω from the permission or obligation equation of Definition 1. If the violation is caused by the **permission equation**, then there exists an event $e \in \omega$ that matches a permission in P but no permissions in P'. Thus $P \not\sqsubseteq P'$. If the violation is caused by the **obligation equation**, then there exist an obligation $\tau' \in O'$ such that no event e exists in ω, such that $match(\tau', e)$ is true. Thus, there cannot exist a $\tau \in O$ such that $\tau \sqsubseteq \tau'$, as this would imply the existence of an e in ω that matches τ and, by inclusion, τ'.

$\Rightarrow$ If either condition (1) or condition (2) of the theorem are true, then we can construct a state of the world ω where p is valid but p' is not (thus proving $p \not\sqsubseteq p'$) as follows. If **the second condition of the theorem is true**, pick an obligation $\tau' \in O'$ such that $\nexists \tau \in O \; (\tau \sqsubseteq \tau')$. Construct ω from an empty set by adding an event e to ω for each $\tau \in O$, such that $match(\tau, e)$ and $\neg match(\tau', e)$ (note that, by the premises, such element must exist). It is easy to see that ω satisfies p but not p', as it violates obligation τ'. If **the second condition of the theorem is false, but the first is true**, construct ω by adding an event e to ω for each $\tau \in O$ (thus every obligation of O is satisfied, and so naturally also every obligation in O', as every obligation in O' is satisfied by the same events that satisfy the obligations from O that they contain), then pick one permission $\tau \in P$ such that $\{\tau\} \not\sqsubseteq P'$, and one event e^* such that $match(\tau, e^*)$ and not exists a $\tau' \in P'$ such that $match(\tau, e^*)$. Such element must exist if $P \not\sqsubseteq P'$. After adding e^* to ω, ω still satisfies p but not longer p'.

5 Related Work

We will now present a brief overview of the evolution of ODRL and its formalisations, and how they compare with the work in this paper. Iannella's work in [8] and [9] presented ODRL's early development and use cases, as well as introduced its XML structure and application to Creative Commons licensing. These studies also discussed the challenges of semantic mapping and policy interpretation. The Open Mobile Alliance and multimedia scenarios provided additional grounding in practical implementation contexts. Serrão et al. [16] demonstrated the role of ODRL in the OpenSDRM architecture for managing streaming media and sensor data, [20] offered a broader review of ODRL's evolution and its comparative position among rights expression languages (RELs), while [13] highlighted the challenges in using ODRL for real-world data-sharing infrastructures.

The problem of providing an ODRL formal semantics has been tackled since the inception of the language. Pucella and Weissman defined a formal semantics of ODRL 1.1 for the problem of defining when a permission (or prohibition) follows from a set of ODRL statements [15]. At the time, the link with RDF had not been established; ODRL was XML based and lacked a proper way to specify instances of Parties and Assets. To the best of our knowledge, the refinement of XML Schemas into RDF and ontologies was first proposed by [6], while [12] proposed the first OWL modelling of ODRL, before the release of ODRL 2.0.

Steyskal and Polleres [18] proposed semantics for ODRL 2.1 supporting reasoning encoded in the functions that matched requests against rules in a policy. Requests contain actions, parties, assets and constraints. The requested actions are matched against rule actions, and return one of six values depending on the relationship between actions (*match*, *no match*, *broader*, *narrower*, *part of*, *required*). Constraints are evaluated as boolean functions, while parties and assets are matched exactly. Based on these matches a policy is found to be *permitted*, *conditionally permitted*, *conditionally prohibited*, *not active* or *not applicable*.

De Vos et al. [4] defined the ODRL Regulatory Compliance Profile (ORCP), a custom ODRL profile, and semantics based on answer set programming where processes (series of events, in our case) are modelled as model states, and ODRL policies, which are in RDF format, are translated into *fluents* –facts that are true if present and false otherwise – following the closed-world assumption. The model and fluents would be compiled into an answer set program and solved to check for compliance between policies and regulatory frameworks. All mentions of reasoning under dependencies are left as future work, but still discussed briefly. Our approach instead saturates policies by materialising new event rules according to dependencies specified in a given ontology.

The work in [3] proposes a novel approach to express ODRL policies that integrates the descriptive ontology terms of ODRL with other languages that allow dynamic data handling or function evaluation. Our evaluation problem can then be cast as the execution of the program in the target language. The framework proposes abstract functions *transformConstraints* and *transformActions* to translate ODRL to a target programming language and provide implementations for Python and Java. They do not propose a formal semantics of the transformation functions, highlighting instead flexibility and practicality.

Recently [17] introduced a Compliance Report Model and test suite to support systematic analysis of policy engines. They also implement an ODRL Evaluator based on the current ODRL formal semantics draft with reasoning capabilities implemented by using the EYE reasoner. However, their evaluator is incomplete due to the incompleteness of the W3C draft.

We now provide a more detailed comparison with the formalisation of ODRL semantics by Bonatti et al. [1]. While their work is an early stage of development, with certain components being left for future work, it is the most comprehensive formalisation of ODRL 2.2 to date, which builds and expands on previous work. They provide a declarative semantics for ODRL which focuses on a variant of the evaluation problem, where policies are used to determine compliance over traces (i.e. sequence of states). A core difference between our semantics and the one presented by Bonatti et al. is that ours is based on query answering and it is (1) our semantics uses a single domain of constants by abstracting away from identifiers (e.g. when two or more syntactical expression refer to the same entity); (2) it deals with inferences (such as dependencies between actions), as a separate materialisation step; (3) it does not require a distinction between "nontemporal" and "temporal" left operands, and between "active" or "inactive" rules.

At a semantic level, our formalisation differs from the one in Bonatti et al. in the following ways. In [1], obligations need to be fulfilled by all possible assignees, while in ours only by one. Our choice simplifies computing compliance with obligations, as determining a complete list of all possible assignees might not always be trivial or even possible. Our semantics also provides an interpretation of obligation with consequences, while in [1] it is not specified under which conditions an obligation can be considered fulfilled in time (thus not requiring consequences), and when instead consequences are required. Moreover, in [1], events that fulfil obligations are automatically permitted. However this is not formalised, and if any one of multiple events fulfils a single obligation, it is unclear if all of them are automatically permitted. Our semantics is more expressive, as it can also capture the case in which some events that could fulfil an obligation are not directly permitted, but might require duties or incur consequences. We also provide an interpretation of constraint operators which avoids redundancy and better aligns with the specification (e.g. differentiating between `isA`, `isPartOf` and `isAnyOf`). Lastly, we extend the study of policy comparison with an asymmetric case and motivate it in a practical scenario.

6 Conclusion

In this paper, we proposed a comprehensive formal semantics for the latest ODRL specification (2.2). Our semantics directly addresses both the problem of policy evaluation (i.e. whether a policy is satisfied in a state of the world) and the problem of policy comparison. For the latter problem, we define both a *strict* symmetric comparison and a *softer* asymmetric one. We motivated the need for both comparison approaches in the context of data usage policies in data marketplaces and Data Spaces, providing an answer to the increasing need for data usage regulation in the age of AI. Our policy comparison semantics is intuitive and easy to interpret, and, due to its grounding in query containment, is easy to implement in any standard query language such as SQL or SPARQL. In particular, we provided a streamlined approach to detect both types of policy comparisons for a core subset of ODRL policies, proved its correctness and provided a reference implementation. Future work will focus on a theoretical analysis of our semantics, studying the decidability and complexity of its core decision problems, and extending its implementation with the aim of making it more usable for the community.

Acknowledgements. This work was supported by the EU Commission and the UKRI Horizon Europe guarantee funding scheme for the Horizon Europe projects UPCAST (101093216), RAISE (101058479) and DataPACT (101189771).

References

1. Bonatti, P.A., Fornara, N., Harth, A.: Towards a Formal Semantics of the Open Digital Rights Language (ODRL 2.2). CEUR Workshop Proceedings, vol. 3977 (2025). https://ceur-ws.org/Vol-3977/OPAL2025-4.pdf
2. Calvanese, D., De Giacomo, G., Lembo, D., Lenzerini, M., Rosati, R.: Tractable reasoning and efficient query answering in description logics: the DL-Lite family. J. Autom. Reason. **39**, 385–429 (2007)
3. Cimmino, A., Cano-Benito, J., García-Castro, R.: Open digital rights enforcement framework (ODRE): from descriptive to enforceable policies. Comput. Secur. **150**, 104282 (2025)
4. De Vos, M., Kirrane, S., Padget, J., Satoh, K.: ODRL policy modelling and compliance checking. In: Fodor, P., Montali, M., Calvanese, D., Roman, D. (eds.) RuleML+RR 2019. LNCS, vol. 11784, pp. 36–51. Springer, Cham (2019). https://doi.org/10.1007/978-3-030-31095-0_3
5. Føllesdal, D., Hilpinen, R.: Deontic logic: an introduction. In: Deontic Logic: Introductory and Systematic Readings, pp. 1–35. Springer (1971)
6. García, R., Gil, R., Gallego, I., Delgado, J.: Formalising ODRL semantics using web ontologies. In: Proceedings of the 2nd International ODRL Workshop, pp. 1–10 (2005)
7. Golpayegani, D., Esteves, B., Pandit, H.J., Lewis, D.: AIUP: an ODRL profile for expressing AI use policies to support the EU AI act. In: Joint Proceedings of Posters, Demos, Workshops, and Tutorials of the 20th International Conference on Semantic Systems co-located with 20th International Conference on Semantic Systems (SEMANTiCS 2024) (2024)
8. Iannella, R.: The open digital rights language: XML for digital rights management. Inf. Secur. Tech. Rep. **9**(3), 47–55 (2004)
9. Iannella, R.: Open Digital Rights Language (ODRL). Sydney University Press (2007)
10. Iannella, R., Villata, S.: ODRL Information Model 2.2. https://www.w3.org/TR/odrl-model/. Accessed 20 Aug 2025

11. Johnson, D.S., Klug, A.: Testing containment of conjunctive queries under functional and inclusion dependencies. In: Proceedings of the 1st ACM SIGACT-SIGMOD Symposium on Principles of Database Systems, PODS 1982, pp. 164–169. Association for Computing Machinery (1982)
12. Kasten, A., Grimm, R.: Making the semantics of ODRL and URM explicit using web ontologies. In: Virtual Goods, pp. 77–91 (2010)
13. Kebede, M.G., Sileno, G., Van Engers, T.: A critical reflection on ODRL. In: Rodríguez-Doncel, V., Palmirani, M., Araszkiewicz, M., Casanovas, P., Pagallo, U., Sartor, G. (eds.) AICOL/XAILA 2018/2020. LNCS (LNAI), vol. 13048, pp. 48–61. Springer, Cham (2021). https://doi.org/10.1007/978-3-030-89811-3_4
14. Pareti, P.: paolo7/ODRL-Conflict-Detection: ODRL Conflict Detection between Policies (2025). https://doi.org/10.5281/zenodo.17077271
15. Pucella, R., Weissman, V.: A formal foundation for ODRL (2006). https://doi.org/10.48550/arXiv.cs/0601085
16. Serrão, C., Dias, M., Delgado, J.: Using ODRL to express rights for different content usage scenarios. In: ODRL 2005 (2005). http://hdl.handle.net/10071/250
17. Slabbinck, W., Rojas, J., Esteves, B., Colpaert, P., Verborgh, R.: Interoperable interpretation and evaluation of ODRL policies. In: Semantic Web–22nd International Conference, ESWC, vol. 2025 (2025). https://doi.org/10.1007/978-3-031-94578-6_11
18. Steyskal, S., Polleres, A.: Towards formal semantics for ODRL policies. In: Bassiliades, N., Gottlob, G., Sadri, F., Paschke, A., Roman, D. (eds.) RuleML 2015. LNCS, vol. 9202, pp. 360–375. Springer, Cham (2015). https://doi.org/10.1007/978-3-319-21542-6_23
19. Yumusak, S., Gheisari, S., Salas, J.O., Ibáñez, L.D., Konstantinidis, G.: Data sharing negotiation and contracting. CEUR Workshop Proceedings, vol. 3828 (2024). https://ceur-ws.org/Vol-3828/paper39.pdf
20. Zhang, Y., Wei, X., Zhao, S.: Research on ODRL connecting with application scenarios. In: 2008 International Seminar on Future Information Technology and Management Engineering, pp. 261–264. IEEE (2008). https://doi.org/10.1109/FITME.2008.43

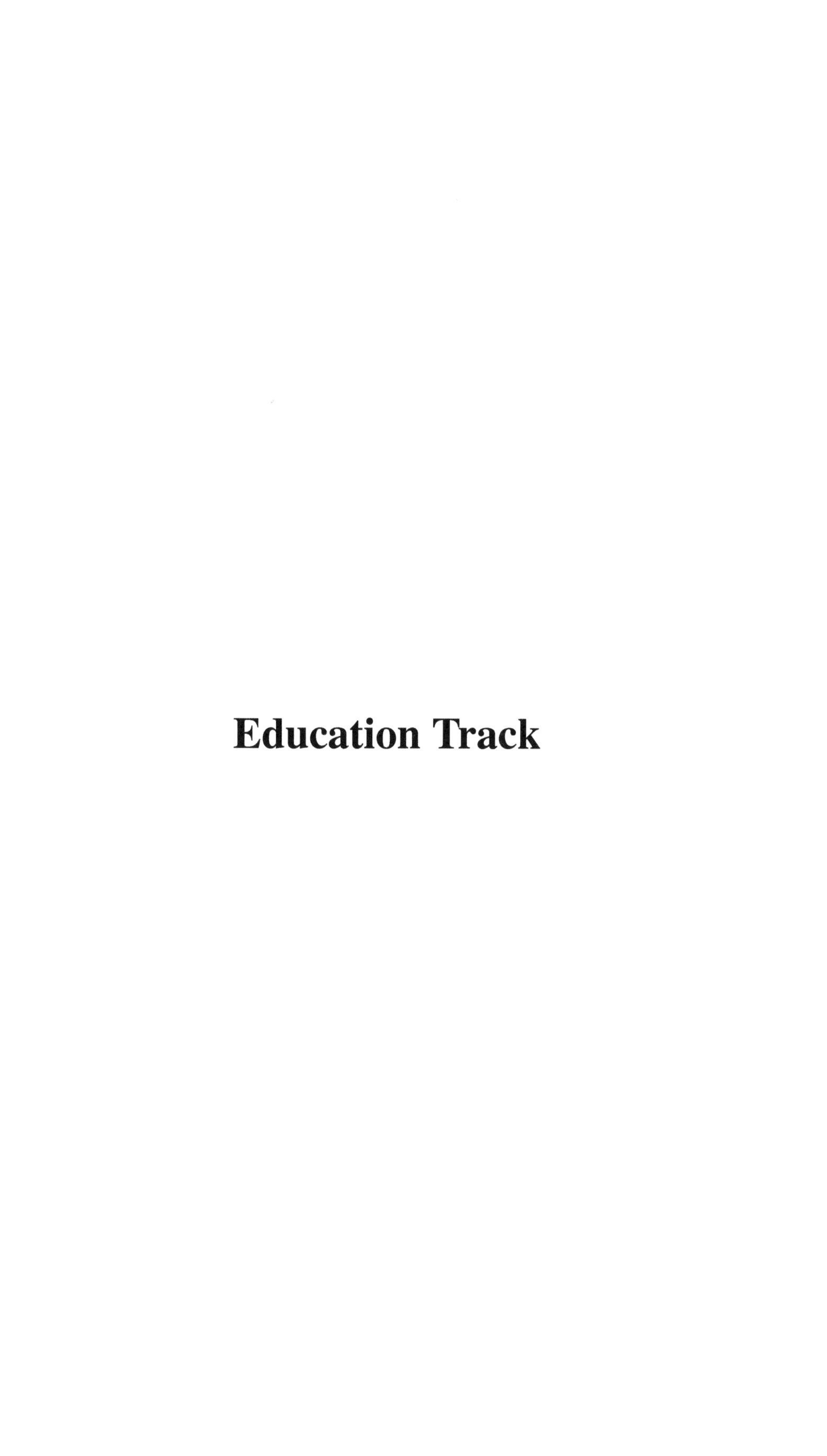

Education Track

An Optimized Pipeline for Automatic Educational Knowledge Graph Construction

Qurat Ul Ain(✉), Mohamed Amine Chatti, Jean Qussa, Amr Shakhshir, Rawaa Alatrash, and Shoeb Joarder

Social Computing Group, Faculty of Computer Science, University of Duisburg-Essen, Duisburg, Germany
qurat.ain@stud.uni-due.de

Abstract. The automatic construction of Educational Knowledge Graphs (EduKGs) is essential for domain knowledge modeling by extracting meaningful representations from learning materials. Despite growing interest, identifying a scalable and reliable approach for automatic EduKG generation remains a challenge. In an attempt to develop a unified and robust pipeline for automatic EduKG construction, in this study we propose a pipeline for automatic EduKG construction from PDF learning materials. The process begins with generating slide-level EduKGs from individual pages/slides, which are then merged to form a comprehensive EduKG representing the entire learning material. We evaluate the accuracy of the EduKG generated from the proposed pipeline in our MOOC platform, CourseMapper. The observed accuracy, while indicative of partial success, is relatively low particularly in the educational context, where the reliability of knowledge representations is critical for supporting meaningful learning. To address this, we introduce targeted optimizations across multiple pipeline components. The optimized pipeline achieves a 17.5% improvement in accuracy and a tenfold increase in processing efficiency. Our approach offers a holistic, scalable and end-to-end pipeline for automatic EduKG construction, adaptable to diverse educational contexts, and supports improved semantic representation of learning content.

Keywords: Educational Knowledge Graphs · Knowledge Modeling · LLM-based Keyphrase Extraction · Natural Language Processing

1 Introduction

Knowledge graphs (KGs) have gained significant attention in recent years due to their ability to structure and integrate diverse information across various domains [14]. Educational knowledge graphs (EduKGs) refer to the application of KGs in the educational domain, structuring knowledge concepts and their relationships to support personalized learning and adaptive education [1]. They play a crucial role in technology-enhanced learning (TEL) systems such as massive open online courses (MOOCs) and learning management systems (LMSs) by

H. Takeda et al. (Eds.): IJCKG 2025, LNAI 16297, pp. 271–287, 2026.
https://doi.org/10.1007/978-981-95-5009-8_18

extracting and visualizing key learning concepts [20]. Structurally, an EduKG is a heterogeneous graph where nodes represent educational entities (e.g., knowledge concepts, learning materials) and edges define various types of semantic relationships between them [20]. EduKGs serve a wide range of purposes in the educational domain, including enhancing learning and scientific discovery [9], educational assessment [11], training KG practitioners [10], learner modeling [5], personalized learning resource recommendation [3], and so on.

With the exponential growth of digital learning content, there is an increasing need for automated and scalable methods to organize and interpret domain knowledge effectively. EduKGs offer a robust framework for modeling such knowledge, yet the automatic construction of high-quality EduKGs remains a significant challenge. Traditionally, EduKGs were constructed manually by domain experts. However, due to the time and effort required for this approach, recent research has shifted towards automating EduKG generation using machine learning (ML) and natural language processing (NLP) techniques [28]. In parallel, the emergence of Generative AI and Large Language Models (LLMs) has sparked interest in their potential for EduKG generation [16,29]. Despite their promise, LLMs face significant challenges in automatic EduKG generation, such as token limits that restrict processing large datasets, inefficient human-machine interaction that either lacks oversight or increases labor costs, and a tendency to generate inaccurate information due to their hallucinating nature. These limitations make LLMs unreliable for fully autonomous EduKG construction [29]. Despite growing research interest in automatic EduKG construction, there remains a pressing need for a holistic, reliable, and scalable end-to-end pipeline that can automatically generate high-quality EduKGs across diverse educational settings. Existing efforts such as EduKG for K-12 [27] are ontology-driven, requiring a predefined schema before extraction rather than fully automatic extraction from unstructured PDFs. As part of our ongoing research efforts to address this challenge [1,4], this study focuses on refining and optimizing the automatic EduKG construction process to develop a unified and robust solution. We first propose a pipeline for automatic EduKG construction that can process any PDF-based learning material, starting from a single page/slide of a PDF, leading to a slide-EduKG per slide which are then combined to form an EduKG to represent the whole learning material. We then evaluate the accuracy of the EduKG generated by the proposed pipeline in our MOOC platform, CourseMapper [2]. The accuracy of the EduKG was not very high which calls for a need for further improvements in the pipeline. Subsequently, we introduce a series of optimizations, including enhancements to the pipeline architecture, improved data preprocessing, and refined modules for text extraction, concept annotation and EduKG expansion. The optimized pipeline is then compared against its initial version in terms of accuracy and efficiency, achieving a 17.5% improvement in accuracy and a tenfold increase in processing efficiency. All experiments were conducted within CourseMapper, to demonstrate the pipeline's applicability in real-world educational environments. The resulting pipeline offers a state-of-the-art, domain-independent solution for automatic EduKG construction and can be easily adapted to a wide range of educational contexts.

2 EduKG Construction Pipeline

The EduKG construction pipeline consists of several steps, as described below.

2.1 Text Extraction

The first step in building EduKG involves extracting text from individual pages/slides of PDF learning materials. This step is particularly challenging due to the diverse and often inconsistent layouts and structures of educational PDFs. To ensure coherent input for the subsequent keyphrase extraction step, it is essential to accurately identify and reconstruct logically connected text segments in the correct reading order. We address this using PDFMiner [21] which applies a layout-aware strategy by detecting text blocks, applying rule-based classification, and merging them to preserve the intended textual sequence.

2.2 Keyphrase Extraction

Once text is extracted from a page/slide of the PDF learning material, keyphrase extraction is performed as a preliminary step to Wikipedia entity linking, optimizing the process by minimizing the amount of text processed by the entity linking service. To determine the most effective keyphrase extraction method, we evaluate both state-of-the-art Large Language Model (LLM) approaches including Zero-Shot, Few-Shot, and KeyBERT-assisted prompts, and traditional techniques, including SingleRank, PatternRank, and SIFRank variants. The Inspec [15] and SemEval [6] datasets serve as benchmarks for performance evaluation, as they contain scholarly texts relevant to the educational domain, making them suitable for this evaluation. Our experiments apply the LLM-based methods using the Mistral-7B-Instruct-v0.1 model [25], while the traditional methods are implemented following their respective original configurations. Mistral-7B, an open-source instruction-tuned LLM with 7 billion parameters, is well-suited for this task due to its strong language understanding, ability to follow prompts, and scalability for large datasets. In the Zero-Shot setting, the LLM extracts keyphrases based solely on its pre-trained knowledge, using prompts such as "Identify key phrases from the following text: [Document Text]". The Few-Shot setting improves performance by including a few example text–keyphrase pairs in the prompt. The LLM-KeyBERT approach combines LLM prompting with KeyBERT [13], which uses BERT embeddings to extract contextually relevant keywords, enhanced by the LLM's nuanced understanding of the input. All methods are evaluated by comparing the top-n extracted keyphrases against expert annotations in the Inspec and SemEval datasets. Effectiveness is measured using Precision, Recall, and F1-score, capturing each method's accuracy, coverage, and balance. The results are presented in Table 1. This comparative analysis shows that LLM-based approaches, particularly with Few-Shot learning, outperform others in keyphrase extraction. This highlights the significance of LLMs' nuanced understanding and adaptability in capturing semantically rich keyphrases. While traditional methods lag behind in performance, they still hold significant value, especially in contexts where access to advanced models is limited. It is worth noting that the reported results for LLM-based methods reflect their best-case performance; in practice, some runs show a 10–15% drop in effectiveness, emphasizing the need for careful validation when deploying these models. Another important dimension of our evaluation is runtime efficiency. LLM-based methods, although accurate, demand significantly more time and computational resources per document than traditional methods (see Table 2). Despite lacking the depth of LLM-based analysis, traditional approaches offer dependable and efficient solutions, particularly when computational constraints or real-time processing are priorities. Among these, $SIFRank_{SqueezeBERT}$ demonstrates a strong balance between accuracy and efficiency, making it a practical choice for keyphrase extraction in our pipeline. The keyphrase extractor extracts n=15 keyphrases from each slide of the PDF material, because in a learning environment it would be overwhelming for the learners to focus on more than 15 concepts within a slide.

Table 1. A comparison of different keyphrase extraction methods

K	Keyphrase Extraction Method	Inspec Dataset			SemEval Dataset		
		P	R	F1	P	R	F1
5	Zero-shot LLM	72.95	75.17	74.04	72.95	75.17	73.2
	Few-shot LLM	**74.28**	**76.14**	**75.20**	**74.28**	**76.14**	**76.33**
	LLM_KeyBERT	71.51	73.81	72.64	71.51	73.81	72.12
	SingleRank	64.78	67.04	67.86	64.75	67.04	67.15
	PatternRank	60.83	62.98	61.89	60.83	62.98	61.45
	SIFRank	65.21	67.49	66.33	65.21	67.49	68.11
	$SIFRank_{SqueezeBERT}$	66.75	67.04	65.89	66.75	69.01	66.24
10	Zero-shot LLM	69.14	**84.98**	76.25	69.14	84.98	75.6
	Few-shot LLM	**72.04**	84.57	**77.81**	**72.04**	**84.57**	**78.18**
	LLM_KeyBERT	66.79	83.00	74.02	66.79	83.00	74.31
	SingleRank	66.11	81.19	73.09	66.11	81.19	73.23
	PatternRank	64.11	79.02	70.79	64.11	79.02	70.91
	SIFRank	62.74	77.06	69.16	62.74	77.06	69.23
	$SIFRank_{SqueezeBERT}$	66.34	81.37	72.87	66.34	81.37	73.86
15	Zero-shot LLM	59.16	**89.08**	71.10	59.16	**89.08**	71.2
	Few-shot LLM	**66.61**	88.15	**75.88**	**66.61**	88.15	**75.68**
	LLM_KeyBERT	55.65	87.27	67.96	55.65	87.27	68.16
	SingleRank	57.12	87.02	69.01	57.12	87.00	69.12
	PatternRank	55.86	85.98	67.72	55.86	85.98	68.09
	SIFRank	54.49	83.05	65.81	54.49	83.05	66.21
	$SIFRank_{SqueezeBERT}$	57.17	87.00	68.96	57.17	87.00	69.33

Table 2. Runtime required to extract (N = 5) keyphrases per document

Keyphrase extraction method	Runtime (min:sec)
LLM-based methods	1:40
SingleRank	<0:01
PatternRank	<0:01
SIFRank	0:03
$SIFRank_{SqueezeBERT}$	0:24

2.3 Concept Identification

The extracted keyphrases are used for concept identification by linking them to external knowledge bases. Following [17], we use DBpedia Spotlight [19] to map keyphrases to DBpedia concepts through spotting, candidate selection, and disambiguation based on their contextual information. The annotated keyphrases are then referred to as Main Concepts (MCs) of the slide. Nevertheless, entity-linking tools like DBpedia Spotlight can occasionally produce incorrect annotations where keyphrases are linked to irrelevant concepts, due to its fully automated nature [17]. To address this issue, we implement a concept-weighting strategy that emphasizes semantic and contextual relevance. We use a transformer-based method (w_{SBERT}), where weights of the MCs (w_{LM}) are computed via cosine similarity between SBERT [22] embeddings of the whole learning material and corresponding Wikipedia articles. Additionally, the similarity score between the embedding of the text of the slide and the embedding of the Wikipedia abstract of the concept is obtained, resulting in a slide similarity score (w_{Slide}). The relevance of the concepts per slide is determined by using the sum of these two similarity scores ((w_{LM}) and (w_{Slide})). After obtaining the slide's MCs and their weights, relations between MCs and the corresponding slide, as well as relations between MCs and the learning material, are established to ensure that concepts exist at both the learning material and slide levels. At this point, the Slide-EduKG containing the current slide, MCs, learning material, and their relations are stored in a Neo4j database and made available for learners, even if the construction process has not yet been completed. By storing the data after annotating each slide, allows learners to explore Slide-EduKGs for completed slides without waiting for the full EduKG construction to finish. Additionally, combining slide-level and learning material-level weights ensures balanced concept importance, reducing bias toward either the slide or the overall learning material. Once all slides have been annotated, the *concept expansion* step is activated.

2.4 Concept Expansion

Since keyphrase extraction and concept identification may overlook some relevant concepts [18], we expand the identified concepts to increase EduKG coverage and diversity to better support knowledge exploration. To achieve this, we apply two expansion strategies based on semantic relationships in DBpedia, as proposed by Manrique et al. [17]: related concept expansion and category-based expansion. In related concept expansion, semantically linked concepts are added to the EduKG using SPARQL queries over the dbo:wikiPageWikiLink property. Whereas, in category-based expansion, hierarchical categories are retrieved via the dct:subject property. After identifying the related concepts (RCs) and categories (Cts), their embeddings are generated using SBERT. For RCs, the embeddings are derived from the abstracts of their corresponding Wikipedia articles, whereas Cts' embeddings are based on their names. Cosine similarity is then calculated between the embeddings of each MC and its associated RCs and Cts to assess their semantic closeness. Additionally, cosine similarity is measured between the RCs and Cts and the whole learning material. These two similarity scores are summed up to compute a unified weight for each RC and Ct. This overall weight is used to rank them allowing the most relevant RCs and Cts to be selected. Specifically, the top 20 RCs and top 3 Cts for each MC are retained and integrated into the EduKG. This method enhances the structure of the EduKG by incorporating and exploring semantically related concepts and categories.

After building the EduKG slide by slide, it is then aggregated to create a LM-EduKG, that is a composite of all the slide-EduKGs. This approach enforces two key constraints: 1) The concepts related to a slide must adequately capture the content of that slide, and, 2) Any concept appearing at the lower level i.e., Slide-EduKG should also be included at the higher level i.e., LM-EduKG.

3 Evaluation

For the evaluation we used CourseMapper, to construct and visualize EduKG from PDF learning materials using the proposed pipeline. We evaluate the accuracy of the generated EduKG using the simple random sampling (SRS) method proposed by Gao et al. [12]. This method evaluates the correctness of samples in a KG, which consist of triples of (subject, predicate, object), where subject and object can be any two nodes in the KG, and predicate is the relation that links them. The evaluation involves two key tasks: entity identification which requires the evaluators to verify the correctness of a node entity by considering contextual information, and relationship validation which assesses the correctness of relations between subject and object nodes. The final accuracy score μ_s is computed as the mean of sample judgments. To ensure statistical reliability, the method follows an iterative sampling process with a predefined margin of error threshold. If the margin of error for the initial sample set exceeds the threshold, additional samples are evaluated until the accuracy stabilizes within the acceptable range. The results are presented using the normal approximation, which is the mean value $\pm$ the standard deviation value σ. The mean value of the evaluated sample set μ_s represents accuracy and falls between 0 and 1, where higher values indicate greater accuracy. Two course experts of the same course were recruited and an EduKG was generated for evaluation using a random PDF material from their course in CourseMapper. The evaluation process took 4 h where the first annotator evaluated 200 random samples from the generated EduKG, while the second annotator evaluated 180 random samples, making sure that each annotator evaluated a different sample set. The mean value of accuracy $\pm$ standard deviation ($\mu_s \pm \sigma$) is found to be 0.4 ± 0.049. This indicates that approximately 40% of the evaluated triples were judged as correct with respect to both entity identification and relationship validation. The standard deviation of 0.049 indicates relatively low variability in the judgments across the sample, implying that the accuracy estimates are consistent and reliable within the evaluated subset. The narrow spread strengthens confidence in the stability of the reported accuracy and suggests that the results are representative of the overall quality of the EduKG.

4 Pipeline Optimization

The observed accuracy of EduKG generated by the proposed pipeline, while indicative of partial success, is relatively low particularly in the educational context, where the reliability of knowledge representations is critical for supporting meaningful learning. This underscores the need for systematic optimizations within the pipeline to ensure the EduKGs are more accurate, efficient, and easily accessible, to support effective learning experiences. In an effort to address these limitations, we propose a series of optimizations across various steps of the initial pipeline, with the anticipation of achieving higher accuracy and efficiency.

4.1 Worker-Based Architecture

One major limitation of the proposed pipeline is its implementation as an API within a Python-Flask application. While generally a reasonable approach, it is unsuitable for the time-intensive EduKG generation process. Handling multiple simultaneous requests can overload the system, especially when users request the EduKGs for multiple materials at once. Additionally, failures at any stage leave the database in an inconsistent state with no straightforward recovery method. The current system also lacks request queuing, forcing users to retry if the server is busy. To address these issues, a worker-based architecture is proposed. This approach utilizes job queues and multiple workers that continuously process queued EduKG construction tasks. Each worker retrieves a job, executes it, stores the results in the database, and updates the user about its status. This architecture enables horizontal scalability by adding more workers as needed, ensures failed jobs are automatically re-queued, and allows request queuing when all workers are busy. Redis is chosen as the message broker due to its simplicity, reliability, and efficiency as an in-memory database.

4.2 Data Preprocessing

A major bottleneck in the proposed pipeline is its reliance on the Wikipedia API particularly in concept annotation, weighting and expansion modules, which introduces significant latency due to a huge number of network requests. Additionally, generating embeddings for identified concepts further impacts performance. To mitigate these issues, an offline data preprocessing step is proposed. This step, executed once a month, extracts relevant data from the Wikipedia XML dump, including article names, abstracts, inter-article links, category associations, redirections, and disambiguation pages. The extracted data is then used to generate embeddings for concept abstracts and category names using the SBERT all-mpnet-base-v2 sentence embedding model. The processed information is stored in a PostgreSQL database, allowing near-instant access to embeddings, categories, and relationships during EduKG generation. While this approach increases storage requirements and preprocessing time, the benefits of eliminating network latency and reducing redundant computations significantly outweigh these costs. Moreover, the database remains accessible to EduKG pipeline workers even if Wikipedia API is down or being updated.

4.3 Optimization of Text Extraction

The current pipeline originally employed a naive text extraction method using PDFMiner. However, this approach lacked any preprocessing, resulting in significant inaccuracies caused by the inclusion of noisy elements such as slide titles, page numbers, references, and footers. To improve the reliability of extracted content, the text extraction module is substantially enhanced to include structured analysis aimed at determining the relevance of extracted elements and ensuring accurate segmentation of sentences and bullet points. The improved approach incorporates several analytical techniques to ensure cleaner and more meaningful output. *Font size analysis* is used to assess whether adjacent text elements should be treated as part of the same sentence. If the difference in font size between two elements exceeds 0.5 points, they are considered separate sentences. In addition, *text distance analysis* is applied to account for vertical and horizontal spacing; when the distance between two adjacent elements is more than 1.5 times the respective line height or width, segmentation occurs at that point. To filter out repetitive or irrelevant content such as footers, sidebars, and page numbers, *text similarity analysis* is also employed. After removing non-alphabetical characters, any text fragment that appears in the same location on more than half of the pages, or on at least five pages in documents with fewer than ten pages, is classified as noise and removed. Furthermore, *bullet point analysis* is integrated to correctly segment list items, addressing inconsistencies introduced by various PDF rendering tools. Bullet points embedded within text naturally serve as sentence separators. However, when bullet points are rendered as separate graphical elements, such as circles or rectangles, additional spatial analysis is required. These elements are identified based on their position relative to adjacent text, under the assumption of a left-to-right writing system. Once a bullet point is detected, the following text is treated as the start of a new sentence. Overall, this enhanced methodology substantially improves the accuracy of the text extraction module and consequently keyphrase identification by filtering irrelevant content and preserving the semantic structure of the learning material.

Evaluation and Results: To evaluate the impact of the proposed changes, we compared the text extracted from the optimized module with the initial one (see Sect. 2.1). Existing datasets like Grotoap2 [26], PubMed and BioMed Central [8] etc., mainly contain text from scientific articles, which do not align with our use case of extracting text from slides and PDF learning materials. Therefore, a different evaluation approach is required. Bast et al. [7] recommend extracting structure from TeX files and evaluating text extraction based on the rendered PDFs, but this method is applicable only to LaTeX-based presentations, which are a small subset of all learning materials. For this evaluation, a new dataset was created with materials from various sources i.e. LaTeX, Microsoft PowerPoint, and Microsoft Word. A total of 24 PDF materials were selected, ensuring variety in layout and content types. The evaluation consisted of three steps. First, text was extracted from the slides using PDFMiner. Second, a human annotator labeled each text block with categories: header, content, footer, page number, navigation, figure, or empty. The final dataset contained 643 annotated text parts from 24 slides across 7 different learning materials. Third, comparing the human annotations to the output of both text extraction modules and calculate precision, recall and F1 scores. We further utilized the text extraction evaluation criteria proposed by Bast et al. [7], which is divided into three groups: newline differences, which capture the quality of the detection of paragraph boundaries (NL^+ is the number of spurious newlines in the output file, whereas NL^- is the number of missing newlines in the output file); paragraph differences, which capture the quality of the distinction between body and non-body text and of the reading order (P^+ is the number of spurious paragraphs, P^- is the number of missing paragraphs, and $P^{\updownarrow}$ is the number of rearranged paragraphs in the output file); and word differences, which capture the quality of the recognition of individual words and their boundaries (W^+ is the number of spurious words, W^- is the number of missing words, and $W^\sim$ is the number of misspelled words in the output file). The results are shown in Table 3 and 4, which clearly demonstrate that our new optimizations outperform the earlier one in nearly every measure.

Table 3. Comparison of the optimized text extraction vs. earlier one

Text Extraction Type	Module Type	P	R	F1
Content and figure	Initial	0.64	0.96	0.77
	Optimized	**1.0**	**0.97**	**0.98**
Non-content	Initial	0.0	0.0	0.0
	Optimized	**0.93**	**1.0**	**0.96**
Beginning of sentence in content	Initial	0.82	0.08	0.14
	Optimized	**0.94**	**0.95**	**0.95**
Beginning of sentence in figure	Initial	0.0	0.0	0.0
	Optimized	**0.36**	**0.38**	**0.37**

Table 4. Results of the evaluation of the initial vs. optimized text extraction using the differences criteria proposed by [7]

Module	NL^+	NL^-	P^+	P^-	$P^\updownarrow$	W^+	W^-	$W^\sim$
Initial	**3**	230	230	**12**	1	0	0	1
Optimized	52	**49**	**1**	17	1	0	0	1

4.4 Optimization of Concept Annotation

In the concept annotation module, extracted keyphrases are sent to the DBpedia Spotlight API for annotation. Multiple requests are processed simultaneously, making the total annotation time dependent on the longest individual request, aside from network overhead. However, this approach has three main limitations: (1) DBpedia Spotlight may return irrelevant or miss relevant annotations, (2) sending only keyphrases omits contextual information crucial for accurate interpretation, and (3) as an online service, DBpedia Spotlight can be inconsistent and unreliable. To mitigate these issues, a *concept weighting* and a *concept disambiguation* modules are integrated to filter irrelevant concepts and enhance annotation accuracy in the EduKG. Earlier we determined the weight of each MC with respect to each slide and to the whole learning material (see Sect. 2.3). We use this same strategy (named as concept weighting) further in concept disambiguation; a process that enhances EduKG accuracy by identifying better alternatives to DBpedia-annotated concepts. It checks whether an annotated concept is the title of or linked from a Wikipedia disambiguation page. A disambiguation page is a Wikipedia page that lists articles sharing the same or a similar title. If this condition is met, all linked pages are considered alternative concepts, which are then weighted by the concept weighting strategy. The highest-weighted concept replaces the original annotation. This step helps improve the quality of the EduKG by providing alternative annotations for keyphrases that may have been incorrectly annotated in the concept annotation step. This is because DBpedia only knows the keyphrase that is passed, but not the context thereof. The next step is *knowledge graph pruning*, i.e., removing irrelevant concepts from the EduKG. Irrelevant concepts are defined as concepts with a low weight as determined by the weighting strategy. All concepts whose weight falls below the weight threshold i.e. 0.192 (based on experiments mentioned below), are removed from the EduKG.

Evaluation and Results: The performance of this module is influenced by three parameters: the sentence embedding model, the weight threshold, and the inclusion of the disambiguation step. The sentence embedding model, used to compute concept-material similarity, significantly impacts concept weighting. Three pretrained transformer models are compared: all-mpnet-base-v2 [24], known for its accuracy but slow performance; all-MiniLM-L6-v2 [24], offering a balance between speed and accuracy; and msmarco-distilbert-base-tas-b [23], optimized for retrieval tasks. The evaluation uses the CCI dataset [17], consisting of transcripts from 96 programming-related learning videos annotated by seven experts. The weight threshold is ignored, and instead, the top-1, top-2, top-3, top-5, and top-10 concepts per resource are selected and evaluated using Precision (see Table 5). Results show that all-mpnet-base-v2 matches expert-annotated concepts highest, followed closely by all-MiniLM-L6-v2 and msmarco-distilbert-base-tas-b, which perform similarly. Due to its superior performance, all-mpnet-base-v2 is selected to be used for embeddings generation.

Table 5. Precision of the sentence embedding models

Model	Top-1	Top-2	Top-3	Top-5	Top-10
msmarco-distilbert	0.18	0.16	0.14	0.10	0.07
all-MiniLM-L6-v2	0.17	0.15	0.13	0.09	0.06
all-mpnet-base-v2	**0.25**	**0.19**	**0.16**	**0.13**	**0.09**

An additional experiment was conducted to determine the optimal weight threshold for knowledge graph pruning. Using the CCI dataset [17], expert-annotated concepts were weighted against transcripts using our weighting strategy. Results show that 50% of concepts have a weight above 0.471, where weight falls in the range of $[-1, 1]$ with all the weights falling between 0.028 and 0.744. However, using such a low threshold would introduce excessive noise into EduKG. Since 95% of weights fall between 0.192 and 0.774, 0.192 is selected as the weight threshold, meaning that all concepts having a weight less than 0.192 will be removed from the EduKG (knowledge graph pruning step).

Lastly, we evaluated the impact of introducing disambiguation step on the generated EduKG accuracy. Using the CCI dataset [17], EduKG was generated twice, once with and once without disambiguation. Precision, Recall, and F1-score were computed (see Table 6). Results confirm that disambiguation enhances EduKG accuracy, emphasizing the importance of considering context rather than just keyphrases during annotation. Consequently, the disambiguation step is included in the optimized EduKG construction pipeline.

Table 6. Concept annotation with and without concept disambiguation

Method	Precision	Recall	F1-Score
Initial Pipeline (Without Disambiguation)	0.13	0.17	0.14
Optimized Pipeline (With Disambiguation)	**0.18**	**0.29**	**0.23**

4.5 Optimization of Concept Expansion

Concept expansion is the process of assigning related concepts (RCs) and Categories (Cts) to the identified main concepts (MCs). In the proposed initial pipeline, these RCs and Cts were retrieved from DBpedia using SPARQL queries. While SPARQL queries offer a straightforward method for retrieving RCs and Cts, they come with practical limitations. Specifically, querying public SPARQL endpoints at scale can lead to performance bottlenecks due to latency, rate limiting, or inconsistent response times, making the approach less suitable for high-throughput or time-sensitive applications. Moreover, relying on external infrastructure introduces risks related to availability and long-term stability, as these endpoints may become temporarily inaccessible or deprecated without warning. To address these issues, we instead retrieve RCs and Cts from our locally hosted Wikipedia dump (Sect. 4.2). This alternative ensures faster and more reliable access to the required data while giving us full control over the extraction process. It also eliminates external dependencies, enabling more scalable and robust integration within our pipeline.

Related-concept expansion consists of three steps: candidate set creation, candidate set weighting, and candidate pruning. It works by creating a set of expansion candidates for each MC. This expansion candidate set contains all Wikipedia pages linking and linked to from the Wikipedia page of the MC. These candidates are then weighted by computing the cosine similarity between the embedding of the abstract of the Wikipedia page of the candidate and the embedding of the text of the whole PDF material. After every candidate RC has been assigned a weight, they are sorted by weight and the top-20 candidates for each MC are chosen as the final set of RCs for that MC. Similarly, category expansion process consists of candidate set creation, candidate set weighting, and candidate pruning. The category expansion candidate set consists of all categories of the all the MCs of the PDF material. The categories of the concepts are the Wikipedia categories to which the Wikipedia article of the concept belongs. This list of categories is retrieved from the Wikipedia dump database which was created in the preprocessing step (Sect. 4.2). The candidate set is then weighted by computing the product of the normalized category weight and the connected concepts weight. The normalized category weight (w_{nc}) is the cosine similarity between the embedding of the category name and the embedding of the text of whole PDF material. The connected concepts weight (w_{cc}) is the inverse of the logarithm of the number of MCs connected to the category plus one. The one is added to avoid zero weights. Using the number of connected MCs to a category allows more relevant categories to the learning material, being assigned a higher weight. The embedding of the category name is precomputed and is simply retrieved from the database. Based on this final weight, the top-5 categories for each MC are then kept and the rest are discarded.

5 Evaluation of the Optimized Pipeline

The optimizations proposed in various steps of the EduKG construction pipeline were evaluated in terms of efficiency and accuracy of the generated EduKG.

5.1 Evaluation of EduKG Efficiency

The runtime of the optimized pipeline was evaluated and compared to the initial version of the pipeline on a machine with the following specifications: CPU: AMD Ryzen 5 5600, RAM: 32GB, Storage: 4TB NVMe M.2 SSD, GPU: None. The GPU was intentionally excluded to reflect performance on a typical commodity server. The speed and memory usage of the pipeline was measured using the following metrics: 1) Mean time per slide, computed as the time required to extract MCs from a PDF learning material divided by the number of slides in the material, and 2) Mean time per concept, computed as the time required to expand the knowledge graph in a learning material divided by the number of MCs in the material. Furthermore, EduKG creation and expansion were evaluated separately. A diverse dataset of learning materials, varying in size, structure, density, and topics, was created. To test scalability, the dataset included one large material with 279 pages, while the remaining materials had 31 to 52 pages. The initial and optimized version of the pipeline were run on each material in the dataset, with DBpedia response caching disabled to ensure results accurately reflect performance on new, unseen data.

The evaluation results (see Table 7) show that the optimized pipeline is over 10 times faster in EduKG construction and more than 100 times faster in EduKG expansion. This substantial improvement in execution time compared to the initial version is primarily due to the use of a preprocessed local Wikipedia database with precomputed embeddings. Leveraging this local dump for both concept annotation and concept expansion significantly reduces computational overhead and enhances the overall reliability of the pipeline.

5.2 Evaluation of EduKG Accuracy

This evaluation aims to assess the accuracy of the EduKG generated with the proposed optimizations in the pipeline and compare it to the EduKG generated with the initial version of the pipeline (Sect. 2). The optimized pipeline is evaluated using the same approach as earlier, employing SRS and expert annotators. To ensure comparability, the same evaluation method, PDF material, metrics, and parameters are used. The process begins with randomly selecting a sample from the EduKG generated by the optimized pipeline, which is then annotated by the same two experts until the stopping criteria are met. The expert annotation part of the evaluation took around 2.5 h in total. The first expert annotated a random set of 200 triples and the second one annotated 203 triples till we met the stopping criteria. The results of the evaluation are shown in Table 7.

The evaluation reveals a 17.5% improvement in EduKG accuracy compared to the initial version of the pipeline, underscoring the effectiveness of the proposed optimizations. Particularly, improvements in the text extraction step played a critical role by significantly reducing noise in text, which previously led to incorrect or redundant annotations. Furthermore, the integration of concept disambiguation and knowledge graph pruning effectively filtered out irrelevant concepts, contributing to a more accurate and semantically coherent EduKG. The adoption of the all-mpnet-base-v2 embedding model further strengthened the pipeline, enabling the generation of high-quality embeddings that substantially improved the performance of the concept weighting step. Despite these advancements, the overall increase in accuracy remains moderate. This can be attributed to the continued dependence on DBpedia as the primary external knowledge base for concept annotation which is an element common to both versions of the pipeline. While the shared knowledge source may have limited the potential for drastic accuracy gains, the proposed optimizations nonetheless yielded meaningful improvements in both the accuracy and efficiency of EduKG construction, demonstrating the robustness and scalability of the optimized pipeline.

Table 7. Evaluation results of EduKG accuracy and efficiency

	Metric	Initial Pipeline	Optimized Pipeline
Efficiency	Avg. Time per Slide (Generation)	24.35 s	**2.3 s**
	Avg. Time per Concept (Expansion)	222 s	**1.89 s**
Accuracy	Accuracy (μ_s)	0.4	**0.47**
	Accuracy ($\mu_s \pm \sigma$)	0.4 ± 0.049	0.47 ± 0.049

6 Conclusion and Future Work

This study presents an end-to-end pipeline for the automatic construction of Educational Knowledge Graphs (EduKGs) from PDF learning materials within the MOOC platform, CourseMapper. The pipeline generates slide-level EduKGs from individual pages/slides of the PDF learning material and subsequently merges them into a comprehensive EduKG representing the entire learning material. Initial evaluation of the EduKG generated by the proposed pipeline indicated moderate performance, emphasizing the importance of further optimization to ensure that EduKGs accurately reflect educational content and support meaningful learning experiences. To address this, we proposed several optimizations across multiple steps of the pipeline and re-evaluated its performance. The optimized version achieved a 17.5% improvement in accuracy, along with significant gains in efficiency where concept annotation became over 10 times faster, and knowledge graph expansion over 100 times faster. These improvements stem from enhanced text extraction that uses sentence boundary detection and filtering of non-content text, the integration of a preprocessed Wikipedia dump database with precomputed embeddings reducing on-the-fly computation, and the use of a stronger transformer model all-mpnet-base-v2 for high quality embedding generation. Furthermore, the disambiguation module improved concept weighting, resulting in only highly relevant concepts being included in the EduKG. The updated worker-queue architecture also contributed to improved stability and reduced resource consumption, trading higher storage use for much lower CPU and memory demands. The pipeline is implemented within our MOOC platform, CourseMapper, to showcase its practical applicability in authentic educational settings. The final pipeline presents a state-of-the-art solution for automatic EduKG construction, with the flexibility to be adapted across various educational settings and use cases.

Despite these advances, some limitations remain. A considerable portion of runtime is spent saving expanded concepts in Neo4j, indicating a need for optimization in database interactions. The current data preprocessing step lacks support for incremental updates of the Wikipedia dump, requiring full reprocessing even for minor changes. Additionally, the text extraction module struggles with ligatures, hyphenated words, umlauts, and accents, and would benefit from more sophisticated layout analysis to better handle complex elements like tables and figures. The evaluation dataset (CCI) is also limited in coverage and outdated, constraining the benchmarking accuracy. A more representative dataset with learning materials actively used in practice and annotated by experts would provide a stronger evaluation framework. As future work, we plan to incorporate a human-in-the-loop mechanism, enabling domain experts to refine, validate, and enhance the automatically generated EduKGs before publishing it for the end users. This integration aims to strike a balance between automation and human expertise, ensuring high-quality EduKGs while significantly reducing the manual effort involved.

Disclosure of Interests. The authors have no competing interests to declare that are relevant to the content of this article.

References

1. Ain, Q.U., Chatti, M.A., Bakar, K.G.C., Joarder, S., Alatrash, R.: Automatic construction of educational knowledge graphs: a word embedding-based approach. Information **14** (2023). https://doi.org/10.3390/info14100526
2. Ain, Q.U., et al.: Learning channels to support interaction and collaboration in coursemapper. In: Proceedings of the 14th International Conference on Education Technology and Computers, pp. 252–260 (2022)
3. Ain, Q.U., Chatti, M.A., Meteng Kamdem, P.A., Alatrash, R., Joarder, S., Siepmann, C.: Learner modeling and recommendation of learning resources using personal knowledge graphs. In: Proceedings of the 14th Learning Analytics and Knowledge Conference, pp. 273–283 (2024). https://doi.org/10.1145/3636555.3636881

4. Ain, Q.U., Chatti, M.A., Shakhshir, A., Qussa, J., Alatrash, R., Joarder, S.: Top-down vs. bottom-up approaches for automatic educational knowledge graph construction in coursemapper. arXiv preprint arXiv:2505.10069 (2025)
5. Alatrash, R., Chatti, M.A., Ain, Q.U., Joarder, S.: Transparent learner knowledge state modeling using personal knowledge graphs and graph neural networks. In: Adjunct Proceedings of the 32nd ACM Conference on User Modeling, Adaptation and Personalization, pp. 591–596 (2024). https://doi.org/10.1145/3631700.3665230
6. Augenstein, I., Das, M., Riedel, S., Vikraman, L., McCallum, A.: Semeval 2017 task 10: scienceie-extracting keyphrases and relations from scientific publications. arXiv preprint arXiv:1704.02853 (2017)
7. Bast, H., Korzen, C.: A benchmark and evaluation for text extraction from PDF (2017). https://doi.org/10.1109/JCDL.2017.7991564
8. BMC: BMC, research in progress (2018). https://www.biomedcentral.com/
9. Chen, P., Lu, Y., Zheng, V.W., Chen, X., Yang, B.: Knowedu: a system to construct knowledge graph for education. IEEE Access **6**, 31553–31563 (2018)
10. Engelbrecht, E., Ilkou, E., Abu-Rasheed, H., Chaves-Fraga, D., Jiménez-Ruiz, E., Labra-Gayo, J.E.: Teaching knowledge graph for knowledge graphs education. Seman. Web Interoperability Usability Appl. (2024)
11. Fettach, Y., Ghogho, M., Benatallah, B.: Knowledge graphs in education and employability: a survey on applications and techniques. IEEE Access **10** (2022)
12. Gao, J., Li, X., Xu, Y.E., Sisman, B., Dong, X.L., Yang, J.: Efficient knowledge graph accuracy evaluation. arXiv preprint arXiv:1907.09657 (2019)
13. Grootendorst, M.: Keybert: minimal keyword extraction with bert (2020). https://maartengr.github.io/KeyBERT
14. Hogan, A., et al.: Knowledge graphs. ACM Comput. Surv. (CSUR) **54**(4), 1–37 (2021)
15. Hulth, A.: Improved automatic keyword extraction given more linguistic knowledge. In: Proceedings of the 2003 Conference on Empirical Methods in Natural Language Processing, pp. 216–223 (2003)
16. Jhajj, G., Zhang, X., Gustafson, J.R., Lin, F., Lin, M.P.C.: Educational knowledge graph creation and augmentation via LLMs. In: International Conference on Intelligent Tutoring Systems, pp. 292–304. Springer, Cham (2024)
17. Manrique, R., Grévisse, C., Mariño, O., Rothkugel, S.: Knowledge graph-based core concept identification in learning resources. In: Ichise, R., Lecue, F., Kawamura, T., Zhao, D., Muggleton, S., Kozaki, K. (eds.) JIST 2018. LNCS, vol. 11341, pp. 36–51. Springer, Cham (2018). https://doi.org/10.1007/978-3-030-04284-4_3
18. Manrique, R., Marino, O.: Knowledge graph-based weighting strategies for a scholarly paper recommendation scenario. In: KaRS@ RecSys, pp. 5–8 (2018)
19. Mendes, P.N., Jakob, M., García-Silva, A., Bizer, C.: Dbpedia spotlight: shedding light on the web of documents. In: Proceedings of the 7th International Conference on Semantic Systems, pp. 1–8 (2011)
20. Qu, K., Li, K.C., Wong, B.T., Wu, M.M., Liu, M.: A survey of knowledge graph approaches and applications in education. Electronics **13**(13), 2537 (2024)
21. Ramakrishnan, C., Patnia, A., Hovy, E., Burns, G.A.: Layout-aware text extraction from full-text PDF of scientific articles. Source Code for Biology and Medicine (2012)
22. Reimers, N., Gurevych, I.: Sentence-bert: sentence embeddings using siamese bert-networks. arXiv preprint arXiv:1908.10084 (2019)
23. Reimers, N., Gurevych, I.: msmarco-distilbert-base-tas-b (2023). https://huggingface.co/sentence-transformers/msmarco-distilbert-base-tas-b

24. Reimers, N., Gurevych, I.: Sentence-bert: pretrained models (2023). https://www.sbert.net/docs/sentence_transformer/pretrained_models.html
25. TheBloke: TheBloke/Mistral-7B-Instruct-v0.1-GGUF (2024). https://huggingface.co/TheBloke/Mistral-7B-Instruct-v0.1-GGUF
26. Tkaczyk, D., Szostek, P., Bolikowski, L.: Grotoap2-the methodology of creating a large ground truth dataset of scientific articles. D-Lib Mag. **20**(11/12) (2014)
27. Zhao, B., et al.: Edukg: a heterogeneous sustainable k-12 educational knowledge graph. arXiv preprint arXiv:2210.12228 (2022)
28. Zhong, L., Wu, J., Li, Q., Peng, H., Wu, X.: A comprehensive survey on automatic knowledge graph construction. ACM Comput. Surv. (2023)
29. Zhu, Y., et al.: LLMs for knowledge graph construction and reasoning: recent capabilities and future opportunities. World Wide Web **27**(5), 58 (2024)

Inferring Prerequisite Knowledge Concepts in Educational Knowledge Graphs: A Multi-criteria Approach

Rawaa Alatrash(✉), Mohamed Amine Chatti, Nasha Wibowo, and Qurat Ul Ain

Social Computing Group, Faculty of Computer Science, University of Duisburg-Essen, Duisburg, Germany
rawaa.alatrash@gmail.com

Abstract. Educational Knowledge Graphs (EduKGs) organize various learning entities and their relationships to support structured and adaptive learning. Prerequisite relationships (PRs) are critical in EduKGs for defining the logical order in which concepts should be learned. However, the current EduKG in the MOOC platform CourseMapper lacks explicit PR links, and manually annotating them is time-consuming and inconsistent. To address this, we propose an unsupervised method for automatically inferring concept PRs without relying on labeled data. We define ten criteria based on document-based, Wikipedia hyperlink-based, graph-based, and text-based features, and combine them using a voting algorithm to robustly capture PRs in educational content. Experiments on benchmark datasets show that our approach achieves higher precision than existing methods while maintaining scalability and adaptability, thus providing reliable support for sequence-aware learning in CourseMapper.

Keywords: Educational Knowledge Graphs · MOOCs · Prerequisite concept identification · Relation extraction · Knowledge structure

1 Introduction

Educational Knowledge Graphs (EduKGs) play a vital role in organizing and representing key entities and their relationships in digital learning environments, supporting knowledge discovery, concept navigation, and learner modeling [1]. In our MOOC platform CourseMapper, EduKGs provide learners with a structured overview of the main knowledge concepts covered in learning materials to enhance understanding and engagement [1–3]. However, a major limitation of current EduKG structures in CourseMapper is the lack of prerequisite relationships (PRs), which define concept dependencies where understanding one concept is necessary before learning another. These relationships determine the logical sequence for mastering topics and support personalized sequence recommendations [15]. Without explicit PRs, learners may struggle to identify which

H. Takeda et al. (Eds.): IJCKG 2025, LNAI 16297, pp. 288–303, 2026.
https://doi.org/10.1007/978-981-95-5009-8_19

concepts should be mastered first, which can hinder self-directed learning and adaptive recommendation systems. Prior research has explored end-to-end learning, feature-based, and hybrid methods for PR extraction [5,6,12,15]. End-to-end learning-based methods depend on domain-specific labeled data and static features, which limits their scalability. Feature-based methods remain widely used and rely on graph structures, text content, or Wikipedia data to capture concept dependencies [7,12]. However, these methods have several limitations. They often overlook how semantic and structural information interact and they often rely only on concept order. Moreover, most feature-based approaches assess each criterion individually instead of combining their contributions for effective PR identification. To overcome these limitations, we propose an unsupervised, multi-criteria approach that combines the contributions of document-based, Wikipedia hyperlink-based, graph-based, and text-based features using a voting algorithm, aiming to effectively extract PRs and enrich the structure of EduKGs in CourseMapper.

2 Background and Related Work

2.1 EduKG Generation in CourseMapper

Educational Knowledge Graphs (EduKGs) represent diverse entities and their relationships within a learning context, such as concepts, instructors, courses, and institutions [1]. In our MOOC platform CourseMapper [2], instructors can upload Learning Materials (LM), for which an EduKG is automatically generated. This includes entities such as LM, Slides (S), Main Concepts (MC), Related Concepts (RC), Categories (Cat), and Learners (L), linked by relationships such as CONTAINS, CONSISTS_OF, RELATED_TO, HAS_CATEGORY, HAS_READ, UNDERSTOOD (U), and DID_NOT_UNDERSTAND (DNU). MCs are identified in each slide using a keyphrase extraction algorithm. These MCs undergo an expansion process via DBpedia Spotlight, which adds additional knowledge concepts as RCs and Cats. The RCs and Cats are then filtered to retain only the most relevant ones using cosine similarity in the EduKG [3]. This structure provides learners with an organized view of key concepts in the LM. The EduKG can be further enhanced by adding prerequisite relationships, which model the learning sequence among concepts and support adaptive learning pathways [15]. Currently, CourseMapper's EduKG does not include PRs, limiting its support for sequence-aware learning.

2.2 Prerequisite Relationship Extraction

PRs describe how concepts depend on each other in learning pathways. Methods for identifying PRs include end-to-end learning-based, feature-based, and hybrid approaches. End-to-end learning-based methods rely on deep learning to detect PRs from large datasets [6,17,19]. Feature-based methods use manually designed features [7,12,15] such as graph-based features [12], which capture structural characteristics of graphs; text-based features, which use descriptions and content [8,23]; and Wikipedia-based features, which exploit hyperlinks, clickstreams, and categories [5,16,18]. Hybrid methods combine both end-to-end learning-based and feature-based methods [5,12]. Many existing end-to-end learning-based methods for PR inference depend on domain-specific labeled data and simplistic features, limiting scalability and adaptability. Manual annotation is time-consuming and inconsistent. Most feature-based methods consider criteria independently, often ignoring the interaction between semantic and structural information. To overcome these issues, we propose an unsupervised multi-criteria approach combining ten distinct criteria document-based, Wikipedia hyperlink-based, graph-based, and text-based features. These criteria integrate semantic, structural, and contextual information through a **voting algorithm** that balances their contributions and mitigates noise from any single criterion. Our method evaluates all concept pairs in the EduKG, which is crucial for unstructured content where structural criteria alone may be misleading. Therefore, incorporating additional criteria that leverage external resources and contextual information can enhance the accuracy of PR inference. Our approach, tested on two public datasets, achieves a precision of 1.0, demonstrating robustness, domain independence, and effective PR extraction.

3 Methodology

To define the likelihood of one concept being a prerequisite of another, we developed and adopted a set of criteria, derived from the literature review and adapted for our context. These criteria are categorized into document-based features, Wikipedia hyperlink-based features, graph-based features, hierarchy-based features, and text-based features. Our multi-criteria approach begins by examining concept pairs against these criteria, each offering a unique perspective. Their outputs are then aggregated using a voting algorithm, followed by a thresholding step to infer whether a PR exists. These criteria are applied to all concept pairs in EduKG in CourseMapper and examined in both directions to ensure comprehensive identification. This study excludes concept synonyms and automatic redirections, as these are managed during the EduKG construction on the platform CourseMapper, ensuring only unique nodes without duplicates are extracted [3]. Before outlining the model structure, we adopt three core assumptions: **Transitivity** [4,5], where if $A \xrightarrow{\text{PR}} B$ and $B \xrightarrow{\text{PR}} C$, then $A \xrightarrow{\text{PR}} C$; **Asymmetry** [4,5], where if $B \rightarrow A$, then $A \nrightarrow B$; and **No Self-Connection** [4], a concept cannot be a prerequisite of itself. Table 1 lists each criterion with its name, abbreviation, and description.

3.1 Prerequisite Relationships Extraction Features

Document-Based Features: This category depends on the document structure to obtain the **temporal order** (TemO) of concepts within an article where the concept introduced first serves as a prerequisite for the subsequent [19]. For instance, in discussing *Clustering*, the term *machine learning* is likely introduced before *Clustering* as understanding the former is essential for grasping the latter.

Table 1. Categories of Criteria with Full Names, Abbreviations, and Descriptions.

Category	Full name	Abbreviations	Description
Document based features	Temporal order	TemO	c_1 precedes c_2 in the learning material or channel
Wikipedia Hyperlinks based features	Hyperlinks on each other's articles	HL-A	c_1 hyperlink is mentioned within c_2's article
	Hyperlinks on each other's abstract	HL-Ab	c_1 hyperlink appears in the abstract of c_2
	Hyperlinks on each other's related concepts' abstracts	HL-RCA	c_1 hyperlink is given on c_2's related concept's abstract
	Reference distance	RefD	c_1 hyperlink is mentioned in the related concepts of c_2
	Inbound links and outbound links ratio	IOLR	c_1 has higher ratio of the number of in/outbound links pointing to c_2
Graph based features	One concept is a category of another concept	CatCon	c_1 is a category of c_2 in DBPedia hierarchy
	One concept is the direct super-category of another concept	SuperCatCon	c_1 is a super category of c_2 in DBPedia hierarchy
Hierarchy based features	CourseMapper Hierarchy	CMH	first occurrence of c_1 in $channel_1$ precedes first occurrence of c_2 in $channel_2$
Text-based features	BERTopic and Shannon's Entropy	BERTropy	c_1 has higher topic entropy than c_2

$TemO(c_1, c_2)$ is assigned the value 1 if the position $P(c_1)$ of concept c_1 appears before the position $P(c_2)$ of concept c_2; otherwise, 0.

$$TemO(c_1, c_2) = \begin{cases} 1, & \text{if } P(c_1) < P(c_2) \\ 0, & \text{otherwise} \end{cases} \tag{1}$$

TemO applies only to MCs, as it relies on their order of appearance in the LM; related concepts are excluded because they are not explicitly mentioned.

Wikipedia Hyperlinks-Based Features: Wikipedia has been exploited to find potential PRs among concepts by analyzing the hyperlink structure of their corresponding Wikipedia articles [5,20].

Hyperlinks on Each Other's Articles (HL-A): In this criteria, we examine whether the hyperlinks for each concept appear in the corresponding Wikipedia articles of other concepts. For example, if the hyperlink for the concept *Artificial Intelligence* (c_1) is found in the Wikipedia article for *Machine Learning* (c_2), it suggests that *Artificial Intelligence* serves as a prerequisite for understanding

Machine Learning. This implies that c_1 is a prerequisite to c_2. These relationships are formalized as follows:

$$HL - A(c_1, c_2) = \begin{cases} 1, & \text{if } c_1 \in H_{c2} \\ 0, & \text{otherwise} \end{cases} \tag{2}$$

where $HL - A(c_1, c_2)$ determines whether the hyperlink for concept c_1 exists in the Wikipedia article of concept c_2, where in this case 1 is assigned, otherwise 0. H_{c2} represents the set of concepts that are hyperlinked within the Wikipedia article for concept c_2. Sometimes, hyperlinks in Wikipedia articles do not indicate a PR especially in long articles or hyperlinks at the end of the article. For example, if *Supervised Learning* appears at the end of the article *Machine Learning* it could misleadingly suggest that *Supervised Learning* is a prerequisite to *Machine Learning*. Therefore, we need to consider criteria focused on the abstracts of articles for such special cases [10].

Hyperlinks on Each Other's Abstract (HL-Ab): This criterion is to find out if c_1 hyperlinks are included in c_2 abstract. The abstract typically provides a brief definition, and if hyperlink for concept c_1 is mentioned in the abstract of concept c_2, it suggests that c_1 is likely a prerequisite for c_2 [10]. For instance, if *Machine Learning* is referenced in the abstract of *Supervised Learning* this implies that understanding *Machine Learning* is necessary before learning *Supervised Learning* [10]. The relationships are formulated as follows:

$$HL - Ab(c_1, c_2) = \begin{cases} 1, & \text{if } c_1 \in AH_{c_2} \\ 0, & \text{otherwise} \end{cases} \tag{3}$$

$HL - Ab(c_1, c_2)$ indicates whether c_1 is a prerequisite to c_2, where 1 is assigned in case hyperlink for concept c_1 belongs to AH_{c_2} which is the set of concepts hyperlinked within the abstract of the Wikipedia article for concept c_2, otherwise 0.

Hyperlinks on Each Other's Related Concepts' Abstracts (HL-RCA): Inspired by the idea of analyzing the importance of related concepts to infer PRs [9] and the role of hyperlinks in abstracts, we developed this criterion. This involves analyzing the abstracts of related concepts for each concept pair. For instance, consider the concepts *artificial intelligence* and *machine learning*. A related concept of *machine learning*, such as *artificial neural network*, is likely to reference a hyperlink for concept *artificial intelligence* in the abstract of its Wikipedia article. In contrast, a concept related to *artificial intelligence*, such as *knowledge representation*, does not mention the hyperlink of *machine learning* concept within the abstract of its article. This pattern arises because related concepts often share common prerequisite concepts. In this example, *artificial intelligence* serves as a prerequisite for both *machine learning* and *artificial neural network* as given in the following formula:

$$HL - RCA(c_1, c_2) = \begin{cases} 1, & \text{if } c_1 \in A_{RC_{c_2}} \\ 0, & \text{otherwise} \end{cases} \tag{4}$$

where $A_{RC_{c_2}}$ is the set of concepts that are hyperlinked within the abstract of the Wikipedia article of the related concepts of c_2. $HL - RCA(c_1, c_2)$ will be 1 in case c_1 hyperlinks are present in $A_{RC_{c_2}}$, otherwise 0.

Reference Distance (RefD): It is defined as the reference distance between concept pair [9]. This measures the proximity of Wikipedia articles based on their reference links and can be an excellent indicator for finding PRs. The idea here is that for each concept pair, we are comparing the asymmetry of c_1's related concept mentioning c_2 and c_2's related concepts mentioning c_1. For example, most of *Data Mining* related concepts mention *Data Science* in their articles. However, none of *Data Science* related concepts mention *Data Mining* in their articles. We adapted this equation to fit our system as follows:

$$RefD_{(c_1,c_2)} = \frac{\sum_{i=1}^{k} r(R_{c_i}, c_2)\, w(R_{c_i}, c_1)}{\sum_{i=1}^{k} w(R_{c_i}, c_1)} - \frac{\sum_{i=1}^{k} r(R_{c_i}, c_1)\, w(R_{c_i}, c_2)}{\sum_{i=1}^{k} w(R_{c_i}, c_2)} \quad (5)$$

where $C = \{c_1, ..., c_k\}$ is the concept space, R_{c_i} is a set of related concepts to c_i, $w(R_{c_i}, c_1)$ is cosine similarity between R_{c_i} and c_1, $r(R_{c_i}, c_1)$ the relationship between related concept of c_i and c_1, where R_{c_i} refers to c_1, meaning whether the related concept of c_i includes c_1. The outcome of RefD is then compared against a threshold to determine whether a PR is indicated as shown in the Eq. 6 below.

$$RefD(c_1, c_2) = \begin{cases} 1, & \text{if } -1 < RefD(c_1, c_2) < -\theta \\ 0, & \text{otherwise} \end{cases} \quad (6)$$

c_1 is considered a prerequisite of c_2 if the RefD result falls between -1 and $-\theta$. Conversely, c_2 is assumed to be a prerequisite of c_1 if the RefD result is between θ and 1. While between $-\theta$ and θ indicates no prerequisite can be inferred, where θ is empirically determined. This is defined as the following:

$$RefD(c_2, c_1) = \begin{cases} 1, & \text{if } \theta < RefD(c_1, c_2) < 1 \\ 0, & \text{otherwise} \end{cases} \quad (7)$$

Inbound Links and Outbound Links Ratio (IOLR): This criterion evaluates a concept's connectivity by comparing its inbound and outbound links [5,10]. Foundational concepts usually receive more inbound links, while advanced ones tend to have more outbound links. To reduce bias from generally popular concepts, we compute the inbound-to-outbound link ratio (IOL), where a higher IOL indicates that a concept is more foundational and likely to be a prerequisite, as given below:

$$IOL(c_i) = \frac{\sum inboundlinks}{\sum outboundlinks} \quad (8)$$

Then, we define the ratio to determine the direction of the relationship, as indicated by the equation $IOLR(c_1, c_2)$.

$$IOLR(c_1, c_2) = \begin{cases} 1, & \text{if } IOL(c_1) > IOL(c_2) \\ 0, & \text{otherwise} \end{cases} \quad (9)$$

Graph-Based Features: This category adopts the hierarchical structure of DBpedia, where the categories in DBpedia represent the hierarchical relationships between concepts.

One Concept is a Category of Another Concept (CatCon): Categories outline the essential topics of concepts, and understanding a concept often requires knowledge of its foundational elements defined by its category [5,20]. Based on the DBpedia hierarchy, if c_2 is classified under c_1, we can infer that c_1 is a category of c_2 and thus a prerequisite for c_2. To formulate this, we define K_{c_2} as the category associated with c_2 and vice versa in case of K_{c_1}. This is given in the equation below:

$$CatCon(c_1, c_2) = \begin{cases} 1, & \text{if } c_1 \in K_{c_2} \\ 0, & \text{otherwise} \end{cases} \quad (10)$$

$CatCon(c_1, c_2)$ indicates that c_1 belongs to the category associated with c_2. In other words, c_2 is categorized under c_1. This, strongly implies that c_1 serves as a prerequisite for c_2.

One Concept is the Direct Super-Category of Another Concept (SuperCatCon): Another useful signal for identifying PRs is the direct super-category of a concept, defined as the category of its category in DBpedia [20]. This follows a transitive logic: if c_1 *(Machine Learning)* is a category of c_2 *(Unsupervised Learning)*, and c_2 is a category of c_3 *(K-Means)*. Thus, c_1 is the super-category of c_3, which means c_1 is inferred as a prerequisite of c_3. Let $K_{c_1}^{up}$ denote the direct super-category of c_1, PR exists when these intersect with the categories of c_2.

$$SuperCat(c_1, c_2) = |K_{c_1}^{up} \cap K_{c_2}| \quad (11)$$

This is also repeated to calculate the existence of prerequisite in both directions $super(c_2, c_1)$. Then, we compare their asymmetry to avoid intertwining categories. We want to prevent scenarios where some super-categories of c_2 are also categories of c_1 for $SuperCat(c_1, c_2)$ or for $SuperCat(c_2, c_1)$, which would imply mutual prerequisites, as follows:

$$SuperCatCon(c_1, c_2) = \begin{cases} 1, & \text{if } len(SuperCat(c_1, c_2)) > len(SuperCat(c_2, c_1)) \\ 0, & \text{otherwise} \end{cases} \quad (12)$$

When $SuperCatCon(c_1, c_2)$ is 1, indicates that more of c_1's super-categories are also categories of c_2 suggesting that c_1 is a prerequisite of c_2.

Hierarchy-Based Features: This category captures the structural organization of educational content, reflecting concept relationships based on their positions within hierarchical formats. This approach can be applied to all educational platforms that follow a hierarchy structure. Inspired by Thareja et al., who modeled conceptual progression using the hierarchical structure of academic textbooks to detect PRs [17]. We applied a similar strategy in our context by leveraging the course hierarchy in CourseMapper, to define the CourseMapper Hierarchy (**CMH**) criterion, which captures PRs among concepts within the course hierarchy. In CourseMapper, courses are structured hierarchically, encompassing multiple topics, each of which consists of several channels that, in turn, contain various LMs. This hierarchical organization can be leveraged to infer PRs, as the arrangement of topics, channels, and LMs follows a sequential order. Specifically, channels that need to be accessed earlier are positioned before those intended for later stages of learning. By exploiting this structured progression, we can systematically derive PRs between different concepts within one course.

$$ConTop(c_1, c_2) = \begin{cases} 1, & \text{if } Top(c_1) > Top(c_2) \\ 0, & \text{otherwise} \end{cases} \tag{13}$$

In this context, $Top(c_1)$ represents the topics or channels within one course in which c_1 appears within a learning material in CourseMapper. $ConTop(c_1, c_2)$ is assigned a value of 1 if the channel in which c_1 is mentioned precedes the channel that mentions c_2. In this case, the likelihood of c_1 being a prerequisite for c_2 increases significantly.

Text-Based Features: This approach utilizes BERTopic combined with Shannon's entropy (**BERTropy**) to infer prerequisite connections [5]. In this context, a topic is defined as a theme discussed in the Wikipedia abstract of a given concept. For example, the abstract of the *Machine Learning* article may include topics such as *Artificial Intelligence* and *Statistical Methods*. The underlying intuition of this approach is that more advanced concepts tend to be more specialized and therefore cover fewer topics, whereas general concepts span a broader set of topics. To operationalize this, BERTopic is employed to extract topics from the Wikipedia abstract of a concept c. The model assigns a probability to each topic, indicating the degree to which the topic is present in the abstract. Let $P(x)$ denote the probability assigned by BERTopic that topic x appears in the abstract of concept c:

$$P(x) = \text{BERTopic}(x, c) \tag{14}$$

Given these topic probabilities, Shannon entropy is used to measure the distribution of topics within the abstract:

$$EntPro(c) = -\sum_{x}^{X} P(x) log \frac{1}{P(x)} \tag{15}$$

where X denotes the set of all identified topics and $P(x)$ represents the probability of topic x for concept c. A higher entropy score indicates that the concept is more general, while a lower entropy score suggests a more advanced and focused concept. This is applied to infer PRs between all concept pairs using entropy rule. A PR is assumed to exist only if the difference exceeds this threshold θ. In our case, the threshold is empirically selected and set $\theta = 1.33$, meaning that between 0 and 1.33, we assume no PR exists.

$$BERTropy(c_1, c_2) = \begin{cases} 1, & \text{if } EntPro(c_1) - EntPro(c_2) > \theta \\ 0, & \text{otherwise} \end{cases} \tag{16}$$

3.2 Voting Algorithm

After applying the criteria mentioned previously to each concept pairs, a **voting algorithm** will take place determine both the presence and direction of the PR between concept pairs. First, we treat all criteria equally by assigning equal weight (1) to all of them. This avoids biasing inference toward any single feature and ensures no criterion dominates the final decision. After that, we constructed two binary arrays (one hot encoded) (A_1, A_2) corresponds to $(c1 > c2)$ and $(c2 < c1)$, each of size 10, to examine how well each concept pair satisfies the set of predefined criteria. Each array element corresponds to a specific criterion and is assigned a value of 1 if the criterion is met or 0 if not. This means that only the criteria that are satisfied contribute to the final computation used to determine the PR, while the criteria that are not met have no influence on the result. We then separately sum the positive values in each array to quantify the total contribution of satisfied criteria for each concept pair. Following that, we compute the difference between the two sums by subtracting the total score of A_1 from that of A_2, as detailed in the following equation.

$$S_{\text{init}} = \sum A_1 - \sum A_2 \tag{17}$$

After that, normalization takes place, which is a common practice in many applications and simplifies setting thresholds for PR inference. To this end, the final score (S_{init}) is rescaled from the original range of $[-10, 10]$ to the standardized range of $[-1, 1]$. This facilitates meaningful aggregation and supporting the inference of PRs.

$$S_{\text{new}} = 2 \cdot \left(\frac{S_{\text{init}} - S_{\text{min}}}{S_{\text{max}} - S_{\text{min}}} \right) - 1 \tag{18}$$

where S_{new} is the normalized score difference between the two arrays, A_1 and A_2. S_{min}, in this case, would be the minimum, which is -10 and S_{max} is the maximum, 10. S_{init} is the current score before normalization a concept pair has. After normalizing the results, we classify the results into three distinct ranges. To do this, we define a threshold on both sides, these are θ and $-\theta$. The first range is if S_{new} or the difference between the arrays is between -1 and $-\theta$ which concludes that c_2 is a prerequisite concept to c_1. When the result is between $-\theta$

and θ, that means there is no PR between the two concepts. The last range is when S_{new} is between θ and 1 then we conclude that c_1 is a prerequisite of c_2. The value of $\theta = 0.28$ is determined empirically through a series of experiments aimed at identifying the optimal parameter setting for our specific application as mentioned in Sect. 4. To illustrate this more, an example in Table 2 is given, after examining all criteria for the concept pair *Machine Learning* (c_1) and *Supervised Learning* (c_2), we use two one-hot encoded arrays (A_1, A_2) to represent how well each concept pair satisfies a set of predefined criteria, in both directions. After that, we calculate the overall sums of the two arrays ($A_1 = 5.00$ and $A_2 = 2.00$), then compute the difference between them and normalize it, as illustrated in Table 3. The final score of $0.3 > 0.28$ falls between θ and 1, indicating that c_1 is prerequisite to c_2.

Table 2. Voting algorithm example

Criteria	TemO	HL-A	HL-Ab	HL-RCA	RefD	IOLR	CatCon	SuperCatCon	CMH	BERTropy
A_1 (c_1->c_2)	1	1	0	0	0	0	1	1	0	1
A_2 (c_2->c_1)	0	0	1	0	1	0	0	0	0	0

Table 3. Example of voting algorithm results

Mechanism	Total A_1 (c_1->c_2)	Total A_2 (c_2->c_1)	$A_1 - A_2$	Normalized difference
Voting algorithm	5.00	2.00	3.00	0.3

4 Experiments and Results

To demonstrate the effectiveness and advantages of our proposed method, we conducted thorough experiments and a comparative analysis using two public datasets, namely the AL-CPL [10,11,18] and Biology datasets, which is a specialized dataset derived from textbook chapters in the Biology curriculum [17].

Threshold Optimization: The threshold is essentially a hyperparameter to control the sensitivity of the model in determining the PR. The thresholds of criteria RefD and BERTopic, as well as the proposed voting algorithm were determined empirically using the same methodology, based on the AL-CPL dataset. To this end, a range of threshold values have been used to run each approach separately and compared the results with the ground truth dataset to compute precision, accuracy, recall, and F1 score. The **RefD** method identifies PRs by analyzing how concepts reference each other within the Wikipedia link structure. Wang et al. [18] originally proposed a threshold value of $\theta = 0.04$. We replicated this in the context of our platform, CourseMapper to generate the PRs. We found out that precision peaks at $\theta = 0.6$ with a value close to 0.9. This aligned with our aim of reliably detecting PRs while minimizing false positives. The higher threshold compared to previous work is attributed to the more domain-specific concept space used in our EduKG. While prior studies operated over the full

Wikipedia concept set, our system limits the scope to concepts and related concepts extracted from AL-CPL, leading to a more focused and specialized semantic space. Similarly, **BERTropy** infers PRs between concepts by analyzing the distribution of topics covered within their respective abstracts. We calculated the distribution of topics covered across all concepts. Then, BERTropy is applied to the entire set of concepts to determine PRs among them. We empirically found that precision peaks at $\theta = 1.4$, reaching approximately 0.9. This suggests that higher thresholds are more conservative, identifying only the strongest relationships, whereas lower thresholds connect more topics but may introduce noise. For the **Voting Algorithm**, we experimentally observed that precision reaches its maximum value of 1.0 at $\theta = 0.28$, indicating complete correctness among positive predictions without false positives at this threshold. In this study, we prioritize precision because minimizing false positives is more critical than capturing all possible PRs. Incorrect prerequisite suggestions could negatively affect learning paths and curriculum design.

Experiment 1: Prerequisite Learning on the AL-CPL Dataset: To evaluate the effectiveness of our proposed model, we conducted experiments using three domains from the AL-CPL dataset: Data Mining, Physics, and Macroeconomics [10,11,18]. It labels concepts with relationships: 1 for "prerequisite," −1 for "reverse prerequisite," and 0 for no relation. This dataset validates all criteria except CMH, which requires a content hierarchy that is lost when materials are uploaded into a single channel, treating all concepts at the same level. Therefore, hierarchical information was only applied in the Biology dataset, where such structure was available. The AL-CPL was uploaded in CourseMapper, and an EduKG was generated including main concepts, related concepts, and categories. This set of related concepts is utilized to infer the likelihood of a PR between concept pairs of the learning materials in CourseMapper.

Evaluation of Individual Criteria: To evaluate the contribution of each criterion, we conducted experiments across three domains—Data Mining, Physics, and Macroeconomics. Results showed that BERTropy achieved (1), HL-Ab obtained (0.84), and RefD accomplished (0.84) notably enhanced precision in Data Mining, while HL-Ab got (1) and CatCon got (1) performed well in Physics, and CatCon excelled in Macroeconomics with value (0.86). In terms of recall, IOLR with value (0.83, 0.84, 0.91) consistently outperformed all other criteria across three domains (Data Mining, Physics, and Macroeconomics), effectively capturing a broader set of prerequisite relations. Conversely, SuperCatCon underperformed in both precision and recall, indicating the need for refinement. To assess the balance between precision and recall, F1 scores were analyzed. IOLR achieved the highest F1 scores with values (0.67, 0.55, 0.59) in all domains (Data Mining, Physics, and Macroeconomics), confirming its effectiveness in maintaining this trade-off. Accuracy was also considered and offered useful trends, despite potential sensitivity to class imbalance. For example, HL-A showed high accuracy (0.65) in Data Mining, while CatCon achieved (0.70, 0.63) strong accuracy in Physics and Macroeconomics but had lower F1 scores, suggesting it may favor majority classes while poor at finding the minority class. The variation in criterion performance across domains can be attributed to differences in concept structure and semantic clarity. For instance, Data Mining and Physics datasets often contain well-defined, hierarchical, and technically specific concepts, which align well with criteria that leverage structural or semantic features. In contrast, the Macroeconomics domain tends to be more abstract and less formally structured, making it harder for some methods to identify clear PRs. In line with prior work, we place greater emphasis on precision, as the reliability of inferred PRs is critical for downstream tasks and system utility.

Comparative Results and Analysis: We compared our proposed model against six baseline methods—RefD [9], Active Learning [11], AdaBoost [22], WikiCPRL [21], Neural Network [14], and a Linguistically-Driven Strategy [13]—to evaluate its effectiveness in capturing PRs without relying on labeled data. As shown in Table 4, we assessed our voting algorithm across three domains: Data Mining, Physics, and Macroeconomics. Compared to existing baselines, our proposed Voting Algorithm consistently achieves higher precision but lower recall and F1 scores, reflecting its focus on reliably extracting only the strongest edges. For example, in Data Mining, our method reaches perfect precision (1.0) while RefD, Active Learning, and AdaBoost achieve lower precision (0.51,0.80, 0.85) but higher recall (0.76, 0.73, 0.85) and F1 scores (0.61, 0.67, 0.85). Similarly in Physics, Bagging-based AdaBoost achieves higher precision (0.87) than our method (0.76), while other baselines such as RefD and WikiCPRL show lower precision (0.49 and 0.43, respectively) compared to our method, but higher recall (0.49, 0.90). The same trend is observed for the Neural Network and the Linguistically-Driven Strategy, which yield higher F1 scores (0.65 and 0.81 in Data Mining, respectively, vs. 0.42 for our method) due to their stronger recall. This pattern shows that while baselines recover more links overall, our method prioritizes the precision and reliability of inferred PRs—critical for building robust learning paths—since, in educational settings, misleading a learner with an incorrect prerequisite (false positive) is more harmful than omitting some prerequisites.

However, baseline models perform better in recall, which we attribute to our precision-focused threshold selection in criteria such as RefD and BERTropy. These thresholds were tuned to prioritize reliability over quantity, resulting in fewer but more accurate inferred relationships. This can be attributed to differences between domains in concept structure and semantic clarity as mentioned earlier. Despite the lower recall, the proposed voting algorithm which integrates various feature-based methods achieves competitive performance without requiring any labeled training data. This makes it particularly well-suited for real-world settings where labeled prerequisite data is limited or difficult to obtain. Although supervised models often outperform unsupervised ones due to access to labeled data, manual labeling is costly, time-consuming, and prone to inconsistency. Most unsupervised methods evaluate criteria independently rather than integrating them for more effective PR identification. Our unsupervised multi-criteria approach addresses this gap with a scalable, flexible voting algorithm for inferring PRs, particularly in diverse educational domains lacking annotations.

Table 4. Performance Comparison Across AL-CPL Datasets

	Data Mining				Physics				Macroeconomics			
Criteria	P	R	F1	Acc.	P	R	F1	Acc.	P	R	F1	Acc.
TemO	0.47	0.31	0.37	0.40	0.36	0.42	0.39	0.47	0.46	0.71	0.56	0.53
HL-A	0.8	0.53	0.64	**0.65**	0.58	0.34	0.43	0.64	0.58	0.34	0.42	0.62
HL-Ab	0.84	0.22	0.35	0.53	**1**	0.13	0.23	0.65	0.62	0.08	0.14	0.60
HL-RCA	0.53	0.53	0.53	0.45	0.42	0.67	0.52	0.49	0.39	0.43	0.41	0.48
RefD	0.84	0.22	0.35	0.53	0.72	0.21	0.33	0.65	0.69	0.15	0.24	0.62
IOLR	0.56	**0.83**	**0.67**	0.53	0.41	**0.84**	**0.55**	0.45	0.44	**0.91**	**0.59**	0.47
CatCon	0	0	0	0.43	**1**	0.25	0.40	**0.70**	**0.86**	0.12	0.21	**0.63**
SuperCatCon	0	0	0	0.43	0	0	0	0.60	0	0	0	0.59
BERTropy	**1**	0.09	0.16	0.48	0.26	0.09	0.13	0.54	0	0	0	0.59
Baselines												
RefD [9]	0.51	0.76	0.61	-	0.49	0.49	0.49	-	-	-	-	-
Active Learning [10]	0.80	0.73	0.76	-	0.8	0.6	0.7	-	-	-	-	-
Bagging-based AdaBoost [22]	0.85	0.85	**0.85**	-	**0.87**	**0.87**	**0.86**	-	-	-	-	-
WikiCPRL [21]	0.59	**0.90**	0.68	-	0.43	0.83	0.57	-	-	-	-	-
Neural Network [14]	0.62	0.68	0.65	-	0.53	0.62	0.57	-	-	-	-	-
Linguistically-Driven Strategy [13]	0.81	0.81	0.81	-	0.77	0.71	0.74	-	-	-	-	-
Proposed - Voting Algorithm	**1**	0.27	0.42	0.58	0.76	0.13	0.22	0.64	0.65	0.16	0.25	0.61

Experiment 2: Prerequisite Learning on the Biology Dataset: To effectively evaluate the proposed model's ability to detect PRs and assess **TemO** criterion more reliably, as well as to measure the contribution of **CMH**, we utilized a specialized dataset derived from textbook chapters in the NCERT[1] Biology curriculum [17]. For our experiments, we selected two conceptually related chapters from NCERT science textbooks: Chapter 7 (Transportation in Animals and Plants, Class 7) and Chapter 6 (Life Processes, Class 10). These were chosen for their natural hierarchical relationship, with foundational content in Class 7 supporting advanced concepts in Class 10. Both chapters were uploaded into CourseMapper under the same course and same topic, but placed in two different sequential channels to preserve the learning hierarchy defined by the platform. We generated the EduKG for each chapter individually by extracting the main concepts from each and then expanding them using DBpedia to identify related concepts. This structure enabled us to evaluate all ten predefined criteria for detecting PRs. It is worth mentioning, when a new learning material is added, the system computes the EduKG for this new material, without needing to reprocess the entire EduKG. In the absence of ground truth in the original dataset, 186 concept pairs were manually annotated by domain experts (PhDs and medical doctors) as either prerequisite or not.

Evaluation of Individual Criteria and Comparative Results: Biology dataset is used to evaluate each individual criterion using four standard metrics: precision, recall, F1-score, and accuracy (Table 5). HL-Ab and HL-A achieved the

[1] https://ncert.nic.in/textbook.php.

highest precision, effectively identifying the most reliable PRs while minimizing false positives. In contrast, IOLR and CMH attained the highest recall, capturing a broader set of dependencies at the expense of tolerating some false positives. Notably, IOLR and CMH also achieved the highest F1-scores, indicating a strong balance between precision and recall. In terms of accuracy, BERTropy performed best (0.75), followed closely by IOLR and CMH, reinforcing their robustness in distinguishing between prerequisite and non-prerequisite pairs. The success of CMH, in particular, highlights the value of structured educational content and hierarchical information in inferring PRs. In this dataset, the placement of materials across channels is well-defined, enabling CMH to leverage this ordering effectively. By contrast, CatCon and SuperCatCon depend on the category hierarchy in DBpedia, and TemO relies on the sequential appearance of concepts in text, both of which are less consistent in the current dataset. Their effectiveness may improve by incorporating additional chapters. In comparison, our proposed voting algorithm achieved competitive results, although it did not surpass the best-performing individual criteria such as CMH and IOLR. With moderate precision and recall, it provided a balanced yet limited performance compared to these criteria when applied independently. The effectiveness of the voting algorithm is largely supported by strong individual contributors like CMH and IOLR. CMH achieved high F1-scores due to its reliance on hierarchical signals, while IOLR benefited from the ratio of in-outbound links—both features being particularly well captured in the Biology dataset. These findings highlight the importance of integrating hierarchical information. The defined criteria are deduced from Wikipedia articles, DBpedia hierarchy, textual information, content order, and platform hierarchy to infer PRs, making it applicable across diverse educational platforms.

Table 5. Performance Comparison Across Biology dataset

	Biology			
Criteria	P	R	F1	Acc.
TemO	0.0	0.0	0.0	0.42
HL-A	0.91	0.32	0.48	0.58
HL-Ab	**1.0**	0.06	0.12	0.45
HL-RCA	0.87	0.23	0.35	0.51
RefD	0.75	0.10	0.17	0.45
IOLR	0.58	**1.0**	**0.74**	0.58
CMH	0.58	**1.0**	**0.74**	0.58
CatCon	0.0	0.0	0.0	0.42
SuperCatCon	0.0	0.0	0.0	0.42
BERTropy	0.55	0.65	0.70	**0.75**
Proposed - Voting Algorithm	0.48	0.35	0.41	0.40

5 Conclusion and Future Work

Educational Knowledge Graphs (EduKGs) capture diverse entities and connections in educational systems, providing a structured view of key concepts from learning materials. Prerequisite relationships (PRs) are essential for defining learning sequences but are often missing in EduKGs. However, automatic methods for PR identification depend on domain-specific labeled data and static features, limiting scalability. Existing feature-based methods often ignore how semantic and structural information interact, rely often on concept order, and assess criteria individually. To address these gaps, we proposed an unsupervised method that combines ten criteria using a voting algorithm to infer PRs without using labeled data. Experiments on two benchmark datasets show that our proposed approach achieves higher precision than baselines and reliably extracts PRs across domains. In future work, we aim to extend our PR identification framework with more criteria. We will also experiment with hybrid and LLM-based approaches for more effective PR identification. Additionally, adaptive thresholding methods, learned per domain and combined with a weighting approach, could be adopted to further enhance the robustness and accuracy of PR inference.

Disclosure of Interests. The authors have no competing interests to declare that are relevant to the content of this article.

References

1. Ain, Q.U., Chatti, M.A., Bakar, K.G.C., Joarder, S., Alatrash, R.: Automatic construction of educational knowledge graphs: a word embedding-based approach. Information **14**(10), 526 (2023)
2. Ain, Q.U., et al.: Learning channels to support interaction and collaboration in coursemapper. In: Proceedings of the 14th International Conference on Education Technology and Computers, pp. 252–260 (2022)
3. Alatrash, R., Chatti, M.A., Ain, Q.U., Fang, Y., Joarder, S., Siepmann, C.: ConceptGCN: knowledge concept recommendation in MOOCs based on knowledge graph convolutional networks and SBERT. Comput. Educ. Artif. Intell. **6**, 100193 (2024). https://doi.org/10.1016/j.caeai.2023.100193. https://www.sciencedirect.com/science/article/pii/S2666920X23000723
4. Bai, Y., Liu, Z., Guo, T., Hou, M., Xiao, K.: Prerequisite relation learning: a survey and outlook. ACM Comput. Surv. (2025)
5. Bai, Y., Zhang, Y., Xiao, K., Lou, Y., Sun, K.: A bert-based approach for extracting prerequisite relations among wikipedia concepts. Math. Probl. Eng. **2021**(1), 3510402 (2021)
6. Gasparetti, F., De Medio, C., Limongelli, C., Sciarrone, F., Temperini, M.: Prerequisites between learning objects: automatic extraction based on a machine learning approach. Telematics Inform. **35**(3), 595–610 (2018)
7. Hu, H., Pan, L., Ran, Y., Kan, M.Y.: Modeling and leveraging prerequisite context in recommendation. arXiv preprint arXiv:2209.11471 (2022)

8. Hu, X., He, Y., Sun, G.: Active learning for concept prerequisite learning in wikipedia. In: Proceedings of the 2021 13th International Conference on Machine Learning and Computing, pp. 582–587 (2021)
9. Liang, C., Wu, Z., Huang, W., Giles, C.L.: Measuring prerequisite relations among concepts. In: Proceedings of the 2015 Conference on Empirical Methods in Natural Language Processing, pp. 1668–1674 (2015)
10. Liang, C., Ye, J., Wang, S., Pursel, B., Giles, C.L.: Investigating active learning for concept prerequisite learning. In: Proceedings of the AAAI Conference on Artificial Intelligence, vol. 32 (2018)
11. Liang, C., Ye, J., Zhao, H., Pursel, B., Giles, C.L.: Active learning of strict partial orders: a case study on concept prerequisite relations. arXiv preprint arXiv:1801.06481 (2018)
12. Manrique, R., Pereira, B., Mariño, O.: Exploring knowledge graphs for the identification of concept prerequisites. Smart Learn. Environ. **6**(1), 1–18 (2019). https://doi.org/10.1186/s40561-019-0104-3
13. Miaschi, A., Alzetta, C., Cardillo, F.A., Dell'Orletta, F.: Linguistically-driven strategy for concept prerequisites learning on Italian. In: Proceedings of the Fourteenth Workshop on Innovative Use of NLP for Building Educational Applications, pp. 285–295 (2019)
14. Moggio, A., Parizzi, A.: Unige se@ prelearn: utility for automatic prerequisite learning from Italian Wikipedia. EVALITA Evaluation of NLP and Speech Tools for Italian-17th December 2020, p. 376 (2020)
15. Pan, L., Li, C., Li, J., Tang, J.: Prerequisite relation learning for concepts in MOOCs. In: Proceedings of the 55th Annual Meeting of the Association for Computational Linguistics (Volume 1: Long Papers), pp. 1447–1456 (2017)
16. Sayyadiharikandeh, M., Gordon, J., Ambite, J.L., Lerman, K.: Finding prerequisite relations using the wikipedia clickstream. In: Companion Proceedings of the 2019 World Wide Web Conference, pp. 1240–1247 (2019)
17. Thareja, R., Garg, R., Baghel, S., Dwivedi, D., Mohania, M., Kulshrestha, R.: Auto-req: automatic detection of pre-requisite dependencies between academic videos. In: Proceedings of the 18th Workshop on Innovative Use of NLP for Building Educational Applications (BEA 2023), pp. 539–549 (2023)
18. Wang, S., et al.: Using prerequisites to extract concept maps from textbooks. In: Proceedings of the 25th ACM International on Conference on Information and Knowledge Management, pp. 317–326 (2016)
19. Xia, J., Li, M., Tang, Y., Yang, S.: Course map learning with graph convolutional network based on AUCM. World Wide Web **26**(5), 3483–3502 (2023)
20. Xiao, K., Fu, Y., Zhang, J., Tianji, W.: A hybrid approach for discovering concept prerequisite relations in Wikipedia. In: 2022 9th International Conference on Behavioural and Social Computing (BESC), pp. 1–5. IEEE (2022)
21. Xiao, K., Li, K., Zhang, Y., Chen, X., Lou, Y.: Wikicprl: a weakly supervised approach for wikipedia concept prerequisite relation learning. In: Asia-Pacific Web (APWeb) and Web-Age Information Management (WAIM) Joint International Conference on Web and Big Data, pp. 177–192. Springer (2023)
22. Zhou, Y., Xiao, K., Zhang, Y.: An ensemble learning approach for extracting concept prerequisite relations from wikipedia. In: 2020 16th International Conference on Mobility, Sensing and Networking (MSN), pp. 642–647. IEEE (2020)
23. Zhu, Y., Zamani, H.: Predicting prerequisite relations for unseen concepts. In: Proceedings of the 2022 Conference on Empirical Methods in Natural Language Processing (EMNLP 2022) (2022)

A Knowledge Graph-Based Approach for Personalized Course and Curriculum Path Recommendation

Constance Jumbo[1], Rashed Al Amin[1(✉)], Hasan Abu-Rasheed[3], Veit Wiese[1], Christian Weber[2], and Roman Obermaisser[1]

[1] Institute for Embedded Systems, University of Siegen, Siegen, Germany
rashed.amin@uni-siegen.de

[2] Medical Informatics and Graph-based Systems, University of Siegen, Siegen, Germany

[3] Studiumdigitale, Goethe University Frankfurt, Frankfurt, Germany

Abstract. The rapid development of educational technology has significantly expanded the range of academic degree programs, associated courses, and corresponding learning resources. This expansion has led to substantial overlaps in course content across programs, making it increasingly difficult for learners to construct coherent and efficient learning paths. Educators must also adapt their materials more effectively to meet the diverse needs of their students. A personalized course and curriculum path recommendation system offers a promising solution by guiding both students and instructors in selecting relevant content and structuring appropriate learning trajectories. Among the techniques available, knowledge graphs (KGs) are particularly effective. By representing information as entities and their interrelationships, KGs provide a structured, semantically meaningful structure for organizing educational data. This conceptual paper introduces a Knowledge-graph driven recommender framework that supports the generation of personalised learning paths aligned with individual learners prior knowledge and goals, while enabling instructors to adapt curriculum design and instructional delivery to identified learner needs. By semantically linking learner's background knowledge to relevant curricular components by leveraging placement assessments, the framework supports individualized assessments and recommends adaptive learning trajectories. The work outlines its theoretical foundations and explores its potential through relevant literature and graph-based design principles, demonstrating how structured semantic representations support goal-aligned, personalized learning. Through this approach, systems based on the framework have the potential to reduce content redundancy, enhance coherence, and improve the quality of teaching and learning, while fostering a more dynamic and responsive educational environment.

Keywords: Knowledge graph · Personalized learning · Path Recommendation

H. Takeda et al. (Eds.): IJCKG 2025, LNAI 16297, pp. 304–319, 2026.
https://doi.org/10.1007/978-981-95-5009-8_20

1 Introduction

The expansion of educational programs and course offerings, both offline and online, has increased significantly in response to technological advancements and the growing demand for accessible, flexible, and lifelong learning opportunities [21,27]. Universities now offer a broad range of interdisciplinary and specialized degree programs. At the same time, online platforms, particularly Massive Open Online Courses (MOOCs), have enabled global access to thousands of courses across various domains. Collaborative learning platforms have emerged as an effective solution to enhance interaction among learners and instructors in distributed environments [5,7,8]. However, while such platforms support engagement and accessibility, they alone cannot address the complexities of navigating overlapping and extensive course content. Moreover, the vast and often unstructured nature of online learning resources makes it increasingly difficult for learners to discern which courses or materials are most relevant to their learning objectives or prior knowledge. Learners may unintentionally repeat content they have already mastered or struggle to identify the next logical step in their academic development. On the other hand, instructors face challenges in maintaining consistency across course materials, updating content in response to technological and pedagogical trends, and adapting instruction to meet the diverse needs of learners. These issues highlight the pressing need for intelligent systems that can effectively organize, integrate, and personalize course content to support both learners and educators in navigating the increasingly complex educational landscape. In this context, personalized learning path recommendation systems become essential for guiding learners through individualized and coherent educational trajectories that align with their goals and prior knowledge while also assisting educators in curriculum planning and content alignment [12].

Personalized learning refers to an educational approach that adapts the content, sequence, and pace of learning activities to match the unique needs, preferences, prior knowledge, and learning goals of each learner [25]. Various approaches have been used, including rule-based [24] and machine learning methods [2,15,18,23,29] but these often fail to capture semantic relationships in curricula. To address this limitation, recent research has increasingly focused on knowledge graph-based approaches, which provide a robust framework for modeling interconnected educational data. Knowledge graphs represent entities, such as courses, topics, skills, assessments, learning resources, and learner attributes, as nodes and the relationships among them as edges. This graph-based structure enables a comprehensive understanding of dependencies and hierarchies within a curriculum, allowing the system to infer optimal paths based on prerequisite relationships, concept progression, and learner profiles [26]. The integration of knowledge graphs into personalized learning path recommendation systems offers multiple benefits. They help to reduce content redundancy, avoid repetitive learning experiences, and improve the alignment between learner goals and instructional content [13]. Additionally, graph-based systems support educators by offering data-driven insights for course sequencing, curriculum refinement, and learner engagement strategies. Promoting a more structured, efficient, and targeted learning experience, knowledge graph-based systems are making significant contributions to improving learning outcomes and ensuring a more adaptive and learner-centered educational environment [16].

This paper proposes a conceptual approach for constructing and utilizing knowledge graphs to provide personalized learning paths. Several scholars have also proposed graph-based structures for learning path recommendations. However, KGs schemas in majority of this line of literature model specific entities, such as courses or concepts, as in [6,17], and [32], leaving out other important entities in higher education, and the need for more granulated definitions of the concepts, missing. To overcome these limitations, this study introduces a knowledge graph-driven and curriculum focused recommendation framework that not only supports personalized course selection for learners but also aids educators in designing adaptive and goal-aligned curricula, since it builds on the analysis of curricula structures in higher education. The framework leverages the interconnected nature of course topics, prerequisites, learner attributes, and academic goals, to construct coherent and individualized learning trajectories. By reducing overlaps and dynamically aligning instructional content with learner needs, the proposed framework aims to create a more streamlined, adaptive, and effective learning experience for both learners and instructors. As part of the funded project IGNITE: "Individual support in STEM teaching", this paper will zoom in specifically on the construction of the knowledge graph and explore, through a use case, how the graph will be used for learning path recommendations.

2 Related Works

Knowledge graphs have emerged as promising foundation for personalized learning, enabling structured reasoning over complex educational data. Prior research shows different directions for KG-based personalization, which has been grouped here into three themes: path optimization and adaptive sequencing via reinforcement learning, semantic enrichment and deep learning for course-specific graphs, and cognitive or learner-centered modeling.

One prominent approach focuses on graph optimization and path planning. Wang et al. [28] applied Steiner Tree Optimization model on a knowledge graph to generate learning paths by minimizing unnecessary intermediate nodes through dynamic programming. Experimental results demonstrated that Portfolio Steiner Tree outperformed baseline methods in terms of scalability and accuracy. Zhao et al. [34] extended this idea to vocational training by constructing subject-specific knowledge graphs. Zhao et al. system visually and structurally maps conceptual relationships, enabling the dynamic adaptation of learning sequences. By analyzing learners' progress, assessment outcomes, and preferences, the model identifies knowledge gaps and updates paths using real-time feedback and Bayesian updates. A related approach was that of Haldar et al. [14], which utilized reinforcement learning to adapt learning sequences dynamically. Haldar et al. [14] introduced a reinforcement learning approach using Q-learning on KGs to derive the optimal sequence between learner's known and desired goals. To address the cold-start problem, Haldar et al. developed an Independent Importance Score, enabling smart initialization. Their model adapted better than static recommenders by generating coherent and efficient paths. Together, these works highlight the potential of opimization and reinforcement learning to create efficient paths. However, they focus on isolated domains or simplified datasets, and they often neglect pedagoogical structures that are central in higher education.

Another significant advancement lies in the semantic enrichment and deep learning for course-specific graphs. Li et al. [20] proposed a CourseKG, a knowledge graph tailored for precision teaching, particularly in the context of C programming. CourseKG enables personalized path planning and supports adaptive instruction, with the potential for multimodal and feedback-integrated extensions. Abu-Rasheed et al. [1] applied semantic text mining to enrich KGs with context edges. The custom Text Mining Pipeline of Abu-Rasheed et al. extracted semantic and contextual relationships from English and German LO descriptions, enhancing the graph with semantic edges for cross-domain recommendations. To address sparsity and scalability, Baig et al. [9] proposed a modular architecture. Their framework included a knowledge graph module to structure relationships, community clustering and a graph embedding module. In order to create a personalised course recommendation Dong et al. [11] combined transformer models and attention mechanisms. Unlike Baig et al., their system constructs a layered course knowledge graph and integrates feature extraction, annotation, and a Transformer model to capture dependencies between learners and course materials. Together, these studies show how semantic enrichment and deep learning approaches expand the flexibility and effectiveness of KG-based personalization, though most remain limited to course-specific contexts. Wei et al. [30] explored a method for personalized learning resource recommendation using a domain-specific knowledge graph constructed for discrete mathematics top-down system. The recommendation algorithm fuses interest similarity and knowledge connection degrees into a weighted score to rank resources. While, Ain et al. [4] proposed a learner modeling and recommendation framework that leverages Personal Knowledge Graphs (PKGs) to enhance personalization and transparency in Educational Recommender Systems (ERS). In the case of Ain et al., Learners explicitly mark concepts they "did not understand", which are integrated into semantically enriched PKGs using sentence embeddings and graph convolutional networks to generate tailored recommendations of external learning resources. These studies significantly outperformed traditional recommendation methods positively impacting learner satisfaction and acceptance underpinning the importance of user input in personalised recommendations.

Taking a cognitive science-inspired approach, Mu et al. [22] introduced the CGLPRS system, which is based on cognitive graphs. Their method visualizes students' cognitive states and incorporates memory decay using Ebbinghaus' forgetting curve. Recommendations are informed by meaningful learning and dual-process theories, utilizing collaborative filtering and TransH embeddings to suggest relevant knowledge units and learning resources. Finally, Zhang et al. [33]

developed KG-PLRR, a model that enhances personalized resource recommendation by propagating student preferences through a course knowledge graph. Their method incorporates student behavior embeddings and learning resource embeddings within the RippleNet and Knowledge Graph Convolutional Network (KGCN) frameworks. These works bring valuable learner-centered perspectives by linking recommendation strategies to cognitive models and behavioural patterns. However, they largely treat learners in isolation and do not address how cognitive personalization integrates with curriculum-level reasoning or institutional structures.

Overall, these studies illustrate the versatility of knowledge graph-based approaches to personalized learning, from structural optimization to semantic enrichment and cognitive modeling. Each method brings a unique perspective that strengthens the foundation for future curriculum and course recommendation systems powered by knowledge graphs. Yet, the majority of approaches remain either domain specific, course-focused, or detached from institutional curriculum structures. Few frameworks address the higher education context, where overlapping domain concepts, modular programs, and cross-domain alignment are critical. Our proposed framework builds on this gap by integrating curriculum, domain, and user models into a unified KG, emphasizing session-level (domain concepts) reasoning as the key mechanism for generating personalized, coherent, and institutionally relevant learning paths.

3 Proposed Framework: A Knowledge Graph-Driven Personalisation of Learning Paths

3.1 Framework Overview

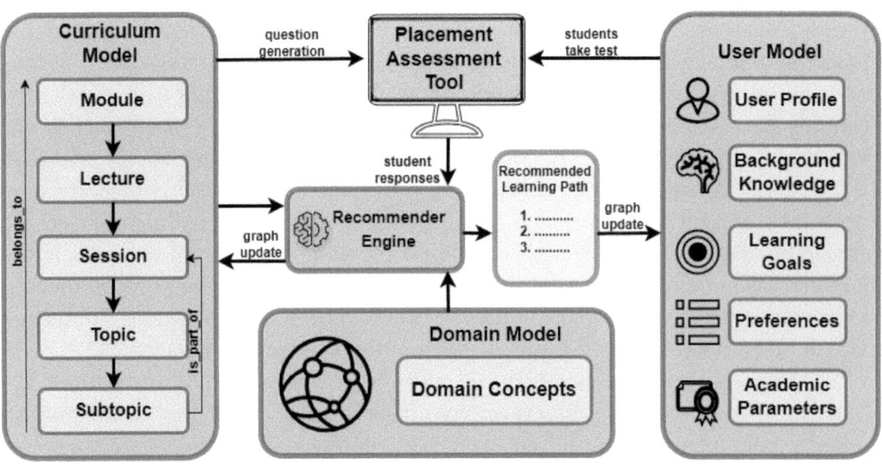

Fig. 1. The IGNITE framework architecture showing the integration of the Curriculum, User, and Domain Models with the Placement Assessment Tool, Graph Update Engine, and Recommender Engine.

We present a semantically-enriched, graph-driven framework designed to generate personalised learning paths within a modular higher education curriculum. Our approach uses a structured knowledge graph to align each learner's current knowledge state with a curriculum's hierarchical and conceptual structure. The framework models educational content as an interconnected semantic network, including course concepts, concept dependencies, and learner performance data. This representation enables systems that are based on the framework to reason over the graph to determine which curriculum components the learner has prior knowledge of, identify knowledge gaps by pinpointing weaknesses, and recommend coherent and context-sensitive learning path. A key contribution of the framework is its dynamic update mechanism, which refines learner profiles based on assessment feedback, enabling intelligent path adaptation. The IGNITE-project framework integrates three essential components as seen in Fig. 1:

1. A curated knowledge model that consists of a User Model, Curriculum Model and Domain Model,
2. A placement assessment tool for diagnosing learners' actual knowledge state, and
3. A recommender engine to generate personalized learning paths.

The process is initiated by a placement assessment tool, which supports the recommender engine to tailor recommendations to individual need of the students. Together, these components create an end-to-end structure for a learning environment where educational decisions are guided by semantic knowledge representation and learning assessment.

3.2 The Knowledge Model

The framework is underpinned by a unified ontology, built on three core knowledge models. These knowledge models represent key elements of a higher education knowledge structure as described in [2]. By integrating user, curriculum, and domain models, the ontological framework connects the user's background and intent with the curriculum and its logic, while utilizing a shared topic-oriented concept of decomposition. The addition of a domain model is done with the intent to have the means to disambiguate the use and meaning of topics which might have different concepts and preconditions in different domains. Furthermore, the formal definition enables a differentiated semantic comparison with similarity methods and is the base for the generation of personalised paths.

The User Model: The user model captures individual learner features and serves as an anchor for personalization in the knowledge graph. This comprises a set of core entities and relationships that capture various aspects of a learner's profile. The User Profile, Personal Info, Preference, Learning Goal, Background

Knowledge State, Academic Parameter, and Log. This is a combination of foundational user input (personal information, learning goals, and academic preferences collected through user input) with dynamically updated attributes, particularly the learner's Background Knowledge State, which is generated from the responses provided in the placement assessments. To maintain a simplified yet flexible schema, entities are connected using an abstract relationship label, *has_a*, which generates different forms of learner-attribute associations. As shown in Fig. 3, learner responses are mapped to Topic nodes and aggregated for model updates. Graph reasoning and updates are triggered when assessment responses are mapped to curriculum nodes and conclude on an update, such as updating the Background Knowledge State. These placement assessment responses propagate through the curriculum and user graphs, utilizing the *includes* relationship labels to update the learner's knowledge state. This information is vital for tailoring learning paths that are both curriculum-aligned and learner-relevant. Additionally, the Learning Goal node connects to the Domain Model via the *includes* relationship, comparing the learner's state to the conceptual requirements of the domain. This dual connection supports reasoning from a structural curriculum perspective and a conceptual domain alignment viewpoint.

The Domain Model: This model comprises domain-specific concepts and serves as the conceptual backbone of the graph. It captures key knowledge components within a subject area, abstracted from specific courses in the program. Its primary function supports semantic alignment between what is taught (curriculum) and what needs to be known (conceptual understanding), thus enabling use-case orientation and reasoning. The domain model comprises Scenario (applied or conceptualised use cases within the domain), Domain (a broad subject area), Subdomain (a narrow focus within the domain), and Domain Topic (atomic knowledge components or concepts) entities. Each Domain Topic node is bidirectionally linked to the Topic and Subtopic nodes in the curriculum model. This creates a cross-module alignment; as a result, if multiple sessions reference the same domain concepts, the model avoids redundant recommendations. The background knowledge is defined through semantic relationships between user nodes and the domain and curriculum models as described in [2]. They consists of lists of curriculum topics, each contextualized by domain descriptions that add conceptual depth. Additionally, the Domain Topic node is linked to the Learning Goal in the User Model, comparing what the learner knows and what they should know, thereby keeping their learning goal current. This facilitates adaptive learning, enabling learners to reason about their progress across related concepts rather than just isolated topics. Reasoning over this model allows the system to measure conceptual proximity and suggest content that best bridges the learner's current and target states.

The Curriculum Model: The curriculum model reflects the hierarchical structure of academic programs, organising module contents into instructional units of varying levels of granularity. As shown in Fig. 2, it models the pedagogical progression of learning by decomposing course materials into semantically annotated units. The curriculum model is structured as follows:

- Module: This represents a full academic course, typically offered over a single semester.
- Lecture: This is a sub-module, representing a thematic grouping of sessions.
- Session: This instructional unit focuses on specific topics and is typically delivered in a single session, e.g., on weekly bases.
- Topic: This is a specific knowledge area within a session.
- Subtopic: This is a finer-grained learning concept.

There is a limit to the quality and complexity of decomposition of human knowledge. Any freely built taxonomy of topics and subtopics will therefore always be uneven in depth and width of decomposition if multiple depth of subtopics are allowed. The decision to allow only a singular level of subtopics in this framework is intended as an endorsement to human and machine-based topic integration to compromise on the expressiveness of decomposition at time of modelling along this limitation, in favor of heavily reducing the need for disambiguation of modelling depths across larger ontologies.

Each entity is modelled as a node in a labelled property graph and annotated with metadata, including titles, IDs, weights, descriptions, and the IDs from the parent node. Two key relationships define the graph structure:

- BELONGS-TO: provides hierarchical links (e.g., Topic BELONGS-TO Session)
- IS-PART-OF: Provides transitive traversal links (Subtopic IS-PART-OF Session)

Although the curriculum is designed in a top-down format, mirroring course design practices and syllabus, the graph reasoning operates bottom-up. Assessment data mapped to Topic nodes is traversed upward via the *belongs_to* relationship label to infer a learner's knowledge state at the Session level, which is the primary unit for recommendation as seen in Fig. 3. In systems based on this framework, such as the one described in [2], graph content is populated in a hybrid manner. Content can be added directly from course materials and instructor-curated syllabus, if the are already fitted into the levels of granularity determined by the ontology. If the course content requires an extraction step of the ontology classes, an algorithmic approach, e.g., via LLM-based topic extraction from lecture media, slides, and manuscripts, has proven a sound ability to populate the KG [2]. In both construction approaches, all outputs are validated by course administrators and domain experts, to ensure pedagogical consistency, and accurate topic extraction.

To connect curriculum topics to their corresponding domain ones, Topic and Subtopic node are connected to one or more Domain Topics via the *equivalent_to* relationship, enabling cross-module reasoning. This facilitates the substitution of curriculum units across modules, avoids redundancy, and enhances the flexibility

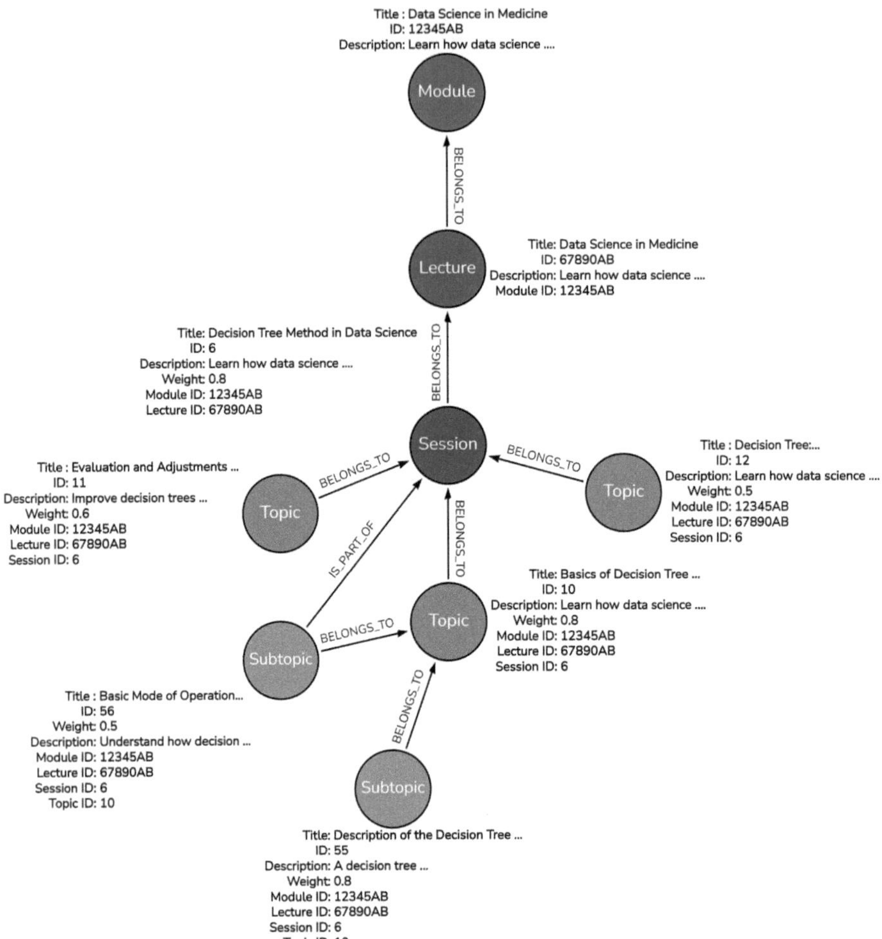

Fig. 2. Curriculum Subgraph illustrating the hierarchical structure from Module to Subtopic. Relationships such as *belongs_to* and *is_part_of* connect modular elements, enabling semantic reasoning.

of the learning path engine. The Curriculum Model is tightly linked to the User and Domain Models via the placement assessment layer, allowing fine-grained detection and recommendation.

3.3 The Placement Assessment Tool

Personalised learning systems often suffer from a mismatch between perceived and actual learner knowledge, resulting in ineffective or disengaging recommendations [19]. To address this, we plan to incorporate a structured placement assessment tool that diagnoses the learner's knowledge using automatically generated questions based on curriculum topics. The assessment items will span multiple cognitive levels, based on Bloom's Taxonomy (recall, understand, apply, analyse, evaluate and create). The graph aggregates learner responses and propagates them upward to Session nodes using traversal paths guided by structural relationships as shown in Fig. 3. This session level reasoning balances granularity with pedagogical coherence.

Session-Level Reasoning and Placement Mapping: Sessions are the smallest complete teaching units with clear learning objectives, while the placement assessment generates evidence at the Topic/Subtopic level, aggregating results to the session level ensures reliable mastery inference and enables actionable, non-fragmented recommendations. It also supports flexible substitution of equivalent sessions across modules while aligning with institutional structures such as credit, assessments, and timetables. To illustrate the session-level reasoning, consider the case of a learner enrolled in a Data Science in Medicine module, which includes a Session Titled "Decision Tree Method in Data Science". This Session contains Topics such as:

- Topic1: Basics of Decision Tree Procedures in Data Science
- Topic2: Evaluation and Adjustments of Decision Trees
- Topic3: Decision Tree: Example Using a Mailing Campaign
- Topic4: Decision Trees for Diagnostic Support in Medicine

Each of these Topics decomposes further into subtopics, which are linked to domain concepts. First, the learner is presented with a placement test consisting of auto-generated questions from the Topic level. The questions are mapped to their Topic nodes and, by upward traversal, connected to Session nodes. Upon answering these questions, the system identifies that the learner correctly answered questions related to Topic 3; however, the learner struggled with questions related to Topics 1, 2, and 4 and did not respond accurately to them as seen in Fig. 3. These incorrect responses are semantically mapped to their corresponding Topic nodes and then rolled up to the session level. Aggregating performance at the Topic level and reasoning upwards, the system concludes that the learner has partial mastery of the session's learning objectives. As a result, the session is flagged as unmet but is annotated with performance details indicating which Topics were successfully mastered. This output is structured as a reasoning result set containing, for each session, a mastery classification and weighed performance scores. This result set is represented as a an annotated sub-graph of the unified knowledge model, enabling further semantic reasoning, visualization and ontology-level queries.

This serves as the primary interface to the recommender engine, which uses this information to select and prioritise sessions for inclusion in a personalised learning path. Finally, the learner is recommended to attend the original session labelled as 'unmet' or an alternative session with overlapping domain coverage that specifically reinforces the Topics missed. If multiple sessions share domain-aligned topics, *equivalent_to* and *includes* links are used to reason over alternative sessions. The reasoning outcomes include:

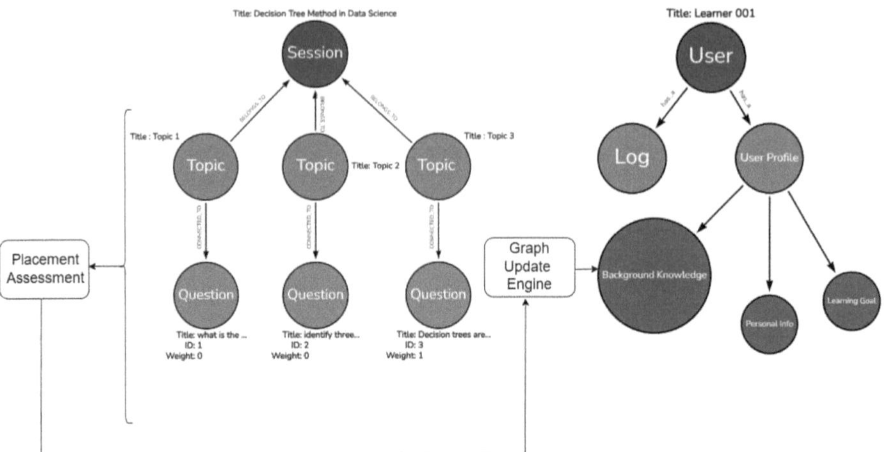

Fig. 3. Graph-based reasoning and model update workflow. Learner responses to questions are mapped to Topic nodes, aggregated to the Session level, and used to update the Background Knowledge in the User Model.

- Localization of knowledge strengths and weaknesses.
- Identification of unmet learning objectives at the Session level.
- Semantic linking to conceptually equivalent sessions.
- Dynamic update of the User Model based on graph-anchored performance.

This ensures the path is pedagogically coherent (session-level recommendation maintains instructional logic), non-redundant (already mastered concepts are not repeated), domain-aligned (the recommended sessions are conceptually equivalent), and personalized (based on actual learner responses). Unlike rule-based recommender systems that operate on fixed logic [31], the graph-based structure supports flexible reasoning, which adapts recommendations as the domain and curriculum knowledge evolve.

3.4 The Recommendation Engine

The knowledge modeling and assessment approaches proposed in this framework share the aim of providing the necessary information for the recommendation system, to fulfill the goal of personalizing learning path recommendations, with respect to the learning context and domain. The recommender system, n this sense, is then the operational core of the framework, utilizing placement feedback and the KG to construct the learning paths. The recommendation draws from three main inputs: aggregated, session-level mastery inferences derived from the assessment tool, network and hierarchical mapping of sessions and topics through shared domain topics, and the learners' attributes from the user model. The output of the recommendation can vary, depending on the requirements of the use case in hand, it can be a ordered list of sessions that a student should attend, list of courses or modules to register for, or otherwise. This flexibility in supporting the implementation of a wider scope of recommendation algorithms is an essential requirement in higher education, among other learning settings. This is due to the diversity of factors that govern the selection of the most suitable recommendation algorithm, which exceeds the end-user preference and extends to factors such as institution-level policies.

In the scope of the IGNITE project, a recommendation engine has been developed to serve the use-case handheld in the scope of the project, which focuses on providing session-level path recommendation for master and bachelor-level students, who join the study programs from diverse, and in many cases international, study backgrounds. The recommender systems is developed for this use case as a context-aware algorithm, following [3], utilizing reinforcement learning over the knowledge graph structure to ensure the pedagogical-alignment and context-awareness of the path recommendation. The recommend system consists of two main parts:

KG Navigation and Learning: This part is responsible for discovering potential paths of sessions and topics that best compliment the students current knowledge state (as determined by the assessment tool). After defining the scope of potential paths to be recommended, a reinforcement-learning virtual agent navigates all paths and ranks them based on user profile and context information. the top-n highest ranking paths are then considered for the final construction of the learning path, which encompasses a time-sensitive element in this use case, as the sessions to be recommended to the learner are time-bounded to predefined timetables of the university modules.

Session Scheduler: The scheduling part of the recommender system is tasked with ensuring that recommended sessions do not conflict with existing commitments by cross-referencing the institutional timetable with the learners schedule, thereby optimising learning continuity. If a conflict exists, the scheduler utilizes the content from the top-n ranked paths, which align timely with the conflict position in the learning path, and determines a replacement that ensures continuity of the learning path and suitability to the learner.

4 Discussion

The knowledge graph-driven IGNITE framework addresses key challenges in higher education: delivering personalised learning paths within an increasingly modular and overlapping curriculum. By linking fine-grained knowledge units, such as subtopics, to broader instructional sessions, the proposed IGNITE framework supports granular diagnosis of learner knowledge and avoids unnecessary repetition. Our framework enforces instructional consistency by modelling conceptual progression, ensuring that learners follow a pedagogically sound path. This structure supports better cognitive load management and knowledge retention. Studies have shown that intelligent sequencing improves performance and contributes to lower dropout rates in learning environments [10]. The IGNITE framework provides instructors with actionable insights into students' knowledge states. By semantically anchoring the assessment results to curriculum nodes, instructors can identify cohort-wide or individual knowledge gaps, which can be used to adjust teaching strategies accordingly. This evidence-based diagnosis facilitates data-informed instruction, improving learning outcomes and teaching relevance [10]. The graph structure furthermore promotes modular content reuse, which is intended to be studied in the next stage of the IGNITE project. Equivalent concepts across different modules are semantically linked, reducing redundancy and enabling efficient content updates and cross-module alignment. This is a valuable framework for designing more flexible, detailed and interconnected curricula or, in broader cases, program-wide instructional planning. At the institutional level, the framework provides a scalable approach to personalised education. By automating the placement assessment, content alignment and recommendation generation, the level of manpower required to carry out these tasks will be reduced. The framework demonstrates how semantic integration and graph-based reasoning can bridge the gap between learner needs, instructional design, and institutional goals, supporting personalised and efficient education delivery.

5 Conclusion and Future Work

We have presented a KG-driven framework that connects curriculum structure, domain concepts, and user profiles into a unified semantic model for personalized learning path generation. The graph design tailors the curriculum to fit the learner's actual background knowledge rather than a one-size-fits-all sequence of classes. The graph structure is designed to be extensible, allowing new modules or updates to be integrated without disrupting existing functions. This enables institutions to provide tailored support to learners while reducing the workload on instructors. Bridging semantic technologies with pedagogical intent, the framework provides a practical model for modernising higher education. Providing learners with tailored paths supports educators in identifying knowledge gaps and enables institutions to design a flexible curriculum that meets the individual needs of learners. Future work will involve evaluating the proposed KG-based recommender system in real academic environments to measure its effectiveness in improving personalized learning pathways. This includes conducting pre-assessment tests with students to analyze their prior knowledge, providing personalized course content suggestions for both students and teachers, and evaluating the usability of the personalized learning path recommendations generated. Further development will focus on integrating the framework with existing learning management systems, ensuring interoperability with diverse educational datasets, and enhancing adaptability through dynamic updates to the KG based on learner progress and changing curricular needs.

References

1. Abu-Rasheed, H., Dornhöfer, M., Weber, C., Kismihók, G., Buchmann, U., Fathi, M.: Building contextual knowledge graphs for personalized learning recommendations using text mining and semantic graph completion. In: 2023 IEEE International Conference on Advanced Learning Technologies (ICALT), pp. 36–40 (2023)
2. Abu-Rasheed, H., et al.: LLM-assisted knowledge graph completion for curriculum and domain modelling in personalized higher education recommendations. In: 2025 IEEE Global Engineering Education Conference (EDUCON), pp. 1–5 (2025)
3. Abu-Rasheed, H., Weber, C., Dornhöfer, M., Fathi, M.: Pedagogically-informed implementation of reinforcement learning on knowledge graphs for context-aware learning recommendations. In: European Conference on Technology Enhanced Learning, pp. 518–523 (2023). https://doi.org/10.1007/978-3-031-42682-7_35
4. Ain, Q.U., Chatti, M.A., Meteng Kamdem, P.A., Alatrash, R., Joarder, S., Siepmann, C.: Learner modeling and recommendation of learning resources using personal knowledge graphs. In: Proceedings of the 14th Learning Analytics and Knowledge Conference, pp. 273–283 (2024)
5. Al Amin, R., Wiese, V., Rahman, M.M., Jacobs, S., Jaschke, S., Obermaisser, R.: Architecture of IoT-based collaborative laboratory for reconfigurable SOC: review and development. In: 2023 IEEE 2nd German Education Conference (GECon), pp. 1–6 (2023)
6. Alatrash, R., Chatti, M.A., Ain, Q.U., Fang, Y., Joarder, S., Siepmann, C.: ConceptGCN: knowledge concept recommendation in MOOCs based on knowledge graph convolutional networks and SBERT. Comput. Educ. Artif. Intell. **6**, 100193 (2024). https://doi.org/10.1016/j.caeai.2023.100193
7. Amin, R.A., Wiese, V., Jacobs, S., Hardebusch, T., Jaschke, S., Obermaisser, R.: Requirement analysis and didactic evaluation of a collaborative remote laboratory for FPGAs. In: 2025 IEEE Global Engineering Education Conference (EDUCON), pp. 1–5 (2025)
8. Amin, R.A., Wiese, V., Rahman, M.M., Jacobs, S., Jaschke, S., Obermaisser, R.: Fpgarlab: leveraging FPGA design and verification using intelligent cyber physical system based collaborative remote laboratory. In: 2024 International Congress on Human-Computer Interaction, Optimization and Robotic Applications (HORA), pp. 1–8 (2024)
9. Baig, D., Nurbakova, D., Mbaye, B., Calabretto, S.: Knowledge graph-based recommendation system for personalized e-learning. In: Adjunct Proceedings of the 32nd ACM Conference on User Modeling, Adaptation and Personalization, pp. 561–566 (2024)
10. Contrino, M.F., Reyes-Millán, M., Vázquez-Villegas, P., Membrillo-Hernández, J.: Using an adaptive learning tool to improve student performance and satisfaction in online and face-to-face education for a more personalized approach. Smart Learn. Environ. **11**(6), 1–24 (2024). https://doi.org/10.1186/s40561-024-00292-y
11. Dong, L., Wang, H., Yang, L.: Personalized learning recommendation method based on knowledge graph. In: 2024 7th International Conference on Advanced Algorithms and Control Engineering (ICAACE), pp. 323–328. IEEE (2024)
12. Frej, J., Shah, N., Knezevic, M., Nazaretsky, T., Käser, T.: Finding paths for explainable MOOC recommendation: a learner perspective. In: Proceedings of the 14th Learning Analytics and Knowledge Conference, pp. 426–437 (2024)
13. Gu, Q.: Personalized learning path recommendation based on knowledge graph. Trans. Soc. Sci. Educ. Humanit. Res. **8**, 11–15 (2024)

14. Haldar, S., Sengupta, S., Das, A.K.: Personalized learning path recommendation using graph reinforcement learning. Procedia Comput. Sci. **258**, 3480–3489 (2025)
15. Hashim, S., Omar, M.K., Ab Jalil, H., Sharef, N.M.: Trends on technologies and artificial intelligence in education for personalized learning: systematic literature. J. Acad. Res. Prog. Educ. Dev. **12**(1), 884–903 (2022)
16. Hou, B., Lin, Y., Li, Y., Fang, C., Li, C., Wang, X.: KG-PLPPM: a knowledge graph-based personal learning path planning method used in online learning. Electronics **14**(2), 255 (2025)
17. Hou, B., Lin, Y., Li, Y., Fang, C., Li, C., Wang, X.: KG-PLPPM: a knowledge graph-based personal learning path planning method used in online learning. Electronics **14**(2) (2025). https://doi.org/10.3390/electronics14020255
18. Julianti, M.R., Heryadi, Y., Yulianto, B., Budiharto, W.: Recommendation system model for personalized learning in higher education using content-based filtering method. In: 2022 International Conference on Information Management and Technology (ICIMTech), pp. 1–6. IEEE (2022)
19. Lee, J., Park, J., Kim, D.: Exploring when learners become aware of their knowledge gaps: content analyses of learner discussions. Instr. Sci. 171–205 (2024)
20. Li, Y., Liang, Y., Yang, R., Qiu, J., Zhang, C., Zhang, X.: Coursekg: an educational knowledge graph based on course information for precision teaching. Appl. Sci. **14**(7), 2710 (2024)
21. Maatuk, A.M., Elberkawi, E.K., Aljawarneh, S., Rashaideh, H., Alharbi, H.: The covid-19 pandemic and e-learning: challenges and opportunities from the perspective of students and instructors. J. Comput. High. Educ. **34**(1), 21–38 (2022)
22. Mu, M., Yuan, M.: Research on a personalized learning path recommendation system based on cognitive graph with a cognitive graph. Interact. Learn. Environ. **32**(8), 4237–4255 (2024)
23. Narvekar, S., Peng, B., Leonetti, M., Sinapov, J., Taylor, M.E., Stone, P.: Curriculum learning for reinforcement learning domains: a framework and survey. J. Mach. Learn. Res. **21**(181), 1–50 (2020)
24. Raj, N.S., Renumol, V.: A rule-based approach for adaptive content recommendation in a personalized learning environment: an experimental analysis. In: 2019 IEEE Tenth International Conference on Technology for Education (T4E), pp. 138–141. IEEE (2019)
25. Shemshack, A., Kinshuk, Spector, J.M.: A comprehensive analysis of personalized learning components. J. Comput. Educ. **8**(4), 485–503 (2021)
26. Steiner, T., Verborgh, R., Troncy, R., Gabarro, J., Van de Walle, R.: Adding real-time coverage to the google knowledge graph. In: 11th International Semantic Web Conference (ISWC 2012), vol. 914, pp. 65–68. Citeseer (2012)
27. Urdaneta-Ponte, M.C., Mendez-Zorrilla, A., Oleagordia-Ruiz, I.: Recommendation systems for education: systematic review. Electronics **10**(14), 1611 (2021)
28. Wang, F., Wu, M., Liu, Y.: A novel personalized learning path mining based on steiner tree and knowledge graph. Procedia Comput. Sci. **242**, 305–312 (2024)
29. Wang, H., Fu, W.: Personalized learning resource recommendation method based on dynamic collaborative filtering. Mob. Netw. Appl. **26**, 473–487 (2021)
30. Wei, Q., Yao, X.: Personalized recommendation of learning resources based on knowledge graph. In: 2022 11th International Conference on Educational and Information Technology (ICEIT), pp. 46–50. IEEE (2022)
31. Wu, S., et al.: A comprehensive exploration of personalized learning in smart education: from student modeling to personalized recommendations. J. ACM **37**(4), Article 111, 1–82 (2024)

32. Yang, Y., Peng, X., Chen, M., Liu, S.: An explainable graph-based course recommendation model based on multiple interest factors. Expert Syst. Appl. **264**, 125889 (2025). https://doi.org/10.1016/j.eswa.2024.125889
33. Zhang, J., Jing, S., Liang, S.: Research on personalized learning resource recommendation based on knowledge graph. In: 2023 3rd International Conference on Electronic Information Engineering and Computer Communication (EIECC), pp. 111–118. IEEE (2023)
34. Zhao, S.: Personalized learning path recommendation strategy based on subject knowledge graph. In: Proceedings of the 2024 3rd International Conference on Artificial Intelligence and Education, pp. 398–402 (2024)

Conceptual Design Considerations of a Knowledge GraphDriven Framework for Scientific Knowledge Exploration

Amir Rajabi(✉), Jan Schneider, and Daniel Schiffner

DIPF, Leibniz Institute for Research and Information in Education, Frankfurt am Main, Germany
{a.rajabi,j.schneider,d.schiffner}@dipf.de

Abstract. In many professions, educated professionals have always needed to stay up-to-date with advances in their field. With the increasing number of published research works, finding relevant content has become more challenging. Traditional search methods often return thousands of papers related to a search query, necessitating significant time for full-text analysis. Conversely, existing AI-based systems also fail to provide accurate responses. To address these challenges, this position paper proposes a system architecture for scientific understanding, based on user preferences identified in a preliminary survey, that leverages a scientific knowledge graph extending beyond abstract analysis to capture the broader context of scholarly works. The framework aims to deliver context-sensitive responses to user queries and recommend relevant academic studies, thereby improving comprehension. Its development entails tackling various considerations, which will also be briefly discussed.

Keywords: Information overload · Scholarly recommendation · Scholarly document analysis · Retrieval-Augmented Generation · Text mining · Automatic Information extraction · Scientific knowledge graph

1 Introduction

Every year, with the expansion of the number of scientific publications and books, obtaining an overview of various topics becomes increasingly overwhelming and time-consuming. This phenomenon can be described as information overload, which refers to the availability of more potentially useful information than an individual can efficiently process and utilize [11]. As a result, students and researchers are often required to invest significant time and effort in identifying appropriate and relevant materials for their topics, making the task of finding pertinent information increasingly challenging. Search results from platforms such as Google Scholar, ACM Digital Library, Elsevier, and other databases often return thousands of results. For instance, a Google Scholar search using the query *"Scientific Information Extraction AND(relation OR Triplets OR keyphrase-extraction) AND("Document-level") AND(Research OR*

H. Takeda et al. (Eds.): IJCKG 2025, LNAI 16297, pp. 320–328, 2026.
https://doi.org/10.1007/978-981-95-5009-8_21

Scholarly works)" yields 18,400 results[1]. Besides, among the numerous AI-based recommender systems, such as ResearchRabbit[2], which retrieves relevant papers based on the citations of the papers collected by users, many suffer from generating an overwhelming number of inaccurate recommendations, still requiring manual revision. Furthermore, reviewing the titles and abstracts of these papers for scientific understanding alone requires a substantial amount of time.

Effectively addressing this primary problem requires tackling three underlying sub-problems. The first sub-problem arises when reviewing various techniques and concepts associated with searched subject. When exploring methods to accomplish a task, individuals are frequently confronted with a wide range of potential techniques. For example, when searching for suitable ML algorithms for classification, such as support vector machines, or neural networks, requires understanding their strengths, limits, and use cases. However, studying all methods in depth is impractical, making a focused comparison crucial for identifying the most appropriate ones. The second sub-problem involves understanding various scenarios contexts and the ability to apply the system in different domains. For instance, applying the same algorithm in healthcare or finance requires understanding the unique challenges and data characteristics in each domain, which can be time-consuming. Lastly, in all stages of literature review or data analysis, the examination of a large volume of data can lead to information overload, making it difficult to distinguish what is truly relevant to the main topic.

To address these problems, a human-centered application that leverages user preferences, recent advancements in large language model (LLM)-driven conversational systems, and the integration of structured knowledge into these models [24] may offer a promising solution. Furthermore, the incorporation of clearly and systematically organized knowledge can enhance the explainability and interpretability of LLM-based systems [25]. This position paper contributes to the field by addressing the following key research question:

- **RQ:** What are the key design considerations and architectural components for building an interactive system that accurately represents and enables exploration of the state of the art in a specific topic?

2 Contextual Knowledge Extraction and Application

An excess of information can be as detrimental as too little, hindering comprehension and wasting time by diverting attention to irrelevant details. A structured overview of information is widely considered essential, and to support this, various approaches from the literature are presented below.

[1] Based on search results retrieved on April 16, 2025.
[2] https://researchrabbitapp.com/home.

2.1 Principles of AI-Driven Conversation System

Innovations in data-driven technologies have made AI-based assistants a major advancement in human-computer communication, enabling meaningful content retrieval through advanced context understanding powered by LLMs [25]. This enables dynamic dialogue systems, including question answering [9], chatbot [2], and text summarization [16]. Turning to scholarly document analysis, which is highly dynamic and requires frequent updates, providing appropriate content is both essential and challenging. Consequently, three main approaches, outlined below, can keep LLM-based models up to date and improve response accuracy:

1. Fine-tuning improves model alignment with context, but is costly and time-consuming, making it less ideal given the fast pace of new publications.
2. Prompt engineering [26] can supply context to LLMs, but often demands many carefully crafted prompts to provide sufficient context and examples.
3. Inference methods, such as Retrieval-Augmented Generation (RAG) [8], leverage external knowledge to efficiently maintain up-to-date information.

2.2 Knowledge Extraction and Representation from Text to Graph

These human-centered dialogue applications, mentioned in Sect. 2.1, have two key features, each of which presents its own set of challenges. The first key aspect addresses the challenge of mitigating information overload and extracting meaningful knowledge from raw data. In light of the rapidly growing volume of scholarly data, to avoid diverting cognitive resources toward constant data processing and searching, rather than reflection and critical thinking, it is essential to manage information overload effectively. Building on the above concept, automated analysis of the content of scientific articles has the potential to significantly support and improve researchers' efforts [23]. This automation begins with the fundamental step of extracting meaningful knowledge and understanding the underlying structure of content [7], utilizing text mining [32] or information extraction techniques [17].

The second feature concerns the presentation and use of this extracted information to enable further analysis or downstream applications to solve complex tasks [15]. Representation can be achieved using techniques such as line plots for text flow analysis, and node-link diagrams [18]. Converting plain text into a mind map can reveal its key semantic structure, making it easier to understand [13]. Mind maps or graphs typically follow a hierarchical and radial structure, with outputs that can vary significantly depending on the user's perspective and personalized strategies. These differences present significant challenges in developing a general application that can address the needs of every individual.

However, the construction of knowledge graphs (KGs) is crucial for the systematic development of the two aforementioned features, effectively turning these challenges into an opportunity to generate well-structured data. Moreover, these non-hierarchical graphs capture multidirectional relationships, making them valuable for information storage, querying, and integration into retrieval systems [25]. Notable domain-specific scientific knowledge graphs (SKGs) include

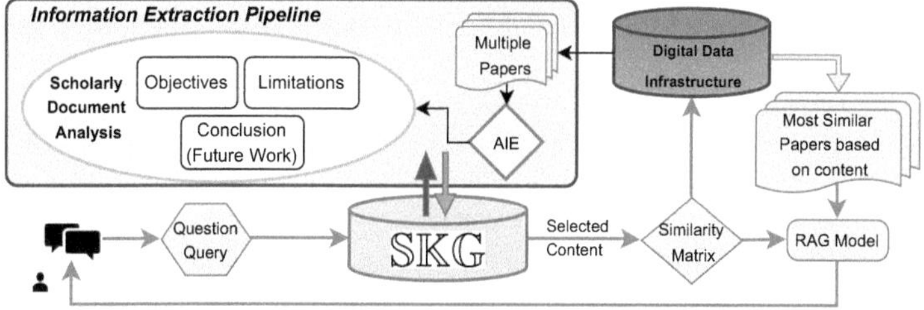

Fig. 1. The proposed framework. AIE stands for Automatic Information Extraction.

MAG [31], ORKG [14], COVID-19 KG [6], and OpenAIRE [22]. The emergence of LLMs, which can encode knowledge within their parameters, has introduced a new paradigm centered on hybrid representations that combine explicit and parametric knowledge [24]. KGs, a well-established form of explicit knowledge representation, are now extensively explored for integration with LLMs, including pre-trained language models (PLMs) such as BERT [5] and GPT-4 [37]. However, this paper does not delve further into the detailed construction of SKGs, as it falls outside its intended scope.

3 Inside the Framework: Design and its Considerations

To address the underlying sub-problems discussed in Sect. 1 and provide a natural language dialogue framework that highlights insights from research papers, a full-text analysis of the articles should be conducted within the defined use case to enable exploration of the state of the art. To operationalize these aims, three key scientific relations, identified through a small case preliminary survey of 14 participating PhD candidates, will be extracted and integrated into the graph. First, a key aspect of the analysis is evaluating the objectives of state-of-the-art papers to determine whether they have achieved and to identify potential future research directions. Second, extracting the limitations of the papers offers valuable insights into their scope. This helps researchers better anticipate potential issues and obstacles and prepare for further developments. Lastly, the discussion of future work provides a clearer outlook on the areas that require more effort and further exploration to answer unsolved questions. These hypotheses imply that systematically analyzing research findings may help individuals to maintain a clear overview of the topic and support the identification of future research.

3.1 The Proposed Framework

Figure 1 depicts the proposed conversational framework, integrating the defined content-aware Scientific Knowledge Graph ($\mathbb{SKG}$) followed by a RAG architecture, where the knowledge provided from the $\mathbb{SKG}$ grounds a LLM in generating contextual responses. Consequently, users receive a direct response alongside a list of content-based recommendations derived from the Digital Data Infrastructure (DDI).

3.2 Utilization of $\mathbb{SKG}$

KGs are defined as explicit knowledge [25], which can be expressed in the form of triples, structured as $\mathbb{SKG} = \boldsymbol{\kappa_m} = \{\boldsymbol{k}_0^T, ..., \boldsymbol{k}_{M-1}^T\} = (h, r, t) \subseteq \mathcal{E} \times \mathcal{R} \times \mathcal{E}$, where $\mathcal{E}$ and $\mathcal{R}$ denote the sets of scientific entities and their scientific relations, respectively, while M represents the number of triples. For example, in NLP research literature, entities such as *Transformer*, *BERT*, and *Tokenization* can serve as nodes, while semantic relationships like *implements*, *enhances*, and *is part of* can define the edges connecting them in a KG. Moreover, the $\mathbb{SKG}$ is continuously connected to the DDI, where all papers are stored, allowing it to update its knowledge dynamically. To generate a response, the query generator encodes the user's request as a triple and retrieves the most semantically similar context from the factual knowledge stored in the $\mathbb{SKG}$, selected based on a similarity metric S. The calculation is performed within the inner product space:

$$S = \cos(\theta) = \frac{\sum_{n=0}^{N-1} q_n \cdot \sum_{m=0}^{M-1} \kappa_m}{\sqrt{\sum_{n=0}^{N-1} q_n^2} \cdot \sqrt{\sum_{m=0}^{M-1} \kappa_m^2}} \geq R, \quad \kappa_m \subset K \in \mathbb{R}^{1\times 3} \tag{1}$$

where q_n and κ_m denote the embedded question and the stored facts, respectively. Since distinguishing similar from irrelevant responses is challenging, the threshold S $\geq R$ requires further investigation. Consequently, triple candidates with the highest similarity scores will be extracted from the $\mathbb{SKG}$ and fed into a RAG system to generate a response and recommend relevant articles. However, its evaluation will be conducted in future work.

3.3 Considerations of Knowledge Acquisition and Integration

The considerations for implementing the framework are listed below:

Consideration 1: *Data Related Difficulties*: The complexity, varying layout, and unique characteristics of scientific data, such as its distribution and continuous evolution, make it challenging to represent the extracted information in a unified graph structure [10]. Moreover, scientific data is highly heterogeneous and available at scale, making the creation of hand-tagged examples both time-consuming and labor-intensive [12,29,35]. In addition, most entity and relation extraction systems employed in the scientific domain operate at the sentence-level [20,21], which is not suitable for full-text analysis. Due to the long-range semantic dependencies in academic papers [36] and the need to uncover structural patterns [30], document-level extraction methods [4] should be utilized.

Consideration 2: *Frequent Knowledge Updates*: Integrating KGs into LLMs can fundamentally enhance their performance by providing domain-specific information [19], and enabling knowledge updates in frequently changing DDIs. *Shirui et al.* [25] have explored the use of KGs to improve the performance of generative systems by verifying their inputs and outputs. However, the applicability of these models is limited by their high computational demands [34].

Consideration 3: *Intelligence-Based Models*: AIE technologies can be categorized into three main approaches: (a) Rule- and dictionary based models, (b) Statistical machine learning-based models, and (c) Deep learning-based models. In contrast to traditional methods, deep learning-based NLP approaches offer a viable solution through supervised and semi-supervised learning. However, they require large amounts of predefined ontologies and manually annotated training datasets. Due to the scarcity of annotated corpora and limited training resources [1,12,36], these models struggle with adaptability in complex domains and demand high costs and extensive labor [29] Additionally, these advanced NLP techniques heavily rely on popular PLMs [5,24,36]. They have become well-established as essential embedding layers within AIE systems, enriched with scientific terminology to enhance contextual understanding [24,27].

Consideration 4: *Human Intervention Level (HIL)*: While researchers widely acknowledge the necessity of human involvement in extraction systems, a fundamental question remains regarding the extent and capacity in which human expertise should be integrated. Scientific works particularly require a high level of domain-specific knowledge, making expert verification crucial for ensuring the quality of the generated data [30]. Furthermore, expert feedback is known to increase the reliability of responses. *Jaradeh et al.* [14] engaged experts to extract essential knowledge from scholarly articles, leading to the development of the Open Research Knowledge Graph (ORKG). They further argue that insufficient expert annotation may cause unsupervised methods to result in blind extraction.

Consideration 5: *Closed- or Open-Domain AIE*: For a comprehensive overview of the main AIE categories, two primary types should be emphasized: *(a) Closed-Domain AIE (supervised or semi-supervised) and (b) Open-Domain or unsupervised AIE* [33]. *Sollaci* and *Pereira* [28] argued that scholarly articles typically follow a structured format, such as Introduction, Methods, Results, and Discussion (IMRaD), emphasizing the importance of this consistent structure in organizing content. It could be assumed here that this might reduce the need for a predefined ontology in AIE system. Building on this assumption and aiming to reduce HIL, Open-Domain AIE, which uses unsupervised techniques to extract content without structural constraints, may offer a promising solution [35]

Consideration 6: *Human Computer Interaction (HCI)*: Over time, HCI has evolved to merge its scientific inquiries with the practical aim of improving the usability of computer systems [3]. Visualizing a SKG is a non-trivial task, necessitating the development of novel representation techniques. Such approaches should generate intuitive and informative visualizations of interconnected facts, potentially improving effectiveness and user satisfaction.

4 Conclusion

This paper aims to analyze the necessary components of a conversational AI-based system built on full-text scholarly analysis. The system is designed to provide users with deep insights, helping them stay up-to-date and explore the state of the art in specific domains, thereby enhancing both comprehension and understanding. Furthermore, a construction considerations list is provided to guide the implementation process. In the next step, we will conduct a larger empirical investigation to identify which types of relations in research articles, beyond the three discussed in this paper, are most relevant to researchers. Consequently, in future work, we will implement and evaluate the proposed framework, employing objective and subjective measures to demonstrate its effectiveness.

References

1. Adnan, K., Akbar, R.: Limitations of information extraction methods and techniques for heterogeneous unstructured big data. Int. J. Eng. Bus. Manage. **11** (2019). https://api.semanticscholar.org/CorpusID:209522992
2. Afzal, A., Kowsik, A., Fani, R., Matthes, F.: Towards optimizing and evaluating a retrieval augmented QA chatbot using LLMs with human in the loop (2024). arxiv:2407.05925
3. Carroll, J.M.: Human–computer interaction: psychology as a science of design. Int. J. Hum.-Comput. Stud. **46**(4), 501–522 (1997). https://doi.org/10.1006/ijhc.1996.0101
4. Dagdelen, J., Dunn, A., Lee, S., et al.: Structured information extraction from scientific text with large language models. Nat. Commun. **15**, 1418 (2024). https://doi.org/10.1038/s41467-024-45563-x
5. Devlin, J., Chang, M.W., Lee, K., Toutanova, K.: BERT: pre-training of deep bidirectional transformers for language understanding. In: Burstein, J., Doran, C., Solorio, T. (eds.) NAACL, pp. 4171–4186. Association for Computational Linguistics, Minneapolis (2019). https://doi.org/10.18653/v1/N19-1423
6. Domingo-Fernández, D., et al.: Covid-19 knowledge graph: a computable, multimodal, cause-and-effect knowledge model of Covid-19 pathophysiology. Bioinformatics **37**(9), 1332–1334 (2020). https://doi.org/10.1093/bioinformatics/btaa834
7. Dridi, A., Gaber, M.M., Azad, R.M.A., Bhogal, J.: Scholarly data mining: a systematic review of its applications. WIREs Data Min. Knowl. Discov. **11**(2), e1395 (2021). https://doi.org/10.1002/widm.1395
8. Fan, W., et al.: A survey on rag meeting LLMs: towards retrieval-augmented large language models. In: Proceedings of the 30th ACM SIGKDD Conference on Knowledge Discovery and Data Mining, KDD 2024, pp. 6491–6501. Association for Computing Machinery, New York (2024). https://doi.org/10.1145/3637528.3671470
9. Feuerriegel, S., Hartmann, J., Janiesch, C., et al.: Generative AI. Bus. Inf. Syst. Eng. **66**, 111–126 (2023). https://doi.org/10.1007/s12599-023-00834-7
10. Gupta, K., Ahmad, A., Ghosal, T., Ekbal, A.: A BERT-based sequential deep neural architecture to identify contribution statements and extract phrases for triplets from scientific publications. Int. J. Digit. Libr. **25**(4), 1–28 (2024). https://doi.org/10.1007/s00799-023-00393-y
11. Himma, K.E.: The concept of information overload: a preliminary step in understanding the nature of a harmful information-related condition. Ethics Inf. Technol. **9**(4), 259–272 (2007). https://doi.org/10.1007/s10676-007-9140-8
12. Hong, Z.: Enabling scientific information extraction with natural language processing. University of Chicago, Physical Sciences Division, p. 131 (2024). https://doi.org/10.6082/uchicago.11351

13. Hu, M., Guo, H., Zhao, S., Gao, H., Su, Z.: Efficient mind-map generation via sequence-to-graph and reinforced graph refinement. In: Moens, M.F., Huang, X., Specia, L., Yih, S.W.T. (eds.) Proceedings of the 2021 Conference on Empirical Methods in Natural Language Processing, pp. 8130–8141. Association for Computational Linguistics, Online and Punta Cana (2021). https://doi.org/10.18653/v1/2021.emnlp-main.641
14. Jaradeh, M.Y., et al.: Open research knowledge graph: next generation infrastructure for semantic scholarly knowledge. In: Proceedings of the 10th International Conference on Knowledge Capture, K-CAP 2019, p. 243–246. Association for Computing Machinery, New York (2019). https://doi.org/10.1145/3360901.3364435
15. Ji, S., Pan, S., Cambria, E., Marttinen, P., Yu, P.S.: A survey on knowledge graphs: representation, acquisition, and applications. IEEE Trans. Neural Netw. Learn. Syst. **33**(2), 494–514 (2022). https://doi.org/10.1109/TNNLS.2021.3070843
16. Jin, H., Zhang, Y., Meng, D., Wang, J., Tan, J.: A comprehensive survey on process-oriented automatic text summarization with exploration of LLM-based methods. ArXiv abs/2403.02901 (2024). https://api.semanticscholar.org/CorpusID:268248096
17. Kaisler, S., Armour, F., Espinosa, J.A., Money, W.: Big data: issues and challenges moving forward. In: 2013 46th Hawaii International Conference on System Sciences, pp. 995–1004 (2013). https://doi.org/10.1109/HICSS.2013.645
18. Kucher, K., Kerren, A.: Text visualization techniques: taxonomy, visual survey, and community insights. In: 2015 IEEE Pacific Visualization Symposium (PacificVis), pp. 117–121 (2015). https://doi.org/10.1109/PACIFICVIS.2015.7156366
19. Lewis, P., et al.: Retrieval-augmented generation for knowledge-intensive NLP tasks. In: Proceedings of the 34th International Conference on Neural Information Processing Systems, NIPS 2020, pp. 9459–9474. Curran Associates Inc., Red Hook (2020)
20. Liu, Y., Li, S., Huang, K., Wang, Q.: AutoIE: an automated framework for information extraction from scientific literature. In: Cao, C., Chen, H., Zhao, L., Arshad, J., Asyhari, T., Wang, Y. (eds.) KSEM 2024, Part II. LNCS, vol. 14885, pp. 424–436. Springer, Cham (2024). https://doi.org/10.1007/978-981-97-5495-3_32
21. Luan, Y., He, L., Ostendorf, M., Hajishirzi, H.: Multi-task identification of entities, relations, and coreference for scientific knowledge graph construction. In: Riloff, E., Chiang, D., Hockenmaier, J., Tsujii, J. (eds.) Proceedings of the 2018 Conference on Empirical Methods in Natural Language Processing, pp. 3219–3232. Association for Computational Linguistics, Brussels (2018). https://doi.org/10.18653/v1/D18-1360
22. Manghi, P., et al.: OpenAIRE research graph dataset (2022). https://doi.org/10.5281/zenodo.7488618
23. Nasar, Z., Jaffry, S.W., Malik, M.K.: Information extraction from scientific articles: a survey. Scientometrics **117**(3), 1931–1990 (2018). https://doi.org/10.1007/s11192-018-2921-5
24. Pan, J.Z., et al.: Large language models and knowledge graphs: opportunities and challenges. Trans. Graph Data Knowl. **1**(1), 2:1–2:38 (2023). https://doi.org/10.4230/TGDK.1.1.2
25. Pan, S., Luo, L., Wang, Y., Chen, C., Wang, J., Wu, X.: Unifying large language models and knowledge graphs: a roadmap. IEEE Trans. Knowl. Data Eng. **36**(7), 3580–3599 (2024). https://doi.org/10.1109/TKDE.2024.3352100
26. Sahoo, P., Singh, A.K., Saha, S., Jain, V., Mondal, S., Chadha, A.: A systematic survey of prompt engineering in large language models: techniques and applications (2024). https://arxiv.org/abs/2402.07927

27. Shaoxiong, J., Pan, S., Cambria, E., Marttinen, P., Yu, P.S.: A survey on knowledge graphs: representation, acquisition, and applications. IEEE Trans. Neural Netw. Learn. Syst. **33**(2), 494–514 (2022). https://doi.org/10.1109/TNNLS.2021.3070843
28. Sollaci, L., Pereira, M.: The introduction, methods, results, and discussion (IMRAD) structure: a fifty-year survey. J. Med. Libr. Assoc.: JMLA **92**, 364–367 (2004)
29. Tchoua, R.B., et al.: Towards a hybrid human-computer scientific information extraction pipeline. In: 2017 IEEE 13th International Conference on e-Science (e-Science), pp. 109–118 (2017). https://doi.org/10.1109/eScience.2017.23
30. Verma, S., Bhatia, R., H.S., et al.: Scholarly knowledge graphs through structuring scholarly communication: a review. Complex Intell. Syst. **9**, 1059–1095 (2023). https://doi.org/10.1007/s40747-022-00806-6
31. Wang, K., Shen, Z., Huang, C., Wu, C.H., Dong, Y., Kanakia, A.: Microsoft academic graph: when experts are not enough. Quant. Sci. Stud. **1**(1), 396–413 (2020). https://doi.org/10.1162/qss_a_00021
32. Xia, F., Wang, W., Bekele, T.M., Liu, H.: Scholarly data mining: a systematic review of its applications. IEEE Trans. Big Data **3**(1), 18–35 (2017). https://doi.org/10.1109/TBDATA.2016.2641460
33. Xiang, W., Wang, B.: A survey of event extraction from text. IEEE Access **7**, 173111–173137 (2019). https://doi.org/10.1109/ACCESS.2019.2956831
34. Zhou, C., et al.: comprehensive survey on pretrained foundation models: a history from BERT to chatGPT. Int. J. Mach. Learn. Cybern. (2024). https://doi.org/10.1007/s13042-024-02443-6
35. Zhou, S., Yu, B., Sun, A., Long, C., Li, J., Sun, J.: A survey on neural open information extraction: current status and future directions. In: Raedt, L.D. (ed.) Proceedings of the Thirty-First International Joint Conference on Artificial Intelligence, IJCAI-22, pp. 5694–5701. International Joint Conferences on Artificial Intelligence Organization (2022). https://doi.org/10.24963/ijcai.2022/793
36. Zhu, X., Gao, W., Yu, Y., Zhang, L., Deng, H.: Syntax-based argument correlation-enhanced end-to-end model for scientific relation extraction. Neurocomputing **586**, 127639 (2024). https://doi.org/10.1016/j.neucom.2024.127639
37. Zhu, Y., et al.: LLMs for knowledge graph construction and reasoning: recent capabilities and future opportunities. World Wide Web **27**(58) (2024). https://doi.org/10.1007/s11280-024-01297-w

ABEL: Artificial Buddy for Effective Learning

T. Y. Emmy Lai[1(✉)], Ann-Kathrin Bernards[2], Dena Baghery[1], Marlena Flüh[2], Tobias Lang[2], Héctor Allende-Cid[1], and Diego Collarana[2,3]

[1] Fraunhofer IAIS, Sankt Agustin, Germany
{emmy.lai,dena.baghery,hector.allende-cid}@iais.fraunhofer.de
[2] Fraunhofer FIT, Sankt Agustin, Germany
{ann-kathrin.bernards,marlena.flueh,tobias.lang,
diego.collarana.vargas}@fit.fraunhofer.de
[3] Universidad Privada Boliviana, Cochabamba, Bolivia

Abstract. This paper presents ABEL (**A**rtificial **B**uddy for **E**ffective **L**earning), a modular, knowledge-graph-driven chatbot designed to enhance and support education in Data Science and Artificial Intelligence. At the core of ABEL, is a hybrid retrieval architecture that integrates a dynamic Knowledge Graph and a Retrieval-Augmented Generation (RAG) pipeline. The knowledge graph, constructed from curated learning resources, enables a concept-driven retrieval of semantically relevant and specific educational content through multi-hop graph queries and embedding-based similarity search. This approach enhances the contextual grounding and supports the generation of personalized, specific, and explainable responses by Large Language Models (LLMs). ABEL is also complemented by a Frequently Asked Questions (FAQ)-based RAG approach, thus offering flexible access to learning content while ensuring traceability and correctness. We present the system's architecture, evaluate its performance using both retrieval and user-based metrics, and discuss the benefits of combining symbolic graph structures. Our results demonstrate that this approach can significantly improve the relevance and adaptability of chatbot-driven learning platforms.

Keywords: Education Chatbots · Large Language Models · Knowledge Graphs · Retrieval-Augmented Generation (RAG) · Effective Learning

1 Introduction

Artificial Intelligence (AI) has shown transformative potential for education in recent years, enabling personalized learning, intelligent tutoring systems, and feedback tailored to individual student needs [2,8]. More recently, through generative models, AI has demonstrated its ability to support education professionals in content creation, evaluation design, and administrative tasks, thereby improving both teaching efficiency and student engagement [13,18].

H. Takeda et al. (Eds.): IJCKG 2025, LNAI 16297, pp. 329–338, 2026.
https://doi.org/10.1007/978-981-95-5009-8_22

Effective learning involves acquiring knowledge and skills that promote long-term retention [3]. Large Language Models (LLMs) can support this process by providing context-aware and personalized explanations [19], adaptive feedback tailored to each learner's rhythm and understanding [12]. When combined with Information Retrieval approaches (such as Retrieval Augmented Generation (RAG)) [1], LLMs can be utilized as personalized educational assistants, dynamically integrating current content with student-specific inquiries [17]. This enables learners to access relevant and crucial information on demand while maintaining critical thinking and engaging in self-directed, self-paced learning.

Within the scope of the ABEL project, we develop a knowledge-graph-driven chatbot to advance education in Data Science and Artificial Intelligence. This educational chatbot integrates LLMs with Knowledge Graphs (KGs), providing both instructor-led and self-directed, adaptive learning modalities. This dual approach addresses the heterogeneous needs of learners and incorporates state-of-the-art technologies to enhance educational efficiency and learner engagement. The system architecture enables adaptation across diverse curricular themes through its modular Knowledge-Graph architecture. We evaluate the chatbot's effect on learning outcomes and participant engagement through a pilot study. The paper is organized as follows: In Sect. 2 we present the System Architecture and its key components. In Sect. 3 we present the results of our proposed solution and validate it with user evaluation. In the last section, we present the conclusions and delineate future work.

2 ABEL: A Modular, Knowledge-Graph-Driven Chatbot

This section presents the architecture of ABEL, detailing its workflow from user input to output delivery. It introduces the core retrieval strategies, namely concept-centric and FAQ-based, explaining how the Knowledge Graph and RAG components interact with the system to support accurate, context-aware answer generation and user interaction.

2.1 System Architecture

The architecture of the ABEL system is illustrated in Fig. 1. The workflow begins when a user submits a request (1), which is then processed as a payload through a handler to the Prompt Service (2). The Prompt Service requests knowledge context based on the user's request via a Digital Knowledge (DK) Hub Connector (3), which interfaces with a Retrieval-Augmented Generator (RAG) and a Knowledge Graph. The DK Hub Connector then forwards the necessary and relevant context to the Prompt Service (4).

Next, the Prompt Service sends all the information (5), including the payload and knowledge context, to an Agent. The Agent processes the request using a suitable LLM (6), which is made available through a Model Hub and accessed via a Model Hub Connector. After the Agent generates a response to the request (7), it is forwarded back to the Prompt Service. The Prompt Service then packages the Agent's response with additional details for traceability and sends it back to the Request Handler (8).

Finally, the Request Handler converts the response provided by the Prompt Service into a human-readable format (9), making it available as a user response. This response is the culmination of the ABEL system's processing, which involves language correction, embedding, similarity search, and knowledge graph queries using two retrieval strategies: an FAQ-based approach and a concept-centric approach. The retrieved context is incorporated through a prompt engineering stage before response generation, validation, and delivery. User queries are initially processed through language correction and embedding modules, and the system utilizes both retrieval strategies to provide accurate and relevant responses.

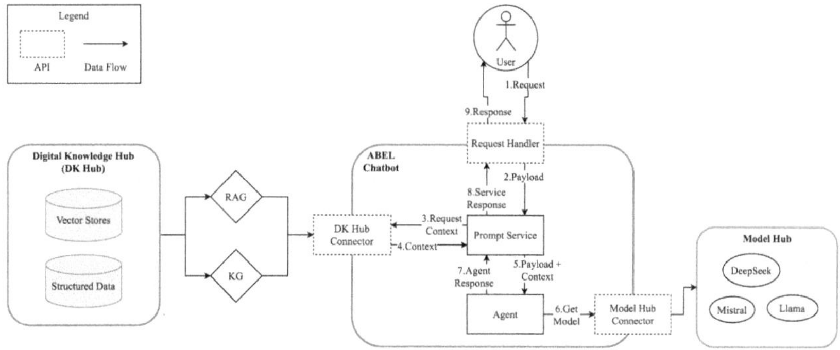

Fig. 1. The ABEL system processes user requests through a sequence of components: Request Handler, Prompt Service, DK Hub Connector, Agent, and Model Hub Connector, to generate and return a response.

2.2 Retrieval Strategies

The system uses two separate retrieval methods to improve answer quality and robustness. This allows each method to be optimized and reduces the chance of missing important information:

1. Concept-Centric Retrieval: This method selects pertinent concepts from the knowledge graph and retrieves associated learning materials in the form of text segments, thereby supplying the LLM with targeted context.
2. FAQ-Based Retrieval: This approach identifies the top-n frequently asked questions (FAQs) most relevant to the user's query and provides these as contextual information to the large language model (LLM).

A REST API connects the LLM chatbot to the knowledge graph, ensuring that the underlying data substantiates generated responses. The API exposes endpoints for both FAQ-based and multi-hop, concept-centric retrieval methods. Additionally, the system prompt directs the LLM to abstain from producing answers when relevant information is not available, thereby reducing the risk of hallucinated content.

Concept-Centric Approach (Knowledge Graph). The knowledge graph used by the ABEL system is constructed from internal Fraunhofer learning resources, including presentations, PDF documents, and curated FAQs. The graph is represented as a property graph within a Neo4j database, comprising nodes, relationships, and associated properties. Each text entity node is enhanced with vector embeddings, enabling semantic search through the application of cosine similarity measures.

FAQ-Based Approach: a RAG System. The FAQ-based approach utilizes a RAG system to provide accurate responses. The process begins with an embedding step, where the user's request is converted into embeddings using a multilingual-e5-large model [16]. This model was chosen for its superior performance in multilingual retrieval tasks, supporting over 100 languages, and outperforming alternative models such as bert-base-multilingual-cased [5], as demonstrated in benchmarks like the Massive Multilingual Text Embedding Benchmark (MMTEB) [6]. The embedded user request is then used to conduct a similarity search on a preprocessed FAQ database, retrieving the top-n most suitable FAQs and their associated context. Finally, an LLM is applied to the top-retrieved answer to conduct a more specific evaluation, ensuring that the provided response is relevant and accurate. Note that the performance of the multilingual-e5-large model is further supported by studies such as [10], which highlights its potential in mitigating the "curse of multilinguality" and achieving state-of-the-art performance in certain language subsets and task categories (Fig. 2).

2.3 Context-Aware Answer Generation and Validation

The answer generation process is initiated when the agent receives the user's request, which is compiled into a payload containing all relevant input information. The agent then retrieves additional context information related to the request. A prompt engineering stage is subsequently applied to carefully craft the prompt that will be used to generate the answer. This prompt is then processed by the system, resulting in the generation of a response. To ensure the quality and accuracy of the generated answer, a second LLM/Agent step is executed, which evaluates the answer against predefined criteria. Only if the generated answer fulfills these specific criteria is it considered valid and provided to the user, thereby guaranteeing a reliable and informed response.

2.4 User Interface

The knowledge-augmented LLM is accessible through a web-based user interface. Users may submit queries by selecting predefined questions or by entering questions in a text input field. The LLM generates responses by submitting requests

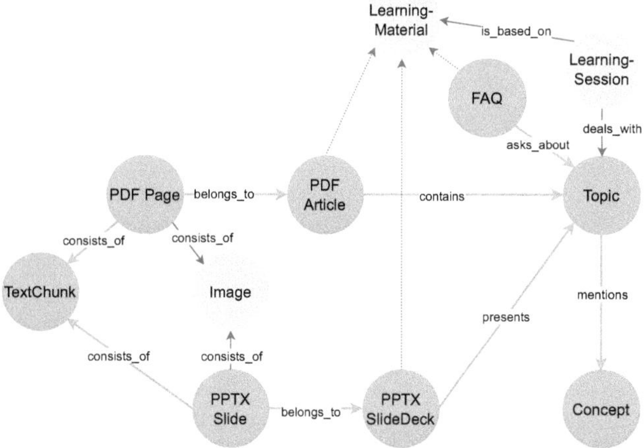

Fig. 2. ABEL KG Schema: The model represents different document types, including their components and frequently asked questions (FAQs), which are related to specific topics. The current ABEL graph is based on the colored nodes. Including images will be the next step. In concept-centric retrieval, RAG mechanics utilize the relationship between Topic and Concept nodes for multi-hop retrieval.

to the API and formulating answers based on the returned information. Each response includes a details option, enabling users to review the answer's underlying information and the data employed by the LLM during response generation.

3 Evaluation

The evaluation of ABEL's performance was conducted through two approaches: a technical assessment using RAGAS and a user-centric evaluation of usability and usefulness.

3.1 RAGAS Evaluation

Our first evaluation procedure was performed with a framework called Retrieval-Augmented Generation Assessment (RAGAS) [7]. This framework was designed as a reference-free evaluation approach that assesses RAG pipelines along three dimensions: faithfulness of the retrieved context, relevance of the generated answer to the query, and focus of the retrieved context. Using this evaluation framework and OpenAI's GPT-4o model (temperature = 0) [11], the system's performance was assessed based on:

- Context Recall: **0.80** – indicating effective retrieval of relevant documents.
- Faithfulness: **0.67** – reflecting moderate consistency between responses and source context.

While the high context recall demonstrates the retriever's efficacy, the faithfulness score suggests the need for enhanced grounding techniques, such as inline citations or post-generation verification.

3.2 User Evaluation

We conducted a user evaluation[1] with 13 participants from Fraunhofer FIT and Fraunhofer IAIS, representing various professional roles. The evaluation took place remotely via Microsoft Teams. Participants first explored ABEL independently for 20 min, asking their own questions and documenting unclear chatbot responses via screenshots.

To assess ABEL, participants used three evaluation tools: a Prompt-Competence checklist (4 items, 5-point scale) [9], an Expert Robustness checklist (7 items, 5-point scale) derived from Nielsen's [14] usability heuristics, and a Technology Acceptance Model (TAM) questionnaire (12 items, 7-point scale) [4,15] covering Perceived Usefulness, Perceived Ease of Use, and Behavioral Intention.

The results showed moderate ratings for prompt competence ($M = 3.08$, $SD = 0.75$), suggesting a need for clearer interactions. Technical robustness was rated positively ($M = 3.72$, $SD = 0.54$), indicating stable system performance. The TAM questionnaire revealed high usability ($M = 6.25$, $SD = 0.80$) and general helpfulness ($M = 4.37$, $SD = 1.63$). However, intent to use ABEL varied ($M = 4.54$, $SD = 2.11$), possibly due to differences in prior experience and individual needs.

Open-ended feedback highlighted strengths in ease of use and thematic fit, but also pointed to limitations in contextual understanding, didactic depth, and terminology sensitivity. Participants emphasized the need for improved response grounding, memory functions, and adaptive user guidance to better accommodate diverse learning contexts. While acknowledging ABEL's potential, users called for refinements in dialogue quality and personalization.

4 Conclusions and Future Work

In conclusion, the ABEL project has demonstrated the potential of leveraging a Knowledge Graph, RAG, and an agent-based chatbot in supporting professional training in Data Science and Artificial Intelligence. The system's utilization of two retrieval strategies, namely Concept-Centric Retrieval and FAQ-Based Retrieval, has shown promise in providing accurate and relevant responses to learners' inquiries. The application of agents in the chatbot has also highlighted the potential for personalized and adaptive support, enabling learners to access relevant information and receive tailored guidance.

Future work will focus on refining the system's capabilities, exploring ways to enhance the effectiveness of retrieval strategies and the agent-based approach,

[1] Instruments used in the user evaluation are listed below, with exact item wordings provided in the appendix.

and improving the overall user experience. By doing so, the ABEL system can become a valuable tool for supporting professional training and education in Data Science and Artificial Intelligence, ultimately helping learners to acquire the knowledge and skills they need to succeed in these fields.

Acknowledgement. This work is supported by the Fraunhofer Entwicklungsfond through the Fraunhofer Academy, which enables the realization of the Project ABEL, by the Fraunhofer ICON grant through the Next Generation Dataspaces Initiative (NGDI), and the German Ministry for Research and Education (BMBF) project WestAI (Grant no. 01IS22094D). We also gratefully acknowledge the additional support provided by Fraunhofer IAIS and Fraunhofer FIT, whose contributions have been instrumental in making this project possible.

A Instruments and Item Wordings Used in the User Evaluation

All questionnaires were administered in German. Below we reproduce the exact item wordings (German), followed by an English translation in brackets. Scale anchors are given for each instrument.

A.1 Prompt-Competence Checklist (4 Items; 5-Point Likert)

Response scale: 1 = trifft überhaupt nicht zu / 5 = trifft voll und ganz zu.

1. Die GenKI interpretiert uneindeutige Prompts korrekt. [The AI interprets ambiguous prompts correctly.]
2. Die Antwort ist kohärent, kontextualisiert und vollständig. [The response is coherent, contextualized, and complete.]
3. Folgefragen werden sinnvoll eingebunden und thematisch passend beantwortet. [Follow-up questions are integrated meaningfully and answered in a thematically appropriate way.]
4. Das Antwortverhalten ist lernförderlich (z. B. Rückfragen, Analogien, Beispiele). [The response behavior supports learning (e.g., counter-questions, analogies, examples).]

A.2 Technical Robustness Expert Checklist (7 Items; 5-Point Likert)

Response scale: 1 = trifft überhaupt nicht zu / 5 = trifft voll und ganz zu. Items were derived from usability-heuristic thinking (e.g., Nielsen) and tailored to LLM chat interactions.

1. Auch bei verschachtelten oder komplexen Prompts bleibt die Antwort logisch und widerspruchsfrei. [Even with nested or complex prompts, the response remains logical and consistent.]

2. Das System erkennt Folgefragen korrekt und integriert den vorherigen Kontext in die Antwort. [The system correctly recognizes follow-up questions and integrates prior context into the response.]
3. Auch bei vagen oder missverständlichen Prompts reagiert das System robust und nachvollziehbar. [The system responds robustly and in a comprehensible manner even to vague or ambiguous prompts.]
4. Das Antwortniveau passt sich dem erkennbaren Vorwissen bzw. der Formulierung des Prompts an. [The response level adapts to apparent prior knowledge and/or the prompt's phrasing.]
5. Die Antwort ist klar gegliedert, sprachlich korrekt und inhaltlich nachvollziehbar. [The response is well-structured, linguistically correct, and substantively clear.]
6. Das System vermeidet unnötige Wiederholungen oder redundante Passagen. [The system avoids unnecessary repetition or redundancy.]
7. Die Ladezeit und Antwortzeit des Systems sind angemessen. [Loading and response times are appropriate.]

A.3 Technology Acceptance Model (TAM) 3 Subscales (12 Items; 7-Point Likert)

Response scale: 1 = stimme überhaupt nicht zu / 7 = stimme voll und ganz zu. Subscales: Perceived Usefulness (PU), Perceived Ease of Use (PEOU), Behavioral Intention (BI).

A3.1 Perceived Usefulness (PU)

1. Die GenKI interpretiert uneindeutige Prompts korrekt. [The AI interprets ambiguous prompts correctly.]
2. Die Antwort ist kohärent, kontextualisiert und vollständig. [The response is coherent, contextualized, and complete.]
3. Folgefragen werden sinnvoll eingebunden und thematisch passend beantwortet. [Follow-up questions are integrated meaningfully and answered in a thematically appropriate way.]
4. Das Antwortverhalten ist lernförderlich (z. B. Rückfragen, Analogien, Beispiele). [The response behavior supports learning (e.g., counter-questions, analogies, examples).]

A3.2 Perceived Ease of Use (PEOU)

1. Ich finde die ABEL-App einfach zu bedienen. [I find the ABEL app easy to use.]
2. Die Interaktion mit der App ist verständlich und unkompliziert. [Interacting with the app is clear and straightforward.]
3. Ich kann die Funktionen der App schnell erlernen. [I can quickly learn how to use the app's features.]
4. Die Nutzung der App erfordert keinen großen mentalen Aufwand. [Using the app does not require much mental effort.]

A3.3 Behavioral Intention (BI)

1. Ich habe vor, die ABEL-App regelmäßig zu verwenden. [I intend to use the ABEL app regularly.]
2. Ich würde die ABEL-App auch in anderen Lernsituationen nutzen. [I would use the ABEL app in other learning situations as well.]
3. Ich kann mir vorstellen, die App Freund:innen oder Kolleg:innen zu empfehlen. [I can imagine recommending the app to friends or colleagues.]
4. Ich plane, die App auch zukünftig zur Unterstützung meines Lernens zu nutzen [I plan to continue using the app to support my learning in the future.]

References

1. Arslan, M., Ghanem, H., Munawar, S., Cruz, C.: A survey on RAG with LLMs. Procedia Comput. Sci. **246**, 3781–3790 (2024)
2. Chen, Y., Jensen, S., Albert, L.J., Gupta, S., Lee, T.: Artificial intelligence (AI) student assistants in the classroom: designing chatbots to support student success. Inf. Syst. Front. **25**(1), 161–182 (2023)
3. Davis, D., Kizilcec, R., Hauff, C., Houben, G.J.: Scaling effective learning strategies: retrieval practice and long-term knowledge retention in MOOCs. J. Learn. Anal. **5**(3), 21–41 (2018)
4. Davis, F., Davis, F.: Perceived usefulness, perceived ease of use, and user acceptance of information technology. MIS Q. **13**, 319 (1989). https://doi.org/10.2307/249008
5. Devlin, J., Chang, M., Lee, K., Toutanova, K.: BERT: pre-training of deep bidirectional transformers for language understanding. CoRR abs/1810.04805 (2018). http://arxiv.org/abs/1810.04805
6. Enevoldsen, K., et al.: MMTEB: massive multilingual text embedding benchmark (2025). https://arxiv.org/abs/2502.13595
7. Es, S., James, J., Anke, L.E., Schockaert, S.: RAGAs: automated evaluation of retrieval augmented generation. In: Proceedings of the 18th Conference of the European Chapter of the Association for Computational Linguistics: System Demonstrations, pp. 150–158 (2024)
8. Hisham, M., Vinod, N., Kuriakose, D., Joshy, M., Syama, S.: Survey on generative AI in education. Int. J. Adv. Eng. Manage. **6**, 615–620 (2024). https://doi.org/10.35629/5252-0611615620
9. Holmes, W., Bialik, M., Fadel, C.: Artificial intelligence in education. Promise and implications for teaching and learning (2019)
10. Huertas-García, Á., Martí-González, C., Muñoz, J., Ambite, E.D.M.: Small language models and large language models in oppositional thinking analysis: capabilities, biases and challenges. In: Conference and Labs of the Evaluation Forum (2024). https://api.semanticscholar.org/CorpusID:271772478
11. Hurst, A., et al.: GPT-4o system card. arXiv preprint arXiv:2410.21276 (2024)
12. Kasneci, E., et al.: Chatgpt for good? On opportunities and challenges of large language models for education. Learn. Individ. Differ. **103**, 102274 (2023)
13. Mittal, U., Sai, S., Chamola, V., Sangwan, D.: A comprehensive review on generative AI for education. IEEE Access **12**, 142733–142759 (2024). https://doi.org/10.1109/ACCESS.2024.3468368

14. Nielsen, J.: Usability Engineering. Morgan Kaufmann, San Francisco (1993)
15. Venkatesh, V., Morris, M., Davis, G., Davis, F.: User acceptance of information technology: toward a unified view. MIS Q. **27**, 425–478 (2003). https://doi.org/10.2307/30036540
16. Wang, L., Yang, N., Huang, X., Yang, L., Majumder, R., Wei, F.: Multilingual E5 text embeddings: a technical report (2024). https://arxiv.org/abs/2402.05672
17. Wong, L.: Gaita: a RAG system for personalized computer science education. Ph.D. thesis, Johns Hopkins University Baltimore, MD, USA (2024)
18. Yu, H., Guo, Y.: Generative artificial intelligence empowers educational reform: current status, issues, and prospects. In: Frontiers in Education, vol. 8, p. 1183162. Frontiers Media SA (2023)
19. Zhang, R., Zhang, Z., Zhu, F., Zhou, J., Rao, A.: Mindalogue: LLM-powered nonlinear interaction for effective learning and task exploration. arXiv preprint arXiv:2410.10570 (2024)

In-use Track

Basketball Question Answering Using Knowledge Graphs and Large Language Models

Michalis Mountantonakis[1,2(✉)], Dimitris Papadopoulos[2], and Yannis Tzitzikas[1,2]

[1] Information Systems Laboratory, FORTH-ICS, Heraklion, Greece
{mountant,tzitzik}@ics.forth.gr

[2] Computer Science Department, University of Crete, Heraklion, Greece
csd4976@csd.uoc.gr

Abstract. There is a recent trend of using LLMs for generating SPARQL queries from natural questions, in order to exploit the generated query for retrieving the answer over a Knowledge Graph (KG). This can be quite challenging for popular domains, such as sports, that include billions of fans worldwide that are not familiar with KGs and query languages, and they typically express their questions in natural language. In this paper, we focus on basketball games which are complex events including several factors and participants, such as teams, players, statistics and others. However, all this information is usually given in separate HTML tables or JSON files, thereby, it is not trivial to be combined. The objective is to show a generic pipeline including the following steps: i) data collection, ii) the definition of competency questions, iii) the creation of an event-based ontology and the KG, and iv) the exploitation of the KG through the LLMs, for enabling natural QA through a text-to-SPARQL method based on ontology and literals path patterns. As a use case, we construct a KG including 5 million triples from the EuroLeague games from 2000–01 until 2024–25, a web application, and an evaluation benchmark with 100 natural questions in 3 languages (English, Greek, Chinese). Finally, we present an experimental evaluation for the QA task; indicatively by using the proposed method, we reached 0.79 accuracy through GPT-4 and 0.71 accuracy through DeepSeek, for the English benchmark, instead of 0.3 for the baseline method.

Keywords: Basketball Analytics · LLMs · Knowledge Graphs · QA

1 Introduction

In several real world domains, such as lifesciences, geography and others [15], Knowledge Graphs (KGs) are constructed for facilitating data analysis and offering advanced services, by answering more complex questions that require to

H. Takeda et al. (Eds.): IJCKG 2025, LNAI 16297, pp. 341–357, 2026.
https://doi.org/10.1007/978-981-95-5009-8_23

combine knowledge from several entities. Towards this direction, the high trend of using Large Language Models (LLMs) has resulted in the creation of several approaches that try to exploit LLMs for generating SPARQL Queries for a given natural question over a desired KG [3,14,18], e.g., see an example for the Cultural Heritage domain [18]. However, most web users are not familiar with KGs and their related technologies. As a result, users typically express their questions in natural language, and even in their native language.

Concerning sports domain, it interests a high number of people worldwide, and regarding research, sports analytics is an emerging topic as it is stated in [5,23]. Here, we focus on Basketball, which is a very popular sport worldwide with over 3 billion fans (http://bit.ly/3QK136X), and over 600 million fans in China. Each Basketball game is a quite complex event including several participants and factors, such as players, coaches, referees, audience, statistics and others. As a result, there is a lot of information that needs to be recorded and there is a high need for updates, since the action never stops; new games are played all the time, new players join the teams, etc. Moreover, there are numerous simple and complex questions, that can be asked either from the fans and journalists for retrieving specific facts, or from the coaching staff for performing a data analysis. For example, to optimize the performance of players and to enable the decision making [27], e.g., a recent study by MIT researchers [29] states the high influence of basketball analytics investment on team performance.

Generally, there are numerous webpages offering traditional and advanced sports statistics in HTML tables and/or in JSON format derived by relational databases. However, there is a lack of RDF KGs for sports domain including such statistics, e.g., the pioneer researcher G. Weikum has stated [30] that "Capturing quantity properties of entities in sports, is important for KG coverage towards analytic tasks. This is a big gap in today's major KGs, and search engines are not a good proxy". Therefore, a major challenge is (a) to construct such RDF KGs for sports for enabling the answer of complex questions, such as those involving relationships across multiple entities (players, coaches, referees, countries), semantic constraints, and numerical statistics, and (b) to offer natural Question Answering (QA) by translating each question into precise SPARQL expressions. Indeed, SPARQL can allow sports analysts to efficiently calculate aggregated metrics such as average, total, max and min directly from large KGs without manually processing individual game records.

As a running example, suppose that a fan of European Basketball desires to find an answer for the question: "Give me the games that each of Kendrick Nunn and Kostas Sloukas, as teammates, scored at least 15 points". To answer this question, the classical approach is to browse every Panathinaikos game boxscore, to find the required information. And this can become even harder for questions that one should combine data from more than one teams, seasons, statistics, players, and others, or/and specific characteristics of them, such as "Is there any game that a player of a Greek team, having height less than 2.00, that had over 10 rebounds and over 10 points". On the other hand, the proposed approach (see Fig. 1) is to exploit a KG where one can find and combine all the information in a structured way, and then to enable the conversion of natural

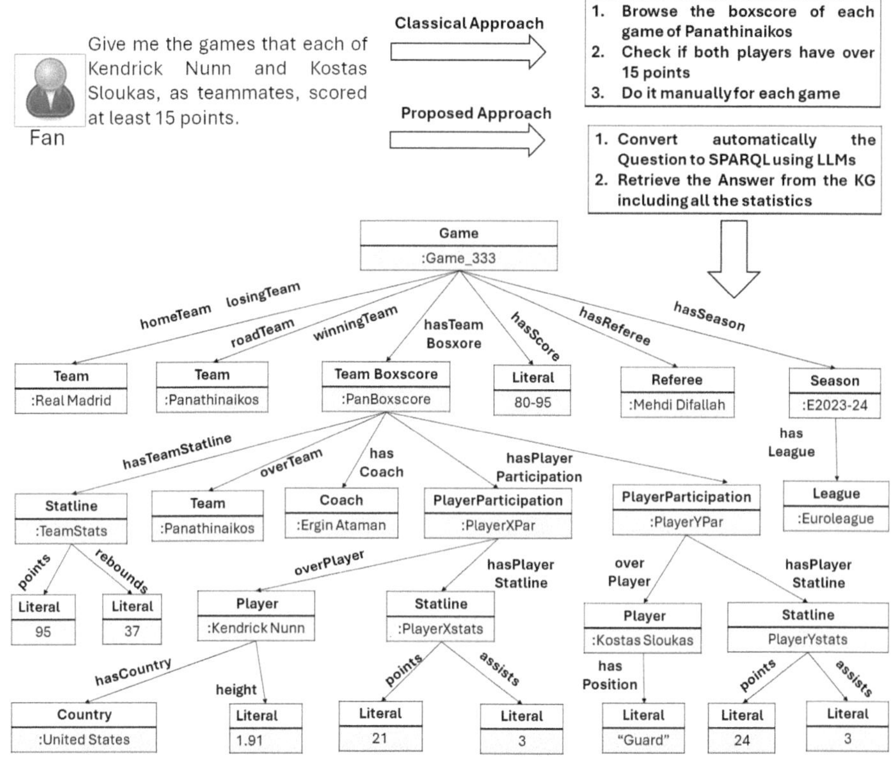

Fig. 1. A Query from a user and the Knowledge Graph for a Basketball Game

language questions to SPARQL, for answering such questions, even in his/her native language. Below we present the research questions:

- *RQ1:* How can we design and construct a domain-specific ontology and KG to support natural QA and complex analytics tasks in basketball?
- *RQ2:* How effectively can LLMs generate accurate SPARQL queries from natural language questions in the basketball domain, when guided by the ontology paths and literal patterns?

Regarding our contribution, in this paper i) we describe a pipeline including all the steps, from the data collection, to the KG exploitation using LLMs, ii) we propose an event-based ontology for basketball games, called *BBall*, and we construct a KG for basketball games by focusing on EuroLeague Basketball, and iii) we use a text-to-SPARQL method based on ontology path and literals patterns that are given to LLMs for enabling natural QA over the KG and we have built an application called *HoopProphets*, which is available online in https://demos.isl.ics.forth.gr/HoopProphets. Finally, iv) we present comparative evaluation results by constructing a benchmark with 100 natural questions (that also includes a paraphrased version in English, a Greek and a Chinese version), by using popular LLMs, i.e., ChatGPT and Deepseek.

To the best of our knowledge, this is the first work proposing such a pipeline, including all the steps from the ontology creation to the KG exploitation from the LLM for enabling the expression of natural language questions. Moreover, we present a KG for the most popular European Basketball league, that can be exploited for QA, data analytics and insights, game simulations and others.

The rest of this paper is organized as follows: Sect. 2 discusses related work, and Sect. 3 describes the pipeline and all the steps. Furthermore, Sect. 4 presents the experimental results based on a QA benchmark, and finally Sect. 5 concludes the paper and discusses the directions for future research.

2 Related Work

Here we discuss the: i) Generation of SPARQL from texts, and ii) KGs and analytics for basketball. Finally, we provide a comparison with our work.

2.1 SPARQL Generation from Text

First, the authors in [32] exploited Chinese KGs for SPARQL generation over LLMs, whereas in [25], example SPARQL queries and question analysis techniques combined for SPARQL generation over the Open Research KG. In [3], several configurations were compared for SPARQL query generation (such as few shows and fine tuning) for the DBpedia [1] and Wikidata [28] KGs. Also, the authors in [2] proposed a hybrid approach for text-to-SPARQL by combining LLMs with RAG techniques, while [14] evaluated the performance of fine-tuned LLMs for the generation of SPARQL queries and analyzed its limitations. Moreover, in [24], a curated dataset with 2,771 unique queries from Wikidata was released, which can be used for fine-tuning LLMs to generate accurate and syntactically valid SPARQL queries from natural text. Finally, in our past work [18], we have proposed ontology path patterns methods, for generating complex queries over the ISO standard CIDOC-CRM for Cultural Heritage domain KGs.

2.2 KGs and Analytics for Basketball

First, the authors in [9], collected data about teams and players from basketball websites and created a KG about NBA, and used BERT for performing QA over the KG. Also, in [10] a QA system based on an NBA KG was proposed, which included a combination of NLP and Machine Learning techniques for answering questions. Another work [13], analyzed and compared statistical data of NBA and EuroLeague Basketball, and they mentioned that EuroLeague "is becoming quantitatively and qualitatively more similar to the NBA". Additionally, there are also works for the prediction of the winner of a basketball game based on machine learning algorithms, for EuroLeague [8] and NBA [26]. Regarding statistics, in [7] a statistical analysis was performed on the historical evolution of EuroLeague. In addition, in [6], LLMs were used for evaluating several tasks for Basketball analytics, such as for analyzing critical statistics in game summaries,

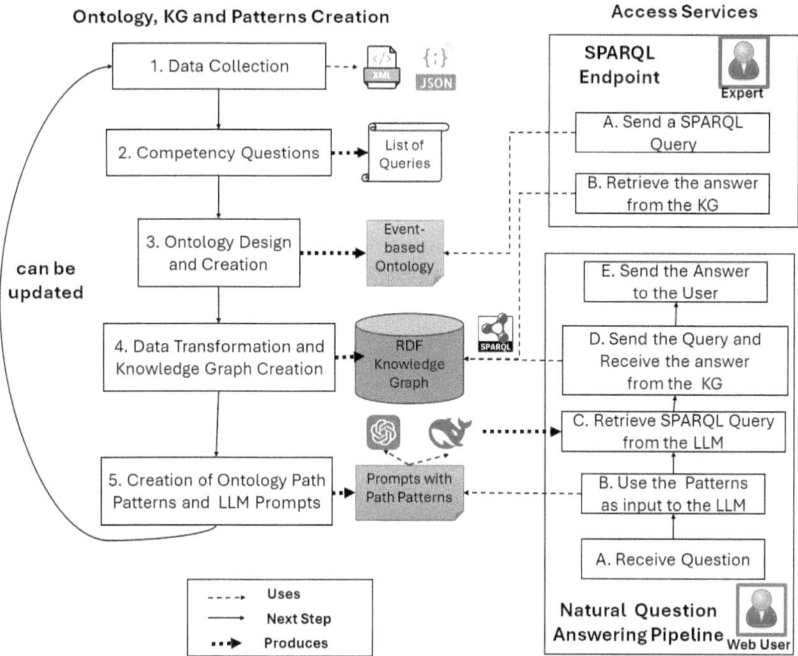

Fig. 2. A Pipeline for Enabling Natural QA with LLMs over Event-Based KGs

whereas [22] used association rule mining techniques for evaluating recovery and economic impacts in NBA basketball. Finally, in [31] the benchmark SportQA was presented, which contains 70K questions for several sports (including basketball) and subjects, such as historical facts and complex scenarios.

2.3 Comparison and Novelty

Compared to the basketball analytics (e.g., [9,10]), this is the first work i) providing a pipeline for basketball games including all the steps from the production of the KG to its exploitation through LLMs for enabling natural QA, and ii) offering a KG and a QA application for the EuroLeague basketball. Regarding text-to-SPARQL, similarly to our past work [18], we also exploit ontology path patterns methods (but here we provide a more compact version by also including literal datatypes), and not fine-tuning or few shots techniques (e.g., [3,14]).

3 The Proposed Steps of the Pipeline

Here we describe all the steps of the pipeline, which are depicted in Fig. 2. As we can see, in the left part, the key steps are the data collection, the creation of Competency Queries, the design and implementation of the ontology, the data transformation and KG creation, and finally the creation of path patterns for constructing prompt templates. On the right part, we can see the steps for exploiting the resulted prompt templates for SPARQL Query Generation over the KG, for offering advanced query services for natural text questions.

3.1 Data Collection

The first step is the data collection, which is important for defining the competency queries and for designing the ontology. Here, we exploit the EuroLeague API[1], where one can export several information about EuroLeague Basketball in JSON or XML format. However, the data are not inherently connected, as they do not follow Linked Data principles, requiring additional processing to establish relationships. Thereby, it is not trivial to ask questions that require information from several API requests. We use 9 endpoints from the API for retrieving in JSON format a) the key metadata of the game, b) the game statistics, and c) metadata for players, coaches, referees, teams, countries and venues.

3.2 Competency Queries

We provide 5 indicative competency questions (in 5 categories) that are not trivial to be answered by the classical boxscores. The full list is available in https://github.com/mountanton/BasketballQA.

Q1. General Question. Give me the winning percentage of Ergin Ataman in 2022–23.
Q2. Filtering Question. Give me all the games of 2023–24 in which the losing team had a higher Performance Index Rating than the winning team.
Q3. Top-K Question. Give me all the top-10 players with the most points in a game of 2023–24 having height less than 1.85.
Q4. Grouping Question. Show Nigel Hayes-Davis's monthly total rebounds in 202324.
Q5. Negation Question. Give me the number of teams that have never won Barcelona at road since 2020.

3.3 Ontology Design and Creation: The BBall Ontology

Although basketball is a very popular sport, we did not manage to find an ontology to cover our competency questions. The most related one is the IPTC Sports Schema[2], however, it mainly contains generic properties for basketball statistics and cannot record the required information of a basketball game. For this reason, we decided to create an ontology called *BBall* that describes the key statistics of a Basketball Game. As we can see in Fig. 3, the key class is the *Game*, which is an *Event* and is connected with the *Teams*, the *Venue*, the

[1] https://api-live.euroleague.net/swagger/.
[2] https://sportschema.org/ontologies/basketball.

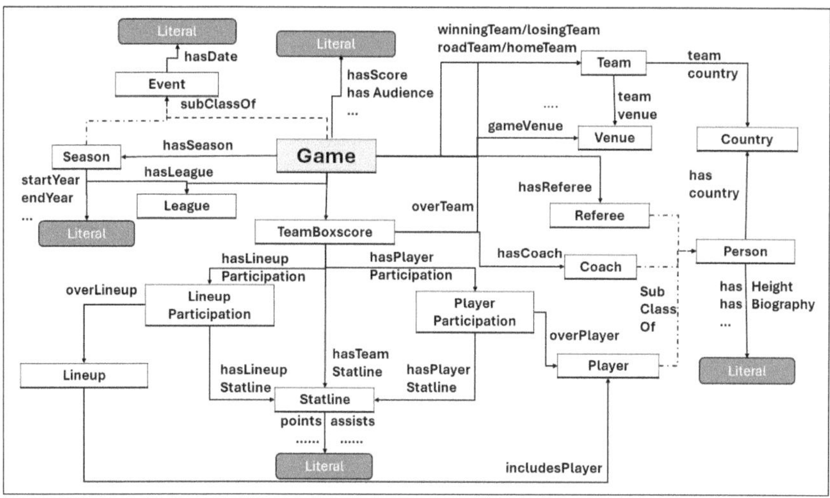

Fig. 3. The BBall Ontology - Key Classes and Relationships

Referees, the *Season*, and contains data properties for the score, the audience, etc. It is also connected with the *TeamBoxscore*, which contains the boxscore of each of the 2 teams of the game. First, each boxscore is connected with a *Team*, and the *Coach* of the team, and also it has a team *Statline*, i.e., for recording the total statistics of the team in the given game. The *Statline* class has one property for each statistic of a boxscore, such as points, rebounds, etc.

Regarding *Players*, we have the *PlayerParticipation* class, which records the participation of each player in the game. For each player participation, we need also to record his *Statline*, i.e., to cover all the statistics of the player. Also, the ontology has another one class related to the statistics, i.e., the *LineupParticipation*, in case that one desires to provide the statistics of a given K-player lineup, i.e., the statistics of the team when a given 5-player lineup is on the court, e.g., see [16]. Finally, for each of the other classes there are several properties, such as for the *Person* (e.g., Players, Coaches) the name, height, birthdate and *Country*; for the season the corresponding *League*, the start year, and so on. An example of the ontology using real data is depicted in Fig. 1, where we start from a game, and we can see the teams, a boxscore of a team, statistics, players, etc.

Validation, Documentation and Publication of the Ontology. For validating the consistency, completeness and conciseness of the BBall ontology we used the tool "Oops!" [21], and the ontology passed the tests without critical errors. Moreover, we use WIDOCO [4] for the documentation of the ontology, which also provides an interactive visualization through the WebVOWL system [12]. Its documentation and the ontology are available online[3]. The ontology contains 16 classes, 25 object properties and 63 data properties (most of them are not depicted in Fig. 3 for reasons of space). Finally, after the ontology cre-

[3] http://users.ics.forth.gr/~mountant/bball.html.

```
SELECT DISTINCT ?class ?p (?object AS ?classOrDatatype)
WHERE {
  ?s ?p ?o . ?s a ?class .
  OPTIONAL { ?o a ?object . }
  OPTIONAL { ?s ?p ?o. FILTER(isLiteral(?o)) .
    BIND(DATATYPE(?o)  AS ?object)
  }
} ORDER BY ?class ?p
```

Fig. 4. The SPARQL query for retrieving the path patterns

ation, an important task is to construct the SPARQL query for each competency question.

Possible Extensions. Certainly, the ontology can be extended to cover more statistics and more sub-events of the game, such as the play-by-play events and more advanced statistics. However, it is worth noting that most of them, such as the number of points per minute, assists/turnover ratio, and others, can already be computed by exploiting the aggregation capabilities of SPARQL (e.g., using COUNT, SUM, AVG, MAX, MIN, etc.). Additionally, more properties can be added for several classes, such as for persons, venues, teams and countries.

3.4 Data Transformation and KG Construction

By using the *BBall* ontology, we need to transform the data and one key objective is the URI policy for having both dereferencable URIs and a single URI for each real world entity. Regarding the steps, they are explained below:

- We retrieve the metadata of all the games for each season. Then we retrieve each game and we create the RDF triples using the *BBall* ontology, and some key properties from the RDF, RDFs, SKOS and FOAF, for the game metadata and the statistics for the players and the team.
- For each person, the teams, venues and countries, we also store their IDs. We exploit the endpoints offered by the API for the persons, teams and venues, for retrieving the data and for transforming them in RDF format.
- In the end, we have created a single ".ttl" file for the games of each season, all the players, teams, referees, coaches, venues and countries, a total of 175 ".ttl" files. Each of the real world entities includes a unique URI having as a suffix a unique code. For most of the entities, we offer dereferencable URIs. The produced files in Turtle format, which have size 322 MB on disc, have been uploaded in a SPARQL endpoints by using the Virtuoso v7 engine.

3.5 QA Access Services: Path Patterns Creation and LLM Prompts

The final step is to create ontology path and literal patterns as input to the LLM for enabling SPARQL Query Generation. The key advantages of such an approach are the following: a) it can be used for non-familiar users to SPARQL, for

```
You are given the classes and their path patterns of a Knowledge Graph
about basketball games.  With the go to, we indicate the class that you
need to visit for constructing a longer path.

#Path Patterns
*bball:Game
 ->bball:homeTeam->bball:Team (go to *bball:Team)
 ->bball:hasTeamBoxscore->bball:TeamBoxscore (go to *bball:TeamBoxscore)
 ->bball:hasAudience-> Literal (xsd:integer)
...
*bball:Team
 ->rdfs:label->Literal
 ->bball:teamCountry->bball:Country (go to *bball:Country)
...
*bball:TeamBoxscore
    ->bball:hasTeamStatline->bball:Statline (go to *bball:Statline)
...

#Extra Patterns
*The URIs include as suffixes identifiers and not names
*The persons names are given by the rdfs:label
with the pattern (Name Surname) in English
*The team names are given through rdfs:label in English

By using properties and classes from the above patterns,
please generate only a  SPARQL query  (without explanation)
for answering the question Q
```

Fig. 5. The prompt template for the Path and Literals Patterns Method

expressing any question that even require to combine information from multiple paths, b) it is a generic method that can be applied in any event-based RDF KG without a need to train the whole KG using an LLM, and c) it can support multilinguality, i.e., the question can be expressed in any language. For enabling the natural QA over the KG, one needs first to create the patterns through a generic SPARQL query. Compared to [18], in this paper we have an ontology which is not so complex as CIDOC-CRM, i.e., in CIDOC-CRM one can use different path patterns to express the same events. For this reason, here we propose to have a single compact prompt template, where we provide the ontology path patterns grouped by each RDF class, and we also include the datatypes of literals (e.g., they can be important for queries that require sums, averages, comparisons, etc.). Figure 4 shows the query to retrieve the path patterns, including patterns having either as an object a class, such as "Game–>homeTeam–>Team", or a literal, such as "Game–>hasHomeTeamScore–>Literal (xsd:integer)".

The next step is to create the prompt templates using the mentioned ontology path and literals patterns, and optionally, to include extra data patterns, such as

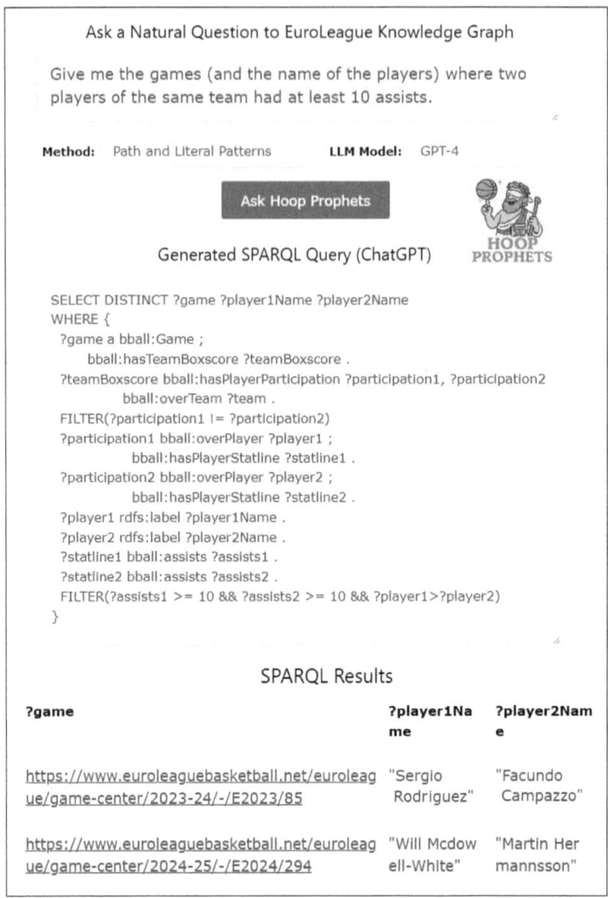

Fig. 6. Screenshots from the HoopProphets

URI patterns about the suffixes, and/or Literals patterns, like patterns related to spaces, capitalization and others. For each triple pattern, we distinguish two types of objects for also enabling the visit of longer paths (e.g., for multi-hop queries): a) for the classes we use the "*go to*" phrase for showing the class that we need to visit for constructing a longer path and b) for literals, we give its datatype. Figure 5 presents such a prompt, where the patterns are grouped by class and extra guidelines are provided for patterns related to the URIs and literals (e.g., person names, team names, URI policy).

Natural QA by using Path Patterns Prompts. The notion is first to retrieve the given question Q, and then to exploit the pre-generated prompt template(s), i.s., right part of Fig. 2. A prompt is sent to the desired LLM, which generates the SPARQL query based on the prompt guidelines. The next step is to send the query to the SPARQL endpoint and to retrieve the answer. In

Table 1. General KG Statistics

Type	Value
Number of Triples	5,288,814
Triples with Object URIs	845,840
Triples with Object Literals	4,442,974
Distinct URIs	309,877
Distinct Literals	29,898
Distinct Subjects	307,750
Distinct Objects	333,573

Table 2. Statistics of the KG Classes

Class	URIs	Triples	avg Triples/URI
Season	25	6,351	254.0
Game	6,201	161,077	25.9
Player	3,234	180,443	55.7
Team	87	37,905	435.6
Venue	239	7,346	30.7
Coach	232	16,157	69.6
Referee	302	20,398	67.5
Statline	140,765	4,170,247	29.6

cases where the SPARQL query returns an empty set, we can repeat the last two steps. Finally, the results can be either shown in HTML tables and/or can be sent to an LLM to create a more natural answer. We can use the same prompt for any language, by sending the question Q expressed in the desired language.

Application. The application *HoopProphets* offers the multilingual natural QA over the KG in https://demos.isl.ics.forth.gr/HoopProphets. A screenshot with an example of a complex query is shown in Fig. 6. Also, a tutorial video can be found in https://youtu.be/X6HCxxFDgXM.

Additional Steps: Updates and Enrichment. Regarding the updates, the steps (or some of the steps) should be repeated in many cases, such as for adding new data like new basketball games, or/and new types of data (such as the play-by-play) which will maybe result in the extension of the ontology. Moreover, in case of having entities that are also available in popular KGs such as DBpedia and Wikidata, they can be linked for enriching the information for the entities. For instance, in our KG the triples of Kostas Sloukas include his code ID in EuroLeague (i.e., 001926), and this information is also covered in his Wikidata page (https://www.wikidata.org/wiki/Q2618541).

4 Experimental Evaluation

We first provide some statistics about the KG, comparative results for the QA over the KG by using ChatGPT and Deepseek and then we provide a discussion and specific research directions. The KG and the application is hosted in a single machine having an i5 core, 4 GB main memory and 50 GB disk space.

4.1 Statistics of the KG

Table 1, provides some general statistics about the constructed KG, including the number of unique URIs, Literals, and others. The KG includes all the 25 seasons from 2000–2001 to 2024–25 and can be easily updated. As we can see, the KG covers over 5 million triples and it has a high number of triples with literals as objects, i.e., 84% of the total triples (they mainly describe the values of statistics). However, since the most literals refer to numerical values, they are repeated, thereby, the distinct literals are only 29 thousand.

Table 2 shows the number of URIs for each class, the number of triples where a URI of the class is either a subject or object and the average triples for the URI of each class. As we can see, the KG includes more than 6 thousand games and 3 thousands players. For each game, there are on average over 25 triples per URI, and for each player over 55 triples. Moreover, the most URIs belong to Statline, i.e., over 140 thousands, whereas the highest number of average triples refer to teams, i.e., 435 triples per team.

4.2 The QA Benchmark over the KG and Experimental Setup

For the given ontology and KG, we have created a benchmark called *EuroLeague-QABench* having in total 100 competency questions. It includes 20 questions in five categories, according to the SPARQL clause needed for each query, i.e., SELECT, FILTER, ORDER BY, GROUP BY, and NEGATION. An example of each category in natural language, with the desired SPARQL query is given in Fig. 7. Moreover, for each natural question, we have asked the LLM to provide a paraphrased version including more informal (and usually less number of) words, say $EuroleagueQABench_{par}$. For example, the question "Which teams missed out on the playoffs in 2023–24 although they won Real Madrid?" was paraphrased as "Who beat Real Madrid but still missed the 202324 Playoffs?".

Moreover, we also provide experiments for the same benchmark in Greek, i.e., $EuroleagueQABench_{gr}$ which was translated by native speakers, and in Chinese i.e., $EuroleagueQABench_{cn}$, which was translated automatically using ChatGPT. As a result, some questions in Chinese may not have been translated accurately.

Experimental Setup. We provide experiments for the above benchmark, by using the following LLMs, Deepseek [11] and ChatGPT [20], for versions 4 and 4-mini. We present comparative results for each LLM, by using i) the path and literals patterns method and ii) a baseline, in which we just give to the LLM a prompt with the list of all the classes and all the properties of the KG.

Metrics. We provide the results of three metrics: *Syntax Score*, i.e., the ratio of syntactically valid SPARQL queries (range $[0, 1]$), the *Accuracy*, i.e., the ratio of SPARQL queries that returned the desired result (range $[0, 1]$), and c) *Triple Patterns Number*, i.e., the number of Triple Patterns per SPARQL query.

GitHub. All the questions/versions of the benchmark, their gold standard, the generated SPARQL queries for each LLM and method, the results and the application code can be found in https://github.com/mountanton/BasketballQA.

Fig. 7. Example with the questions of the benchmark

Table 3. Accuracy scores for the English version of the benchmark

Rank	Model	Method	SELECT	FILTER	ORDER	GROUP	NEG.	All Qs
1	GPT-4	Patterns	**0.90**	**0.75**	**0.90**	0.80	0.60	**0.79**
2	DeepSeek	Patterns	0.70	0.70	0.65	**0.85**	**0.65**	0.71
3	GPT-4mini	Patterns	0.55	0.15	0.50	0.30	0.30	0.36
4	DeepSeek	Baseline	0.40	0.25	0.40	0.25	0.20	0.30
5	GPT-4	Baseline	0.45	0.10	0.25	0.25	0.20	0.25
6	GPT-4mini	Baseline	0.15	0.05	0.10	0.05	0.00	0.07

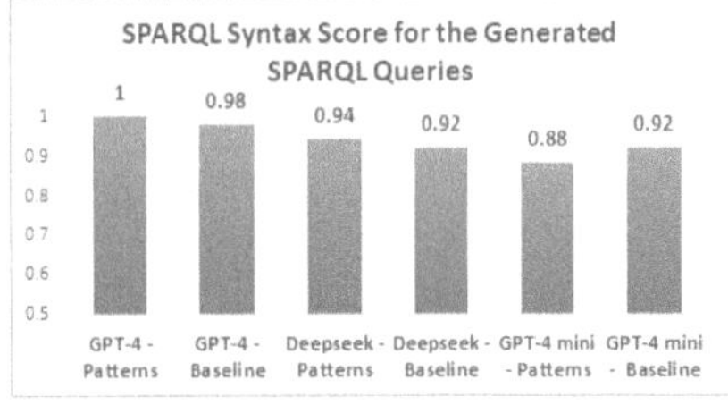

Fig. 8. SPARQL Syntax Score for the Generated SPARQL Queries

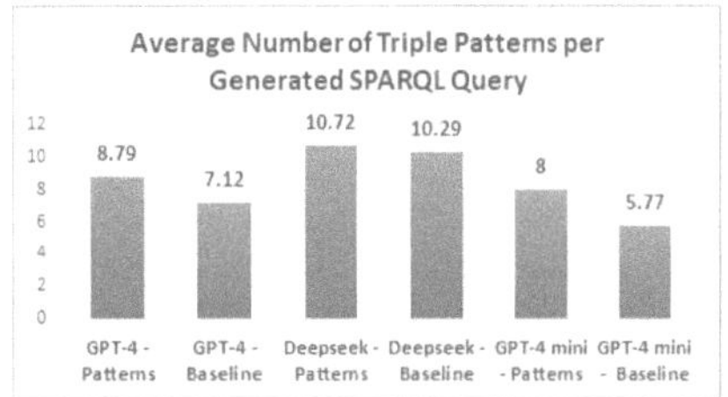

Fig. 9. Average Number of Triple Patterns per SPARQL Query

4.3 Effectiveness for Natural QA Using the LLMs

First, for the English benchmark, for each LLM we compare the results for the path and literals patterns and the baseline, and for the paraphrased and multilingual ones, we provide the results for the path and literals patterns method.

Effectiveness Results on English Version. As we can see in the last column of Table 3, the patterns method outperforms the baseline for all the LLMs, i.e., the accuracy for the baseline is less than 0.3 in all the cases. The LLM having the best results is GPT-4, with accuracy 0.79, whereas *Deepseek* also reaches high accuracy, i.e., 0.71. Regarding the query types, the results were better for SELECT, ORDER BY and GROUP BY queries, whereas the lowest accuracy obtained for the NEGATION queries. Figure 8 shows that *Deepseek* is more vulnerable to syntax errors than *GPT-4*, whereas in Fig. 9, it is clearly that *Deepseek* constructs on average SPARQL queries with more (and usually redundant) triple patterns than the two *GPT* models.

Common Errors. In most cases, both *GPT-4* and *Deepseek* selected the correct patterns (which was not the case for *GPT-4 mini*). Concerning the most common errors, for the *GPT-4* were the following: i) it did not use the DISTINCT in some cases (so it returned duplicates), ii) wrong calculation of some statistics requiring average values, iii) in the negation, it used FILTER instead of FILTER NOT EXISTS. On the other hand for the *Deepseek* the main problems were: a) syntax errors, b) wrong calculation of some statistics requiring average values and c) extra filters and triple patterns that were either redundant or erroneous.

Effectiveness Results on Paraphrased and Multilingual version. Figure 10 shows the results for the other versions of the benchmark. First, for the paraphrased version, there is a decrease in the accuracy, approximately −0.1 for both models. Generally, the questions were more informal and contained fewer words, which resulted in many erroneous queries especially for the NEGATION questions. For the other categories the accuracy remained high. Concerning the multilingual versions, for the Greek version, GPT-4 had a high accuracy (i.e., 0.76) and in most cases, it correctly translated the Greek names, e.g., of the players, to their English name. On the other hand, for the *Deepseek* the accuracy for Greek was low (it failed in many cases to translate correctly the names), but it achieved higher accuracy for the Chinese version compared to *GPT-4*.

4.4 Efficiency and Monetary Cost for Natural QA Using the LLMs

Figure 11 shows the execution time for each LLM. *GPT-4* needs on average 1.8 s for generating the SPARQL query, i.e., it is approximately 8× faster compared to *Deepseek*, which needs on average 14.6 s for each query. Concerning the monetary cost, *GPT-4* is 10× more expensive, i.e., for 100 queries the price was 0.34$, whereas for *Deepseek* and *GPT-4 mini*, it was less than 0.03$.

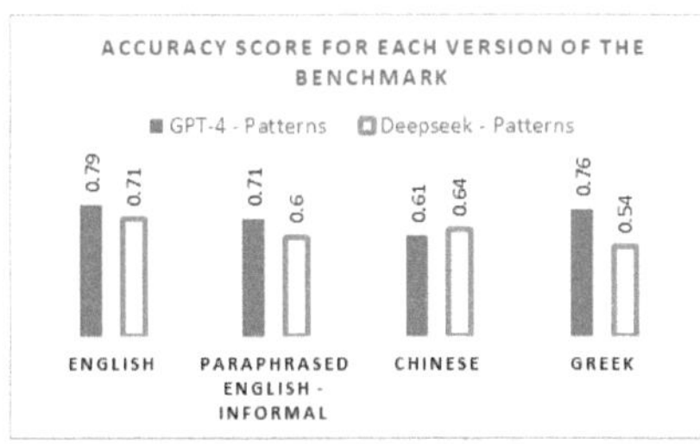

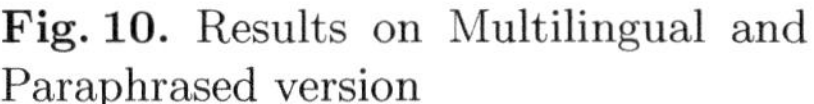
Fig. 10. Results on Multilingual and Paraphrased version

Fig. 11. Average Execution Time for each LLM with the patterns method

4.5 Discussion and Limitations

As regards *GPT-4*, it generally produces correct queries, it has a high performance on the English and Greek questions, and it is quite fast. Some key limitations are that it forgets the distinct in some cases, and it can be expensive as the number of questions and/or patterns increases. Concerning *Deepseek*, it usually produces correct queries, it has high effectiveness for the English and Chinese version, and it is quite cheaper than GPT, i.e., more than 10×. On the other hand, it adds more redundant triple patterns, it is more vulnerable in syntax errors and it is quite slower than the GPT models. Finally, the GPT-4mini is both fast and cheap, however, it produces much more erroneous SPARQL queries.

Regarding other limitations, the users tend to search with different names for the entities of a KG, e.g., they can use abbreviations for the teams (e.g., Barca instead of Barcelona) only the surnames of persons (Lessort instead of Mathias Lessort), and so forth. This is also clear for the results of the paraphrased version of the benchmark. One solution is to exploit autocomplete techniques, e.g., when a user types a surname, say Nunn, to send to the LLM the question including the full name of the player (e.g., Kendrick Nunn) or even the URI of the player.

5 Conclusion

In this paper, we proposed a pipeline including the steps of data collection, competency questions, ontology design, KG construction, and QA services using LLMs. We focused on the sports domain and specifically on basketball, and we used the steps of the pipeline to construct an ontology about basketball games and a KG for the EuroLeague Basketball. The KG includes 5 million triples for all the games (over 6 thousands games) from the season 2000–01 until 2024–25. Afterwards, we described an ontology paths and literals patterns method, and an application for generating SPARQL queries from natural questions (even multilingual ones). Regarding the experiments, we used a dedicated benchmark with 100 questions in three languages. For the english version we achieved accuracy score 0.79 using GPT-4 and 0.71 using Deepseek for the path and literals patterns method (instead of less than 0.3 for the baseline). For the multilingual, GPT-4 achieved for the Greek version accuracy 0.76 and for the Chinese 0.61.

As a future work, we plan to a) create a more interactive dialogue-based web application for enabling users to ask their own questions in any language, by using the proposed QA approach (such as [19]), b) extend the ontology for enriching the KG with more data, like the play-by-play of each game and more data for the rest entities, e.g., players, teams, by finding the same entities in external KGs (e.g., by using LODChain [17]), and c) create a service for automatically evaluating the results of SPARQL query generation over a benchmark.

References

1. Auer, S., Bizer, C., Kobilarov, G., Lehmann, J., Cyganiak, R., Ives, Z.: DBpedia: a nucleus for a web of open data. In: Aberer, K., et al. (eds.) ASWC/ISWC -2007. LNCS, vol. 4825, pp. 722–735. Springer, Heidelberg (2007). https://doi.org/10.1007/978-3-540-76298-0_52
2. Avila, C.V.S., Vidal, V.M., Franco, W., Casanova, M.A.: Few-shot learning or RAG in LLM-based text-to-SPARQL? Why not both? (2025)
3. Diallo, P.A.K.K., Reyd, S., Zouaq, A.: A comprehensive evaluation of neural SPARQL query generation from natural language questions. IEEE Access (2024)
4. Garijo, D.: WIDOCO: a wizard for documenting ontologies. In: d'Amato, C., Fernandez, M., Tamma, V., Lecue, F., Cudré-Mauroux, P., Sequeda, J., Lange, C., Heflin, J. (eds.) ISWC 2017. LNCS, vol. 10588, pp. 94–102. Springer, Cham (2017). https://doi.org/10.1007/978-3-319-68204-4_9
5. Ghosh, I., Ramasamy Ramamurthy, S., Chakma, A., Roy, N.: Sports analytics review: artificial intelligence applications, emerging technologies, and algorithmic perspective. Wiley Interdisc. Rev.: Data Min. Knowl. Discov. **13**(5), e1496 (2023). https://doi.org/10.1002/widm.1496
6. Hu, Y., et al.: SportsMetrics: blending text and numerical data to understand information fusion in LLMs. arXiv preprint arXiv:2402.10979 (2024)
7. Katris, C.: Exploring Euroleague history using basic statistics. arXiv preprint arXiv:2301.02443 (2023). https://doi.org/10.48550/arXiv.2301.02443
8. Lampis, T., Ioannis, N., Vasilios, V., Stavrianna, D.: Predictions of European basketball match results with machine learning algorithms. J. Sports Anal. **9**(2), 171–190 (2023). https://doi.org/10.3233/JSA-220639
9. Li, A., Li, X., Zhu, B., Li, G.: Knowledge-graph-driven question-answer system for basketball. In: In: 5th International RICAI Conference, pp. 757–761. IEEE (2023)
10. Li, Y., Cao, J., Wang, Y.: Implementation of intelligent question answering system based on basketball knowledge graph. In: 4th IAEAC, pp. 2601–2604. IEEE (2019)
11. Liu, A., et al.: DeepSeek-v3 technical report. arXiv preprint arXiv:2412.19437 (2024)
12. Lohmann, S., Link, V., Marbach, E., Negru, S.: WebVOWL: web-based visualization of ontologies. In: Lambrix, P., et al. (eds.) EKAW 2014. LNCS (LNAI), vol. 8982, pp. 154–158. Springer, Cham (2015). https://doi.org/10.1007/978-3-319-17966-7_21
13. Mandić, R., Jakovljević, S., Erčulj, F., Štrumbelj, E.: Trends in NBA and Euroleague basketball: analysis and comparison of statistical data from 2000 to 2017. PLoS ONE **14**(10), e0223524 (2019)

14. Mecharnia, T., d'Aquin, M.: Performance and limitations of fine-tuned LLMs in SPARQL query generation. In: GenAIK Workshop Proceedings, pp. 69–77 (2025)
15. Mountantonakis, M.: Services for Connecting and Integrating Big Numbers of Linked Datasets, vol. 50. IOS Press, Amsterdam (2021)
16. Mountantonakis, M.: Efficient statistical computation for k-player basketball lineups using semilattice structures. Electronics **14**(11) (2025)
17. Mountantonakis, M., Tzitzikas, Y.: LODChain: strengthen the connectivity of your RDF dataset to the rest LOD cloud. In: Sattler, U., et al. (eds.) ISWC 2022. LNCS, vol. 13489, pp. 537–555. Springer, Cham (2022). https://doi.org/10.1007/978-3-031-19433-7_31
18. Mountantonakis, M., Tzitzikas, Y.: Generating SPARQL queries over CIDOC-CRM using a two-stage ontology path patterns method in LLM prompts. ACM J. Comput. Cult. Heritage **18**(1), 1–20 (2025)
19. Mountantonakis, M., Tzitzikas, Y.: TCRMQ: an application for generating CIDOC-CRM SPARQL queries from text by using large language models (2025)
20. OpenAI: ChatGPT (2025). https://chat.openai.com
21. Poveda-Villalón, M., Gómez-Pérez, A., Suárez-Figueroa, M.C.: OOPS! (OntOlogy pitfall scanner!): an on-line tool for ontology evaluation. Int. J. Semant. Web Inf. Syst. (IJSWIS) **10**(2), 7–34 (2014)
22. Sarlis, V., Papageorgiou, G., Tjortjis, C.: Leveraging sports analytics and association rule mining to uncover recovery and economic impacts in NBA basketball. Data **9**(7), 83 (2024)
23. Singh, N.: Sport analytics: a review. Learning **9**(11) (2020)
24. Strappazon, A., Granitzer, M., Egyed-Zsigmond, E., Mitrovic, J., Amor, M.B.: Instruct-to-SPARQL: a text-to-SPARQL dataset for training Wikidata agents. In: ACM SIGIR Conference on Human Information Interaction and Retrieval. ACM (2025)
25. Taffa, T.A., Usbeck, R.: Leveraging LLMs in scholarly knowledge graph question answering. arXiv preprint arXiv:2311.09841 (2023)
26. Thabtah, F., Zhang, L., Abdelhamid, N.: NBA game result prediction using feature analysis and machine learning. Ann. Data Sci. **6**(1), 103–116 (2019)
27. Vinué, G., Epifanio, I.: Archetypoid analysis for sports analytics. Data Min. Knowl. Disc. **31**(6), 1643–1677 (2017). https://doi.org/10.1007/s10618-017-0514-1
28. Vrandečić, D., Krötzsch, M.: Wikidata: a free collaborative knowledgebase. Commun. ACM **57**(10), 78–85 (2014)
29. Wang, H., Sarker, A., Hosoi, A.: The Effect of Basketball Analytics Investment on National Basketball Association (NBA) Team Performance. JSE, SAGE (2025)
30. Weikum, G.: Knowledge graphs 2021: a data odyssey. Proc. VLDB Endow. **14**(12), 3233–3238 (2021)
31. Xia, H., Yang, Z., Wang, Y., Tracy, R., et al.: SportQA: a benchmark for sports understanding in large language models. arXiv preprint arXiv:2402.15862 (2024)
32. Yang, S., Teng, M., Dong, X., Bo, F.: LLM-based SPARQL generation with selected schema from large scale knowledge base. In: Wang, H., Han, X., Liu, M., Cheng, G., Liu, Y., Zhang, N. (eds.) CCKS 2023. CCIS, vol. 1923, pp. 304–316. Springer, Singapore (2023). https://doi.org/10.1007/978-981-99-7224-1_24

PG2RDF: Schema-Guided Transformation of Property Graphs to RDF

Sophia Sideri[1(✉)], Vasilis Efthymiou[1,2], Dimitris Plexousakis[1], and Haridimos Kondylakis[1]

[1] Institute of Computer Science, FORTH and Computer Science Department, UOC, Heraklion, Greece
{sophisid,dp,kondylak}@ics.forth.gr

[2] Department of Informatics and Telematics, Harokopio University of Athens, Athens, Greece
vefthym@hua.gr

Abstract. Interoperability between heterogeneous graph data models remains a significant challenge in knowledge graph construction. While both Property Graphs (PG) and RDF models are widely used for data representation, their structural and semantic differences hinder seamless integration and data exchange. In this work, we present PG2RDF, a fully automated and schema-aware pipeline for transforming PG data into RDF. Our method relies on an XML Schema Definition (XSD) to formalize the PG structure and can generate both X3ML or RML mapping rules for deterministic RDF production. By integrating schema discovery, XML annotation, and RDF materialization through the X3ML Engine and RML Mapper, we demonstrate the feasibility of our approach across heterogeneous datasets and their integration into an interoperable Knowledge Graph. PG2RDF ensures URI consistency and accurately represents the PG information.

Keywords: Data Integration · Graph Databases · Property Graphs · RDF · Data Transformation

1 Introduction

Modern applications in various domains increasingly rely on graph-based representations to manage complex and interconnected data [27]. Among the most prominent Knowledge Graph (KG) data models are Resource Description Framework (RDF) [43] and Property Graphs (PGs) [1]. Both share fundamental graph concepts (using nodes and edges), but differ in structure and query languages [2].

RDF is described in triples of subject-predicate-object structure [46], while PGs have multi labeled nodes and edges along with their key-value pairs [34]. RDF enables semantic interoperability through ontologies [7,24,25], while PGs [22] are mostly used in industry, due to their flexible structure. These

H. Takeda et al. (Eds.): IJCKG 2025, LNAI 16297, pp. 358–373, 2026.
https://doi.org/10.1007/978-981-95-5009-8_24

differences pose challenges when trying to integrate data across the two models [6,19]. In particular, direct transformation between the two models remains an open research problem, due to the difficulty of preserving their semantics and structure [33].

An effective and accurate transformation must consider not only node and edge mappings, but also inference rules, multi-valued attributes, diverse edge types, and schema constraints [23]. Furthermore, the lack of a standard schema language in PGs complicates even more the automated process of mappings to RDF, especially in scale. Several existing approaches for PG-to-RDF transformation rely on manual or schemaless mappings, often resulting in loss of information, ambiguity, and limited scalability [9,10,15,21,41]. In addition, most research is focused on RDF-to-PG [5,8,11,13,18,33] or try to define a unified data model [4,29,47].

In this work, we propose a schema-driven and fully automated pipeline for transforming PG data into RDF. Assuming the availability of the structural schema, we present it using a custom XML Schema Definition (XSD), distinguishing node types, edge types, their properties, and constraints. Based on this XSD format, we then automatically generate both X3ML [28,30] or RML [14] mapping rules, which are executed using the X3ML Engine and RML Mapper, respectively, in order to produce the RDF triples. We create intermediate nodes for edge properties, enabling RDF semantics and extensibility. Finally, we evaluate our method on three synthetic datasets, demonstrating feasibility, integration, and transformation reproducibility, and we then compare the two engines in terms of scalability. Our approach ensures consistency between data and mappings and enables batch transformations that can easily adapt to evolving datasets. Therefore, it can be used in various data integration scenarios.

Outline. The rest of the paper is organized as follows. In Sect. 2, we cover the background of PGs, RDF, transformation languages, and their integration. In Sect. 3, we present the methodology, in Sect. 4 our key findings, and finally Sect. 5 concludes the paper and presents future work.

2 Background and Related Work

Recently, considerable effort has been made to support the interoperability, reconciliation, and data exchange between RDF [43] and PGs [1].

RDF and PGs. While RDF and PGs share foundations, their structural and semantic differences complicate the integration process [33]. RDF represents data as triples and is based on open-world semantics, and queried using SPARQL [44], whereas PGs have with key-value properties, and use languages like Cypher [31], Gremlin [40] and nGQL [17]. These differences make transformation challenging.

Definition 1 (RDF Graph). *An RDF graph is a tuple* $G_{rdf} = (I, B, L, T)$*, where* I *is a set of IRIs,* B *is a set of blank nodes,* L *is a set of literals,* T *is a finite set of RDF triples* (s, p, o)*, with* $s \in I \cup B$ *(the subject),* $p \in I$ *(the predicate), and* $o \in I \cup B \cup L$ *(the object).*

Definition 2 (Property Graph). *A property graph is a tuple* $G_{pg} = (V, E, \rho, \lambda, \pi)$, *where*

- V *and* E *are disjoint finite sets of nodes and edges, respectively,*
- $\rho : E \rightarrow (V \times V)$ *is a total function mapping edges to ordered pairs of nodes,*
- $\lambda : (V \cup E) \rightarrow 2^{\mathcal{L}}$ *is a partial function assigning to each node and edge a finite set of labels, each label belonging to a type, and*
- $\pi : (V \cup E) \times K \rightarrow R$, *is a partial function assigning to each node and edge a set* K *of key-value properties* R.

From Definitions 1 and 2, we observe that in RDF, edges are atomic, i.e., a triple (s, p, o) has no identifier and cannot natively carry metadata. On the other hand, in PGs, edges $e \in E$ are first-class entities with identifiers and can carry properties via $\pi(e, k)$. Therefore, formally, there exists no standard bijective function $f : G_{pg} \rightarrow G_{rdf}$ that preserves both structure and semantics when edge properties are present. This structural mismatch implies that any transformation $T : G_{pg} \rightarrow G_{rdf}$ must introduce auxiliary constructs to represent edges with properties.

Transformation Languages. Several mapping languages have been developed to support RDF graph construction. RML [14], a superset of R2RML [32], is the W3C standard for KG construction. X3ML [28,30] is an XML-based mapping language supporting consistent URI generation and human-readable rule definitions. Other tools like YARRRML [42] and SPARQL-Generate [26] offer flexible RDF generation from different sources. However, these are not natively designed for PGs and require workarounds which limit scalability and semantic precision.

Related Work. Most approaches examine the transformation of RDF-to-PG [5,8,11,13,18,33] or unified models [4,29,47]. A key challenge is semantic preservation; mixed property values (e.g., IRIs and literals under the same predicate) are often mishandled. Other difficulties include handling schema rules, multi-labeled nodes, and nested structures. Many methods ignore the graph schema or rely on manual mappings, which is not efficient. Some tools like G2GML [11], PGO [41], and PREC [10] use templates for mappings, but require human effort. Other approaches try to automate the process using rules [12], but produce large and imprecise RDF outputs [21]. Some other tools like PRSC [9] and ProGOMap [15] use the schema but depend on manual input and do not support schema constraints. A useful integration requires dynamic mapping, structural validation, and needs to support schema updates. Our work proposes a simple approach, addressing most of the aforementioned gaps by automatically generating X3ML and/or RML mapping rules from given or inferred XSD schema describing a PG dataset that is also automatically discovered using our tools.

3 Methodology

In this work, we define a simple, yet straightforward, mapping transformation T from a PG G_{pg} to an RDF graph G_{rdf} as:

$$T(V, E, \rho, \lambda, \pi) = \{(s, p, o)\} \cup \{(e_i, p_k, v) \mid \pi(e_i, k) = v\},$$

where e_i are introduced as RDF resources (URIs) for edge instances to retain metadata, as intermediate nodes.

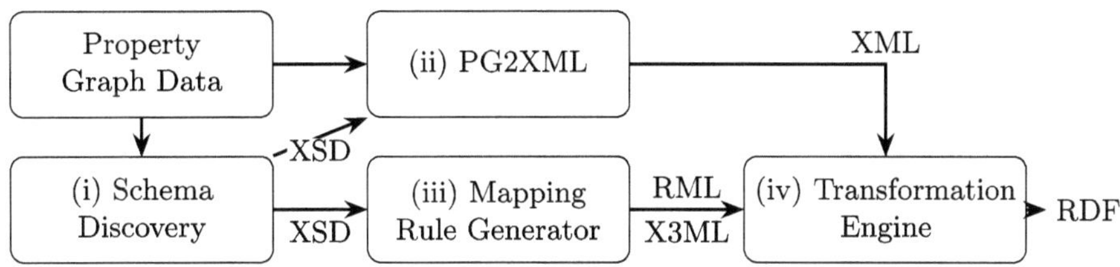

Fig. 1. PG2RDF pipeline: from property graph data to RDF via schema discovery, mapping rule generation, XML transformation, and the Transformation engine.

Following, we analyze our methodology for the automatic transformation of a PG dataset into RDF, under the assumption that the structural schema of the graph has been discovered or provided in XSD. Our approach follows the steps shown in Fig. 1: (i) schema extraction and formalization, (ii) structural annotation of PG instances in XML, (iii) automatic generation of X3ML or RML mapping rules, and (iv) materialization of PG data into RDF with the X3ML Engine [28,30] or the RML Mapper [14]. The goal is to automatically transform property graph data into RDF, while preserving structure and instance information.

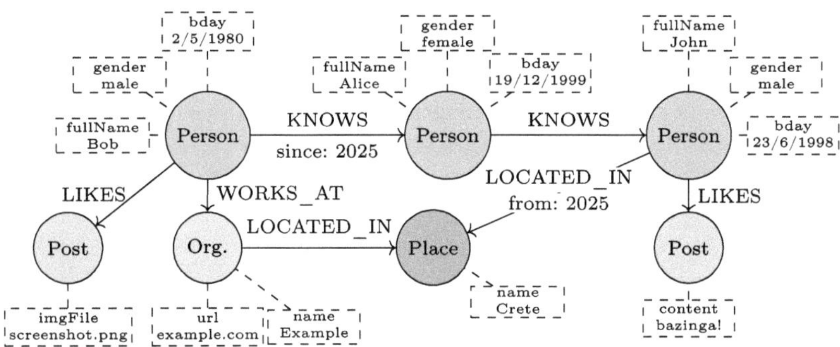

Fig. 2. Example Property Graph.

Property Graph Input. The input is a property graph G_{pg} (see Definition 2) consisting of nodes and edges with key-value properties. Nodes N may have a label λ and sets of properties P_n, while edges E represent relationships and also have properties P_e. This model is flexible but lacks structural constraints.

For presentation purposes, from now on, we assume that we have at most one label on each node and edge; however, extending them to multiple labels is straightforward. Further, we assume that there is already a schema present or discovered (more on this later) that does not change during data insertions. Figure 2 shows a small example of a property graph, illustrating the heterogeneity of PGs. Each entity (e.g., `Person`, `Post`, `Organisation`) is connected with edges (e.g., `KNOWS`, `LIKES`, `WORKS_AT`), and both nodes and edges have attributes. Not all instances of a type share the same schema – for example, some `Post` nodes may contain only a `content` field while others provide an `imgFile`. This structural variety poses challenges for RDF conversion, as the RDF datasets are often, but not always, accompanied by well-formed ontologies.

(i) Schema Extraction and Formalization. To infer the underlying structure, we use **PG-HIVE**[1], a schema discovery tool designed to extract the schema from property graphs. PG-HIVE clusters graph entities based on shared structural features and labels, using Locality-Sensitive Hashing (LSH) [16,45]. The result is a concise and accurate graph schema both in PG-schema [3] and XSD, capturing node types, edge types, properties, and constraints. For our approach in this paper we use the extracted XSD schema.

An **XSD PG-Schema** [35] S for a property graph G_{pg} is a tuple $S = (N_s, E_s, P)$. N_s is a finite set of *node types* where each $n \in N_s$ is represented as a complex type with an associated set of properties P_n. Similarly, E_s is a finite set of *edge types*, each $e \in E_s$ is defined as a complex type that includes a source node type $s_e \in N_s$, a target node type $t_e \in N_s$, and a set of properties P_e. The set P denotes the union of all properties defined across the nodes and edges, i.e., $P = \bigcup_{n \in N_s} P_n \cup \bigcup_{e \in E_s} P_e$. Each property $p \in P$ is a tuple: $p = (name, type, minOccurs, maxOccurs)$, where, $name$ is the property key, $type$ is the XSD datatype (e.g., `xs:string`, `xs:date`), $minOccurs \in \{0, 1\}$ denotes the minimum number of times the property may appear and $maxOccurs \in \mathbb{N}_{\geq 1} \cup \{\texttt{unbounded}\}$ denotes the maximum. Additionally, each node and edge type has `id` and `label` attributes.

This schema serves as a structural template for validating property graph data and generating XML documents that conform to the set XML_S, the collection of XML documents or fragments valid under S. Based on the output of PG-HIVE and the availability of the original schema, the discovered structure is given in a custom XSD, inspired by [35] with key modifications to support edge properties.

Each **node type** is defined as an <xs:complexType> and includes a sequence of <property> elements, one for each property that instances of this node type may carry. Each property contains metadata such as name, type (e.g., `xs:string`, `xs:date`), and cardinality constraints through `minOccurs` and

[1] https://github.com/sophisid/PG-HIVE.

`maxOccurs` attributes. They also carry `id` and `label` attributes, where both are used to generate unique URIs for each instance.

Similarly, **edge types** are defined as a separate <`xs:complexType`>, also with `id` and `label` attributes. In addition to the elements <`property`> that describe the properties of the edges, the types of edges include two <`choice`> elements: <`source`> and <`target`>. These indicate the types of nodes allowed for the origin and destination of the edge. This design enables flexibility in modeling heterogeneous relationships while ensuring compliance with a structured schema.

This meta-schema acts as a structural template, used for validating the PG data and for guiding the automatic generation of RDF mappings in the following steps. It explicitly encodes which properties are mandatory with the `minOccurs="1"` and which are optional `minOccurs="0"`, skipping missing values without violating schema constraints.

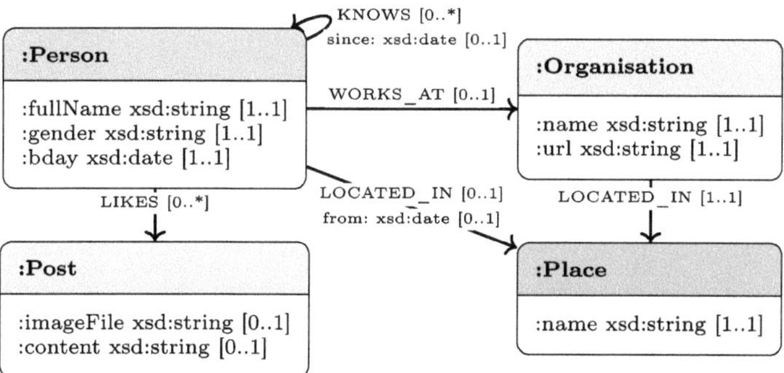

Fig. 3. XSD-based schema for the example PG.

The extracted schema from the graph in Fig. 2 is shown in Fig. 3. It captures the main types, along with their properties, relationships and constraints.

(ii) XSD Annotation of PGs (PG2XML). The generated XSD schema is used for the annotation of the graph instances. Each `complexType` entry captures the node N_s or edge E_s type, specifying the expected attributes. Assume that the PG is stored in a Neo4j[2]. For performing the serialization, for each node and edge instance, the property values are represented as text nodes under the corresponding XML element, while missing values are omitted.

Given a property graph $G_{pg} = (V, E, \rho, \lambda, \pi)$ and an XSD schema $S = (N_s, E_s, P)$, the **XSD annotation** of G_{pg} is a partial function $\alpha : V \cup E \rightarrow \text{XML}_S$, where XML_S denotes the set of XML documents or fragments that are valid with respect to the schema S. For every $x \in V \cup E$, there exists a type $t \in N_s \cup E_s$ such that $t \in \lambda(x)$. Each element x is serialized as an XML element corresponding to type t. For each property $p = (\mathit{name}, \mathit{type}, \mathit{minOccurs}, \mathit{maxOccurs}) \in$

[2] https://neo4j.com/.

P_t, the value is extracted through the function $\pi(x, name)$. Properties with $minOccurs = 1$ must be present in the XML serialization, while those with $minOccurs = 0$ are optional and may be omitted.

The resulting XML document $\alpha(x)$ is valid under the schema S. The parser processes the XSD and produces a set of XML documents that represent the PG instances that align with the schema. Each XML item encodes the node or edge instance accompanied with its properties of the corresponding PG entry. Also, for edge instances, the XML serialization includes elements that identify the source and target nodes to preserve the relation direction. We provide a simple example of the serialization in Fig. 4. Instances are grouped by the type into files `<NodeType>Group_k.xml` where k is the chunk index, having a batch size of 10000 instances by default, and each XML document is structured around grouped instances (e.g., PersonGroup, WorksAtGroup).

(iii) Creating Mapping Rules from Schema (XSD2X3ML, XSD2RML). Each node in the graph must be mapped to an RDF resource with a unique URI. Its properties become RDF literals, and its label is mapped to an RDF class. Similarly, each edge must be transformed into a meaningful relation. As edges can carry properties, the relation is modeled with an intermediate node. This design ensures schema-awareness, semantic interoperability, aligning with Linked Data best practices [20].

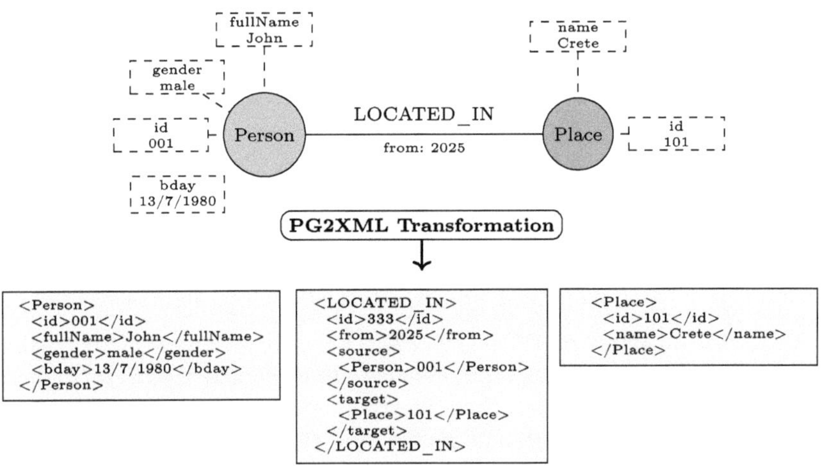

Fig. 4. XSD-based XML serialization of a simple PG instance.

Mapping Rule Generation. Let $S = (N_s, E_s, P)$ be an XSD schema, and let XML be a set of annotated PG instances that conform to S. The mapping generation process is a function $\mu : S \rightarrow \mathcal{M}$ that produces a set of mapping rules $\mathcal{M}$, such that:

- For every node type $n \in N_s$, $\mu(n)$ contains a domain mapping with XPath target //n, mapped to RDF class :<n> and for each property $p \in P_n$, a link mapping to predicate :<name>.
- For every edge type $e \in E_s$, regardless of whether it carries properties (P_e), $\mu(e)$ contains a domain mapping with XPath target //e/source/<SourceType>, of type :<SourceType>, an additional block linking the subject to an intermediate node i_e of class :intermediate, a nested additional entity from i_e to the target node via predicate :hasTarget, finally, for each property $p \in P_e$, another entity from i_e to a literal via predicate :<e>_<p> using XPath ../../p/text().
- Each URI is generated using a deterministic URI policy:

$$\texttt{URI}(x) = \texttt{base} + \texttt{/type/} + \lambda(x) + \texttt{/id/} + \pi(x, \texttt{id}),$$

  where '+' represents concatenation.

This ensures consistent and unique URIs across all documents. Given the schema in XSD format, our system automatically generates the respective X3ML or RML mapping definitions without manual effort. This process enables structural consistency between XML-annotated PG instances and RDF output.

In X3ML, each complexType results in a <mapping> block. For nodes, a <domain> is defined with XPath //<NodeType>, the RDF class is set to :<NodeType>, and each nested property is linked as a literal using a <link> block. For edges, we use an intermediate node of type :intermediate, identified by the edge's id, which connects the source and target entities via nested <additional> blocks. Any edge properties are also attached to this intermediate node, ensuring a uniform and expressive structure.

In the case of RML, the same logic is expressed declaratively through rr:TriplesMap definitions in Turtle syntax. Each node or edge type becomes a separate mapping block with an XPath iterator and subject URI template. For edges, the intermediate node is treated as a first-class RDF resource, linked to source and target via ex:source and ex:target, while edge properties are attached via additional rr:predicateObjectMap statements.

The intermediate representation design has several advantages. First, it ensures a standard structure, as all edges are treated the same, resulting to consistent query patterns. Second, it allows edge metadata to be expressed as RDF properties of the intermediate node, without changing the structure. Third, it aligns with best practices for RDF modeling, where relations are treated as first-class entities when additional context is needed [20].

To explain further the use of intermediate node, in our example of Fig. 2, the PG relation **LOCATED_IN** with the property **since:2025** cannot be represented with one triple like **:John :LOCATED_IN :Crete.**. To support this, the intermediate node represents the edge instance and we can reconstruct the previous triple as in Fig. 5. This ensures that the edge's properties are not lost during the transformation process. The intermediate node acts as a proxy of the original edge and additional statements about the relationship can be attached.

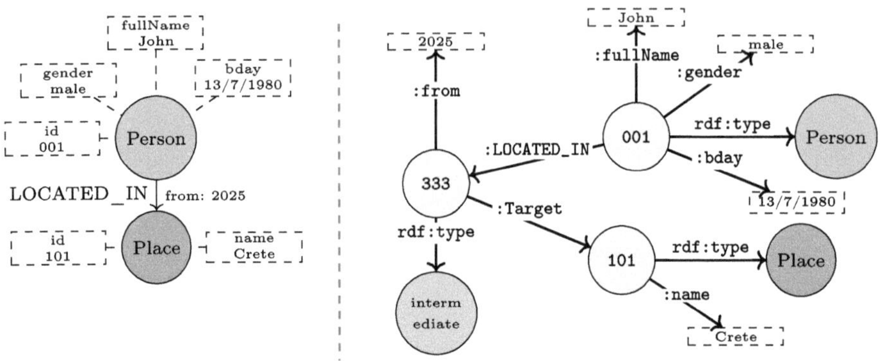

Fig. 5. Representation of a Property Graph (left) and its RDF (right) transformation using PG2RDF.

(iv) Automatic Transformation and RDF output. Following, we need to transform the data using either X3ML Engine [28,30] or RML Mapper [14]. For scalability, we run the materialization process in batches. Each XML file is processed using the previously generated mappings and the standard transformation policies. These policies ensure that the generated URIs include type-specific identifiers, which describe the type of instances and guarantees consistent linking.

This process is schema-aware and ensures the generation of deterministic mapping rules from the inferred or provided schema. Since the URIs of instances are uniquely constructed based on their `id` and `type`, the process supports batch materialization. RDF triples generated from multiple XML files are linked without duplication when loaded in the Blazegraph [39]. If an entity reappears in another chunk, its URI remains the same and is further enriched rather than duplicated.

The engine produces RDF output in TriG format, encoding both the semantic types, entities, and their relations. This pipeline is fully automated, making it suitable for large datasets through batch processing. All artifacts (XSD, XML, X3ML, RML, TriG) can be regenerated deterministically from the same input. So, the methodology is dataset-agnostic and can support data insertions, if the schema remains unchanged, to the RDF graph, without the need for manual intervention.

X3ML Time Complexity. Internally, the engine performs a series of XPath evaluations and applies the mappings to each XML element. Let N_i denote the number of nodes in the input XML, M the number of mappings, L the number of links per mapping, and P the number of XPath expressions per link. The transformation process iterates over all XML elements and evaluates all rules. In the worst case, where each rule applies to every element, the time complexity is $\mathcal{O}(M \cdot L \cdot P \cdot N_i)$. Although X3ML natively uses caching to avoid unnecessary recalculations, it still processes the transformation in sequence. This creates a bottleneck when dealing with large XML files. In our case, where nodes and edges are millions and have many properties, this complexity might incur significant overhead to the whole process. To mitigate this, in our approach we leverage Apache Spark to distribute the transformation workload across multiple workers, significantly improving scalability for large datasets.

RML Time Complexity. RML Mapper also iterates all XML elements and apply rules defined as TriplesMaps. Each rule has an XPath and one or more predicate-object maps. Assuming M mapping rules and N_i input elements, the complexity is $\mathcal{O}(M \cdot N_i)$, assuming one rule per matching element. However, if the RML mapping includes joins or nested object maps (e.g., parent-child or reference lookups), the complexity may increase to $\mathcal{O}(M \cdot N_i \cdot J)$, where J is the average number of join evaluations per rule. In our approach, mappings are flat, so we do not have joins, keeping complexity close to linear to the number of elements.

Overall Time Complexity of PG2RDF. We analyze the complexity of the post-schema discovery stages described in Fig. 1. Let n be the number of graph nodes, e the number of graph edges, p the average number of properties per node or edge, T_n the number of discovered node types, T_e the number of discovered edge types, $M = T_n + T_e$ the total number of mapping rules (one per type), L the average number of links per mapping rule, P the average number of XPath expressions per link. Then, following the complexity of each stage:

- **PG2XML:** $\mathcal{O}((n + e) \cdot p)$ – serialization of each instance into XML.
- **XSD2X3ML:** $\mathcal{O}(M \cdot p)$ – rule creation per type and property.
- **X3ML Engine :** $\mathcal{O}(M \cdot L \cdot P \cdot (n + e))$
- **RML Mapper:** $\mathcal{O}(M \cdot (n + e))$ – assuming no joins.

Thus, the total complexity of the transformation pipeline is $\mathcal{O}((n + e) \cdot p + M \cdot L \cdot P \cdot (n + e))$, with M, L, and P typically much smaller compared to the number of instances $n+e$, the dominant cost remains in the RDF materialization step, especially when using X3ML. In contrast, RML has lower complexity due to flatter mapping rules.

4 Evaluation

In this section, we present the experimental evaluation of PG2RDF, focusing on transformation performance, scalability and the integration quality.

Table 1. PG Dataset Statistics.

Dataset	Nodes	Edges	Node Types	Edge Types	Node Patterns	Edge Patterns	Max Properties	Avg. Properties
LDBC	3,181,724	17,256,038	7	17	9	23	10	~5
FIB25	802,473	1,625,428	4	5	31	9	11	~6
MB6	486,267	961,571	4	5	52	9	11	~6

Setup. All experiments were conducted on a 4-node Spark Standalone cluster (Ubuntu 20.04.2 LTS, 38 cores/machine, SSDs, Gigabit Ethernet), using Apache Spark 3.4.1, Scala 2.12.10, Neo4j (Community) 4.4.0 and Blazegraph 2.1.5. All configurations and code are publicly available for reproducibility[3].

Datasets. To evaluate the performance of the PG2RDF pipeline, we conducted experiments on three synthetic graph datasets with different structure and scale: LDBC Social Network [36], FIB25 [37], and MB6 [38]. These are widely used benchmark datasets in graph data management community and come with known schemas. They were selected to capture both realistic graph workloads and heterogeneous schema complexity. The characteristics of the dataset are summarized in Table 1. We then compared our approach in terms of scalability of X3ML Engine [28] and RML Mapper [14].

Metrics. For each dataset, we measured (a) the total number of RDF triples, (b) transformation time (post-schema extraction) using X3ML Engine and RML Mapper across 1, 2, and 4 worker nodes to assess scalability, and (c) the number of RDF classes and relations produced to evaluate integration quality.

Table 2. Number of triples, transformation time, classes, and relations per dataset.

Dataset	Triples	X3ML Time (s)			RML Time (s)			Cls.	Rel.
		1 node	2 nodes	4 nodes	1 node	2 nodes	4 nodes		
LDBC	49,562,383	301,299	266,982	130,976	251,345	191,122	98,563	8	37
FIB25	4,799,554	29,411	26,376	21,675	23,241	18,331	13,563	6	41
MB6	2,985,543	20,906	18,012	12,543	15,371	11,005	8,325	6	59
Integrated	55,342,378	362,205	289,165	162,732	279,241	215,412	121,037	14	93

Results. A summary of the results for each individual dataset and the integrated graph, is given in Table 2. Our experimental results indicate that PG2RDF can handle different sizes of property graphs accurately; however, it requires significant time to transform large datasets. For example, transforming a graph into RDF with X3ML using one worker, that contains 486K nodes and 961K edges, such as MB6, requires about 6 h. FIB25 requires approximately 8 h, and LDBC, being larger and having more properties, needed 3 and a half days to transform. The integrated graph, containing more than 55 million triples required more than four days. RML performs slightly faster, but still requires a significant amount of time. For example, using one worker, LDBC took more than 2 and a half days, FIB25 6.5 h, MB6 4.3 h and the integrated graph over 3 days. These results show that the process had significant bottlenecks due to the sequential evaluation of XPath expressions over a large number of nodes, edges, and their properties, which can be justified from the time complexity as discussed in Sect. 3.

[3] https://github.com/sophisid/PG2RDF.

Scalability via Parallel Execution. To improve performance, we experimented with a distributed setup by running the X3ML Engine and RML Mapper on four separate machines using Apache Spark. Figure 6 shows the speedup achieved with 1, 2 and 4 nodes for both X3ML and RML across all datasets. This resulted in significant reductions in processing time across all datasets. For example, LDBC's transformation time dropped by over 56%, completing in 1.5 days. As MB6 and FIB25 also had a 40% improvement, demonstrating that using distributed mechanisms can effectively improve computational bottlenecks of X3ML. RML consistently outperformed X3ML, with up to 60% faster transformations. The speedup is sublinear due to overheads in distributed coordination, but the results confirm that parallel execution significantly mitigates the computational cost of processing large graphs.

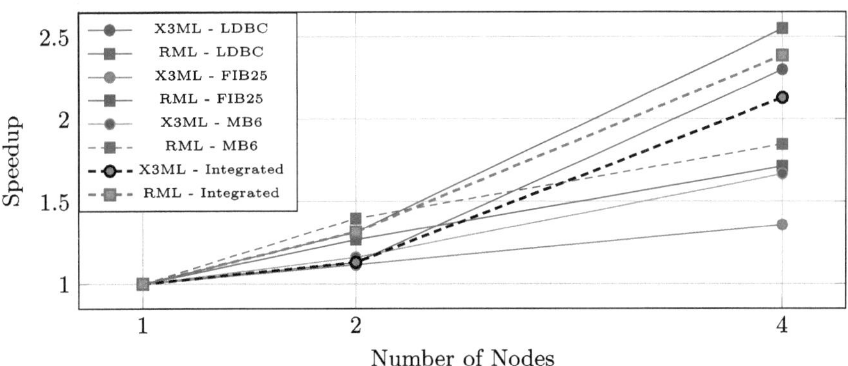

Fig. 6. Speedup of PG2RDF transformation with X3ML and RML using 1, 2, and 4 nodes across individual and integrated datasets.

Integrated Dataset. To evaluate PG2RDF's integration capabilities, we combined the three datasets into a single graph, loaded into a shared Neo4j instance, and applied the PG2RDF pipeline. Table 2 shows the results. This was performed by loading all datasets into a shared Neo4j instance and then applying the same PG2RDF pipeline to the integrated graph. Table 2, shows the statistical results. Unlike each individual dataset measurements, the values in the integrated dataset **(LDBC+FIB+MB6)** are not computed as a direct sum of each dataset's values. The total number of triples in the integrated graph is less than the sum of the individual triples. This reduction occurs due to the overlap in schema and instances across datasets. Specifically, RDF resources such as properties (e.g., `:hasTarget`, `rdf:type`) and ontology classes reused in multiple datasets are deduplicated during integration. Similarly, entities (nodes or edges) that appear in more than one dataset are merged based on their identifiers or matching features.

The result number of classes and relations shows the PG2RDF's ability to perform semantic integration, by unifying equivalent resources while avoiding redundancy. As a result, the output RDF graph is more compact, containing 14 unique classes and 93 unique relations, compared to the summary of the datasets from the separate transformation processes. So we can assume that, $T_{\text{int}} \leq T_{\text{LDBC}} + T_{\text{FIB25}} + T_{\text{MB6}}$ Formally, for a set of n input datasets $D_1, D_2, \ldots, D_n$, the total number of triples in the integrated RDF graph can be expressed as:

$$\mathrm{T}_{\text{integrated}} = \left(\sum_{i=1}^{n} \mathrm{T}_{D_i}\right) - \text{Redundancy}_{\text{schema}} - \text{Redundancy}_{\text{instances}},$$

where T_{D_i} is the number of triples generated from dataset D_i, $\text{Redundancy}_{\text{schema}}$ accounts for reused classes and properties across datasets, $\text{Redundancy}_{\text{instances}}$ accounts for overlapping or equivalent entities that are merged into a single RDF resource.

In terms of execution time, the integration process remains computationally intensive. On a single node the pipeline requires ~100 hours to complete and on four nodes ~45 h, though the speed-up is not linear.

Recoverability and Losslessness. This transformation workflow is structurally lossless; each node and edge is deterministically mapped into RDF, preserving their type, properties, and identifiers. The use of intermediate nodes for edges ensures that no relationship metadata is lost. Each node v with ID id_v, label $\lambda(v)$, and properties $\pi(v)$ maps to an RDF entity with unique URI $: id_v/\lambda(v)$, type $rdf : type : \lambda(v)$, and triples $(: id_v/\lambda(v), : k_i, r_i)$. Each edge e with ID id_e, label $\lambda(e)$, source/target (s, t), and properties $\pi(e)$ maps to an intermediate node $: id_e/intermediate$ with triples linking s, t, and properties. This preserves all PG information (losslessness). The inverse mapping reconstructs v from its URI and triples, and e from its intermediate node, ensuring $T^{-1}(T(G_{PG})) = G_{PG}$ (recoverability), assuming a static schema and unique IDs. Although we do not use semantic vocabularies, and ontology alignment is not guaranteed, the transformation is lossless in terms of instance-level information, and can be reversible under the given schema assumptions.

5 Conclusion and Future Work

In this work, we proposed a fully automated pipeline for transforming PGs into RDF by using the XSD-based schema and generating executable X3ML and RML mappings. Our method ensures consistency, supports batch processing, and reduces manual effort. While our approach establishes a foundational pipeline, remaining challenges include (a) scalability, (b) semantic enrichment, (c) ontology matching, and (d) schema compatibility.

The transformation of a PG to RDF is executed sequentially with the X3ML Engine or RML Mapper. Both are single threaded, and load all data in memory, introducing performance bottlenecks at scale. Integrating it with distributed frameworks could address this. Additionally, our current implementation assumes a standardized XSD schema, whereas, in practice, the schemas might be more complex, having nested types, mixed content, or ambiguous structures. Future work should handle any type of XSD or PG schema to support more integration scenarios. Furthermore, while our system generates syntactically correct RDF triples, it does not incorporate ontologies. Future versions could integrate existing vocabularies like CIDOC CRM[4], FOAF[5], Schema.org[6], and enable semi-automated matching with PG elements. Incorporating vocabulary matching would increase expressiveness and interoperability.

This work provides a promising baseline for automated PG-to-RDF transformation, and vice versa. We envision this framework as part of a broader ecosystem for constructing knowledge graphs from different sources. Integrating with existing KG construction engines, supporting mapping validation and enabling both PG to RDF and back transformations, with provenance and versioning support would make this approach as a general purpose solution for semantic data integration.

Acknowledgments. The work reported in this paper is implemented in the framework of H.F.R.I call "Basic research Financing (Horizontal support of all Sciences)" under the National Recovery and Resilience Plan "Greece 2.0" funded by the European Union NextGenerationEU (H.F.R.I. TALE Project, Number: 16819).

References

1. Angles, R.: The property graph database model. In: AMW (2018)
2. Angles, R., Arenas, M., Barceló, P., Hogan, A., Reutter, J., Vrgoč, D.: Foundations of modern query languages for graph databases. ACM CSUR **50**(5), 1–40 (2017)
3. Angles, R., et al.: PG-schema: schemas for property graphs. Proc. ACM Manag. Data **1**(2), 198:1–198:25 (2023). https://doi.org/10.1145/3589778
4. Angles, R., et al.: Multilayer graphs: a unified data model for graph databases. In: GRADES-NDA (2022)
5. Angles, R., Thakkar, H., Tomaszuk, D.: Mapping RDF databases to property graph databases. IEEE Access **8**, 86091–86110 (2020)
6. Angles, R., Tomaszuk, D., Thakkar, H.: RDF and property graphs interoperability: status and issues (2019)
7. Antoniou, G., Van Harmelen, F.: A Semantic Web Primer. MIT press (2004)
8. Brandizi, M., Singh, A., Hassani-Pak, K.: Getting the best of linked data and property graphs: rdf2neo and the KnetMiner use case. In: SWAT4LS (2018)
9. Bruyat, J., Champin, P.A., Médini, L., Laforest, F.: PRSC: from PG to RDF and back, using schemas. Semant. Web **15**(6), 2555–2595 (2024)
10. Bruyat, J., Champin, P.A., Médini, L., Laforest, F.: PREC: semantic translation of property graphs. arxiv:2110.12996 (2021)

---

[4] http://www.cidoc-crm.org/.
[5] http://xmlns.com/foaf/spec/.
[6] https://schema.org/.

11. Chiba, H., Yamanaka, R., Matsumoto, S.: G2GML: graph to graph mapping language for bridging RDF and property graphs. In: ISWC, pp. 160–175 (2020)
12. Das, S., Srinivasan, J., Perry, M., Chong, E.I., Banerjee, J.: A tale of two graphs: property graphs as RDF in oracle. In: EDBT, pp. 762–773 (2014)
13. Di Pierro, D., Ferilli, S., Redavid, D.: LPG-based knowledge graphs: a survey, a proposal and current trends. Information **14**(3), 154 (2023)
14. Dimou, A., Sande, M.V., Colpaert, P., Verborgh, R., Mannens, E., de Walle, R.V.: RML: a generic language for integrated RDF mappings of heterogeneous data. In: Workshop on Linked Data on the Web, vol. 1184 (2014)
15. Fathy, N., Gad, W., Badr, N., Hashem, M.: ProgoMap: automatic generation of mappings from property graphs to ontologies. IEEE Access **9**, 113100–113116 (2021)
16. Gionis, A., Indyk, P., Motwani, R.: Similarity search in high dimensions via hashing. In: VLDB, pp. 518–529 (1999)
17. Graph, N.: nGQL - nebula graph query language (2024). https://docs.nebula-graph.io/. Accessed 27 June 2025
18. Haihong, E., Han, P., Song, M.: Transforming RDF to property graph in hugegraph. In: ICEMIS (2020)
19. Hartig, O.: Reconciliation of RDF* and property graphs. arXiv:1409.3288 (2014)
20. Heath, T., Bizer, C.: Linked Data: Evolving the Web into a Global Data Space. Morgan & Claypool Publishers (2011)
21. Khayatbashi, S., Ferrada, S., Hartig, O.: Converting property graphs to RDF: a preliminary study of the practical impact of different mappings. In: GRADES-NDA, pp. 1–9 (2022)
22. Kondylakis, H., et al.: Property graph standards: state of the art & open challenges. In: VLDB (2025)
23. Kondylakis, H., Efthymiou, V., Troullinou, G., Ymeralli, E., Plexousakis, D.: Property graphs at scale: a roadmap and vision for the future (short paper). In: CAiSE Workshops, vol. 521, pp. 180–185 (2024)
24. Kondylakis, H., Plexousakis, D.: Exelixis: evolving ontology-based data integration system. In: SIGMOD, pp. 1283–1286. ACM (2011)
25. Kondylakis, H., Plexousakis, D.: Ontology evolution: assisting query migration. In: Atzeni, P., Cheung, D., Ram, S. (eds.) ER 2012. LNCS, vol. 7532, pp. 331–344. Springer, Heidelberg (2012). https://doi.org/10.1007/978-3-642-34002-4_26
26. Lefrançois, M., Zimmermann, A., Bakerally, N.: Flexible RDF generation from RDF and heterogeneous data sources with SPARQL-generate. In: Ciancarini, P., et al. (eds.) EKAW 2016. LNCS (LNAI), vol. 10180, pp. 131–135. Springer, Cham (2017). https://doi.org/10.1007/978-3-319-58694-6_16
27. Lin, S., et al.: Prototypical graph contrastive learning. IEEE Trans. Neural Networks Learn. Syst. **35**(2), 2747–2758 (2024)
28. Marketakis, Y., Minadakis, N., Samaritakis, G., Doerr, M.: X3ML mapping framework for information integration in cultural heritage and beyond. IJDL **18**(4), 301–319 (2017)
29. Marx, M., Krötzsch, M., Thost, V.: Logic on mars: ontologies for generalised property graphs. In: IJCAI, vol. 2017, pp. 1188–1194 (2017)
30. Minadakis, N., et al.: X3ML framework: an effective suite for supporting data mappings. In: Ronzino, P. (ed.) Workshop on Extending, Mapping and Focusing the CRM, vol. 1656, pp. 1–12 (2015)
31. Neo4J: Cypher query language (2023). https://neo4j.com/developer/cypher/

32. Pinkel, C., Binnig, C., Haase, P., Martin, C., Sengupta, K., Trame, J.: How to best find a partner? An evaluation of editing approaches to construct R2RML mappings. In: Presutti, V., d'Amato, C., Gandon, F., d'Aquin, M., Staab, S., Tordai, A. (eds.) ESWC 2014. LNCS, vol. 8465, pp. 675–690. Springer, Cham (2014). https://doi.org/10.1007/978-3-319-07443-6_45
33. Rabbani, K., Lissandrini, M., Bonifati, A., Hose, K.: Transforming RDF graphs to property graphs using standardized schemas. SIGMOD **2**(6), 1–25 (2024)
34. Rodriguez, M.A., Neubauer, P.: The graph traversal pattern. In: Graph Data Management: Techniques and Applications, pp. 29–46. IGI Global (2012)
35. Sarkar, A.: Conceptual level design of semi-structured database system: graph-semantic based approach. CoRR arxiv:1202.4532 (2012)
36. Szárnyas, G., et al.: The LDBC social network benchmark: business intelligence workload. Proc. VLDB Endow. **16**(4), 877–890 (2022)
37. Takemura, S.Y., et al.: Synaptic circuits and their variations within different columns in the visual system of drosophila. Proc. Natl. Acad. Sci. **112**, 13711–13716 (2015)
38. Takemura, S.Y., et al.: A connectome of a learning and memory center in the adult drosophila brain. eLife **6** (2017)
39. Thompson, B., Personick, M., Cutcher, M.: The bigdata® RDF graph database. In: Linked Data Management, pp. 221–266. Chapman and Hall/CRC (2016)
40. TinkerPop, A.: Gremlin: A graph traversal language (2023). https://tinkerpop.apache.org/gremlin.html
41. Tomaszuk, D., Angles, R., Thakkar, H.: PGO: describing property graphs in RDF. IEEE Access **8**, 118355–118369 (2020)
42. Verborgh, R., Herwegen, J.V., de Meester, B.: YARRRML: yet another RDF mapping language (2018). https://rml.io/yarrml/. Accessed 11 June 2025
43. W3C RDF Working Group: Resource description framework (RDF) (2014). https://www.w3.org/RDF/. W3C Recommendation
44. W3C SPARQL Working Group: SPARQL 1.1 overview (2013). https://www.w3.org/TR/sparql11-overview/. W3C Recommendation
45. Wang, H., Cao, J., Shu, L., Rafiei, D.: Locality sensitive hashing revisited: filling the gap between theory and algorithm analysis. In: CIKM, pp. 1969–1978 (2013)
46. Wood, D., Lanthaler, M., Cyganiak, R.: RDF 1.1 concepts and abstract syntax. W3C Recommendation, W3C (2014)
47. Zhang, R., Liu, P., Guo, X., Li, S., Wang, X.: A unified relational storage scheme for RDF and property graphs. In: WISA, pp. 418–429 (2019)

Construction, Access, and Application of a Knowledge Base for Sustainable Aquafood Communication

Yannis Marketakis[1,2(✉)], Yannis Tzitzikas[1,2], Alessandro Petrocelli[3], Sara Pittonet Gaiarin[4], Siân Astley[5], Christine Absil[6], Michelle Boonstra[6], Tim Huntington[7], Tracy Murai[7], and Nicolas Baily[8]

[1] Institute of Computer Science, FORTH-ICS, Heraklion, Greece
{marketak,tzitzik}@ics.forth.gr
[2] Computer Science Department, University of Crete, Heraklion, Greece
[3] Communication Platforms and Online Solutions - COMMpla, Pisa, Italy
a.petrocelli@trust-itservices.com
[4] Trust-IT services, Pisa, Italy
s.pittonet@trust-itservices.com
[5] European Food Information Resources - EuroFIR AISBL, Brussels, Belgium
sa@eurofir.org
[6] Clupea Consultancy, Nijmegen, The Netherlands
Michelle@catchwelfareplatform.com
[7] Poseidon Aquatic Resource Management Ltd., Lymington, UK
{tim,tracy}@consult-poseidon.com
[8] Quantitative Aquatics Inc., Los Baños, Philippines
n.bailly@q-quatics.org

Abstract. In the information age, making informed decisions about food should be straightforward. Consumers increasingly seek trustworthy information on sustainability, nutrition, provenance, and the risks and benefits associated with the products they consume. However, in the case of aquafood, the intersecting complexities of health considerations, environmental sustainability, and climate impacts create a challenging landscape for both consumers and industry stakeholders. To address this, the VeriFish project introduces a robust framework of verifiable indicators designed to support informed and responsible choices among retailers, producers, fishers and citizens. This paper presents the data-driven methodology for constructing a semantic knowledge base that underpins the VeriFish indicator framework by integrating heterogeneous data sources. We outline the key challenges and requirements in building this knowledge base, describe its technical implementation, and detail the mechanisms developed to support discovery and access. Finally, we showcase dedicated applications built on top of the knowledge base that enable practical use of the indicator framework.

Keywords: aquafood sustainability · verifiable indicators · aquafood knowledge base

H. Takeda et al. (Eds.): IJCKG 2025, LNAI 16297, pp. 374–389, 2026.
https://doi.org/10.1007/978-981-95-5009-8_25

1 Introduction

In today's information-rich society, consumers are increasingly expected to make informed decisions about the food they purchase and consume. This demand goes beyond taste and price to include deeper concerns about sustainability, nutrition, and provenance, and the health implications of food products as shown in [14] and [16]. Public awareness of environmental degradation, climate change, and ethical production practices have grown, prompting consumers to seek transparency and accountability from the food industry. As a result, food systems are under pressure to provide reliable, accessible, and meaningful information to support responsible consumer choices.

However, when it comes to aquafood, a category encompassing fish and other aquatic organisms used for food, this task becomes especially complex. Despite its great advantages, aquafood is frequently disregarded or lumped together with meat [2]. Aquafood occupies a unique position at the intersection of human health, environmental sustainability, and economic livelihoods. On the one hand, it is praised as a source of high-quality nutrition and as a low-carbon alternative to terrestrial meat. On the other hand, concerns around overfishing [6], bycatch [7], supply chain opacity [13], and carbon emissions complicate sustainability assessments. The resulting information landscape is fragmented and often inaccessible to end users. In response to these challenges, the VeriFish project[1] proposes a solution: a structured framework of verifiable indicators designed to integrate diverse sustainability, nutritional, and provenance data into a coherent knowledge infrastructure. The framework aims to empower a range of stakeholders, such as retailers, producers, fishers, and consumers, to make choices aligned with the principles of environmental stewardship, human health, and socioeconomic transparency.

The developed framework [9] seeks to provide clarity on different aspects of interest to communicate sustainability to citizens. By integrating three foundational pillars: (a) environmental, (b) nutrition and health, and (c) social and economic, it aims to provide a robust, multi-dimensional approach to assessing and communicating sustainability. The framework is built on a tiered data system to accommodate diverse needs and capacities of aquafood actors. Tier 1 comprises publicly available data from global or regional repositories, and contains highly accessible, free, and scientifically validated data sources. Tier 2 contains data provided by the fisheries and aquaculture production value chain, specifically relating to their production methods, products, and processes.

To facilitate the development of the indicator framework and support the implementation of applications on top of it requires constructing a knowledge base that will semantically integrate data from the different data sources (under Tier 1) and expose them in a uniform manner. In this paper, we describe the details for constructing such a knowledge base, and also elaborate on the main challenges and how they were addressed. We provide the details of the workflow that has been developed for semantically integrating heterogeneous data sources,

[1] Horizon Europe, GA number: 101156426 https://verifish.info/.

as well as the implemented facilities for searching and accessing them. Finally, we discuss the VeriFish-related applications, namely, the factsheets and the mobile application, that utilize the contents of the knowledge base, and report on two indicative use cases.

The paper is organised as follows: Sect. 1 introduces the paper's objectives and describes the main motivation of our work, Sect. 2 provides details about the VeriFish Indicator Framework, Sect. 3 discusses the main requirements, challenges, and related works, Sect. 4 provides the details for the construction and use of the knowledge base, Sect. 5 describes the applications that use the knowledge base, and Sect. 6 concludes and identifies directions for future work.

2 The VeriFish Sustainability Indicator Framework

VeriFish project kicked off in May 2024, planning to develop, in the subsequent 24 months, a series of different outcomes including a prototype mobile application and a series of media products and awareness campaigns, the establishment of a community of practice of organizations and stakeholders committed to promoting sustainable aquafood, an EU Good Practice recommendation for how to efficiently organize sustainable aquafood consumption campaigns, published as a CEN Workshop Agreement (CWA). At the core of these outcomes lies a dynamic framework for the improvement of communication and understanding of sustainable aquafood production and consumption.

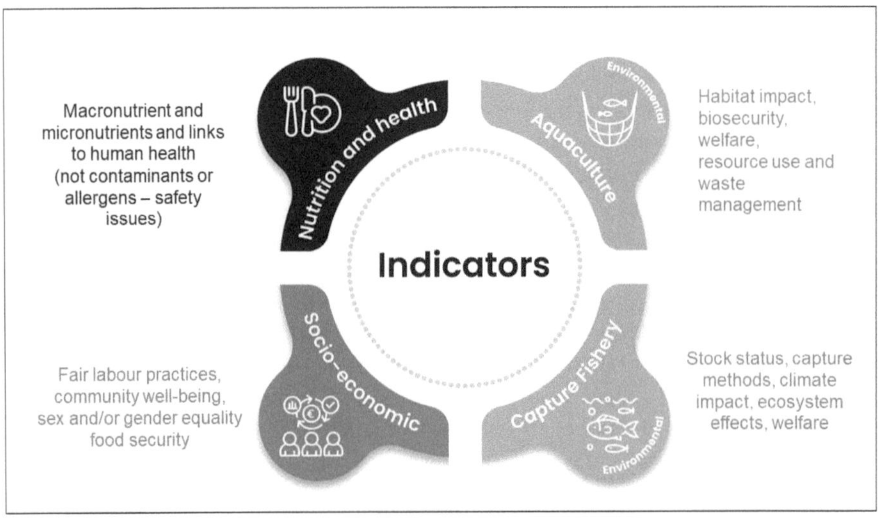

Fig. 1. The VeriFish Indicator Framework.

The indicator framework (shown in Fig. 1) is built on top of three core pillars: (a) environmental, (b) nutrition and health, and (c) socio-economic. The first pillar is further divided into capture fisheries (i.e., fish harvested from their natural habitats) and aquaculture (i.e., farmed fish). This distinction was made because of their distinct ecological impacts, management practices, and sustainability challenges. Below, we describe the main contents and indicators of the three pillars. The complete list of indicators and their relevant data sources can be found in the corresponding deliverables of the project [1,9].

The environmental pillar focuses on minimizing ecological harm and promoting sustainable resource use within capture fisheries and aquaculture, leveraging scientifically validated indicators to measure and communicate key aspects, including stock status, climate and ecosystem impacts, and waste and pollution. This pillar emphasizes ecological aspects, such as stock availability, species biology, habitats and adaptability, allowing consumers to make more informed decisions that protect aquatic ecosystems while maintaining productivity. Overall, this pillar consists of 34 distinct indicators, such as stock status, fishing mortality, bycatch, habitat impact, GreenHouse Gas (GHG) emissions, maximum stocking density, food conversion ratio, etc.

The nutrition and health pillar highlights the critical role of aquafood in supporting human health and addressing dietary needs. This pillar integrates data on the nutritional composition of aquafood products enabling actors to understand and communicate health benefits and consumers to make more informed decisions about consumption. Key areas of focus include nutrient quality and dietary impact on health. This pillar supports the development of guidelines emphasizing the importance of aquafood in achieving balanced and sustainable diets. Overall, this pillar consists of 36 indicators such as omega-3 fatty acids levels, sodium, saturated fat, micronutrients, etc.

The socio-economic pillar highlights the importance of fair and responsible practices in production of aquafood, focusing on people, communities, and governance systems that underpin supply chains. This pillar integrates data on labor rights, working conditions, governance, and economic contributions to provide a clearer picture of social risks and opportunities in both wild-capture and aquaculture systems. It supports users in identifying potential human rights concerns, promoting equitable employment, and recognizing the role of aquafood in supporting livelihoods. Key areas of focus include decent work, social inclusion, transparency, and community well-being. This pillar underpins the development of responsible sourcing guidance and enhances the credibility of sustainability claims across the sector. Overall, this pillar consists of 14 distinct indicators such as illegal, unregulated, and unreported fishing vessels, wages and working conditions, workers' rights, prevention of child labor, etc.

To support the development of the indicator framework, a wide array of publicly available data sources can be leveraged. These include scientific databases, regulatory reports, sustainability assessments, datasets, statistics reports, and many more. Despite their abundance, these data sources are often heterogeneous in format, scope, and terminology, making them difficult to integrate and use efficiently. To harness their full potential, a unifying semantic layer is needed that can organize, align, and interconnect the relevant information.

3 Requirements and Challenges

A fundamental requirement for constructing the knowledge base is ensuring access to publicly available and credible data sources. The reliability and impact of the VeriFish indicator framework heavily depend on the quality and transparency of the underlying data. This means prioritizing data that is not only openly accessible but also sourced from authoritative institutions and well-established scientific repositories. Such sources enhance the traceability and verifiability of the indicators, which is crucial for building trust among stakeholders and enabling reproducibility of results. Moreover, leveraging open data aligns with the FAIR principles [20], facilitating broader reuse of the knowledge base in both research and policy contexts.

Developing a robust knowledge base to support the VeriFish indicator framework presents a series of technical and conceptual challenges arising from the nature of the data sources and the complexity of the indicators. A primary challenge lies in handling the heterogeneity of data formats. Information relevant to the framework originates from a wide range of sources, including datasets, databases, spreadsheets, APIs, reports, etc., each with its own structure, schema, and access mechanism. Harmonizing them into a coherent, machine-readable form requires the definition of schema mappings for transforming them with respect to a target ontology.

A second challenge is addressing the thematic diversity of the data. The indicator framework encompasses multiple dimensions of sustainability, grouped into the three core pillars. Each pillar is associated with distinct conceptual models, data standards and domain-specific vocabularies. This diversity introduces the need for careful semantic modeling and data integration, ensuring that data from different domains can coexist within a unified, interoperable structure while preserving their domain-specific meanings.

Another critical requirement is ensuring that the knowledge base remains current and valid over time. The data sources that inform the sustainability indicators are inherent dynamic and subject to scientific advancements, regulatory updates and revisions. Consequently, the knowledge base must be equipped with mechanisms to support its continuous evolution and integration of timely data resources in the knowledge base.

Finally, the constructed knowledge base must be effectively accessed and utilized by external applications and services. This entails the provision of standardized access mechanisms such as SPARQL endpoints and RESTful APIs. Such interfaces will support flexible querying and retrieval of relevant information, while abstracting the underlying complexity of the knowledge representation.

3.1 Related Work

Several efforts in the literature have addressed the construction of knowledge bases relevant to aquafood sustainability. Regarding nutrition, initiatives such as FoodOn [3] and WikiFCD [15] demonstrate how ontologies can harmonize food composition data across repositories, while systems like NutriBase [19] integrate nutrient information. For aquatic foods, FishBase has recently expanded[2] to include empirical nutrient datasets for thousands of fish species, providing an empirical foundation for linking aquafood to health and nutrition. On the environmental side, there are tools like Seafood Carbon Emissions Profiling Tool[3] that provide structured data on carbon and ecosystem impact.

Closest to our work is SeafoodWatch[4] that guides consumer choices through an indicator-based scoring. The sustainability assessment is carried out in terms of the targeted species, the fishing or farming method, the location, and other related concerns and classifies aquafood into three categories: (a) Best choices, (b) Good Alternatives, and (c) Avoid. In contrast, our approach adopts a broader perspective, integrating environmental, nutritional and health, and socio-economic dimensions, while also ensuring transparent linkage to trustworthy, publicly available data sources.

4 The Knowledge Base

In this section, we provide the details for constructing, maintaining and accessing the knowledge base. First, we identify the data sources that will be used (in Sect. 4.1), then, we discuss the adopted ontology and the definition of the schema mappings (in Sect. 4.2), subsequently we elaborate on the workflow for the construction and evolution of the knowledge base (in Sect. 4.3), and finally we discuss the means for accessing it (in Sect. 4.4).

4.1 Exploited Data Sources

The complete list of relevant data sources (Tier 1) that can be utilized is identified in [9]. The first version of the knowledge base relied on data sources from the environmental and nutrition, and health pillars. Subsequent versions, which will be developed throughout the VeriFish project duration, will incorporate more data sources, also from the social and economic pillar. In this paper, we describe the ones that have been used for the implementation of the first version of the knowledge base. The data sources that were used are:

- The Global Record of Stocks and Fisheries (GRSF) [18] is an international initiative aimed at improving transparency and traceability in global fisheries and aquaculture. It serves as a centralized database that provides standardized and scientifically validated information on fish stocks and fisheries worldwide. GRSF integrates data from four different data sources, owned by different stakeholders in a knowledge base. These data sources are: (a) Fisheries and Resources Monitoring System (FIRMS)[5], (b) RAM Legacy Stock

[2] https://www.seaaroundus.org/fish-nutrient-profile-now-available-on-fishbase/ (accessed on Sep 1, 2025).
[3] https://seafoodcarbonprofilingtool.seafish.org/.
[4] https://www.seafoodwatch.org/.
[5] https://firms.fao.org/firms/en/home.

Assessment Database[6], (c) FishSource[7], and (d) records from the FAO SDG 14.4.1 Questionnaire[8].

- FishBase[9] [5] is the world's most widely accessed specialized database on marine and freshwater fishes. It provides information on life history, population dynamics, ecology, distribution, taxonomy, human uses, and other key aspects of fish biology. It is widely used by scientific databases, research institutions, and organizations as a trustworthy source of ichthyological knowledge. Fishbase integrates data from a range of sources, including peer-reviewed literature, fisheries surveys, and expert contributions, resulting in a robust and continuously updated repository. It is worth mentioning that the transformation of FishBase into a semantic web knowledge started already through the project SemantyFish[10] and the results of this project were consumed for the development of the VeriFish knowledge base as well.
- The Global Biodiversity Information Facility (GBIF)[11] provides access to biodiversity data. Through a network of national and thematic nodes, it provides an infrastructure enabling the publication of and access to large volumes of data gathered over centuries about the Earth's species. GBIF integrates data from widely varying sources including museum and herbarium specimens, ongoing wildlife monitoring programmes, and citizens' observations. All records are freely available to consult and download from a global platform.
- The IUCN Red List of Threatened Species[12] [8] is an inventory of the global conservation status of plant, animal, and fungi species. It is maintained by the International Union for Conservation of Nature and is one of the most authoritative sources of information on species' extinction risk. It categorizes species into various threat levels, ranging from "Least Concern" to "Extinct", based on rigorous scientific assessments, and includes detailed information on species' population trends, distribution, habitats, threats, and conservation actions.
- FoodEXplorer [4] is an online platform developed by EuroFIR to provide access to harmonized food composition data from various national food databases across Europe and beyond. The platform integrates detailed information about the nutritional content of foods, including macro- and micronutrients, bioactive compounds, and food classifications. Its primary goal is to support research, policy-making, industry applications, and consumer health initiatives by providing reliable, standardized food composition data. FoodEXplorer allows users to search data simultaneously from most EU mem-

[6] https://www.ramlegacy.org.
[7] https://www.fishsource.org/.
[8] https://www.fao.org/sustainable-development-goals-data-portal/data/indicators/1441-fish-stocks-sustainability/.
[9] https://fishbase.org/.
[10] https://semantyfish.github.io/ open science project funded under the OSCARS 1st Call for Open Science Projects and Services.
[11] https://www.gbif.org/.
[12] https://www.iucnredlist.org/.

ber states and an increasing number of countries outside Europe, as well as access to standardized value documentation. For the initial development of the knowledge base, we relied on static datasets compiled from FoodEXplorer, that contain up-to-date and curated nutritional data for various aquafood items. These datasets were constructed and utilized in the context of the projects FishEUTrust[13] and SEAFOOD^{TOMORROW}[14].
Table 1, demonstrates the data sources and the main entities collected from each one of them.

Table 1. Data sources and the entities collected from them

Source	Entities
GRSF	stocks, fisheries, species, water areas, fishing gears, flag states, stock status categories, timeseries (i.e., fishing pressure, abundance level, catches, landings, etc.)
FishBase	species names, pictures, habitats, harmless indicators, biological information
GBIF	species common names, taxonomies
IUCN	endangered species identification, threat levels
FoodEXplorer	nutritional data

4.2 Target Ontology and Schema Mappings

For the construction of the knowledge base, we started from GRSF, as it was already agreed to do so in the project description. Moreover, GRSF is already a knowledge base that has been derived from the semantic integration of other data sources, and already contains information that is valuable for the indicator framework. The conceptual backbone GRSF is MarineTLO [17], an ontology for the marine domain. It is a top-level ontology, generic enough to provide consistent abstractions or specifications of concepts included in all data models or ontologies of marine data sources and provide the necessary properties to make this distributed knowledge base a coherent source of facts relating to observational data with the respective spatiotemporal context and categorical (systematic) domain knowledge and is adequate for integrating information from the marine domain.

MarineTLO could be used as is (i.e., without requiring any major extensions), for modeling the majority of the data sources described in Sect. 4.1. To properly model the resources from FoodEXplorer, it was required to introduce new classes and properties related to the food and nutrition domain. For this reason: (a) we

[13] Horizon Europe, GA number: 101060712 https://fisheutrust.org/.
[14] H2020, GA number: 773400 https://seafoodtomorrow.eu/.

reused particular classes and properties from FoodOn ontology [3] (i.e., we reused the class FOODON_00002403[15]), (b) we introduced new classes and properties whenever it was required, and (c) we extended MarineTLO to support them. Figure 2 demonstrates a part of the operational ontology we have used.

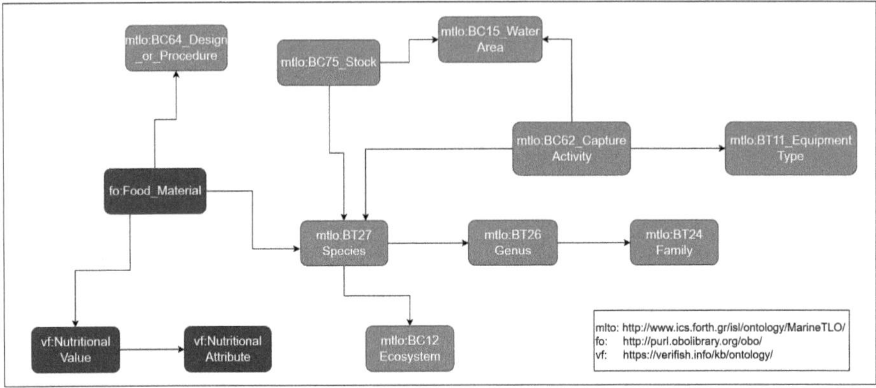

Fig. 2. Part of the extended ontology that was used for constructing the Knowledge Base (using namespaces mtlo:http://www.ics.forth.gr/isl/ontology/MarineTLO/, fo:http://purl.obolibrary.org/obo/, vf:https://verifish.info/kb/ontology/)

Based on the conceptual model, we defined a set of schema mappings to enable the transformation of the collected data into ontological instances aligned with the target ontologies. These mappings bridge the gap between the original data formats, and the semantic representation that is orchestrated by the target ontologies. To implement this, we used X3ML Framework [10], which offers a declarative language for specifying schema mappings along with a suite of tools that facilitate the transformation workflow. Using the X3ML mapping language, we managed to define the correspondences between the elements in different source schemata and their equivalent concepts in the target ontologies, in a human-readable manner. Once the mappings are defined, they can be used by the corresponding tools of the framework (i.e. X3ML Engine[16]) for the generation of RDF triples.

4.3 Construction and Evolution Workflow

We have designed a configurable workflow that handles the construction of the knowledge base. The workflow uses as input the different data sources that are used and collects their data. This was needed since data sources expose their contents using different technologies and formats; ranging from entire databases (e.g., MS Access database for FishBase), APIs (e.g., for GBIF and IUCN Red List

[15] Class: Food Material http://purl.obolibrary.org/obo/FOODON_00002403.
[16] https://github.com/isl/x3ml.

of Threatened Species), knowledge graphs (e.g., for GRSF), and spreadsheets (e.g., for nutritional information from curated FoodEXplorer datasets).

The workflow is responsible for two basic tasks: (a) collecting data and storing them locally as structured datasets, and (b) transforming data with respect to the defined schema mappings and adding them to the knowledge base. This architecture also supports the evolution of the knowledge base, as it can be executed periodically to include up-to-date information from the underlying data sources, in a similar manner as already been carried out for the evolution of GRSF KB [11]. Figure 3 visualize the workflow, and Table 2 provides a few statistics about the actual contents of the first version of the knowledge base. Just indicatively, the current version of the knowledge base consists of more than 6.7 million RDF triples.

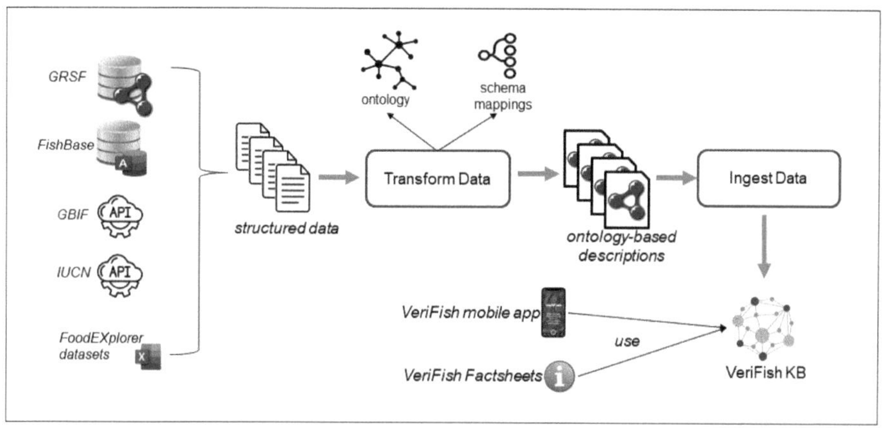

Fig. 3. The construction workflow of the knowledge base.

Table 2. Statistics about the contents of the knowledge base

Entity	Total	Entity	Total
Stocks	3,745	Fisheries	17,173
Species	13,602	Water areas	1,680
Fishing gears	82	Flag states	169
Management entities	268	Timeseries (stocks & fisheries)	573,733
Scientific names (species)	13,602	Common names (species)	16,157
Species IDs	71,568	Species IUCN data	5,282
Food Items	141	Nutritional attributes	77

The proposed workflow supports the evolution of the knowledge base by enabling its periodic reconstruction, either partial or complete, based on updates in the original data sources. This is a critical requirement, as many of the publicly available data sources integrated into the knowledge base are regularly revised to reflect new findings or corrections. Incorporating these updates ensures that the knowledge base provides current, reliable, and accurate information. Additionally, the data sources follow different update cycles; for instance GRSF is revised multiple times per year, while FishBase is updated on an annual basis. To accommodate this, the workflow is designed to refresh the contents of its data source independently, allowing the knowledge base to evolve in a modular and scalable manner.

To illustrate the benefits of the knowledge base, Fig. 4 demonstrates how heterogeneous data sources are semantically integrated into a unified structure. In particular, it shows how resources from different data sources are interconnected, allowing the combination of information about fish stocks (i.e., GRSF), fish species (i.e., FishBase, GBIF, IUCN), and nutritional data (i.e., FoodEXplorer). This interconnected structure ensures complex query answering as well as providing a solid foundation for the indicator framework.

4.4 Accessing the Knowledge Base

To facilitate discovery and access to the knowledge base, we have implemented two distinct facilities: (a) a SPARQL-2-CSV feature and (b) the VeriFish API[17].

The former is a lightweight and user-friendly interface that enables the execution of precompiled SPARQL queries and the delivery of results in CSV format, which are easily consumable by external applications. SPARQL-2-CSV does not simply serve pre-existing resources; it dynamically computes the result of each SPARQL query at the time of request[18]. This ensures that the output reflects the most up-to-date state of the knowledge base, combining relevant data from multiple sources into a coherent response. This functionality is natively supported by Virtuoso triplestore[19], which serves as the underlying storage and querying engine of the knowledge base.

The latter is the official access method for VeriFish KB and its landing page, and documentation is shown in Fig. 5. It has been implemented as a RESTful API offering methods for accessing particular resources (i.e., fisheries, stocks, species, etc.) by specifying the corresponding parameters. It was implemented as an extension of the API that is used for accessing GRSF. It now includes dedicated methods for retrieving information about various species-related resources. Notably, through the API, users can query for the fishing gears used to harvest particular species, the water areas where those species are found, their status in terms of over-exploitation or overfishing, and their nutritional profiles, such as protein content, micronutrient levels and fat composition. For example, the results of the method `searchNutrientsUsingSpecies` for mackerel (with scientific name *Scomber scombrus*), reveals all the available nutritional information on

---

[17] https://demos.isl.ics.forth.gr/verifish/verifish-api/#/VeriFish.

[18] A CSV list of species common names in English, with pictures and harmless indicators, can be collected from https://tinyurl.com/m8vkpha7.

[19] https://virtuoso.openlinksw.com/.

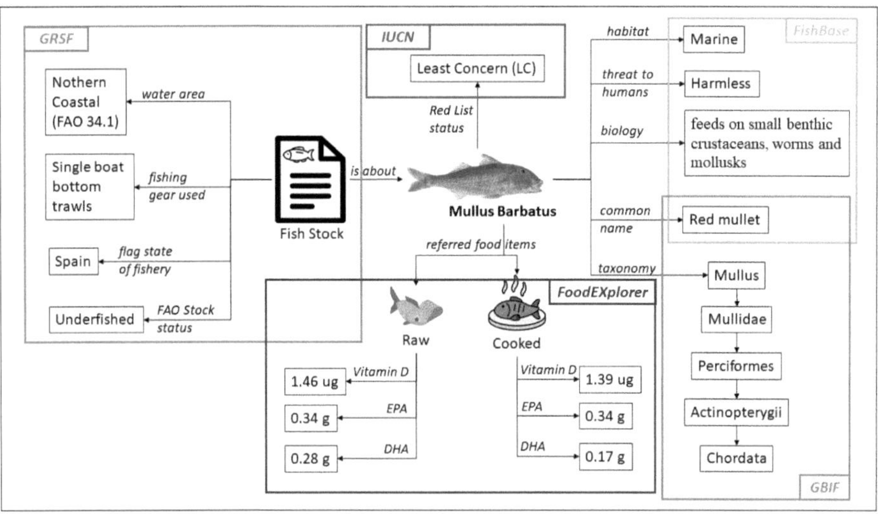

Fig. 4. Integrating heterogeneous data sources into the knowledge base, connecting species, stocks, fisheries, and nutritional data.

that species, in various processing states such as raw, cooked, smoked, or canned. The results are available in JSON format from https://tinyurl.com/38p6xza5.

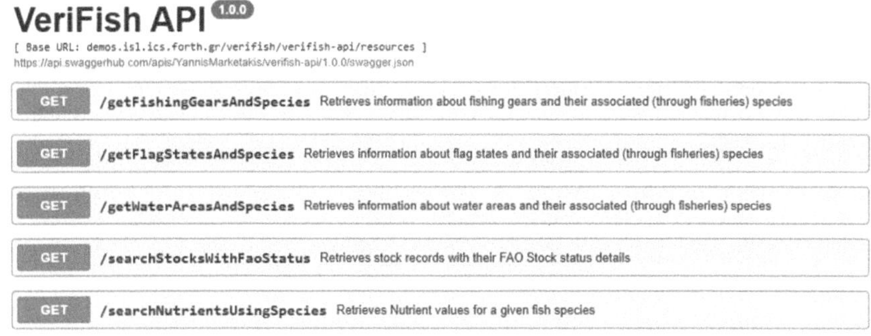

Fig. 5. The landing page of the VeriFish API providing discovery and access to the knowledge base.

5 Applications

To exploit the wealth of information that is integrated in the knowledge base and support the efficient communication of sustainability indicators, VeriFish will produce a series of media products, campaigns, and applications. In this section, we focus on two of them, that are directly related to the knowledge base, since they utilize its contents to provide various information and facilities. In the sequel, we provide more information about the VeriFish mobile application and the online factsheets.

The mobile application [12] responds directly to the project's mission of enabling citizens, producers, retailers, and stakeholders across the aquafood value chain to make informed consumption decisions. It offers seamless access to complex and diverse datasets in an accessible digital product, by exploiting the underlying knowledge base. It is designed with a hybrid data architecture that balances robust, authoritative data sources with real-time responsiveness and scalability. Information is collected from the knowledge base, through the different access mechanisms that are described in Sect. 4.4.

To ensure fast data access and smooth performance, particularly under a high user load, the mobile application implements Firebase[20] as a high-performance caching layer on top of the main knowledge base. It is used to cache frequently accessed records, enabling quicker retrieval, reducing latency and minimizing repeated calls to VeriFish API. Firebase stores and manages data using a NoSQL structure, organized into collections and documents. This schema-less flexibility allows Firebase to efficiently manage the wide range of entities and relationships inherent to the VeriFish domain, such as aquafood products, sustainability indicators, factsheets, and more. The application also supports customized search and filtering facilities to allow users to browse or narrow down resources with respect to different attributes.

In addition to interactive features, the application provides users with access to digital factsheets, which compile key information related to both species and associated sustainability indicators. These factsheets present a structured overview of relevant data, and the sustainability aspects derived from the indicator framework and the knowledge base. Designed for clarity and accessibility, they facilitate users better understand the context and sustainability implications of different aquafood products. While they are currently accessible through the mobile application, plans are underway to make them available through the official project website, broadening access and supporting transparency across the aquafood value chain.

Some indicative screenshots of the mobile application and the factsheets are depicted in Fig. 6. Currently, the mobile application is available in a testing-beta version[21] for Android and iOS devices.

5.1 Use Cases

To demonstrate the practical value of the knowledge base and its integration into real-world decision-making processes, we outline two representative use cases

[20] https://firebase.google.com/.
[21] https://verifish.info/verifish-web-app/.

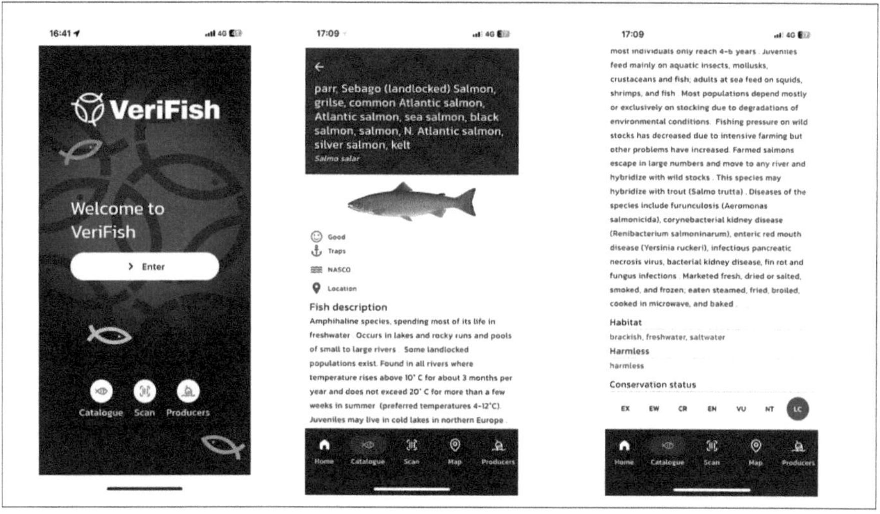

Fig. 6. Screenshots of the VeriFish mobile application, exposing the contents of the knowledge base.

that will be supported by the VeriFish mobile application. These scenarios highlight how different stakeholder groups can interact with the application to access information derived from the knowledge base for different purposes.

Use case 1: informed aquafood purchasing decisions

In this scenario, the mobile application will be used to empower consumers with reliable information to support informed and responsible aquafood choices. It targets everyday aquafood shoppers, whether at supermarkets, fishmongers, or at home, who seek clear, trustworthy, and actionable data on the sustainability and the nutritional value of the products they consume. By scanning a product's QR code or manually searching for a fish species within the mobile app, users gain access to detailed, verifiable information drawn directly and lively from the knowledge base. This includes indicators such as stock health, fishing or farming methods, environmental impact, and nutritional composition.

Use case 2: fishers verifying and sharing sustainable catch data

In this use case, the mobile application serves as a tool for fishers to record, verify, and communicate the sustainability attributes of their catch. More specifically, a fisher uses the app to log essential catch details, including the species, the fishing methods, the geographic location, and any relevant sustainability certifications, such as those issued by the Marine Stewardship Council (MSC)[22]. Then, this input is linked to VeriFish knowledge base, which cross-references submitted data with existing sustainability indicators and regulatory standards. Then the app performs an automated assessment based on the different dimensions of the indicator framework, evaluating the environmental impact and compliance of

[22] https://www.msc.org/.

the reported catch data. Following this assessment, the app generates a unique QR code, representing a digital catch factsheet, that can be easily shared with wholesalers, retailers, or even end-consumers. This process facilitates the end-to-end traceability and enhances market transparency, building trust throughout the aquafood supply chain.

6 Conclusion and Next Steps

In this paper, we described the requirements and the process for constructing a knowledge base for sustainable aquafood communication, an activity carried out in the context of the Horizon Europe project VeriFish. The knowledge base was constructed by semantically integrating heterogeneous and thematically diverse data sources. More specifically, we elaborated on the main challenges and described the implementation of a technical workflow to automate its construction and evolution. To facilitate access, dedicated software components were developed, that exposes information in a structured manner. Moreover, we described the rationale and functionality of a mobile application that utilizes the contents of the knowledge base and exposes them to mobile users.

In the future, we plan to enhance the contents of the knowledge base, with the inclusion of more data sources, including as well datasets related to the socio-economic pillar of the indicator framework. Moreover, we plan to interconnect particular knowledge base resources with other external and publicly available data sources, enhancing its interoperability.

Acknowledgments. This work has received funding from the Horizon Europe project VeriFish (Grant Agreement No: 101156426) and SemantyFish (Horizon Europe OSCARS 1st Call for Open Science Projects and Services).

References

1. Astley, S.: Verifish indicator framework defined - d2.1 (1.0) (2024). https://doi.org/10.5281/zenodo.14384281
2. de Boer, J., Schösler, H., Aiking, H.: Fish as an alternative protein-a consumer-oriented perspective on its role in a transition towards more healthy and sustainable diets. Appetite **152**, 104721 (2020)
3. Dooley, D.M., et al.: Foodon: a harmonized food ontology to increase global food traceability, quality control and data integration. npj Sci. Food **2**(1), 23 (2018)
4. Durazzo, A., et al.: Food composition data and tools online and their use in research and policy: Eurofir aisbl contribution in 2022. Nutrients **14**(22), 4788 (2022)
5. Froese, R., Pauly, D., et al.: Fishbase (2010)
6. Hilborn, R.: Overfishing: What Everyone Needs to Know®. Oxford University Press, Cambridge (2011)
7. Lewison, R.L., Crowder, L.B., Read, A.J., Freeman, S.A.: Understanding impacts of fisheries bycatch on marine megafauna. Trends Ecol. Evol. **19**(11), 598–604 (2004)

8. List, I.R., et al.: The iucn red list of threatened species. Di sponí vel em:<http://www.iucnredlist.org/info/categories_criteria2001.html. Acessoem 12 (2004)
9. Marketakis, Y.: Indicator framework developed - d2.2 (1.0) (2025). https://doi.org/10.5281/zenodo.15358656
10. Marketakis, Y., et al.: X3ml mapping framework for information integration in cultural heritage and beyond. Int. J. Digit. Libr. **18**, 301–319 (2017)
11. Marketakis, Y., Tzitzikas, Y., Gentile, A., van Niekerk, B., Taconet, M.: On the evolution of semantic warehouses: the case of global record of stocks and fisheries. In: Garoufallou, E., Ovalle-Perandones, M.-A. (eds.) MTSR 2020. CCIS, vol. 1355, pp. 269–281. Springer, Cham (2021). https://doi.org/10.1007/978-3-030-71903-6_26
12. Petrocelli, A.: Verifish web app requirements & specification document - d3.2 (1.0) (2025). https://doi.org/10.5281/zenodo.15359008, version 1.0
13. Sodhi, M.S., Tang, C.S.: Research opportunities in supply chain transparency. Prod. Oper. Manag. **28**(12), 2946–2959 (2019)
14. Talati, Z., Egnell, M., Hercberg, S., Julia, C., Pettigrew, S.: Food choice under five front-of-package nutrition label conditions: an experimental study across 12 countries. Am. J. Public Health **109**(12), 1770–1775 (2019)
15. Thornton, K., Seals-Nutt, K., Matsuzaki, M.: Introducing wikifcd: Many food composition tables in a single knowledge base. In: JOWO (2021)
16. TIME, O.B.A.A.: Consumers and the transition to sustainable food (2020). https://www.beuc.eu/publications/one-bite-timeconsumers-and-transition-sustainable-food/. Accessed 25 Jan 2022
17. Tzitzikas, Y., et al.: Unifying heterogeneous and distributed information about marine species through the top level ontology marinetlo. Program **50**(1), 16–40 (2016)
18. Tzitzikas, Y., et al.: Towards a global record of stocks and fisheries. In: HAICTA, pp. 328–340 (2017)
19. Valenčič, E., Beckett, E., Bucher, T., Collins, C.E., Koroušić Seljak, B.: Nutribase-management system for the integration and interoperability of food-and nutrition-related data and knowledge. Front. Nutr. **11**, 1503389 (2025)
20. Wilkinson, M.D., et al.: The fair guiding principles for scientific data management and stewardship. Sci. Data **3**(1), 1–9 (2016)

To Automate or Not to Automate the Transcription of Ancient Earthquakes: Toward a Global Knowledge Graph About Ancient Earthquakes

Sophia Sideri[1,2(✉)], Emmanouil Patronakis[1,2], Evangelos Nikiforos[2], Isidoros Chatzichrysos[2], Apostolos Baniotis[1,2], Valantis Zervos[1,2], Stavros Tzormpatzakis[1,2], Michalis Saridakis[1,2], Iosif Oikonomakis[1], Michalis Mountantonakis[1,2], Pavlos Fafalios[1,3], and Yannis Tzitzikas[1,2]

[1] Institute of Computer Science (ICS), Foundation for Research and Technology – Hellas (FORTH), Heraklion, Greece
{sophisid,patronakis,baniotis,vzervos,stzorba,saridakism, sifisoik,mountant,fafalios,tzitzik}@ics.forth.gr

[2] Computer Science Department, University of Crete, Heraklion, Greece
{csd4966,csd4338}@csd.uoc.gr

[3] School of Production Engineering and Management, Technical University of Crete, Chania, Greece

Abstract. The transcription of ancient earthquakes remains a significant challenge due to the ambiguity and uncertainty of historical descriptions. The vision is to document this information, found in books and manuscripts in various languages, in a structured, semantically rich, and interoperable manner, enabling a complete picture of past seismic activity. While ontologies like CRM-EQ offer semantically expressive models for capturing seismic events, the manual transcription of such information from historical sources remains time-consuming and requires expertise. In this paper we propose and evaluate a hybrid and collaborative workflow for transcribing earthquakes from historical texts. We compare five different methodologies from manual curation to LLM-based pipelines, assessing them in terms of effort, accuracy, and scalability. Our findings show that semi-automated approaches are practical, balancing effort and accuracy. We captured 248 earthquake events and produced over 24,500 RDF triples, demonstrating the feasibility of semi-automated transcription for historical seismic data.

Keywords: Knowledge Graph Construction · Historical Data · Earthquakes · Ontology-based Data Extraction · Large Language Models (LLMs)

S. Sideri, E. Patronakis and E. Nikiforos —These authors contributed equally to this work.

H. Takeda et al. (Eds.): IJCKG 2025, LNAI 16297, pp. 390–405, 2026.
https://doi.org/10.1007/978-981-95-5009-8_26

1 Introduction

Recently, the use of semantic technologies and knowledge graphs to represent historical and uncertain data has gained significant traction, especially in the domain of cultural heritage, history, and natural disasters [7,13,28]. Many descriptions of ancient earthquakes can be found in various books [35,39] which contain information expressed at various levels of detail and precision: dating at century level, statements with magnitude intervals, relative magnitude, statements about the provenance of information, uncertainties and beliefs. Such textual representation does not facilitate the machine-readable data necessary for researchers and the public to assess the frequency of catastrophic earthquakes [37].

Particularly, it would be very useful if all data that are now described in various books or manuscripts could be recorded in machine readable formats along with their semantics. That would provide a more comprehensive understanding of what happened in the past and would help seismologists discover patterns in the frequency of the occurred earthquakes. The vision would be to construct a global knowledge graph about all ancient earthquakes. However, the manual transcription of such historical data is laborious and time-consuming.

To the best of our knowledge, no prior work has systematically evaluated the range of solutions from entirely manual to fully automated methods for this problem. To address this, in this paper we explore a hybrid and collaborative workflow that combines Large Language Model (LLM) services with human transcription, aiming to reduce human effort while ensuring accuracy and semantic consistency.

The key contributions of this paper are: (a) the design and empirical evaluation of a collaborative workflow for the transcribing of historical earthquakes according to CRM-EQ [30] (b) a comparative analysis of the methodologies used (c) the enrichment and instance matching of the transcribed data with authorities, and (d) the development of a user-interface for semantic exploration of the resulting Knowledge Graph (KG). The transcription was based on the book "Oi Seismoi tis Kritis" by Platakis [39], which describes earthquakes in Crete (Greece). This work was designed following key principles for semantic content management systems, introduced by Hyvönen et al. [22], including consistent URI naming, modular data structures, validation steps and usability publishing practices.

Outline. The rest of the paper is organized as follows. In Sect. 2, we cover some background on semantic technologies and KGs from archival data. In Sect. 3, we present the collaborative methodology and approaches that we followed. In Sect. 4, we present our insights of each method and step. Finally, Sect. 5 concludes the paper and presents directions for future work.

2 Background and Related Work

Recently, there has been a growing interest in using semantic technologies and knowledge graphs for representing historical and uncertain data, such as in the domain of cultural heritage [11,17,28], biodiversity [1], Virtual Research Environments (VREs) [26,40], and natural disasters [27,30,38]. For our use case, we utilised the CRM-EQ [30] ontology, to represent the uncertainty in earthquake data. It is based on two CIDOC-CRM [8] extensions, CRMsci [9] and CRMinf [10], and supports the representation of provenance, approximate dates, multiple sources, and inconsistent or incomplete information.

Uematsu and Takeda [45,46] suggest the creation of an Earthquake Linked Open Data (LOD) dataset utilizing Ontology Oriented Design Patterns. A part of this approach was the development of a new Earthquake Ontology that represents specialized domain concepts including hypocenters, seismic motion, and the seismic intensity scale of the Japan Meteorological Agency, reusing existing ontologies such as SOSA [23] for sensor data.

Several works have explored workflows for transcribing and curating historical data. Fafalios et al. [15] proposed a provenance-aware workflow model where transcription and curation is separated, to preserve original data. Their workflow was applied to the SeaLiT project [6,14] demonstrating how semantic technologies can be used in large-scale historical research. A similar work in the field of maritime history developed a collaborative platform for historians to curate archival sources like ship logbooks [16,36]. While these approaches focus mainly on manual transcription by domain experts, our goal is to explore whether this transcription task can be automated, or at least partially. A relevant automated approach from Piantanida et al. [38] provides a pipeline for extracting geological knowledge from reports using annotators and machine learning. Similarly, Mastoras et al. [27] built a KG for modern earthquakes using Natural Language Processing (NLP) modules and satellite data.

Recent works have shown the promising usage of LLMs to automate the process of digitizing historical records. In geosciences, Ma et al. [25] suggested a workflow that uses models offered in their public Llama 2 [44]. Wei et al. [49] propose a chain-of-thought strategy to derive location and depth information from archives. Similarly, a proof-of-concept study in the social sciences led by Nicole Schwitter [41] applied Cohere Labs Command R+ model [4] to enhance the efficiency and accuracy of extracting information from 19th-century newspapers.

In conclusion, most related approaches focus either on structured modelling or on partial automation of historical data transcription. No prior work evaluates the different approaches in terms of quality and time. This constitutes the central focus of our approach. Our work examines a hybrid workflow that combines manual assertions, automated information extraction from books using LLM, and rule-based post-processing steps to ensure data consistency.

3 The Steps of Transcribing Ancient Earthquakes

3.1 Collaborative Methodology

We investigated a collaborative approach, involving 23 persons with different backgrounds: 10 postgraduate students, 9 undergraduates, and 4 professionals

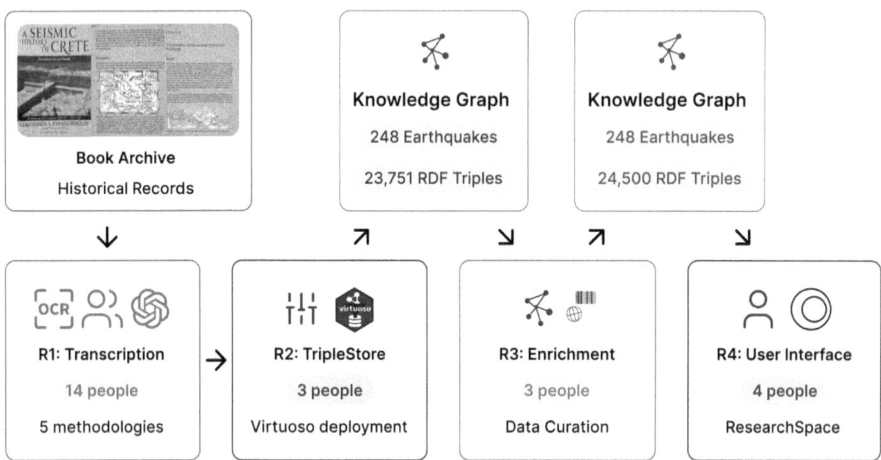

Fig. 1. Project Workflow.

with expertise in semantic technologies. We formed teams with particular roles and asked all members to record the steps and the time they spent on each task. In this workflow, also shown in Fig. 1, we identified four main roles: **R1** for producing RDF triples from the book (14 people), **R2** for organizing the produced data (3 people), **R3** for quality checking and enrichment (3 people), and **R4** for deploying various data access services (4 people).

Transcription Methodologies. Converting historical earthquake descriptions into structured RDF representations aligned with the CRM-EQ ontology is a foundational element of our knowledge graph construction. Given the collaborative nature of this project, we explored various methodologies, ranging from fully manual transcription to automated pipelines using the latest LLMs. Through analysis of the approaches adopted by the 14 participants in the R1 group, we identified five distinct methodological patterns for RDF production:

1. Manual Transcription using Protégé: In this approach we manually created RDF triples using the Protégé editor [29]. This required deep understanding of the ontology but provided the highest accuracy. We noted that after an initial learning curve of 1–2 hours of transcription, the process became more efficient.

2. OCR-Assisted Manual Transcription: We employed Google OCR (Optical Character Recognition) [19] to digitize the scanned pages, then manually authored the RDF/TTL using Protégé [29]. This method reduced the time needed for text extraction while maintaining human oversight for semantic accuracy.

3. LLM-Supported RDF Generation: We utilized LLMs, such as ChatGPT-4 [32], GPT-o1 [33], Gemini 2.0 [5], and DeepSeek [3] to generate RDF triples from OCR-extracted text. To help guide the models, we provided the ontology along

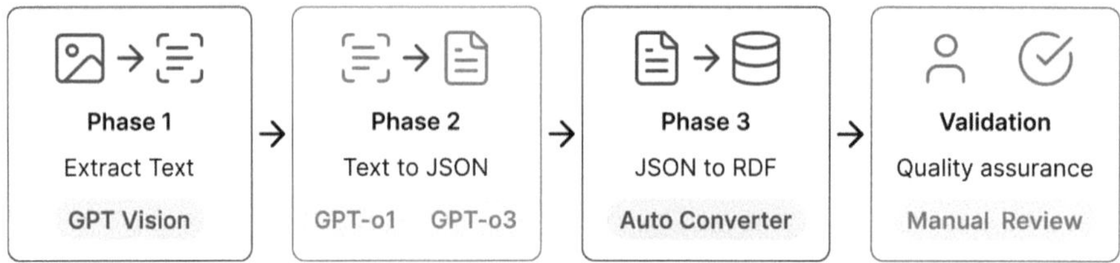

Fig. 2. Custom Pipeline with Intermediate Representation.

with example triples. However, all participants who used this method found that the output needed manual corrections due to various errors in the LLM output.

4. Direct Vision-Language Model Processing: Another approach involved using multimodal LLMs (GPT-4o) to directly process page screenshots, bypassing traditional OCR. This method needed approximately one minute per page for text extraction. However, manual verification remained essential.

5. Custom Pipeline with Intermediate Representation: We developed a custom pipeline to improve accuracy and consistency. As shown in Fig. 2, the process involves: (a) multimodal LLM to extract text directly from page screenshots (b) applying reasoning models (GPT-o1 [33], GPT-o3 [34]) with specific instructions to transform text into standardized JSON schemas containing earthquake information such as dates, locations, seismic data, references and related phenomena, and (c) a dedicated tool, the CRM-EQ JSON Converter[1], to transform the JSON into CRM-EQ compliant RDF triples.

Analysis of the Custom Pipeline. Among the aforementioned methods, the custom pipeline stood out due to its modular architecture and its potential for scalable RDF generation. We introduce an intermediate representation (IR) as a JSON that mediates between historical text (images or OCR output) and RDF triples. By constraining the extraction to a schema whose fields have defined semantics, the pipeline reduces the ambiguity and errors of the transformation.

The IR organizes earthquake information into groups. We have spatiotemporal information, seismological details, source data, associated phenomena, and uncertainties or beliefs indicating when something is approximate or disputed, and we also keep the exact original phrase. Writing things this way keeps the original meaning, but removes random wording that would confuse the mapping into RDF. To ensure accurate extraction, we used schema-based prompting. Each prompt instance is composed of four XML-labeled templates plus a final emission directive. Instead of treating this prompt in an ad hoc manner, we formalize it as a higher-order function that constructs a well-formed extraction request and specifies an expected contract for the model output.

Formal Prompt Construction. Figure 3 illustrates the formulation of the prompt used as input to the LLM. It shows how main components are combined

[1] https://github.com/EmmanouilPatronakis/crm-eq-json-converter.

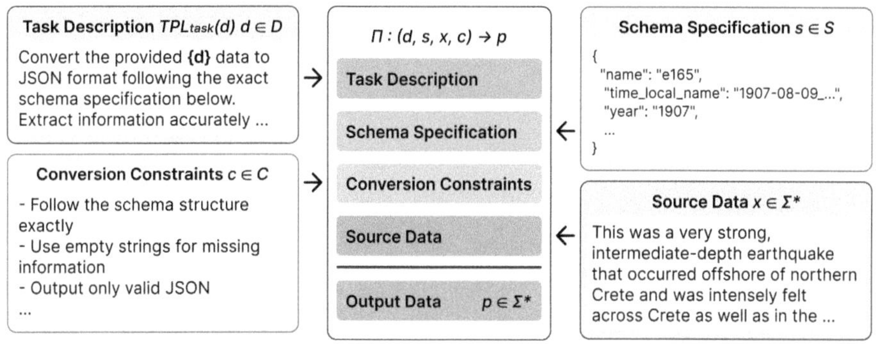

Fig. 3. Formal Prompt Construction.

into a structured prompt string $p \in \Sigma^*$. Let $\mathcal{D}$ denote the set of supported domain labels (e.g., `EARTHQUAKE_HISTORY`). Let $\mathcal{S}$ be the set of JSON Schema specifications for the IR, and Σ^* the set of Unicode strings (source transcriptions). We define the function:

$$\Pi : (d, s, x, c) \mapsto p \quad \text{with} \quad d \in \mathcal{D},\ s \in \mathcal{S},\ x \in \Sigma^*,\ c \in \mathcal{C}$$

where $\mathcal{C}$ is the set of finite lists of conversion constraints. The element $p \in \Sigma^*$ is the prompt string supplied to the reasoning LLM. Specifically:

$$\begin{aligned} \Pi(d, s, x, c) = {} & \texttt{"<task\_desc>"} + \mathrm{TPL}_{\mathrm{task}}(d) + \texttt{"</task\_desc>"} \\ & + \texttt{"<schema\_specs>"} + s + \texttt{"</schema\_specs>"} \\ & + \texttt{"<constraints>"} + c + \texttt{"</constraints>"} \\ & + \texttt{"<source>"} + x + \texttt{"</source>"} \\ & + \texttt{"Output the converted JSON"} \end{aligned} \tag{1}$$

The task $\mathrm{TPL}_{\mathrm{task}}(d)$ is parametrized as, *"Convert the provided"* $+\ d\ +$ *"data to JSON format following the exact schema specification below. Extract information faithfully; do not infer or fabricate."* The constraint list c typically includes (but is not limited to): *Follow the schema exactly; Use empty strings for missing atomic values; Preserve numeric precision; No comments; Output only valid JSON*. We define $\mathcal{J} \subseteq \Sigma^*$ as the set of strings that parse (under a strict JSON parser) to an object instance conforming to schema. The model is considered successful on input tuple (d, s, x, c) iff its emission is valid JSON conforming to the schema. Any deviation, like, malformed JSON, schema violations, needs human intervention.

Transformation to RDF. Figure 4 illustrates each field mapping during transformation from valid JSON input data $p \in \Sigma^*$ to RDF. Each input field is processed by a dedicated mapping function (e.g., `ensure_time_span`, `create_dimension`) which creates the triples, contributing to the final RDF output. Let $f : \mathcal{J} \to \mathcal{R}$ map valid JSON IR documents to the set $\mathcal{R}$ of CRM-EQ

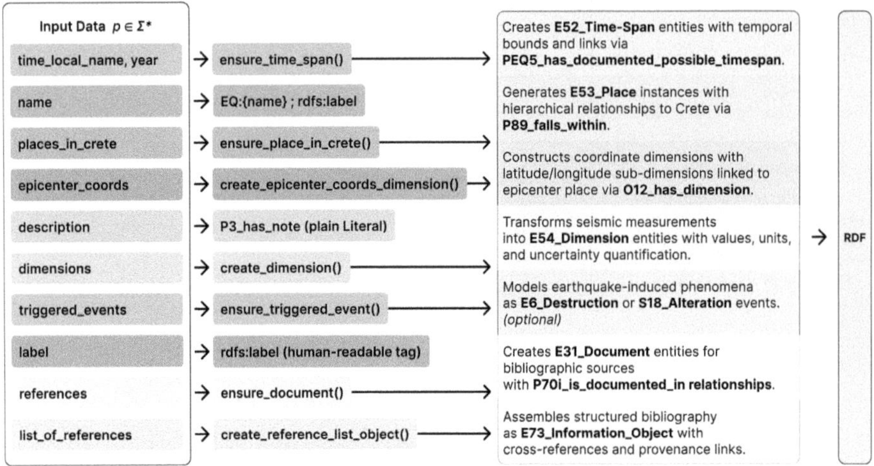

Fig. 4. Transformation to RDF.

compliant RDF triple sets. We decompose f as a composition of field-level deterministic mappings $f = g_n, g_{n-1}, \ldots, g_1$. Each g_i consumes one semantic entity (e.g. temporal, spatial, magnitudes) and emits a new set of triples.

For a JSON document j, we write $f(j) = (T, E)$, where T is the (multi)set of triples and E a (possibly empty) list of validation exceptions (e.g. missing mandatory path `event.time.begin`). A document is *ontology-valid* if $E = \varnothing$. Each mapping employs URI templates of the form, $\text{URI}(\text{entity}, \alpha) = b/\text{entity}/h(\alpha)$, with base IRI b and hash function h over a canonical serialization ensuring trivial regeneration of the RDF entities.

Common Transcription Challenges. Participants faced several challenges in all methodologies. The most significant were *LLM hallucinations and errors*, including missing properties, syntax issues, and incomplete date representations. Furthermore, the *learning curve of the CRM-EQ ontology* delayed the process. The uncertainty structure and conflicting sources demanded a long learning time period. Most overcame this through collaborating and sharing strategies, gradually learning the ontology. *Ambiguous historical data* required a careful interpretation. Uncertain dates, conflicting locations, and different magnitude scales were modelled using CRM-EQ's belief entity to retain the original source that had the claims and their uncertainties for each earthquake. Finally, *OCR inaccuracies* remained problematic despite modern tools. Historical typography and notation, often led to recognition errors. Participants spent considerable time on manual correction, highlighting the need for human oversight.

3.2 Uploading the Files To the TripleStore

Following the transcription phase, the team R2, created a collaborative workflow, to upload the Terse RDF Triple Language (TTL) files into a Virtuoso triplestore [12]. More specifically, each member of R1 uploaded their TTL files to a shared cloud repository which were then imported into Virtuoso to formulate the RDF graphs. Each graph was assigned a URI, using the pages of the book assigned to each person, and the student's registration number, making it unique and consistent based on the following format:`http://csd.uoc.gr/<starting page>-<ending page>-<student regist. number>`. For instance, the TTL file http://csd.uoc.gr/022-051-1234 represents content from pages 22 to 51 of the book and was curated by the student with registration number 1234. Because the KG was part of a group project, we used the unique IDs of each student, for ease of use, and to ensure unique URIs within the triplestore. Following upload, a validation step was performed to ensure that the triples were syntactically correct, and aligned to the ontology. Once the RDF graphs were successfully integrated, the SPARQL endpoint of the Virtuoso Triplestore was shared with Teams R3 and R4.

3.3 Enrichment and Instance Matching

The next step, implemented by the R3 team, was associated with improving the data and linking it with external authoritative resources such as Wikidata [47] and Geonames [18]. This enables interoperability within the Linked Open Data [2] ecosystem and follows the FAIR principles [50]. We used the standard OWL [20] relation, `owl:sameAs`, to indicate entities as identical. Since OWL does not provide a relation for approximate matches we introduced a custom property `custom:closeMatch` to capture such semantics. It is important to note that we do not incorporate all external information into our knowledge graph. Instead, we keep basic information, and query the rest on demand. In our approach, we start by enriching the data with the external resources and use them to match local data.

Enrichment. We perform different enrichment approaches on each entity type, querying external sources to retrieve additional metadata. The **Places** (`E5_Place`) were enriched using the GeoNames API [18], retrieving coordinates, administrative divisions, and alternative labels. We identified candidate matches by their geographic coordinates, if available, otherwise, we used the label to perform a keyword search. We cached GeoNames responses to avoid redundant API requests, as geographic information is stable. **Persons** (`E21_Person`) were enriched using Wikidata, incorporating fields such as occupations, birth and death years, and external identifiers. However, due to incomplete data, especially missing birth or death dates, exact matching was often not feasible. To improve the ranking of Wikidata candidates, we introduced a weighted scoring prioritizing relevant occupations (e.g., geologists, seismologists, historians) and temporal consistency

$$S(e) = \sum_{i=1}^{n} w(o_i) + \delta_{\text{birth}}(e) \cdot w_b + \delta_{\text{death}}(e) \cdot w_d \quad (2)$$

where $S(e)$ is the total score of candidate entity e, $o_1, \ldots, o_n$ are the occupations of e, $w(o_i) \in [0, 5]$ is the weight assigned to occupation o_i based on relevance, with $w(o_i) = 0$ if o_i is irrelevant, $\delta_{\text{birth}}(e) = 1$ if the birth year matches the reference and $\delta_{\text{death}}(e) = 1$ if the death year matches the reference (both 0 otherwise), and $w_b = w_d = 5$ are the weights assigned for birth and death year matches. All weights in the scoring function, including those for occupations ($w(o_i)$), birth year (w_b), and death year (w_d), were empirically chosen. Occupations considered highly relevant to seismic data, such as geologist, seismologist, and historian were assigned higher weights[2].

In many cases, names were incomplete (e.g., only surname was available), and associated dates were ambiguous or found along with their names (e.g., Troulinos (1900)), possibly representing birth or death year or unrelated facts. We utilized these dates as possible birth/death dates for the purpose of scoring. To address incomplete name labels, we searched for all individuals who shared the same family name and used the scoring system to identify the most relevant. As a result, person entities were typically linked using the custom relation `custom:closeMatch`.

Instance Matching. Following the enrichment, we performed instance matching to identify duplicate or equivalent entities across both the corpus and external linked data. We preserve all local data and use `owl:sameAs` or `custom:closeMatch` to record equivalence, depending on the confidence level. **Places.** We define two places Pl_1, Pl_2 as equivalent if they share the same GeoURI, or if they have highly similar labels and are located near each other:

$$
\begin{aligned}
Pl_1 \equiv Pl_2 \iff \; & \text{GeoURI}(Pl_1) = \text{GeoURI}(Pl_2) \\
& \vee (\text{sim}(L_1, L_2) \geq \tau \;\wedge\; d(C_1, C_2) \leq \delta)
\end{aligned}
\tag{3}
$$

where, $\text{sim}(L_1, L_2)$: label similarity between names L_1 and L_2, computed using Levenshtein distance [24], $\tau = 0.95$: similarity threshold, $d(C_1, C_2)$: Haversine distance [42] between coordinates C_1 and C_2, $\delta = 0.2\,\text{km}$: spatial proximity threshold. When these criteria are not satisfied but close matches are observed (e.g., $\tau \geq 0.8$, $\delta \leq 5\,\text{km}$), we assign `custom:closeMatch`. Earthquakes. To identify equivalent earthquake events, we rely on a combination of label similarity, temporal overlap, and spatial proximity. The equivalence conditions are defined as:

$$
\begin{aligned}
E_1 \equiv E_2 \iff \; & (\text{sim}(L_1, L_2) \geq \tau \;\wedge (\Delta T(B_1, B_2) \leq \theta \;\vee\; \Delta T(E_1, E_2) \leq \theta)) \\
\vee \; & (\text{sim}(L_1, L_2) \geq \tau \;\wedge\; d(C_1, C_2) \leq \delta) \\
\vee \; & ((\Delta T(B_1, B_2) \leq \theta \;\vee\; \Delta T(E_1, E_2) \leq \theta) \;\wedge\; d(C_1, C_2) \leq \delta)
\end{aligned}
\tag{4}
$$

[2] The full list of weights is available at: https://github.com/sophisid/instance_matching_EQ/blob/master/occupations_weights.json.

where, E_1, E_2 two earthquakes, $\text{sim}(L_1, L_2)$: label similarity (e.g., Levenshtein distance) between L_1 and L_2, $\tau = 0.95$: similarity threshold, $\Delta T(B_1, B_2)$, $\Delta T(E_1, E_2)$: time difference between begin and end dates, $\theta = 1\,\text{day}$: temporal threshold, $d(C_1, C_2)$: Haversine distance between coordinates, and $\delta = 50\,\text{km}$: spatial threshold. If none of the strict equivalence conditions are met, but weaker ones apply (e.g., $\tau \geq 0.80$), we assign a `custom:closeMatch`. Persons. We define

two persons P_1, P_2 as equivalent if they share a Wikidata URI through the `owl:sameAs` relation, or if they have very similar labels and matching years of birth and death.

$$\begin{aligned} P_1 \equiv P_2 \iff \quad & \text{WikidataURI}(P_1) = \text{WikidataURI}(P_2) \\ \vee \ & (\text{sim}(L_1, L_2) \geq \tau) \\ \vee \ & (\text{YOB}_1 = \text{YOB}_2 \ \wedge \ \text{YOD}_1 = \text{YOD}_2) \end{aligned} \tag{5}$$

where, $\text{WikidataURI}(P)$: external identifier of person P in Wikidata, $\text{sim}(L_1, L_2)$: Levenshtein similarity between labels L_1 and L_2, $\tau = 0.95$: label similarity threshold, YOB: year of birth, YOD: year of death. If $\text{sim}(L_1, L_2) \in [0.80, 0.95)$ or if one label is a strict subset of the other, we assign a weaker equivalence relation, `custom:closeMatch`, instead of `owl:sameAs`.

Enriching the data not only enhances its semantic expressiveness, but also facilitates integration with external data sources. Upon completion of this phase, the graph had transformed from a basic transcription archive to a rich semantic knowledge graph. Ideally, we would like to extend instance matching to cover `E5_Event` and `E7_Activity`, by comparing actors, time spans, and locations. Another direction is to explore how to connect beliefs (`I2_Belief`) together by using LLMs to match uncertain information.

3.4 Access Services

To ensure the accessibility of the transcribed and enriched data to both technical and non-technical users, we developed a user interface built on top of the ResearchSpace (RS) framework [31], and a SPARQL endpoint.

SPARQL Endpoint. The underlying knowledge graph is provided through a SPARQL endpoint[3]. For domain experts, the SPARQL endpoint allows direct querying and interaction with the raw RDF data, facilitating tasks such as validation and enrichment.

User Interface. For non-expert users we also created a public interface[4], to navigate the KG through interactive tools, including a structured search component that helps build complex queries by selecting entities and relations from a predefined catalogue, a keyword search, and a map navigation. RS was selected for its native support for semantic data exploration. The UI has been customized

[3] http://demos.isl.ics.forth.gr/earthquakes-rs/sparql.
[4] http://demos.isl.ics.forth.gr/earthquakes-rs/.

around the main entities of our dataset: Earthquakes, Places, Persons, and Documents. Each entity page includes a dynamic RDF catalogue showing incoming and outgoing relations. In addition, it queries information from the external authorities linked in the enrichment process.

4 Evaluation

To ensure data quality, we implemented several validation strategies. To verify the RDF triples' syntax we used the W3C RDF Validation Service [48]. To ensure ontology compliance, we checked property domains and ranges using Protégé's [29] reasoning capabilities and OntoGraf [43] for visual consistency checking. Additionally, we did a peer review process, where we cross-checked the triples to identify inconsistencies or errors. Finally, we tested the correctness of the generated RDF data by executing predefined SPARQL queries derived from a group of competency questions.

Competency Questions. Team R1 contributed a set of natural language competency questions, which were manually translated into SPARQL queries by R2. More specifically, the set included questions such as:

- How many / Which earthquakes occur between a specific time period?
- How many / Which earthquakes occur in a specific place?
- How many / Which earthquakes had more than a specific Richter number?
- Which places have the most earthquakes?

We formulated queries based on the previous questions, and used them during the early stages of development to test the integrity of the graphs and validate the endpoint's functionality.

Empirical Observations. Our findings reveal that while automated approaches using LLMs are scalable, they require human oversight to achieve acceptable quality levels. Participants who started with manual methods reported improved efficiency as they became familiar with the ontology, often achieving comparable or better throughput than those using automated methods when factoring in correction time. The most effective approach appeared to be a hybrid methodology that combines automated text extraction with human-guided RDF generation.

The time investment in the transcription process varied across methods and participants, ranging from 7 to 36 h for completing the assigned pages ($\sim$22 pages per person). Table 1 summarizes the key characteristics observed across different methodologies, including average time per page, error rate, scalability, and the level of ontology expertise required. As shown, manual transcription has the lowest error rate but is the least scalable and demands expertise. In contrast, automated or semi-automated methods are faster and scalable, with moderate error rates, emphasizing the trade-offs between efficiency and accuracy.

Table 1. Comparison of different RDF production methods.

Method	Time/Page	Error Rate	Scalability	Ont. Expertise
Manual (Protégé)	30–45 min	Very Low	Low	High
OCR + Manual	20–30 min	Low	Medium	High
OCR + LLM	15–25 min	Medium	High	Medium
Custom Pipeline	10–20 min	Low-Medium	High	Medium
Vision-LLM	5–10 min	Medium	Very High	Low

Table 2. Triples per entity and source added after the enrichment.

Entity	Named Graphs	Triples	Source
EQ1 Earthquake	custom:earthquake	337	Manuscript
E53 Place	custom:places, custom:geonames	342	Manuscript & Geonames
E21 Person	custom:people, custom:wikidata	70	Manuscripts & Wikidata

Effort Spent. The collaborative transcription effort involved 14 participants in the role of R1: 1 expert supervisor, 9 post-graduate students, and 4 undergraduate students. The total time investment reached 8,644 min (144 h), with each participant spending an average of 11 h on their assigned sections. This time included the initial learning curve for learning the CRM-EQ ontology, familiarizing with the book content, formalizing effective LLM prompts, and the actual transcription work.

The R2 team, comprising one post-graduate, one undergraduate, and an expert supervisor, spent approximately one week setting up the Virtuoso triple store. Similarly, the R3 team, including a post-graduate, an undergraduate, and an expert supervisor, devoted two weeks to formulate the scoring system and implement the corresponding code. Finally, R4 with the same composition, developed the user interface in approximately two weeks.

Post Processing. As shown in Table 2, the post-processing phase contributed a total of 749 additional triples across three entity types. The number of triples increased from 23,751 to 24,500, which cover all entities, incoming and outgoing relations. In total, the project led to the documentation of 248 ancient earthquakes, fully described according to the CRM-EQ ontology, covering diverse locations and time periods as described in the book.

Knowledge Graph Statistics. Table 3 shows some key statistics for the resulted KG, such as the number of distinct triples, literals, URIs, and others. Table 4 offers a more detailed view of some of the main classes in the KG. For instance, the **E53 Place** class has the most triples (6,234), reflecting the large number of spatial references in the corpus. Earthquake instances (**EQ1 Earthquake**) are associated with an average of 14.89 triples each. Other classes, such as **E54 Dimension** and **E73 Information Object**, also show many measure-

Table 3. Statistics about the KG.

Type	Value
Triples	24,500
Triples with Object URIs	14,262
Triples with Object Literals	10,130
Distinct URIs	3,085
Distinct Literals	5,895
Distinct Subjects	3,040
Distinct Objects	8,545

Table 4. Statistics of some key KG Classes.

Class	URIs	Triples (T)	avg T/URI
E54 Dimension	673	4,235	6.29
EQ1 Earthquake	248	3,695	14.89
E52 Time Span	205	1,188	5.79
E53 Place	146	6,234	42.69
E73 Information Object	104	1,200	11.53
I12 Adopted Belief	83	728	8.77

ments and textual information material. This highlights the semantic richness and expressivity of the knowledge graph.

Reproducibility. To ensure reproducibility, we provide open access to all relevant resources. The CRM-EQ JSON configuration and conversion code is available at https://github.com/EmmanouilPatronakis/crm-eq-json-converter, accompanied by working examples. The enrichment and instance matching components are available at https://github.com/sophisid/instance_matching_EQ, and the source code of the ResearchSpace application is accessible at https://github.com/sophisid/EarthquakeRS.

5 Concluding Remarks

This paper presented a collaborative workflow for transcribing historical earthquake information into a knowledge graph. The process resulted in 248 earthquake entities, producing over 24,500 RDF triples, and showed that semi-automated pipelines reduced time by over 50% compared to manual methods.

Lessons Learned. The collaborative process revealed several important insights that can inform future efforts in similar domains:

- **Hybrid approaches are most effective:** Combining automated extraction using LLMs and OCR with structured intermediate representations and manual validation achieves high-quality and scalable results.
- **Ontology learning curve matters:** Although the ontology learning was initially challenging, the use of CRM-EQ became more efficient over time.
- **Intermediate schemas help control quality:** Using intermediate representations restricted LLM hallucination outputs and improved the pipeline.
- **Collaboration fosters consistency:** Well-defined roles, peer feedback and validation, significantly improved data quality and participant engagement.

The results of this work can be useful for enabling the community to use semi-automated workflows, aiming to create a knowledge graph that contains as complete information as possible about all ancient earthquakes. Moreover, this approach can be easily applied in other domains. Future work can explore how

LLMs and OCR can accelerate the transcription process, with strict validation and corrections of the data, to ensure semantic accuracy, or, try other ontologies like OMIT [21]. In addition, more advanced prompt engineering strategies could further reduce human effort by improving the reliability and consistency of LLM outputs.

Acknowledgments. Many thanks to all students who participated in this project, namely: Dimitris Andronikou, Pavlos Ziotas, Georgios Panagiotis Mellios, Christos Konstantinos Panorios, Vasileia-Eirini Pitikaki, Emmanouil Troulis, Petros Tsalikis, Torbjorn Fries, Evaggelos Mixelioudakis, Ilias Kapsis, Maria Foukaki and Despoina Athanasiou.

References

1. Basset, A., Los, W.: Biodiversity e-science: lifewatch, the european infrastructure on biodiversity and ecosystem research. Plant Biosystems-An Int. J. Dealing Aspects Plant Bio. **146**(4), 780–782 (2012)
2. Bauer, F., Kaltenböck, M.: Linked open data: the essentials. Edition mono/monochrom, Vienna **710**(21) (2011)
3. Bi, X., et al.: Deepseek LLM: scaling open-source language models with longtermism. arXiv preprint arXiv:2401.02954 (2024)
4. Cohere: command r+ (2024). https://cohere.com/blog/command-r-plus
5. DeepMind, G.: Gemini 1 technical report (2023). https://deepmind.google/technologies/gemini/
6. Delis, A.: Seafaring lives at the crossroads of mediterranean maritime history. Int. J. Maritime History **32**(2), 464–478 (2020)
7. Dirgahayu, T., et al.: Semantic web in disaster management: a systematic literature review. In: IOP Conference Series: Materials Science And Engineering. vol. 803, p. 012043. IOP Publishing (2020)
8. Doerr, M.: The CIDOC conceptual reference model: an ontological approach to semantic interoperability of metadata. AI Mag. **24**(3), 75–75 (2003)
9. Doerr, M., et al.: Definition of the CRMsci - an extension of CIDOC CRM to support scientific observation. Tech. rep. https://cidoc-crm.org/crmsci
10. Doerr, M., et al.: Definition of the CRMinf - an extension of CIDOC-CRM to support argumentation (2023). https://cidoc-crm.org/crminf
11. Doerr, M., Theodoridou, M.: Mapping archaeological databases to CIDOC-CRM. In: CAA (2011). https://publications.ics.forth.gr/_publications/DOERR_MappingCIDOC_1_wFig.pdf
12. Erling, O.: Virtuoso, a Hybrid RDBMS/Graph column store. IEEE Data Eng. Bull. **35**(1), 3–8 (2012)
13. Fafalios, P., et al.: Towards semantic interoperability in historical research: documenting research data and knowledge with synthesis. In: ISWC, pp. 682–698 (2021)
14. Fafalios, P., Kritsotaki, A., Doerr, M.: The SeaLiT ontology-an extension of CIDOC-CRM for the modeling and integration of maritime history information. ACM J. Comput. Cultural Heritage **16**(3), 1–21 (2023)
15. Fafalios, P., Marketakis, Y., Axaridou, A., Tzitzikas, Y., Doerr, M.: A workflow model for holistic data management and semantic interoperability in quantitative archival research. Digit. Scholarship Humanities **38**(3), 1049–1066 (2023). https://doi.org/10.1093/llc/fqad018

16. Fafalios, P., et al.: Fast cat: collaborative data entry and curation for semantic interoperability in digital humanities. J. Comput. Cultural Heritage (JOCCH) **14**(4), 1–20 (2021)
17. Felicetti, A., Gerth, P., Meghini, C., Theodoridou, M.: Integrating heterogeneous coin datasets in the context of archaeological research. In: Proceedings of the Workshop on Extending, Mapping and Focusing the CRM co-located with 19th International Conference on Theory and Practice of Digital Libraries (2015), Poznań, Poland, September 17, 2015. CEUR Workshop Proceedings, vol. 1656, pp. 13–27 (2015). https://ceur-ws.org/Vol-1656/paper2.pdf
18. GeoNames team: geonames (2024). https://www.geonames.org
19. Google cloud: cloud vision api: optical character recognition (OCR) (2024). https://cloud.google.com/vision/docs/ocr
20. Group, W.O.W.: Owl 2 web ontology language document overview (second edition). https://www.w3.org/TR/owl2-overview/
21. Huang, J., et al.: Omit: dynamic, semi-automated ontology development for the microrna domain. PLoS ONE **9**(7), e100855 (2014)
22. Hyvönen, E., Tuominen, J., Alonen, M., Mäkelä, E.: Linked data finland: a 7-star model and platform for publishing and re-using linked datasets. In: ESWC, pp. 226–230 (2014)
23. Janowicz, K., Haller, A., Cox, S.J., Le Phuoc, D., Lefrançois, M.: Sosa: a lightweight ontology for sensors, observations, samples, and actuators. J. Web Semantics **56**, 1–10 (2019)
24. Levenshtein, V.I., et al.: Binary codes capable of correcting deletions, insertions, and reversals. In: Soviet physics doklady. vol. 10, pp. 707–710. Soviet Union (1966)
25. Ma, Z., E. Santos, J., Lackey, G., Viswanathan, H., O'Malley, D.: Information extraction from historical well records using a large language model. Sci. Rep. **14** (2024). https://doi.org/10.1038/s41598-024-81846-5
26. Martin, P., Remy, L., Theodoridou, M., Jeffery, K.G., Zhao, Z.: Mapping heterogeneous research infrastructure metadata into a unified catalogue for use in a generic virtual research environment. Future Gener. Comput. Syst. **101**, 1–13 (2019). https://doi.org/10.1016/J.FUTURE.2019.05.076
27. Mastoras, V., et al.: Towards a framework for seismic data. In: Iberoamerican Knowledge Graphs and Semantic Web Conference, pp. 106–119 (2023)
28. Meroño-Peñuela, A., et al.: Semantic technologies for historical research: A survey. Semantic Web **6**(6), 539–564 (2015)
29. Musen, M.A.: The protégé project: a look back and a look forward. AI matters **1**(4), 4–12 (2015)
30. Oikonomakis, I., Fafalios, P., Mountantonakis, M., Tzitzikas, Y.: Modeling modern and historical data as a knowledge graph: a case study for earthquake data. In: Proceedings of the 18th Metadata and Semantics Research Conference (MTSR) (2024)
31. Oldman, D., Tanase, D.: Reshaping the knowledge graph by connecting researchers, data and practices in researchspace. In: ISWC, pp. 325–340 (2018)
32. OpenAI: GPT-4 (large language model). https://openai.com/research/gpt-4
33. OpenAI: Openai o1 (2024). https://openai.com/o1/
34. OpenAI: Openai o3 (2025). https://openai.com/index/introducing-o3-and-o4-mini/
35. Papadopoulos, G.A.: A Seismic History of Crete. The Hellenic Arc and Trench - Earthquakes and Tsunamis: 2000 BC - 2011 AD (2011)

36. Petrakis, K., et al.: Digitizing, curating and visualizing archival sources of maritime history: the case of ship logbooks of the nineteenth and twentieth centuries. Drassana, pp. 60–87 (2021). https://doi.org/10.51829/Drassana.28.649
37. Philips, J.P., Tabrizi, N.: Historical document processing: a survey of techniques, tools, and trends (2020). https://arxiv.org/abs/2002.06300
38. Piantanida, M., et al.: Using knowledge graphs to navigate through geological concepts extracted from documents (2021)
39. Platakis, E.: Oi seismoi tis kritis. Kritika Chronika **4**, 494–496 (1950)
40. Remy, L., et al.: Building an integrated enhanced virtual research environment metadata catalogue. Electron. Libr. **37**(6), 929–951 (2019)
41. Schwitter, N.: Using large language models for preprocessing and information extraction from unstructured text: A proof-of-concept application in the social sciences. Methodolog. Innov. 18(1), 61–65 (2025). https://doi.org/10.1177/20597991251313876
42. Sinnott, R.W.: Virtues of the haversine. Sky and telescope **68**(2), 158 (1984)
43. Stanford center for biomedical informatics research: ontograf - protégé wiki (2025). https://protegewiki.stanford.edu/wiki/OntoGraf
44. Touvron, H., et al.: Llama 2: open foundation and fine-tuned chat models (2023). https://arxiv.org/abs/2307.09288
45. Uematsu, H., Takeda, H.: Earthquake LOD: seismic dataset construction with ontology oriented design patterns. In: 12th International Joint Conference on Knowledge Graphs (IJCKG 2023) (2023)
46. Uematsu, H., Takeda, H.: Earthquake ontology and LOD. In: ISWC (Posters/Demos/Industry) (2023)
47. Vrandečić, D., Krötzsch, M.: Wikidata: a free collaborative knowledgebase. Commun. ACM **57**(10), 78–85 (2014)
48. W3C: W3C RDF validation service (2025). https://www.w3.org/RDF/Validator/
49. Wei, J., et al.: Chain-of-thought prompting elicits reasoning in large language models (2023). https://arxiv.org/abs/2201.11903
50. Wilkinson, M.D., et al.: The fair guiding principles for scientific data management and stewardship. Scientific data **3**(1), 1–9 (2016)

Knowledge Graph-Driven Policy Enforcement in Urban Dataspace

Dessislava Petrova-Antonova[1(✉)], Paolo Pareti[2], Petar Tomov[1], Semih Yumusak[2], Christopher Maidens[2], Shanshan Jiang[3], George Konstantinidis[2], and Dumitru Roman[3,4]

[1] GATE Institute, Sofia University, Sofia, Bulgaria
dessislava.petrova@gate-ai.eu
[2] University of Southampton, Southampton, UK
[3] SINTEF AS, Trondheim, Norway
[4] Bucharest University of Economic Studies, Bucharest, Romania

Abstract. With the diffusion of interconnected sensors, systems, and stakeholders, smart cities are evolving into complex, data-intensive environments where urban data has become a critical asset. New challenges emerge in urban dataspaces regarding enforcing multi-actor, heterogeneous, dynamic and real-time policies. To explore the potential of applying Knowledge Graph (KG) technologies in the urban context, we identified a comprehensive list of real-world policy challenges related to data governance through a novel urban dataspace testbed and an illustrative use case. Among the key challenges are new categories of policy conflicts in urban data, complexity in multilateral data governance, and real-time sensor compliance scenarios. Key features are proposed to extend an existing Open Digital Rights Language (ODRL)-based policy engine to address these challenges, exploiting KG technologies and capabilities of large language models, including a concise formal semantics for ODRL, and enhancements to key components of the policy engine. We also provide concrete examples to illustrate the challenges and how this enhanced ODRL-based framework addresses them in the smart cities setting.

Keywords: Urban Dataspace · Knowledge Graph · ODRL · Data Governance · Policy Enforcement

1 Introduction

The rapid expansion of interconnected sensors, systems and stakeholders is transforming smart cities into highly complex, data-intensive environments. Urban services now rely heavily on data-driven systems to manage resources, optimise operations, and improve citizens' quality of life. An urban dataspace is a shared environment where data flows among diverse actors and systems. In such context, urban data has become a critical asset, demanding appropriate governance to ensure its quality, security, privacy, and ethical use.

D. Petrova-Antonova and P. Pareti—Equal contribution.

H. Takeda et al. (Eds.): IJCKG 2025, LNAI 16297, pp. 406–421, 2026.
https://doi.org/10.1007/978-981-95-5009-8_27

Urban data is generated rapidly and continuously from heterogeneous sources, often in varying formats, granularities, quality, and update frequencies. Moreover, these data are typically shared and reused by multiple stakeholders with divergent objectives, responsibilities, and enforcement powers. This raises critical concerns related to multi-actor, dynamic and real-time policy enforcement, and the negotiation of usage rights and constraints across organisational boundaries.

The heterogeneity of data sources is a fundamental challenge in urban dataspace contexts. Knowledge Graph (KG) technologies can help address this by enabling semantic interoperability through shared vocabularies, flexible data models, and reasoning, thereby supporting accurate, trustworthy, and collaborative data sharing across domains [20]. In addition, policy enforcement—ensuring that data usage complies with regulations and expectations—represents another critical challenge. While policy frameworks such as the Open Digital Rights Language (ODRL) offer a foundation for expressing usage policies, current implementations fall short in supporting the emerging challenges specific to urban contexts, such as the spatio-temporal, IoT-aware, and dynamic policy enforcement. Existing tools lack adequate mechanisms to manage heterogeneous, fine-grained, context-aware constraints. There is also a lack of evidence-based research reporting the challenges and lessons learned from real-life urban dataspaces.

To explore the potential of applying KG technologies in urban contexts, this paper introduces the Urban Dataspace, initiated by GATE Institute – the International Data Space Association (IDSA) Hub in Bulgaria, and presents key policy challenges identified from real-world urban use cases. The challenges can be covered by ecosystem mechanisms or policy engines. While existing dataspace solutions, such as IDS-RAM [18], provide generic guidelines or mechanisms related to the ecosystem (roles, responsibilities, templates, etc.), they lack guidelines and adaptations to concrete use cases. In this paper, we focus on the challenges that can be addressed by policy engines, in particular, adapted to the specific requirements that emerged from the urban context, and we propose key extensions to an existing ODRL-based policy framework [14,32] in the smart city context. The extensions enable spatio-temporal and dynamic policy enforcement, covering crucial support related to data heterogeneity, fine-grained temporal conditions, actor dynamism, and sensor-context awareness.

The main contributions of this work are:

- A detailed analysis of the challenges for data governance regarding policy enforcement in urban dataspaces derived from real-life smart city use cases.
- A design proposal for extensions to an ODRL-based policy engine supporting spatio-temporal, multi-actor, and IoT-aware policy enforcement, and a semantic grounding that enables comparison between policies.

The rest of the paper is structured as follows: Sect. 2 describes background and related work. Section 3 presents a novel urban dataspace and policy challenges derived from associated urban use cases. Section 4 presents the ODRL policy engine extensions to address the identified challenges and an illustrative example. Finally, Sect. 5 concludes the paper and outlines future work.

2 Background and Related Work

2.1 Dataspace and Dataspace Protocol

A dataspace is an "interoperable framework, based on common governance principles, standards, practices and enabling services, that enables trusted data transactions between participants" [8]. A dataspace enables data sharing between various trusted participants, providing both a multi-organisational agreement and a supporting technical infrastructure. It answers the technology challenges of data sharing related to publishing and finding data, maintaining control over the shared data, negotiating the agreements between participants, etc. Sharing data between participants requires a metadata provision to facilitate the transfer of datasets through a data transfer protocol. The Dataspace Protocol [9] specifies how the metadata is delivered, defining how datasets are offered as Data Catalogue Vocabulary (DCAT) *catalogues* and how usage control is expressed as *Policies* with ODRL. It also prescribes how data usage *Agreements* are syntactically expressed and electronically negotiated, and how datasets are accessed based on Transfer Process Protocols. The primary purpose of the Dataspace Protocol is to support interoperability on several levels, like technical and semantic interoperability for participants, as well as interoperability on trust, organisational and legal levels. The Participants interact in the Dataspace through so-called Participant Agents, which are *Connectors*, implementing the Dataspace Protocol. The Connectors can implement additional internal functionalities, like monitoring or policy engines. The identity management is performed by the Identity Provider, which also validates additional claims. It is responsible for the provision of a Trust Framework and can be implemented as a centralised system, a decentralised system or a federated system.

DCAT enables a data provider to describe datasets and data services in a catalogue based on a standard model and vocabulary, enabling the consumption and aggregation of metadata from multiple catalogues and increasing the discoverability of datasets and data services. A decentralised approach to publishing data catalogues can be implemented, allowing federated search for datasets across catalogues in multiple locations. The interaction between a data provider and a data consumer starts with a *Contract Negotiation*, which is tracked through a series of states such as requested, offered, accepted, agreed, verified, finalised and terminated. After the contract agreement is accepted, the *Transfer Process* between the data provider and data consumer can be initiated.

2.2 Existing Approaches for Policy Enforcement in Urban Contexts

Policy enforcement implements mechanisms, rules, and technologies that are put in place to ensure that data is used in accordance with established regulations and expectations [7]. This represents one of the significant challenges for urban dataspaces [4].

High-level approaches to policy enforcement in urban dataspaces typically span four interrelated dimensions: governance frameworks, technical mechanisms, regulatory compliance, and trust-building practices [15].

Organisational and data governance frameworks define organisational structure, processes, and compliance, e.g. defining clear policies and rules (e.g. access, usage, and consent) [7], legal/ethical foundations, approaches to multi-stakeholder governance models (e.g. centralised, decentralised or federated) [22].

Technical enforcement approaches automate elements of policy enforcement through access [24] and data usage [17] control mechanisms. Additional techniques include privacy-enhancing technologies (PETs) [1], blockchain-based (e.g., smart contracts), metadata governance and auditing/monitoring solutions.

Legal and regulatory compliance is essential to align enforcement with frameworks like the GDPR, Data Governance Act [11], and specific contractual agreements (e.g., data transaction or participation agreements).

Trust and accountability support legitimacy and transparency through auditability, independent oversight, ethical guidelines and engagement [23].

Prominent initiatives and frameworks providing foundations for orchestration of enforcement implementation include [21]: International Data Spaces Association (IDSA) [16], FIWARE Foundation [12], Gaia-X [13,25], Data Spaces Support Centre [6], BSI [3], iSHARE [19].

Open Digital Rights Language (ODRL) [26] is a flexible, extensible W3C standard that allows users to define machine-readable policies that may control usage of content and services. This is expressed through permissions (actions allowed), prohibitions (actions forbidden), duties/obligations (actions that must be performed) and constraints (conditions under which the former apply). Its application has expanded significantly within environments like urban dataspaces.

While ODRL describes policies, it doesn't inherently enforce them. Enforcement relies on data governance through a separate "policy execution engine" that interprets ODRL policies and translates them into actionable controls. Challenges and limitations exist, including precision of semantic definitions of how to process and evaluate policies [5], limited expressiveness for complex scenarios, lack of standardised implementation specification for operands and operators, limited capabilities for conflict resolution and integration complexities.

3 Urban Dataspace and Challenges

This section presents the Urban Dataspace, the first dataspace in Bulgaria, which is initiated and deployed by the Big Data for Smart Society (GATE) Institute.

3.1 Architecture and Components

The initial deployment of the Urban Dataspace is based on the IDS reference architecture model (IDS-RAM) [18], including the following building blocks in addition to the IDS Connector: Identity Provider, App Store, Metadata Broker, Clearing House and Vocabulary Hub.

With the release of the Dataspace Protocol, the Urban Dataspace has been migrated to the Eclipse Specification, which provides a comprehensive framework as a set of Eclipse Dataspace Components (EDC) with a basic set of features (functional and non-functional) for the implementation of dataspaces. It supports interoperability by design, leveraging the framework's defined APIs. The framework's components, shown in Fig. 1, provide functionality that is mandatory for dataspaces as follows: (1) *Catalogue service*, publishing and securing assets that can be shared with other organisations; (2) *Identity service*, managing and verifying organisational credentials using, for example, OAuth2 tokens; (3) Control plane services, including the creation and processing of data usage agreements that grant access to data; and (4) *Data plane and monitoring services*, initiating and managing data transfers using the HTTP protocol.

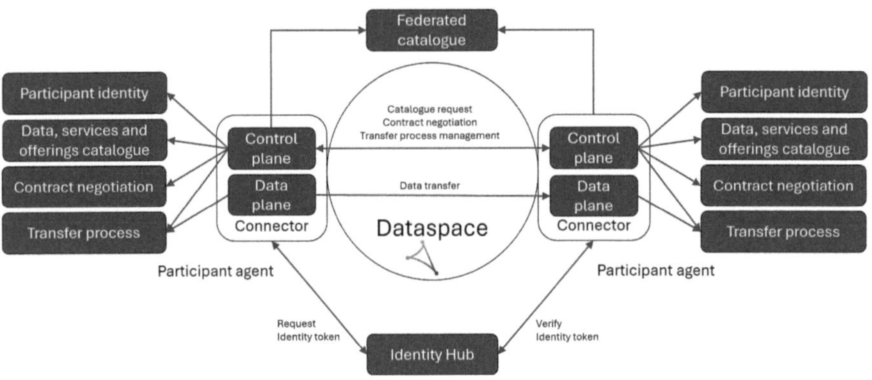

Fig. 1. Dataspace architecture and components.

The *Connector* consists of components for controlling data sharing and executing data transfer, namely the *Control Plane* and the *Data Plane*. The contract agreements that grant access to data, manage data transfers, and monitor usage policy compliance are managed by the Control Plane. The crawling and caching of data catalogues is performed by the *Federated Catalogue*. To determine whether it trusts and should grant data access to a counterparty, a participant connector uses verifiable credentials presented by the Identity Hub.

3.2 Real-World Policy Challenges in an Urban Dataspace

Urban Dataspace Context. Challenges in any dataspace, and particularly that we faced in an urban context, stem from the heterogeneity of data sources, providing data at varying spatial and temporal resolutions, the prevalence of incomplete or low-quality datasets, data sparsity, and the diversity of users with differing aims and roles. Data streams range from high-volume remote sensing sources, such as LiDAR scanning, Earth observation, and airborne sensing, producing data at rates of several gigabytes per minute, to low-volume, asynchronous sensors, including air quality and noise monitoring stations, operating at variable reporting intervals. Static spatial datasets from cadastral systems and diverse dynamic data flows, such as mobility feeds, parking occupancy, and pedestrian movement patterns, contribute to the overall information complexity. From the user perspective, typical urban stakeholders include data providers (e.g., municipal departments, utilities, private operators), data consumers (e.g., urban planners, researchers, startups), service providers and their clients (e.g., analytics vendors and transport operators), and governing authorities (e.g., municipal administrations, regulators). Although these actors operate within a shared technical infrastructure and regulatory framework, their objectives and priorities diverge. These differences influence access rights, permissible use contexts, temporal and spatial constraints, applicable obligations, and the manner in which rights and restrictions extend to derivative products. The deployment burden also differs in kind and weight across stakeholders. Not only do teams bring uneven legal, data, and IT literacy, but they also have different levels of process maturity, tooling, and capacity. For instance, stakeholders often face procurement and DPIA/security reviews that impose formal approvals and clearances (often including certification) which can force design changes, constrain data scope and contracts, and extend schedules. As data providers, they must translate licenses into approved templates, attach rich metadata to data products, and ensure policy inheritance for derived products. As data consumers, they must express intended use in machine-readable terms, while IT operations integrate connectors, identity, and audit logging with existing systems without disruption. These tasks are demanding and often block adoption due to limited staff, expertise, or tooling; however, many steps can be automated with the policy engine, proposed in Sect. 4, lowering barriers and enabling broader participation in urban dataspaces.

Use Case Development. Use cases are specific scenarios where participants are able to create economic, social, or environmental value through data sharing. The Use Case Development building block provides an operational environment for shaping new business models. This process unfolds in a series of elements:

1. Identification and Monitoring of Use Case Scenarios: Potential use case scenarios are brainstormed by leveraging various sources like participant needs, competitor analysis, and other data spaces to gather ideas.

2. Refinement of Use Case Scenarios: Promising use case scenarios are specified in detail, including the purpose, participants, data flows, and value propositions for each participant. The Data Cooperation Canvas, Use Case Playbook or self-created templates can support this refinement process.

3. Brokering and Ensuring Synergies: The data space connects the right participants and resources for use cases, fostering synergies between use cases. For example, multiple use cases can use common data products or infrastructure, reducing costs and boosting efficiency.

4. Support for Value Sharing in Use Cases: A critical element for successful use cases is to ensure the costs generated by it for the data space infrastructure are properly covered. Subscription fees or data exchange-based payments, for example, can facilitate this value distribution.

5. Ensuring Adherence to Principles: Use cases must also adhere to dataspace principles like data sovereignty and trust, along with any additional rules established at the dataspace or individual use case level.

6. Use Case Implementation: Once a use case scenario is refined, it is transformed into a functioning use case. This might involve further infrastructure development, data product creation, participant onboarding, agreements, including access and usage policies, and technical configurations.

7. Use Case Continuous Improvement: The performance of the use case is continuously analysed to identify possibilities for improvement. Changes are planned and implemented systematically throughout the life cycle of a use case.

Problems and Challenges. Consideration of the policy enforcement challenges from the use case development has led to the identification of problem areas (**Px**) and related policy challenges (**Cx**).

Heterogeneous Temporal and Spatial Granularity (P1): Datasets have different update frequency, causing *temporal mismatches* (e.g., sensors may report per minute, hour or day). Fine-grained policies may not apply consistently due to *spatial resolution differences* (e.g., point, street segment, neighbourhood, citywide). Hence, policy constraints may not map cleanly to what the data can support, risking under- or over-enforcement. **C1:** *Defining and enforcing consistent policies across uneven resolutions without losing critical context or introducing unfair bias.*

Schema and Semantic Heterogeneity (P2): Data providers describe the same phenomena with *different schemas, labels and units* (e.g. time may appear as "time", "t", or "timestamp", measured in hours, seconds, or ISO 8601 strings) while data consumers and service providers express intended use in their own terms and baseline vocabularies. Policy rules that reference such concepts and units depend on correct alignment; when mismatches occur, checks on usage conditions or licensing constraints may yield false denials or unsafe permissions. **C2:** *Policies must be semantically aware to avoid misenforcement due to schema mismatches.* This requires normalisation or semantic mediation layers.

Data Reliability and Trustworthiness (P3): Datasets *differ widely in quality and origin* (e.g. sensory data may come from certified measurement stations or community-maintained devices), while key metadata is often incomplete, non-comparable, or entirely missing (e.g., calibration dates, accuracy/error bounds). Policy constraints often rely on such quality indicators (e.g., "allow access only to data with known error margins" or "train models only on recently calibrated sensors"), but in their absence or inconsistency, such policies may be misleading, overly restrictive, or silently bypassed. **C3:** *Enforcing policies which rely on data confidence is difficult without standardized reliability metrics.*

Policy Conflict and Ambiguity (P4): Datasets covering similar domains, such as traffic flows, air quality, or mobility, may come with *diverging access rules:* one provider allows open reuse, another enforces strict time-bounded conditions. Users or services consuming or aggregating such datasets face unclear or incompatible obligations, especially when the data is republished or combined. This undermines legal certainty and hinders automation. **C4:** *Need mechanisms to detect, reconcile, or prioritise conflicting policies across sources.*

Dynamic and Context-Aware Requirements (P5): Certain policies are stateful and situational: clauses depend on geofences (e.g., in the EU only), time windows (rush hour only), and events (accidents, maintenance). Context may be specified by providers, yet consumers need timely decisions while data arrives as a mix of live streams, delayed batches, and backfills. It is difficult to determine consumer compliance as context shifts. **C5:** *Enforcing spatiotemporal and event-triggered policies in heterogeneous, sometimes delayed, data flows.*

Licensing and Usage Constraints (P6): Datasets arrive under *varied licensing models* (open, non-commercial, purpose-limited) with terms in heterogeneous templates or plain prose. Service outputs also carry policies that must be declared and inherited. Stakeholders (providers, consumers, service providers, governance) must author, compare, negotiate, and combine terms, decide access, and propagate rights and obligations to derivatives. When terms are non-standard or only human-readable, machines cannot reliably assess intended use, verify obligations, or detect cross-asset incompatibilities, leading to overblocking or non-compliant reuse. **C6:** *Automating the enforcement of legal and policy terms when licenses are non-standard or human-readable only.*

Multilateral Data Governance (P7): *Shared data assets* span public authorities, private operators, and citizen groups, each with its own governance model, accountability, and risk posture. As data and services cross boundaries, joint obligations, tiered access, and decision rights must be aligned to avoid conflicts and stalemates. **C7:** *Coordinating policy across stakeholders with divergent goals and enforcement powers.*

Data Ownership (P8): Europe *lacks a clear and universally accepted definition of data ownership*, and the way different existing ownership models should interact remains uncertain. The Data Act does not establish rules on data ownership itself. Instead, it centres on the concept of data holders, rather than data owners, and sets out regulations governing how third parties may access and use data in the possession of these holders, regardless of any actual or asserted ownership claims. **C8:** *Allocation and enforcement rights in an ecosystem where ownership concepts are ambiguous and potentially conflicting.*

Some of the challenges we faced in practice are inherently human-factor and outside any technical stack: ambiguous or shifting policies, inconsistent/missing metadata, mis-calibrated sensors, misaligned incentives, and legal review (i.e., issues needing governance/training/contracting rather than code). Where automation is viable, we separate ecosystem and engine duties. The ecosystem, mostly realised by the dataspace's control/identity planes, defines baseline rules and templates, negotiates exceptions, certifies connectors/providers, and handles dispute resolution and tiered access. The policy engine, proposed in Sect. 4) provides: (i) a formal layer for policy equality/containment/merge to detect conflicts and ensure stricter-than inheritance; (ii) an ontology/mediation layer (DCAT, SSN/SOSA, QUDT, PROV-O) to normalise schemas, units, and provenance; (iii) automated evaluation of credentials, provenance completeness, calibration/accuracy signals, and spatiotemporal clauses; and (iv) decisions and audit logs for runtime usage control via the connector.

3.3 Air Quality Use Case

The Urban Dataspace provides a secure digital infrastructure that enables trusted and sovereign data and service sharing among various stakeholders. It supports the implementation of the Urban Digital Twin (UDT) of Sofia city, covering use cases related to urban planning, air pollution and climate change mitigation, mobility and transportation, energy efficiency and assessment of building solar potential. To avoid complexity, this section presents the air quality use case as a sample scenario for showcasing the Urban Dataspace.

Dataspace Participants. The air quality data is shared as Data Offers by four main data providers shown in Fig. 2.

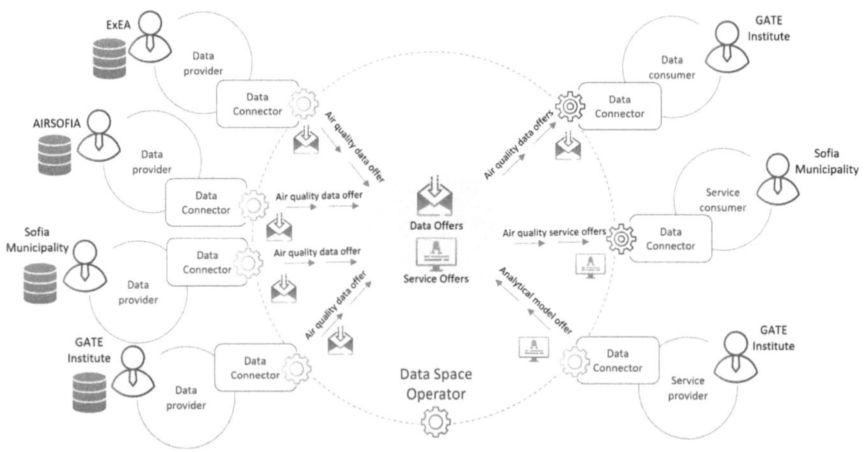

Fig. 2. Air quality scenario within the Urban Dataspace.

Bulgarian Executive Environment Agency (ExEA)[1] is an administration under the Minister of Environment and Water and a National Coordination Centre of the European Environment Agency. As such its data is considered the most reliable. *AIRSOFIA*[2], part of the Sensor.community's projects about Fine Particulate Matter (PM2.5) measurement in the environment. The data originate

[1] https://eea.government.bg/en.
[2] https://airsofia.info.

from low-cost community sensors, not subject to official calibration or regulatory standards. As a result, the measurements may exhibit higher uncertainty and variability compared to certified monitoring stations, limiting their reliability for precise air quality assessments or policy enforcement. *Sofia Municipality*[3], collecting air quality data from 21 air quality stations that, while not officially certified like those of the ExEA, offer a higher degree of reliability compared to community-based sensors. These stations typically employ more robust hardware and maintenance protocols, providing more consistent and accurate measurements than low-cost, crowd-sourced alternatives. *GATE Institute*[4], collecting air quality data through its City Living Lab, equipped with 12 air quality stations, similar to those of the municipality.

Given that participants in the dataspace may assume multiple roles, the GATE functions as both a data consumer and a service provider. Air quality data is utilised to develop predictive models for spatial and temporal prediction, employing a Gaussian auto-regressive approach and various machine learning models. The predictive models are offered as services within the dataspace and accessed by the Sofia Municipality to support evidence-based decision-making.

Problems and Challenges. The problems and challenges outlined for the Urban Dataspace in Sect. 3.2 can be illustrated through the air quality use case as follows:

Heterogeneous Temporal and Spatial Granularity (P1): Sensors report per-minute, hourly, or daily; locations vary city-wide, and some sensors sit near sensitive sites (e.g., school areas), where policies often require fine-grained geofences (e.g., "no access within 500 m of schools").

Schema and Semantic Heterogeneity (P2): Providers expose similar concepts with different schemas and units; PM2.5 might be labelled differently across data providers as pm25, PM_2_5, fine_particles. Policies referencing fields or units require mediation.

Data Reliability and Trust (P3): ExEA stations use certified, precise sensors with reliable uptime and rich metadata, but the raw feeds still contain gaps, unnormalised values, and artefacts, requiring substantial cleaning and validation before analysis; AIRSOFIA includes low-cost sensors, some out of calibration. Accuracy/calibration-dependent policies must account for this variance.

Policy Conflict and Ambiguity (P4): ExEA and Sofia Municipality data are open for reuse, whereas AIRSOFIA's terms are ambiguous; combining them creates uncertainty about downstream obligations.

Dynamic and Context-Aware Requirements (P5): Private providers can apply conditional access (e.g., "allow during pollution spikes"), requiring evaluation over time, place, and event state.

---

[3] https://platform.airthings-project.com.

[4] https://citylab.gate-ai.eu.

Licensing and Usage Constraints (P6): Where licenses exist, they are often human-readable only, preventing automated assessment and inheritance of obligations.

Multilateral Data Governance (P7): ExEA, AIRSOFIA, the Municipality, and GATE have distinct operational models, necessitating tiered access and clear responsibilities.

Data Ownership (P8): The problem of ownership arises from overlapping contributions and unclear rules on how rights and responsibilities are shared when datasets are integrated and shared as a common data product.

4 Proposal for a Policy Engine for Urban Dataspaces

In this section we present the proposal of an architecture of an enhanced policy engine and show how it addresses the urban dataspace challenges identified in Sect. 3.2 (referenced in text with code **(C1-C7)**). This proposal outlines key directions for advancing research and implementation efforts. Although no comprehensive solution currently exists for the challenges of urban dataspaces, several of the approaches reviewed in Sect. 2.2 could be adapted and integrated into our proposed framework. The architecture has been designed with scalability and usability in mind; however, a thorough evaluation of these dimensions—both theoretically and in practice—remains an important avenue for future work. Figure 3 illustrates the overall architecture, showing how its five key components align with the challenges.

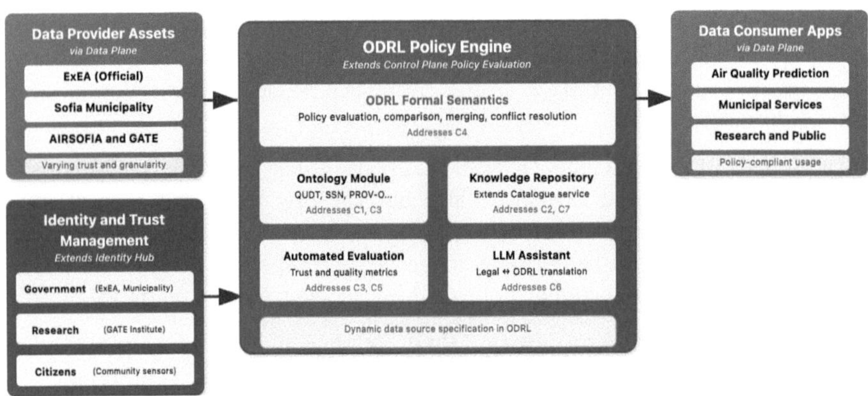

Fig. 3. ODRL-based Policy Engine Architecture for Urban Data Space

4.1 ODRL Formal Semantics (C4)

A formal semantics of ODRL is needed to provide a consistent approach to evaluate and compare policies, as required by challenge **(C4)**. However, such semantics is currently not available. While a number of partial formalisations exist, such as [10] and [2], these are focussed on particular features of ODRL, do not consider the latest ODRL version, namely version 2.2, or are preliminary works. To address the needs of an urban data-space, a suitable ODRL semantics should specify the semantics of multiple decision problems, such as (1) the evaluation of a policy on a state of the world, or on a data access request (e.g. does policy X allow me to download and share derivative work from dataset Y?), (2) the comparison of policies (e.g. are two policies semantically equivalent? Do the permissions of one imply the permissions of the other?) and (3) the merging of multiple policies to characterize the conditions of use of data derived from multiple sources. Moreover, a formal semantics should be simple, intuitive, easily implementable and computationally efficient. For example, in a data sharing scenario, a practical and cautious approach is to state explicitly what is permitted, rather than listing everything that is forbidden. Thus the chosen semantics can assert that everything not explicitly permitted is forbidden.

4.2 Ontology Module (C1, C3, C5)

Interoperability in an Urban Dataspace requires concepts and policies to be expressed in a mutually understandable way. While no single ontology covers all use cases, reuse of a core set is essential. In this regard, spatial and temporal resolution in policies can be expressed using DCAT properties such as `dcat:temporalResolution`, while sensor data and units of measurement are expressed with QUDT.[5] Custom vocabulary for common granularities (e.g., "street level") enable greater expressivity **(C1)**. Data reliability depends on quality and provenance: Sensor, Observation, Sample, and Actuator Ontology (SOSA/SSN) models sensor accuracy, the Provenance Ontology (PROV-O) describes data derivation, and the Verifiable Credentials Data Model (VC) [31] provides certifiable claims, linking datasets to trusted issuers **(C3)**.

To regulate policies, ODRL constraints compare operands (e.g., PM2.5 values $> 55.5\ \mu g/m^3$). However, specifying trusted data sources is critical. We extend ODRL with `leftOperandSource`, which points to specific providers, APIs, or models. Provenance and versioning are expressed with the Provenance, Authoring and Versioning (PAV) ontology [28], ensuring that when multiple sources exist, the chosen one is transparent and verifiable **(C5)**.

4.3 Knowledge Repository Module (C2, C7)

This module introduces repositories for both ontologies and policies. The ontology repository extends IDS-RAM's Vocabulary Hub by not only storing vocabularies but also mappings across them, supporting RDFS [29], SKOS [30] and

[5] https://qudt.org.

OWL [27] terminology. These can specify when two concepts are the same, or when one is a more specific/general version of another. Adoption statistics will also promote convergence toward widely reused ontologies **(C2)**.

The policy repository stores ODRL agreements linked to regulated datasets, ensuring transparency of data usage. Stakeholders can review how datasets are being shared and their restrictions. For confidentiality, repositories allow selective visibility, e.g., granting access only to verified IDs. Redacted versions may be made public, while sensitive details remain private. This balance of openness and control supports accountability while protecting sensitive data **(C7)**.

4.4 Automated Evaluation Module (C3)

The Automated Evaluation Module provides scalable mechanisms to certify datasets and enrich them with quality and trust indicators. By generating and validating standardised metadata, it strengthens dataset discoverability and reliability across the dataspace **(C3)**. We envision this as a set of components, such as the following ones. The *ontology compliance* component can analyse datasets and their metadata to extract information about which ontologies it reuses. This information can be published alongside the dataset to enhance discoverability and provide quality measures related to its standardisation. The *granularity compliance* component analyses a dataset and, if it contains temporal and location data defined in a standard format, evaluates its temporal and/or spatial granularity. This information can then be used to annotate the dataset with this quality metric, or, if this quality metric already exists, to verify its correctness. The *digital certification* component is an extension of the Certificate Authority component, which allows for the verification of a wide array of facts, including dataset metadata, using the VC ontology. Together, these automate the evaluation of compliance, quality, and trust, reducing manual overhead.

4.5 LLM Assistant Module (C6)

The LLM Assistant connects machine-readable ODRL policies with human-readable legal text **(C6)**. Its functions are twofold: explaining ODRL policies in natural language for non-technical users, and converting between ODRL and legal text for expert review. For general users, it provides clear policy explanations and supports conversational queries (e.g., "Can I publish derived data if I remove location fields?"). For legal contexts, it proposes draft translations between ODRL clauses and legal formulations, though expert validation remains necessary due to the impossibility of guaranteeing correctness. This capability fosters transparency by enabling stakeholders to compare policies in different formats across the dataspace, while also highlighting potential conflicts or violations for further investigation. By serving as an interface between legal and technical domains, it ensures policies are both enforceable and understandable. Besides the task of policy translation, LLMs could also assist with secondary tasks, such as extracting ontological information from legal texts.

4.6 Example Scenario

We illustrate the policy engine with a practical case. Here, a policy permits use of air quality data only if the spatial resolution is up to 500 m (e.g. district level) and the temporal resolution is hourly or finer. If these conditions are not met (e.g., faulty sensor output), the data may still be used, but only as background for trend analysis. The **Ontology Module** and **ODRL Formal Semantics** are shown in the following policy formalization. The first rule grants permission to use the dataset, while the second restricts usage to trend analysis when spatial or temporal limits are exceeded. Other actions (e.g., transfer of ownership) remain prohibited by default since they are not explicitly allowed (Fig. 4).

```
ex:airQualityPolicy a odrl:Policy ;
 odrl:permission [
  odrl:target ex:AirQualityDataset ;
  odrl:action odrl:use ;
  ] ;
 odrl:prohibition [
  odrl:target ex:AirQualityDataset ;
  odrl:action odrl:use ;
  odrl:constraint [
   odrl:leftOperand dcat:spatialResolutionInMeters ;
   odrl:operator odrl:gt ;
   odrl:rightOperand "500"^^xsd:decimal
   ] ;
  odrl:constraint [
   odrl:leftOperand dcat:temporalResolution ;
   odrl:operator odrl:gt ;
   odrl:rightOperand "PT1H"^^xsd:duration
   ] ;
  odrl:constraint [
   odrl:leftOperand odrl:purpose ;
   odrl:operator odrl:neq ;
   odrl:rightOperand ex:backgroundTrendAnalysis
   ]
  ] .
```

Fig. 4. Example ODRL policy for the sharing of Air Quality Data.

When a consumer requests a dataset subset, the **Automated Evaluation Module** verifies spatial/temporal resolution and communicates usage permissions via the **LLM Assistant Module**. This ensures utility, maintains compliance, and provides graceful fallbacks in heterogeneous sensor networks, directly addressing the granularity mismatch problem (**C1**).

5 Conclusion and Future Work

In this paper, we identified eight problem areas and the associated key challenges for policy enforcement that emerged in urban dataspaces and the limitations in the existing ODRL solutions. To bridge this gap, we proposed the design of an enhanced policy engine grounded on a concise ODRL formal semantics for policy definition and validation, and the extensions to core modules in this context. Specifically, our proposal demonstrates how KG technologies (such as ontologies and knowledge repositories) and LLMs can be utilised to address the real-life urban challenges for policy enforcement and support for non-technical users.

While this paper focuses on urban policy challenges that can be addressed technically through policy engines, we also acknowledge human and organisational factors related challenges as highlighted in Sect. 3.2. For example, the challenge related to data ownership requires an agreed definition of data ownership and harmonised data ownership models, while multilateral data governance needs organisational coordination in addition to technical solutions.

For future work, we will extend the formalisation of ODRL semantics, implement the policy engine extensions and validate them in real-life use cases using the Urban Dataspace testbed. We also plan to integrate the policy engine with the Dataspace Connector to support the replication of the solution.

Acknowledgments. The work is funded through the projects DataPACT (HE 101189771), DataTwin (BG16RFPR002-1.009-0005), CauseFinder (PNRR 760049), and UPCAST (HE 101093216).

Disclosure of Interests. The authors have no competing interests.

References

1. Auñón, J., et al.: Evaluation and utilisation of privacy enhancing technologies–a data spaces perspective. Data Brief **55**, 110560 (2024)
2. Bonatti, P.A., Fornara, N., Harth, A.: Towards a formal semantics of the open digital rights language (ODRL 2.2). In: Proceedings of the OPAL 2025 (2025)
3. BSI: Smart cities framework for sharing data and information services - pas 183. https://www.bsigroup.com/en-GB/insights-and-media/insights/brochures/pas-183-smart-cities-framework-for-sharing-data-and-information-services. Accessed 18 Aug 2025
4. Choenni, S., Bargh, M.S., Busker, T., Netten, N.: Data governance in smart cities: challenges and solution directions. J. Smart Cities Soc. **1**(1), 31–51 (2022)
5. Cimmino, A., Cano-Benito, J., García-Castro, R.: Open digital rights enforcement framework (ODRE): from descriptive to enforceable policies. Comput. Secur. **150**, 104282 (2025)
6. Data Space Support Centre: Data space support centre. https://dssc.eu. Accessed 18 Aug 2025
7. Data Spaces Support Centre: Access & usage policies enforcement. https://dssc.eu/space/BVE/357075567/Access+&+Usage+Policies+Enforcement. Accessed 18 Aug 2025
8. Data Spaces Support Centre: Data spaces blueprint v2.0. https://dssc.eu/space/BVE2/1071251781/1+Key+Concept+Definitions. Accessed 30 Aug 2025
9. Data Spaces Support Centre: Dataspace protocol. https://docs.internationaldataspaces.org/ids-knowledgebase/dataspace-protocolhttps://dssc.eu/space/BVE/357075567/Access+&+Usage+Policies+Enforcement. Accessed 18 Aug 2025
10. De Vos, M., Kirrane, S., Padget, J., Satoh, K.: ODRL policy modelling and compliance checking. In: Fodor, P., Montali, M., Calvanese, D., Roman, D. (eds.) RuleML+RR 2019. LNCS, vol. 11784, pp. 36–51. Springer, Cham (2019). https://doi.org/10.1007/978-3-030-31095-0_3
11. European Commission: Data act explained. https://digital-strategy.ec.europa.eu/en/factpages/data-act-explained#:~:text=In%20addition%2C%20it%20introduces%20safeguards,the%20European%20economy%20and%20society. Accessed 18 Aug 2025

12. FIWARE: Fiware. https://www.fiware.org. Accessed 18 Aug 2025
13. Gaia-X: Gaia-x. https://gaia-x.eu. Accessed 18 Aug 2025
14. Gheisari, S., Yumusak, S., Salas, J.O., Ibáñez, L.D., Konstantinidis, G., Roman, D.: Towards modular data marketplaces. In: Proceedings of the 8th International Joint Conference on Rules and Reasoning (2024)
15. International Data Spaces: Guiding principles. https://docs.internationaldataspaces.org/ids-knowledgebase/idsa-rulebook/idsa-rulebook/2.-guiding-principles. Accessed 18 Aug 2025
16. International Data Spaces: International data spaces. https://internationaldataspaces.org. Accessed 18 Aug 2025
17. International Data Spaces: Usage control in the international data spaces. https://internationaldataspaces.org/wp-content/uploads/dlm_uploads/IDSA-Position-Paper-Usage-Control-in-the-IDS-V3..pdf. Accessed 18 Aug 2025
18. International Data Spaces Association: Ids-ram 4.0. https://github.com/International-Data-Spaces-Association/IDS-RAM_4_0. Accessed 18 Aug 2025
19. iSHARE: Trust framework for data rights. https://ishare.eu. Accessed 18 Aug 2025
20. Lam, A.N., et al.: Towards a toolkit for semantic interoperability in data spaces. In: 2025 IEEE 8th International Conference on Industrial Cyber-Physical Systems (ICPS), pp. 1–7 (2025). https://doi.org/10.1109/ICPS65515.2025.11087848
21. Martella, A., Martella, C., Longo, A.: Designing data spaces: navigating the European initiatives along technical specifications. arXiv:2503.1599 (2025)
22. Micheli, M., Ponti, M., Craglia, M., Berti Suman, A.: Emerging models of data governance in the age of datafication. Big Data Soc. **7**(2) (2020)
23. Perucci, F., Swanson, E.: Building trust and facilitating use of data. Stat. J. IAOS **40**(1), 71–79 (2024)
24. ProdataKey: Understanding access control systems: what is access control? https://www.prodatakey.com/single-post/what-is-an-access-control-system. Accessed 18 Aug 2025
25. Siska, V., Karagiannis, V., Drobics, M.: Building a dataspace: technical overview. https://www.gaia-x.at/wp-content/uploads/2023/04/WhitepaperGaiaX.pdf. Accessed 18 Aug 2025
26. W3C: ORDL information model. https://www.w3.org/TR/odrl-model. Accessed 18 Aug 2025
27. W3C: OWL web ontology language. https://www.w3.org/TR/owl-features. Accessed 18 Aug 2025
28. W3C: Pav - provenance, authoring and versioning. https://pav-ontology.github.io/pav. Accessed 18 Aug 2025
29. W3C: RDF schema 1. https://www.w3.org/TR/rdf-schema. Accessed 18 Aug 2025
30. W3C: SKOS simple knowledge organization system namespace document. https://www.w3.org/2009/08/skos-reference/skos.html. Accessed 18 Aug 2025
31. W3C: Verifiable credentials data model v2.0. https://www.w3.org/TR/vc-data-model-2.0. Accessed 18 Aug 2025
32. Yumusak, S., Gheisari, S., Salas, J.O., Ibáñez, L.D., Konstantinidis, G.: Data sharing negotiation and contracting. In: Proceedings of ISWC 2024 (2024)

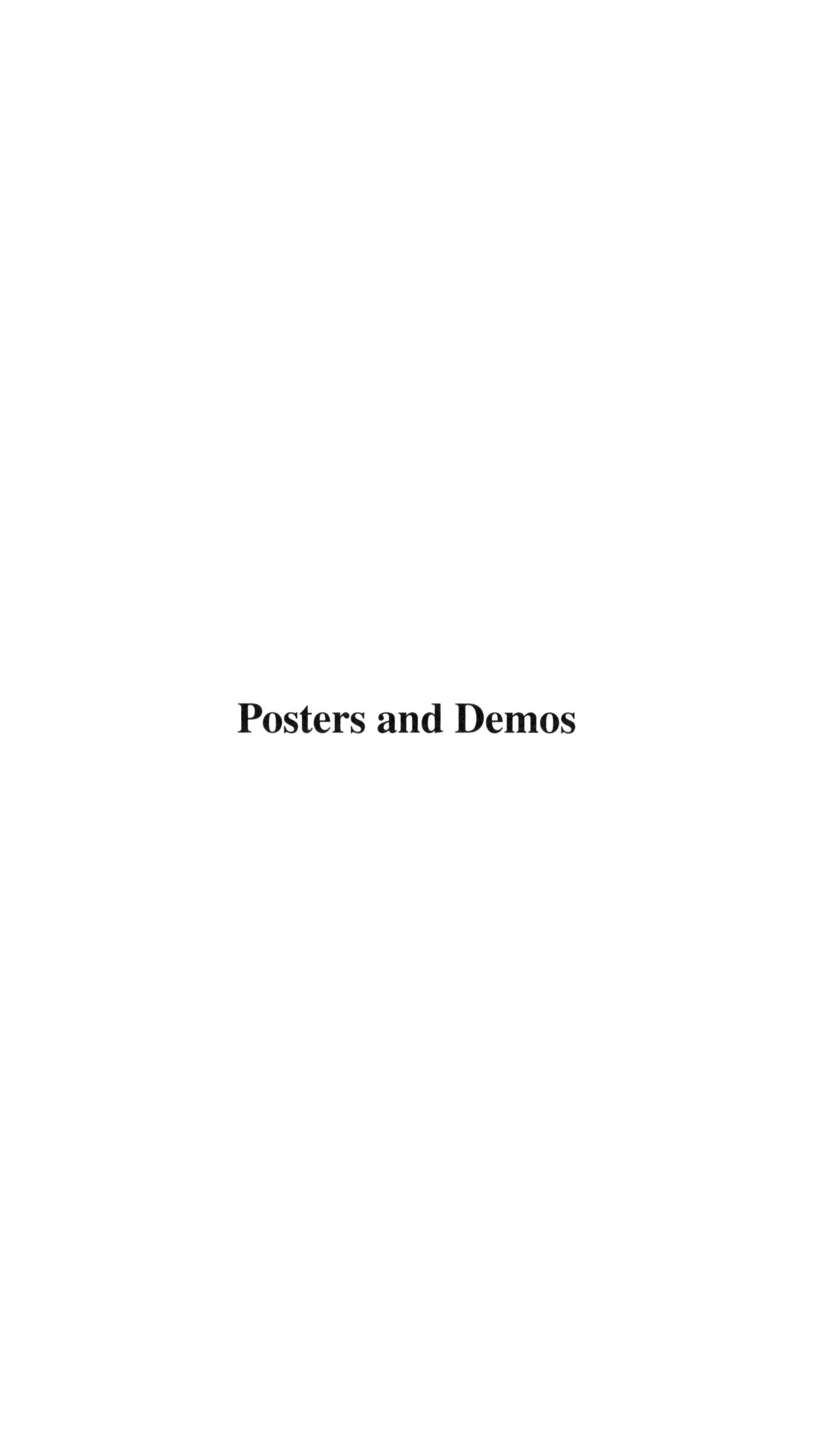

Posters and Demos

ULYSSES: Demonstrating Automated FreqUentLY ASked QueStions for KnowlEdge GraphS

Iakovos Giannakos[1], Giannis Vassiliou[1], Georgia Eirini Trouli[1,3], Nikolaos Papadakis[1], and Haridimos Kondylakis[2,3](✉)

[1] Department of Electrical and Computer Engineering, HMU, Heraklion, Greece
[2] Computer Science Department, UOC, Rethymno, Greece
[3] Institute of Computer Science, FORTH, Heraklion, Greece
kondylak@ics.forth.gr

Abstract. The exponential growth of knowledge graphs necessitates effective and efficient methods for their exploration and understanding. Frequently Asked Questions (FAQ) is a list of questions and answers related to a specific topic intended to help people understand a particular subject. In this demonstration, we present ULYSSES, the first system for automatically constructing a FAQ for large Knowledge Graphs. Our method consists of three key steps. First, we select the most frequent queries by exploiting available query logs. Next, we answer the selected queries using the original graph and finally, we construct textual descriptions of both the queries and the corresponding answers by exploring state-of-the-art transformer models, i.e., ChatGPT and Gemini. In the demonstration, we explain the various steps and allow conference participants to test the system in real time and identify its usefulness.

Keywords: Frequently Asked Questions · Knowledge Graphs · Query Logs

1 Introduction

Due to the complex structure of RDF Knowledge Graphs (KGs), their heterogeneity, and the massive size of the information they contain, users find it difficult to comprehend and explore them [6,7,11]. As a result, there is an increasing need to develop methods and tools that facilitate the quick understanding and exploration of the information they contain [2,8,9].

Frequently Asked Questions (FAQ) is an essential service for users' self-assistance, offered by several websites. FAQs typically present a list of questions, each associated with an answer, enabling users to rapidly identify key questions/answers in the context of the information provided by the website. Existing works in the domain work on a given textual corpus and focus on auto-generating questions and answers [5], or on answering FAQs questions by

H. Takeda et al. (Eds.): IJCKG 2025, LNAI 16297, pp. 425–429, 2026.
https://doi.org/10.1007/978-981-95-5009-8_28

building a knowledge graph to train the pre-trained model [3,10,12,13]. However, although FAQs have proved their usefulness on the web [4], they remain completely unexplored in the context of big knowledge graphs now available online.

Of course, FAQs unavailability for KGs comes from several reasons. First, since the KGs contain a vast amount of information, it is not clear which is the most interesting information to be presented to the users. Second, even if we could identify the most interesting information, users are accustomed to seeing textual descriptions of both the questions and the answers and not structured queries or subgraphs of the original graph.

In this paper, we demonstrate ULYSSES, the first system enabling the construction of high-quality FAQs over knowledge graphs. To address the problem of selecting the relevant information to be presented to the users, ULYSSES exploits query logs available from the SPARQL endpoints of the corresponding KGs. ULYSSES identifies the most frequent SPARQL queries in the logs and uses the corresponding SPARQL endpoints for retrieving their answers. Then, in order to transform both queries and answers into text, it exploits transformer models, i.e., the ChatGPT and the Gemini LLMs, using appropriate prompts. We demonstrate our approach on the DBpedia KG and we show the interesting results achieved by ULYSSES.

To the best of our knowledge, this is *the first approach for automated construction of FAQs for large knowledge graphs*, exploiting existing user logs and transformer models.

2 Architecture

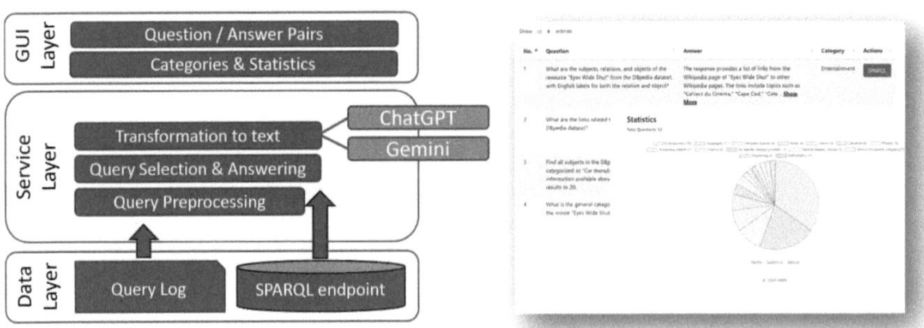

Fig. 1. ULYSSES high-level architecture (left) and GUI (right).

To generate an FAQ for a KG, ULYSSES uses a three-layer architecture, shown on the left of Fig. 1. At the bottom, the data layer includes the sources upon which the FAQ is built, i.e., the query log and the SPARQL endpoint of the selected KG. On top, the service layer constructs the FAQ, using three discrete and sequential steps. First, existing query logs are preprocessed and cleaned. Then the most frequent queries are identified and the corresponding KG is used to answer them. Eventually, both the queries and answers are transformed into text using transformer models. In the GUI layer, the constructed FAQ is presented. A screenshot of our system is shown on the right of Fig. 1. In addition, the GUI allows switching between the two language models to see their results and also organizes the extracted information into distinct categories, generated using the transformer models. In the sequel, we describe the service layer in detail.

Query Preprocessing and Cleaning. Initially, ULYSSES preprocesses the SPARQL log file, eliminating queries that were malformed or erroneous. Then it excludes *DESCRIBE*, *ASK*, and *CONSTRUCT* queries as they would not contribute to a FAQ. It also drops queries that are too generic and exploratory and would not contribute to a useful FAQ (e.g., *SELECT * WHERE ?a <property> ?b.*). Finally, it excludes queries that do not return any results.

Query Selection & Answering . After the preprocessing of the queries, ULYSSES selects the next most frequent ones to use for the construction of the FAQ. The selected queries are next answered using the corresponding SPARQL endpoint. The results are collected to be provided for the transformation step.

Transformation to Text. In order to generate textual FAQs we translate in the sequel both queries and answers to text using large transformer models. The aim is to unlock the potential for a seamless and intuitive translation of intricate SPARQL queries and their corresponding answers, paving the way for a more user-friendly and inclusive knowledge-sharing experience. ULYSSES uses two modules for the translation, specifically, ChatGPT and Gemini. GPT-3.5 is a state-of-the-art language model developed by OpenAI, capable of understanding and generating human-like text across a wide range of topics and styles. It uses a deep neural network architecture with 175 billion parameters to achieve its language processing abilities. Google's Gemini on the other hand, is a more recent entrant, enabling even broader capabilities (e.g., multimodal AI), integrating language processing with image processing and potentially other modalities [1].

First, the models translate the most frequent SPARQL queries into plain text. Then, based on the KG responses and the textual question, ULYSSES requests a plain-language translation for the responses from the two LLMs. In the response conversation within those models, the translated query is included in order to ensure coherence and context continuity.

3 Demonstration Scenario and Conclusion

The purpose of this demonstration is primarily to understand the procedure for constructing an FAQ for large Knowledge Graphs, exploring two state-of-the-art transformer models (LLMs). The code of the ULYSSES system, along with the dataset and workload, is available in our GitHub repository[1] while the system is accessible online[2].

In the demonstration, we will use a real-world RDF dataset, specifically, DBpedia v3.8, which occupies 103GB of storage, for which we have available a query log with 58,604 user queries. The demonstration will proceed in three phases:

(a) *Queries/FAQ.* Initially, we will demonstrate the most frequently asked questions for DBpedia in their SPARQL form. Conference participants will be able to explore the queries and submit them to the corresponding SPARQL endpoint for inspecting the returned results.

(b) *Query Translation and Filtering.* In this phase, the SPARQL queries will be transformed into text using a selected LLM and categorized into different domains, such as music, physics, history, etc., based on their content using the LLMs. Users can visualize all the queries of the FAQ or filter questions based on their category. Statistical information about the number of queries in the various categories is also available.

(c) *Answer Translation.* This phase allows visualizing the query answer into text and verifying the results. We will discuss the differences in the FAQs from the two LLMs, understanding the deficiencies and capabilities of both pipelines, and comparing their results with those from experts. Overall, ChatGPT performs best in providing the FAQ with the highest ROUGE scores.

4 Conclusions

In essence, to the best of our knowledge, there is no other work currently available, generates FAQ for knowledge graphs. As such, this study demonstrates the unique potential of AI-driven techniques in managing KGs and opens a new direction of research, laying the groundwork for further study and advancement. As future work, we intend to explore personalized FAQs for users of FAQs for specific domains/categories.

Acknowledgments. The work reported in this paper is implemented in the framework of H.F.R.I call “Basic research Financing (Horizontal support of all Sciences)” under the National Recovery and Resilience Plan “Greece 2.0” funded by the European Union âĂŞ NextGenerationEU (H.F.R.I. TALE Project, Number: 16819).

References

1. Anil, R., et al.: Gemini: a family of highly capable multimodal models. CoRR arxiv:2312.11805 (2023)

[1] https://github.com/giannisvassiliou/KGFaq/.

[2] https://clinquant-belekoy-781b9f.netlify.app/.

2. Bonifati, A., Dumbrava, S., Kondylakis, H., Troullinou, G., Vassiliou, G.: Progressive querying on knowledge graphs. In: Simitsis, A., Kemme, B., Queralt, A., Romero, O., Jovanovic, P. (eds.) Proceedings 28th International Conference on Extending Database Technology, EDBT 2025, Barcelona, Spain, 25–28 March 2025, pp. 106–118. OpenProceedings.org (2025). https://doi.org/10.48786/EDBT.2025.09
3. Liu, A., et al.: BB-KBQA: bert-based knowledge base question answering. In: CCL, vol. 11856, pp. 81–92 (2019)
4. de Oliveira, E.C.C., et al.: Extracting and searching useful information available on web faqs. In: XXI Simpósio Brasileiro de Banco de Dados, pp. 102–116 (2006)
5. Raazaghi, F.: Auto-faq-gen: automatic frequently asked questions generation. In: Advances in Artificial Intelligence, vol. 9091, pp. 334–337 (2015)
6. Trouli, G.E., Papadakis, N., Kondylakis, H.: Constructing semantic summaries using embeddings. Information **15**(4), 238 (2024). https://doi.org/10.3390/INFO15040238
7. Trouli, G.E., Pappas, A., Troullinou, G., Koumakis, L., Papadakis, N., Kondylakis, H.: Summer: structural summarization for RDF/S kgs. Algorithms **16**(1), 18 (2023). https://doi.org/10.3390/A16010018
8. Troullinou, G., Agathangelos, G., Kondylakis, H., Stefanidis, K., Plexousakis, D.: DIAERESIS: RDF data partitioning and query processing on SPARK. Semant. Web **15**(5), 1763–1789 (2024). https://doi.org/10.3233/SW-243554, https://doi.org/10.3233/SW-243554
9. Troullinou, G., Stefanidis, K., Plexousakis, D., Kondylakis, H.: DIAERESIS: knowledge graph partitioning for efficient query answering. In: Chua, T., Ngo, C., Lee, R.K., Kumar, R., Lauw, H.W. (eds.) Companion Proceedings of the ACM on Web Conference 2024, WWW 2024, Singapore, Singapore, 13–17 May 2024. pp. 987–990. ACM (2024). https://doi.org/10.1145/3589335.3651231
10. Tseng, W., et al.: FAQ retrieval using question-aware graph convolutional network and contextualized language model. In: APSIPA ASC 2021, pp. 2006–2012 (2021)
11. Vassiliou, G., Alevizakis, F., Papadakis, N., Kondylakis, H.: isummary: Workload-based, personalized summaries for knowledge graphs. CoRR arxiv:2403.02934 (2024). https://doi.org/10.48550/ARXIV.2403.02934
12. Xie, R., et al.: Faq-based question answering via knowledge anchors. In: NLPCC, vol. 12430, pp. 3–15 (2020)
13. Zhao, H., et al.: Knowledge graph based question pair matching for domain-oriented FAQ system. In: SMC, pp. 2103–2108 (2022)

Unsupervised Linking of Unstructured Data Entities to a Knowledge Graph

Gabriel Dumont[1(✉)], Pedro Merino-Laso[2,3], and John Puentes[4,5]

[1] Chair of Naval Cyber Defense, French Naval Academy, Brest, France
gabriel.dumont@ecole-navale.fr
[2] French Maritime Academy (ENSM), Brest, France
[3] French Naval Academy Research Institute (IRENav), EA 3634 Brest, France
[4] IMT Atlantique, Brest, France
[5] Lab-STICC, CNRS UMR 6285, Brest, France

Abstract. Recently, Large Language Models (LLM) have been used to generate structured graph data from unstructured text. However, parts of the extracted nodes lack information and retraining is necessary whenever the graph changes. We propose an unsupervised approach based on embeddings similarity, to link entities extracted by a LLM from unstructured text data to nodes of a knowledge graph, generated according to an ontology. Tested on three datasets from a maritime cyber security use case, results indicate an accuracy improvement from 86.2% up to 91,4%, depending on the number of neighbors and context candidate nodes.

Keywords: Knowledge Graph · Ontology · Embedding Similarity · Context · Free Text

1 Introduction

Root cause analysis (RCA) permits to interpret data and information within a broad system perspective. To this end, a knowledge graph (KG) represents system behavior, enabling to trace cause-effect chains, as well as pattern recognition from historical and real-time data. Moreover, in the domain of cyber security, RCA applying one or several KGs, is likely to provide a dynamic global and local representation of complex, interconnected system components, to detect, analyze, and mitigate threats. In this case, KGs serve as reference to instantiate entities derived from unstructured data like free text. While a KG can be relatively stable, derived entities from unstructured data like free text, lack semantic consistency, are ambiguous, and/or difficult to generalize. It remains therefore unclear how to verify in an unsupervised manner, candidate linkages to a KG, of entities extracted from unstructured data. This complex issue makes necessary to design mechanisms capable of resolving incoherences of entities, with respect to a predefined structure of known nodes connections, i.e., a KG. Our work proposes a hybrid approach based on embeddings similarity, to link reliably entities

H. Takeda et al. (Eds.): IJCKG 2025, LNAI 16297, pp. 430–436, 2026.
https://doi.org/10.1007/978-981-95-5009-8_29

extracted by a large language model (LLM) from unstructured data to nodes of a KG, generated according to an ontology[1].

2 Related Works

Entity linking (EL) encompasses identifying and disambiguating elements of data and information found in unstructured text and linking them to their corresponding entries in a KG. Mapping of extracted entities to the right concepts in the KG, allows examining the coherence, completeness, and pertinence of analyzed documents.

LLMs and Knowledge Graphs: Since LLMs can process vast amounts of unstructured textual data, it is feasible to generate and refine KGs. Integration of LLMs into KG construction workflow intends to automate tasks like nodes extraction, relationship identification, and semantic link prediction [11]. Likewise, LLMs application for EL [6] has emerged as an alternative to named entity recognition systems [9]. An advantage of the LLM approach is that entity recognition can be done as a generative task, without any training or LLM fine-tuning [1,11]. However, resulting entities are not directly usable, making necessary the inclusion of additional information to resolve structure and semantic conflicts.

Graph Embeddings: Elements of a graph like nodes, edges, or entire subgraphs, can be represented as text embeddings[2]. Despite its interest, few works have developed unsupervised embeddings to link unstructured text entities and KG nodes. Even if an embedding is trained on a KG that uses different types of contexts [3], or two embeddings are trained in parallel, one on a knowledge base and another on a text corpus [10], data from raw text cannot be linked. Given that training on a dataset is mandatory for these algorithms, the resulting KG cannot be evolutive, except at the cost of embedding retraining, whenever the graph is modified. We propose therefore a general-purpose text-embedding approach, to remain as flexible as possible, avoiding multiple trainings.

Entity Linking to a Knowledge Graph: Among numerous approaches to achieve EL on a graph [8], a common approach is to combine two techniques, one to select preliminary entity candidates and then another to identify the best ones. Using text embeddings and weighted similarity scores for a candidate node and its neighbors, candidate nodes selection can be done with a pipeline trained on a KG and the respective documents [5]. Additional elements like the context of entities, may be also taken into account in the training. However, state-of-the-art approaches require annotated data, which is not necessarily available.

Unsupervised Entity Linking: In cases where annotated data is lacking, unsupervised approaches become valuable. Such methods use general-purpose

[1] An ontology means here the structure of a KG. It describes classes of nodes and relationships.

[2] An embedding can mean either a representation of data as a low-dimensional vector, or the vectorization process itself.

models to achieve EL, using for instance Abstract Meaning Representation [7] or a LLM [2]. Nevertheless, none of these methods use simple (not fine-tuned) text embedding models for this task.

3 Data

We conceived an easily interpretable dataset of reduced size, consisting on geographic locations of maritime places with ports. Our experimental framework includes two main components (Fig. 1): a KG and a dataset of named entities. The KG is based on an ontology (Fig. 2), which integrates geographic information from Wikipedia[3,4] and maritime information from the International Hydrographic Organization, available on the Searates site[5].

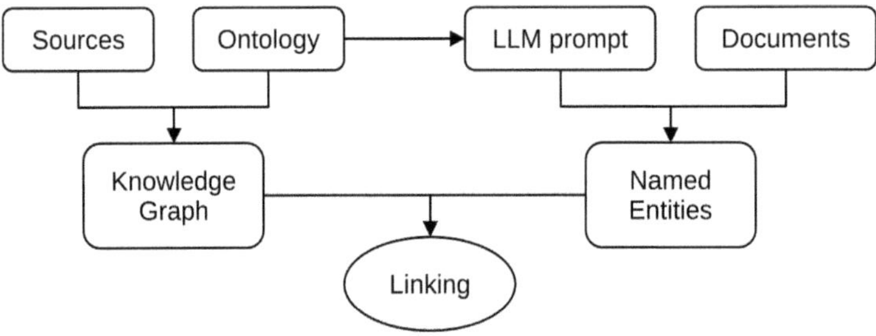

Fig. 1. Block diagram of the experimental framework.

The entities dataset was extracted from 48 articles of year 2020, identified in the ADMIRAL - Advanced Dataset of Maritime cyber Incidents ReleAsed for Literature - database [4], which documents disclosed cybersecurity incidents in the maritime sector. Generally, cybersecurity incidents reports have different forms, with variable degrees of detail, written in different languages, missing oftentimes some data and information. Such characteristics are those of a suitable dataset to identify entities in unstructured text, as well as to evaluate linking performance.

[3] Wikipedia. List of sovereign states and dependent territories by continent https://en.wikipedia.org/wiki/List_of_sovereign_states_and_dependent_territories_by_continent.
[4] Wikipedia. List of countries and territories by land and maritime borders. https://en.wikipedia.org/wiki/List_of_countries_and_territories_by_land_and_maritime_borders.
[5] Searates. Sea Ports in the World. https://searates.com/maritime.

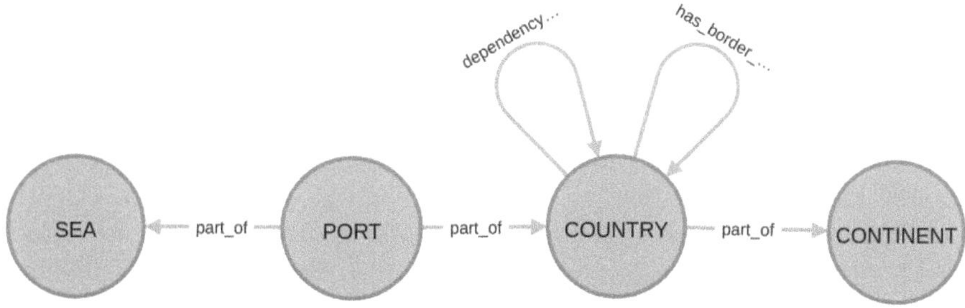

Fig. 2. Ontology of the maritime geographic KG.

Entities were extracted using the structured output feature of MistralAI's API[6]. The used prompt includes a natural language description of the ontology. Entities were then matched manually to identifiers of target nodes in the KG. These identifiers are used to compute the linking accuracy. Three versions of the 210 entities dataset, named with the common root label "MaLin", for Maritime Linking, are utilized: bilingual (original French & English dataset), fully translated to English, and translated to French. These three datasets are referred to as MaLin-bi, MaLin-en, and MaLin-fr, respectively. Below is an example of a MaLin-en entity, structured as *(class, entity, context, corresponding node id)*:

```
COUNTRY
Iran
Microsoft announced that Iran had been "rapidly accelerating"
    cyberattacks since mid-2022.
55
```

While the KG is extracted from a verified source, the test dataset serves a different role. Its noisy nature reflects real-world challenges that models must navigate. As a result, measuring traditional quality aspects is out of scope and does not detract from the usefulness of the test dataset.

4 Method

Candidates Preselection: Before EL, each node in the KG is embedded (represented as a vector). Then, textual entities to be linked are processed with the same embedding model. For each entity, the closest nodes are selected, according to a similarity computed on the embeddings. The used measure is cosine similarity. Selected nodes are kept as candidates for the second step.

Linked Nodes Selection: Similarities could be computed initially only for the simplest version of extracted entities. Still, this calculation is also possible when the associated context the part of the document around the entity is embedded and included in the similarity evaluation [5]. In that case, even if the number of

[6] Mistral API documentation: https://docs.mistral.ai/.

context elements is highly variable, a maximum pooling is applied to similarities between the enhanced entity elements and each neighbor of the candidate node.

Similarity can be computed as well, between an entity's context and the neighborhood of a KG node, producing four distinct similarities: entity/candidate; entity/neighborhood; context/candidate; and context/neighborhood. Once obtained, these four indicators are weighted to have a single score. On the other hand, keeping some weights constant allows testing specific values, for instance with or without neighbors, to select the nodes with the best score for linking.

5 Results

We tested our approach applying two different open-source generic-purpose text embeddings approaches of the same size, namely nomic and mxbai, provided by Nomic and Mixedbread respectively. As a baseline, neither context embedding nor node neighborhood embedding are used. Only the similarity between the entity itself and a candidate node is considered. The performance of the two embeddings is shown in Table 1. Results show the percentage of correctly linked entities. Given these baseline results, the model used in all other tests is mxbai.

Table 1. Performance of the initial embeddings

Embedding	Dimension	MaLin-bi	MaLin-en	MaLin-fr
nomic	1024	80.5%	85.2%	62.9%
mxbai	1024	**81.4%**	**86.2%**	**65.2%**

Table 2 presents obtained accuracies when applying our approach to the three datasets. Tests compare the performance depending on the usage of neighbors embedding on the candidate node side and/or context embedding on the entity side. Five candidates are used as a baseline for these tests. A random grid search is used to find the best weighting values between the 4 similarities mentioned in the proposed approach's second step.

Table 2. Effect on linking accuracy of neighbors and context

Neighbors	Context	MaLin-bi	MaLin-en	MaLin-fr
No	No	81.4%	86.2%	65.2%
Yes	No	83.8%	88.6%	68.1%
No	Yes	81.9%	87.1%	65.7%
Yes	**Yes**	**84.8%**	**89.5%**	**68.6%**

Results suggest that combining neighbors and context improves the linking process. Their outcome appears to be different, with neighbors having a higher impact than context, when considered alone, while the best results are obtained making use of both. Regarding datasets, the best linking occurs when it is applied to a monolingual dataset, for which content corresponds to the same language as the KG, i.e., English. An equivalent global behavior with better accuracy for each higher size set of candidates, is observed when the number of candidates increases gradually. Figure 3 shows the relation between needed time to compute accuracy on each dataset, accuracy itself, and the corresponding number of candidates. The best result reached on MaLin-en with 30 candidates is 91.4%.

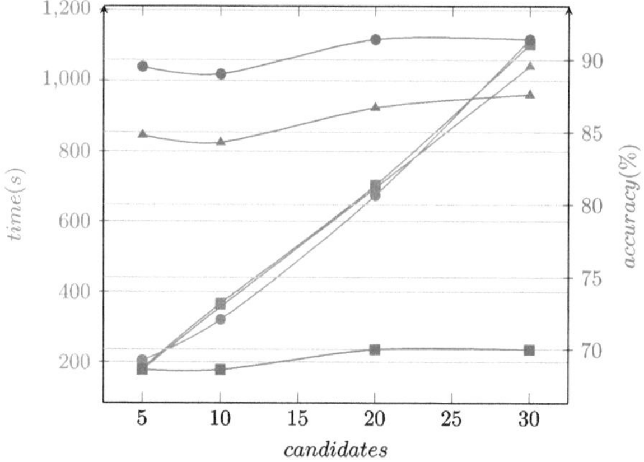

Fig. 3. Impact of the number of candidates on computation time and accuracy. ●: Malin-en, ▲: Malin-bi, ■: Malin-fr

6 Conclusion and Perspectives

Linking of unstructured text entities and KG nodes is a complex problem for which learning-based approaches have been examined, relying on embedding retraining whenever the graph is modified. A flexible general-purpose text embedding approach, of linear complexity, was defined and tested to link in an unsupervised manner, entities extracted from unstructured data to nodes of a KG. The highest accuracy was measured combining both candidates' nodes neighbors and entities context, with candidates' nodes neighbors having a higher impact than entities context, if applied alone. Our approach will be operable in a maritime cybersecurity testing use case, for RCA. Since the algorithm behavior may change depending on the datasets, future works will include evaluation on other datasets, as well as different ontologies to build the KG. Furthermore, in spite of linear complexity, optimization is required to reduce calculation time.

Acknowledgements. This work was possible thanks to the Chair of Naval Cyber Defense, supported and funded by the French Naval Academy, IMT Atlantique, ENSTA, ENSM, Naval Group, and Thales.

References

1. Chen, H.: Large knowledge model: perspectives and challenges. Data Intell. **6**(3), 587–620 (Aug2024). https://doi.org/10.3724/2096-7004.di.2024.0001
2. Cho, Y.M., Zhang, L., Callison-Burch, C.: Unsupervised entity linking with guided summarization and multiple-choice selection. In: Proceedings of the 2022 Conference on Empirical Methods in Natural Language Processing, pp. 9394–9401 (2022). https://doi.org/10.18653/v1/2022.emnlp-main.638
3. Feng, J., Huang, M., Yang, Y., Zhu, X.: GAKE: graph aware knowledge embedding. In: Proceedings of COLING 2016, the 26th International Conference on Computational Linguistics: Technical Papers (2016)
4. Jacq, O.: Advanced dataset of maritime cyber incidents released for litterature, admiral dataset. https://gitlab.com/m-cert/admiral
5. Nedelchev, R., Chaudhuri, D., Lehmann, J., Fischer, A.: End-to-end entity linking and disambiguation leveraging word and knowledge graph embeddings (2020). https://doi.org/10.48550/arXiv.2002.11143
6. Pan, S., Luo, L., Wang, Y., Chen, C., Wang, J., Wu, X.: Unifying large language models and knowledge graphs: a roadmap. IEEE Trans. Knowl. Data Eng. **36**(7), 3580–3599 (2024). https://doi.org/10.1109/TKDE.2024.3352100
7. Pan, X., Cassidy, T., Hermjakob, U., Ji, H., Knight, K.: Unsupervised entity linking with abstract meaning representation. In: Proceedings of the 2015 Conference of the North American Chapter of the Association for Computational Linguistics: Human Language Technologies, pp. 1130–1139 (2015). https://doi.org/10.3115/v1/N15-1119
8. Shen, W., Wang, J., Han, J.: Entity linking with a knowledge base: issues, techniques, and solutions. IEEE Trans. Knowl. Data Eng. **27**(2), 443–460 (Feb2015). https://doi.org/10.1109/TKDE.2014.2327028
9. Wang, S., et al.: GPT-NER: named entity recognition via large language models (2023). https://doi.org/10.48550/arXiv.2304.10428
10. Zhong, H., Zhang, J., Wang, Z., Wan, H., Chen, Z.: Aligning knowledge and text embeddings by entity descriptions. In: Proceedings of the 2015 Conference on Empirical Methods in Natural Language Processing, pp. 267–272 (2015). https://doi.org/10.18653/v1/D15-1031
11. Zhu, Y., et al.: LLMs for knowledge graph construction and reasoning: recent capabilities and future opportunities (2024). https://doi.org/10.48550/arXiv.2305.13168

Process Copilot: Speeding Up Process Digitization in the German Public Sector

Birte Caesar(✉), Solveig Peter, Celsa Diaz-Tejada, Matthias Werner, Oskar Krafft, and Michél Mahler

PD-Berater der öffentlichen Hand GmbH, Friedrichstr. 149, 10117 Berlin, Germany
{birte.caesar,solveig.peter,celsa.diaz-tejada,matthias.werner, oskar.krafft,michel.mahler}@pd-g.de
https://www.pd-g.de/

Abstract. Public institutions seldom document the informal, undocumented processes that shape everyday work. As institutions digitize and modernize, such tacit knowledge is easily lost, particularly amid demographic change and limited process modeling expertise. To address this problem, we present Process Copilot, a generative AI system that converts natural language process descriptions into BPMN 2.0 models using large language models (LLMs). To improve the quality and reliability of generated process models, we incorporate domain-specific ontologies into the prompting process. We evaluate two prompting strategies, with and without ontologies, across ten real-world processes, using syntactic and structural similarity metrics alongside stability analysis based on output variance. Results indicate that ontology integration enhances semantic and structural performance, though at the cost of stability.

Keywords: Process Modeling · BPMN2.0 · Generative AI · Knowledge Graphs · Domain Ontology

1 Introduction

The ongoing modernization of public administration compels institutions to digitize and streamline their internal processes [18]. Yet process knowledge remains fragmented across individuals, hindering transparency and documentation [22]. This problem is further exacerbated by demographic shifts and the loss of experienced staff [4]. While formal process modeling automation using standards like BPMN 2.0[1] offers a potential solution, many public institutions lack the modeling expertise to standardize and implement such methods effectively.

Prior to LLMs, automation of knowledge graph generation relied on natural language processing (NLP) techniques like embedding-based and pre-trained language models (e.g., PG-BERT), which struggled with domain ambiguity and required large, annotated datasets, limiting scalability [12,23]. **LLM-based**

[1] https://www.omg.org/bpmn/.

H. Takeda et al. (Eds.): IJCKG 2025, LNAI 16297, pp. 437–444, 2026.
https://doi.org/10.1007/978-981-95-5009-8_30

knowledge graph construction offers a promising alternative, enabling the joint extraction of entities and relations in a single step through methods such as in-context learning and fine-tuning [6], though fine-tuning is often impractical due to resource constraints [1]. In-context learning, while more accessible, faces challenges such as hallucinations, lack of implicit knowledge, and poor reproducibility [13]. Recent advances incorporate domain-specific ontologies as semantic guardrails to improve output quality, consistency, and reduce hallucinations in LLM-generated knowledge graphs [5,17]. **LLM-based process modeling** is closely linked to LLM-driven knowledge graph construction, as both require domain-specific knowledge and produce structured outputs like graphs or diagrammatic languages. To improve quality and reproducibility, Kourani et al. [14] proposed a prompting strategy combining role definition as a BPMN 2.0 expert and Chain-of-Thought prompting to extract intermediate POWL graphs, later converted to BPMN code outside the LLM. Neuberger et al. [19] experimented with embedding ontology-like information in prompts to structurally guide outputs, while Köpke et al. [16] introduced a meta-model intermediate representation similar to ontologies to reduce prompt complexity during iterative refinements. However, and the best of our knowledge, no process modeling approach to date integrates formal ontologies with LLMs.

Ontology-based Process Modeling aims to reduce errors in manual process design by enabling the creation of semantically precise, BPMN-compliant models using formal ontologies. While prior approaches either focus on modeling BPMN syntax or embedding domain-specific knowledge, they typically rely on expert users and do not leverage LLMs for automatic generation from unstructured text. To address this limitation, we extend **Process Copilot** – a generative AI-based system that transforms natural language process descriptions into BPMN 2.0 models – by integrating domain-specific ontology information into the prompting pipeline. This enables non-experts to produce formal models via a guided interface and LLM-powered transformation process. Building on prior work in LLM-guided modeling [9,10], evaluate whether ontology integration improves the quality, consistency, and stability of the generated models. Our evaluation compares prompting strategies – with and without ontology integration – across ten real-world processes using structural and syntactic similarity metrics, as well as output variance.

2 Process Copilot

Process Copilot is an application to transform natural language process descriptions into BPMN 2.0-compliant process models using LLMs. The generated process models are presented in an interactive interface, allowing users to review, edit, and refine the diagrams as needed. Additionally, it includes functionality to export the resulting models in a format compatible with Camunda,[2] enabling seamless integration into existing workflow management systems.

[2] https://camunda.com/platform/modeler/.

LLMs for Process Modeling. As our work is conducted for the German public administration, we prioritize open-source LLMs that can be deployed on premise and perform reliably in German. We therefore tested a range of recent open-source models *Qwen 2.5 72B, LLaMA 3.3 70B, DeepSeek-R1 32B/8B, Qwen 2.5 32B/7B, Qwen 2.5-Coder 32B/7B, Mistral 3.1 24B, and Pharia-1 7B* for generating process diagrams in Mermaid syntax. Only the Qwen and Mistral LLMs consistently produced grammatically correct German Mermaid graphs without graph syntax errors (results are not part of this paper). While these analyses are ongoing, for the purpose of testing our central hypothesis, whether ontologies improve model outputs, we used OpenAI's *GPT-4o*, chosen for its stability, reproducibility, and strong performance under controlled hyper-parameters (*max_tokens* = 1500, *temperature* = 0.1, *top_k* = 50, *top_p* = 0.9). We excluded OpenAI's *o3* model because of lower reproducibility and a higher hallucination rate under comparable settings (results not published).

Prompt Engineering. We developed a hybrid approach that integrates established prompting techniques with domain-specific knowledge and empirical observations from previous work on administrative processes. Our strategy includes the following prompting methods: **(1) Role prompting**, assigning distinct responsibilities to the LLM across different phases to simulate expert behavior and improve task-specific reliability. **(2) Least-to-most prompting**, where simpler process models are generated and later integrated into more complex structures to support gradual model construction. **(3) Input enhancement**, where the initial prompt focuses on improving the quality of the user-provided natural language description to avoid inconsistencies during information extraction. **(4) Use of JSON as an intermediate representation**, serving as a bridge between text and BPMN 2.0. This choice leverages the overrepresentation of JSON in LLM training data (especially via web graph data such as JSON-LD) and facilitates conversion using open-source BPMN tools. **(5) Knowledge injection** via structured guidelines, including Chain-of-Thought prompting to guide the translation from text to JSON-based graph representations. **(6) Ontology-guided modeling**, using a custom RDF ontology to define the elements and evaluation criteria for process models.

The prompting methods outlined above are incorporated into the experimental prompts used to generate the process models (see Fig. 1), whose results are presented in the following section. The design of the methodology has been refined through multiple internal experiments using example cases tailored to our real-world public administration client needs. Nonetheless, we have identified potential areas for improvement, such as adding a dedicated BPMN 2.0 syntax correction role or a preprocessing step to handle inputs containing multiple processes. These points are discussed further in Sect. 4.

Ontology Usage. In this work, we investigate whether using an ontology as context in in-context learning improves model performance and stability compared to unstructured input. We adopt the same chain-of-thought prompting

```
1  role_definition: |
2  You are an expert in BPMN 2.0 and system architecture. Your task is to design a business
3  process model based on a textual process description.
4
5  instruction:
6  Task 1. Creation of a HIGH-LEVEL model by connecting suitable BPMN 2.0 elements.
7  Task 2. Creation of a DETAILED model by extending and configuring the elements.
8
9  requirements:
10 - The model MUST be created in JSON format that complies with the **JSON SCHEMA** specified
11   below.
12 - The process MUST begin with a `Start` event and end with one or more `End` events.
13 - ...
14
15 context:
16     component_library: ${Component Libary}
17     json_schema: ${JSON Schema}
18
19 input_data: ${Corrected Process Description}
```

Fig. 1. Excerpt of the used prompt

approach described previously but replace the component library with an ontology in RDF/Turtle syntax, which previous work has shown to be well-handled by LLMs. Instead of reusing an existing BPMN 2.0 ontology [21], we constructed a simplified version tailored to our application, omitting unnecessary details. As described above, we use a JSON-schema-based intermediate format, which later is transformed into BPMN XML format. The JSON schema's limited expressiveness restricts useful ontology modeling options, so we limited the ontology to classes, subclasses, and rdfs:comment annotations. The ontology was developed in Protégé by extracting all classes, subclasses, and their hierarchical relationships from the BPMN 2.0 specification, following the ontology engineering methodology proposed by [11]. We incorporated definitions from the BPMN 2.0 specification as rdfs:comment annotations to capture concept semantics. Using the BPMN 2.0 standard as the core ontology facilitates integration with domain-specific ontologies via ontology alignment techniques, enabling future applications that respect the modeling guidelines of individual public institutions. Figure 2 depicts the class hierarchy of the BPMN ontology, which was included in the prompt as a component library (see Fig. 1, line 16) using RDF/Turtle syntax.

3 Quantifying the Impact of Ontology Integration

To assess the impact of ontology integration, we compare process models generated by the **Process Copilot** using manually created ground-truth models as a reference. The ground-truth models are based on anonymized administrative processes. The LLM used is the GPT-4.0 model. We selected ten processes of varying categories and complexity levels. For both quality and stability evaluations, we assess two dimensions of similarity: node similarity and structural similarity.

Node Similarity. Node similarity is computed based on entity comparison. In a BPMN diagram, both activities and events are represented as nodes. For each node pair across two models, we compute the labels' string edit distance normalized by the maximum label length; the syntactic similarity. If the syntactic similarity exceeds a given threshold of 0.5, the nodes are considered equivalent. We use the threshold suggested in the corresponding paper [8]. The overall node similarity between two models is then calculated as the normalized sum of the syntactic similarities across matched node pairs.

Structural Similarity. To assess structural similarity, the process models are converted into graphs, where nodes represent activities, events, and gateways, and are labeled accordingly. The structural similarity is then given by the normalized graph edit distance by the total number of nodes [8].

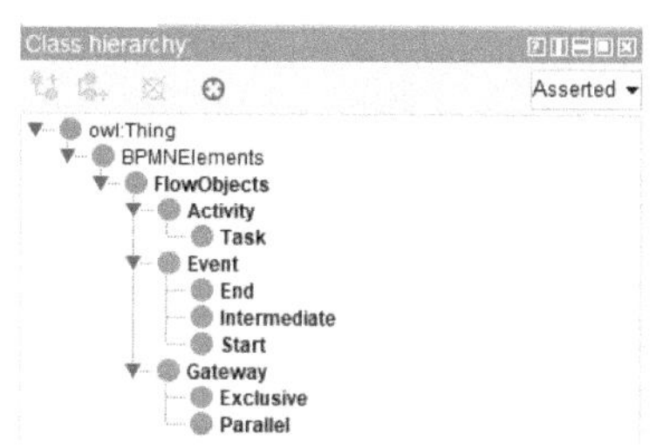

Fig. 2. Ontology class hierarchy.

Ontology Integration Improves Performance. To evaluate performance, we ran each method ten times on each of the ten selected set of ten process examples. The results are presented in Fig. 3, which shows dot plots for both metrics across ten examples. The comparison reveals that ontology integration consistently results in higher average similarity scores. Based on a two-sided MannWhitney U-test, there is only weak evidence of a performance difference for node similarity ($p = 0.053$), whereas ontology integration significantly improves performance for structural similarity ($p = 0.007$).

Stability may be Affected by Increased Complexity. The performance evaluation indicates a greater variance in model outputs when the ontology is incorporated. To quantify stability, we compute pairwise similarities between all outputs within each process individually. The mean node similarity was 0.835 with ontology and 0.866 without ontology usage respectively with a p-value of 0.00009 indicating statistically significant difference. Similarly, the mean structural similarity was 0.866 and 0.875 with and without ontology usage respectively, with a p-value of 0.004. In general, the similarities within each process remain high, indicating relatively strong output stability across all processes and both prompting strategies.

Discussion. Despite performance improvements, the absolute similarity scores remain moderate (0.3–0.6) reflecting the inherent variability and subjectivity in process modelling, where the lack of objectivity is a recognized challenge [2]. In addition, the ground-truth models include BPMN elements such as pools, message flows, catch events and text annotations, which are not represented in our component library or ontology, potentially penalizing structural mismatches outside the expressiveness of our generation system. However, stability is slightly lower with ontology integration, suggesting that the added semantic complexity introduces more variation in how the models are generated. This effect may stem from the expanded search space introduced by the ontology, especially in complex scenarios. Nevertheless, this increase in variance appears modest and does not outweigh the performance gains in model quality.

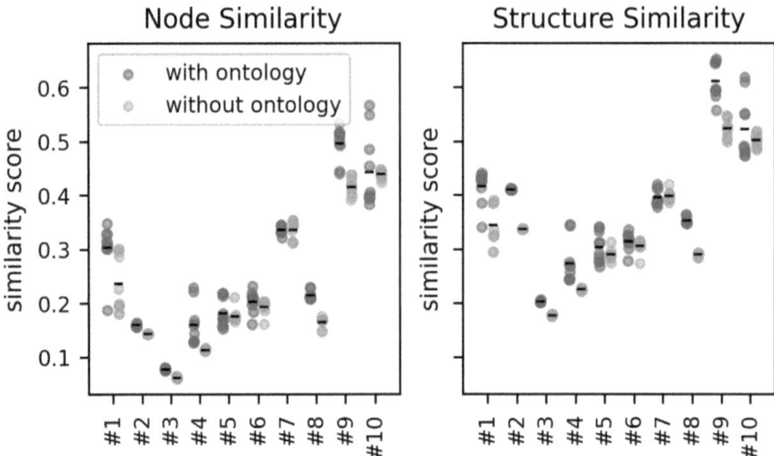

Fig. 3. Including the ontology improves model performance. The figure shows two side-by-side dot plots comparing node similarity and structural similarity between generated and ground-truth process models across ten different example processes (#1 to #10). Each dot represents the result of one run (10 runs per method per example), black lines indicate the mean similarity.

4 Summary and Outlook

This work highlights the potential of ontology-based knowledge integration to improve the semantic and structural quality of LLM-generated process models. While results show benefits, stability remains a concern.

Future work will explore the benefits of incorporating domain-specific ontologies. In particular, we aim to investigate algorithmic integration through constraints or validation to ensure consistent and accurate outputs. We also plan to assess the performance of open source LLMs for BPMN models. Additionally, we will evaluate Process Copilot using the PET dataset [3] to enable comparison with related tools. The PET dataset is a corpus of business process descriptions that are known to experts and considered relevant. This ensures comparability across different methods. Preliminary validation indicates that Process Copilot can already generate a wider range of BPMN element types than ProMoAI [15], supporting more complex process models, and we intend to extend its capabilities further by enabling BPMN subprocesses, frequently requested by our clients.

We will scale evaluations to larger datasets and may include hallucination metrics such as GraphEval [20] for more rigorous assessment. Techniques such as syntax correction with negative prompting, hallucination pruning [7], and agent-based restructuring might further enhance quality. Additional directions include interactive model editing, ontology-guided modeling hints, and integration of open-source models to support use in privacy-sensitive domains. Overall, unlocking the full potential of this approach will require deeper technical integration of knowledge and user-centric system design.

References

1. Bai, J., et al.: AutoSchemaKG: autonomous knowledge graph construction through dynamic schema induction from web-scale corpora (2025). https://doi.org/10.48550/arXiv.2505.23628
2. Beerepoot, I., et al.: The biggest business process management problems to solve before we die. Comput. Ind. **146**, 103837 (2023). https://doi.org/10.1016/j.compind.2022.103837
3. Bellan, P., van der Aa, H., Dragoni, M., Ghidini, C., Ponzetto, S.P.: PET: an annotated dataset for process extraction from natural language text tasks. In: Cabanillas, C., Garmann-Johnsen, N.F., Koschmider, A. (eds.) Business Process Management Workshops - BPM 2022 International Workshops, Münster, Germany, 11–16 September 2022, Revised Selected Papers. Lecture Notes in Business Information Processing, vol. 460, pp. 315–321. Springer (2022). https://doi.org/10.1007/978-3-031-25383-6_23
4. Buck, H.: Demographic change and the public sector (2007)
5. van Cauter, Z., Yakovets, N.: Ontology-guided knowledge graph construction from maintenance short texts. In: Proceedings of the 1st Workshop on Knowledge Graphs and Large Language Models (KaLLM 2024), pp. 75–84 (2024). https://doi.org/10.18653/v1/2024.kallm-1.8
6. Chalkidis, I., et al.: LEGAL-BERT: the muppets straight out of law school. In: Findings of the Association for Computational Linguistics: EMNLP 2020, pp. 2898–2904. Association for Computational Linguistics (2020). https://doi.org/10.18653/v1/2020.findings-emnlp.261
7. Chen, H., et al.: SAC-KG: exploiting large language models as skilled automatic constructors for domain knowledge graphs. In: Proceedings of the 62nd Annual Meeting of the Association for Computational Linguistics (Volume 1: Long Papers), pp. 4345–4360 (2024). https://doi.org/10.18653/v1/2024.acl-long.238
8. Dijkman, R., et al.: Similarity of business process models: metrics and evaluation. Inf. Syst. **36**(2), 498–516 (2011). https://doi.org/10.1016/j.is.2010.09.006
9. Duesterwald, E., et al.: FLOW-BENCH: towards conversational generation of enterprise workflows (2025). https://doi.org/10.48550/arXiv.2505.11646
10. Grohs, M., et al.: Large language models can accomplish business process management tasks. In: International Conference on Business Process Management, pp. 453–465. Springer (2023). https://doi.org/10.1007/978-3-031-50974-2_34
11. Hildebrandt, C., et al.: Ontology building for cyber–physical systems: application in the manufacturing domain. IEEE Trans. Autom. Sci. Eng. **17**(3), 1266–1282 (2020). https://doi.org/10.1109/TASE.2020.2991777
12. Jiang, P., et al.: Text-augmented open knowledge graph completion via pre-trained language models. In: ACL (Findings), pp. 11161–11180 (2023). https://doi.org/10.18653/v1/2023.findings-acl.709
13. Kaplan, A., et al.: Combining knowledge graphs and large language models to ease knowledge access in software architecture research. In: SemTech4STLD 2024, co-located with the ESWC 2024, vol. 3697, pp. 76–82 (2024)
14. Kourani, H., et al.: Process modeling with large language models. In: International Conference on Business Process Modeling, Development and Support, vol. 511, pp. 229–244. Springer (2024). https://doi.org/10.1007/978-3-031-61007-3_18
15. Kourani, H., et al.: ProMoAI: process modeling with generative AI (2024). https://doi.org/10.24963/ijcai.2024/1014

16. Köpke, J., Safan, A.: Introducing the BPMN-chatbot for efficient LLM-based process modeling. In: Proceedings of the International Conference on Business Process Management (2024)
17. Mehenni, G., et al.: Ontology-constrained generation of domain-specific clinical summaries. In: International Conference on Knowledge Engineering and Knowledge Management, pp. 382–398. Springer (2025). https://doi.org/10.1007/978-3-031-77792-9_23
18. Mergel, I.: Digital transformation of the German state. In: Kuhlmann, S., Proeller, I., Schimanke, D., Ziekow, J. (eds.) Public Administration in Germany. GPM, pp. 331–355. Springer, Cham (2021). https://doi.org/10.1007/978-3-030-53697-8_19
19. Neuberger, J., et al.: A universal prompting strategy for extracting process model information from natural language text using large language models. In: International Conference on Conceptual Modeling, pp. 38–55. Springer, Cham (2024). https://doi.org/10.1007/978-3-031-75872-0_3
20. Sansford, H., et al.: GraphEval: a knowledge-graph based LLM hallucination evaluation framework (2024). https://doi.org/10.48550/arXiv.2407.10793
21. Singer, R.: An ontological analysis of business process modeling and execution (2019). https://doi.org/10.48550/arXiv.1905.00499
22. Weerakkody, V., Janssen, M., El-Haddadeh, R.: The resurgence of business process re-engineering in public sector transformation efforts: exploring the systemic challenges and unintended consequences. IseB **19**(3), 993–1014 (2021). https://doi.org/10.1007/s10257-021-00527-2
23. Yao, L., Mao, C., Luo, Y.: KG-BERT: BERT for knowledge graph completion (2019). https://doi.org/10.48550/arXiv.1909.03193

Building Geospatial Knowledge Graphs for Cultural Heritage with the GeoLinks API

Nicolò Pratelli[1(✉)], Claudio De Martino[1], Emanuele Lenzi[1,2], and Valentina Bartalesi[1]

[1] Institute of Information Science and Technologies (ISTI), National Research Council of Italy (CNR), Pisa, Italy
{nicolo.pratelli,claudio.demartino,emanuele.lenzi,valentina.bartalesi}@isti.cnr.it

[2] Department of Information Engineering (DII), University of Pisa, Pisa, Italy

Abstract. The Cultural Heritage domain is rich in contextual information, often found in unstructured texts. These sources frequently reference places but typically lack the semantic structure required for geospatial integration and analysis. We introduce the GeoLinks API, a lightweight web service that extracts and links geospatial entities from narrative texts to support the automatic construction of structured geospatial knowledge graphs. The API performs two core functions: (i) extracting geometries from GeoNames IRIs, and (ii) recognising and linking place entities to Wikidata. Outputs are expressed in GeoSPARQL, enabling spatial reasoning, semantic interoperability, and integration with Linked Data infrastructures. We demonstrate the performance of the GeoLinks API on a dataset of contextual narratives related to traditional craft production, illustrating how it transforms unstructured cultural texts into machine-readable, semantically enriched geospatial knowledge. This approach advances Cultural Heritage preservation and research by making spatial context explicit and supporting new forms of digital analysis.

Keywords: Geospatial Knowledge Graphs · Named Entity Linking · GeoSPARQL · Cultural Heritage Crafts · Linked Open Data

1 Introduction

Cultural heritage data are rich in contextual, spatial, and historical dimensions. Much of this information is embedded in unstructured textual sources, such as descriptions, notes, and historical or ethnographic narratives [9]. These texts frequently reference places (such as villages, regions, landmarks, and landscapes) that are essential for understanding the significance of heritage practices, material culture, and historical events. However, such references are often implicit and lack the semantic structure required for integration into geospatial

H. Takeda et al. (Eds.): IJCKG 2025, LNAI 16297, pp. 445–452, 2026.
https://doi.org/10.1007/978-981-95-5009-8_31

data infrastructures or formal knowledge representations. Furthermore, they frequently omit critical geospatial information (i.e. coordinates, bounding boxes, or polygons) needed to accurately define a place's location and spatial extent.

This paper introduces the GeoLinks API, a lightweight and domain-adaptable web service designed to retrieve geometries associated with specific IRIs and to extract place entities from unstructured texts. The API outputs a graph expressed in GeoSPARQL [1], enabling integration into any semantic ecosystems. By facilitating the construction of geospatial knowledge graphs from narrative sources, GeoLinks supports novel forms of spatial analysis, data enrichment, and digital storytelling within the Cultural Heritage domain.

In the sections that follow, we present the implementation of the GeoLinks API and validate its application using a dataset of ethnographic narratives related to traditional craft production, drawn from the European project Craft Understanding, Education, Training, and Preservation for Posterity and Prosperity (CRAEFT) [13].

2 Related Works

Several tools have been developed for named entity recognition and linking, such as BLINK [12], REL (Radboud Entity Linker) [11], Genre [3] and mGenre [4], which support general-purpose entity extraction from text. More recent systems like ReLiK [10] offer improved disambiguation and support for linking entities to Wikipedia, whereas the current development of modules such as spaCy Entity Linker plugin[1] and Flair entity linker[2] promise accuracy, efficiency, and flexibility in training. Furthermore, there is Wikifier [2] that is linked to the Wikipedia Knowledge Base (KB) However, these tools are not specifically tailored for geospatial entities and typically lack integration with geographic resources like Geonames[3] or OpenStreetMap[4]. There are some tools such as Irchel Geoparser[5] and Mordecai 3 [5] that are Python libraries and they use spaCy for the Named Entity Recognition (NER) task and retrieve candidates from the GeoNames gazetteer. Furthermore, they do not produce output in geospatial RDF formats such as GeoSPARQL [1]. In parallel, efforts within the semantic web community have focused on modeling spatial data using GeoSPARQL and on developing spatial knowledge graphs, yet these approaches often rely on structured datasets rather than extracting knowledge directly from text. In the Cultural Heritage domain, most enrichment initiatives have concentrated on metadata, with limited attention given to the semantic linking of narrative content. Projects such as Europeana [6] and Pelagios [7] have demonstrated the value of linking heritage data to places, but they do not offer automated pipelines for processing

[1] https://github.com/egerber/spaCy-entity-linker.
[2] https://flairnlp.github.io/docs/tutorial-basics/entity-linking.
[3] https://www.geonames.org/.
[4] https://www.openstreetmap.org/.
[5] https://github.com/dguzh/geoparser.

unstructured texts. In contrast, the GeoLinks API addresses this gap by combining geospatial entity recognition, linking to both Geonames and Wikidata[6], and the automatic generation of GeoSPARQL-compliant knowledge graphs directly from Cultural Heritage narratives.

3 GeoLinks API

The GeoLinks API was developed with two main objectives: (i) enriching geospatial gazetteer IRIs with geometries, and (ii) extracting and geospatially enriching place entities from unstructured text. These functions support the transformation of heterogeneous Cultural Heritage data into geospatial Linked Open Data, using GeoSPARQL for interoperability and semantic reasoning.

Figure 1 shows the pipeline of the GeoLinks API based on the type of request. The two different modules are invoked depending on the type of request, as described in Sects. 3.1 and 3.2. The pipeline follows different paths but shares common submodules for entity linking, geometry retrieval and knowledge graph creation.

Key design requirements included API accessibility for integration into digital heritage workflows, automatic recognition of geographic entities in text, and retrieval of corresponding geospatial data in line with Linked Open Data principles. Additionally, the system had to be lightweight and efficient, capable of running on standard hardware without reliance on high-performance GPUs.

The API is implemented in Python using the FastAPI framework[7]. Testing is performed through various endpoints to streamline and automate the process. The API can be accessed at the provided endpoint[8] and the code is available for download or testing via the GitHub repository[9].

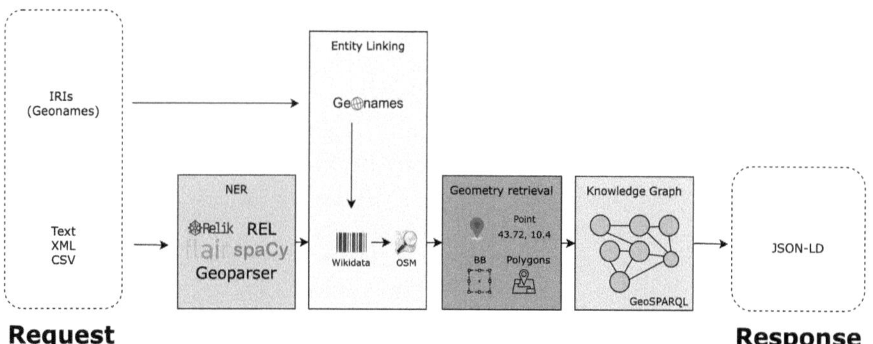

Fig. 1. Diagram of GeoLinks API pipeline.

6 https://www.wikidata.org.
7 https://fastapi.tiangolo.com.
8 https://gel.isti.cnr.it.
9 https://github.com/AIMH-DHgroup/geometry-retrieving-api.

3.1 Geospatial Enrichment of Gazetteer IRIs

In many cases, Cultural Heritage records already contain references to places, often using identifiers from sources like GeoNames. For instance, in the CRAEFT project, several place names in metadata were annotated with GeoNames IRIs. However, the public GeoNames API[10] imposes strict rate limits, particularly problematic in high-volume or batch-processing contexts, and it only provides point coordinates and bounding boxes—lacking full geometrical representations.

To overcome these limitations, we developed a module within the GeoLinks API that enriches GeoNames references with detailed geometries (e.g., coordinates, bounding boxes, polygons). Following a Linked Open Data approach, we leveraged existing links from GeoNames to Wikipedia, Wikidata, and OpenStreetMap to access richer spatial representations. By traversing these connections, we retrieved detailed geometries from OpenStreetMap via Wikidata, effectively bypassing GeoNames' constraints and enhancing the available spatial data for integration.

This step enriches existing metadata with explicit geospatial representations, which are critical for spatial queries and visualization.

To validate the effectiveness of the module for geometry extraction from semi-structured data, we conducted an analysis on a subset of Cultural Heritage records from the CRAEFT project that had been manually annotated with GeoNames identifiers. We began by extracting all distinct GeoNames IRIs referenced in the KB. For each IRI, our API queries the GeoNames public API to retrieve associated geospatial information, including point geometries (latitude and longitude) and bounding boxes. It then uses the internal GeoLinks pipeline to retrieve polygonal representations of the places.

The evaluation measured the percentage of retrieved point coordinates, bounding boxes, and geometries for the list of GeoNames IRIs present in the CRAEFT dataset. The focus was on determining the proportion of GeoNames IRIs for which these three types of geospatial data could be successfully retrieved via the GeoLinks API. Preliminary results indicate that, while the API successfully retrieved all point coordinates for every IRI by exploiting the GeoNames API, only 93% of bounding boxes for the tested IRIs were available through the GeoNames API alone. Since the GeoNames API does not provide full geometries, the GeoLinks API leverages the described pipeline to retrieve them and was able to obtain geometries for 77% of the IRIs. This demonstrates that a substantial proportion of annotated places were effectively enriched with additional spatial information.

These findings highlight the module's capability to enhance metadata with explicit spatial representations, emphasizing the benefits of a federated data integration approach. Rather than relying exclusively on a single service, the GeoLinks API leverages the interoperability of Linked Data sources—specifically, the connections between GeoNames, Wikidata, and OpenStreetMap—to provide more robust, scalable, and sustainable access to geospatial data.

---

[10] https://www.geonames.org/export/web-services.html.

3.2 Place Entity Recognition and Geospatial Linking

The second and more complex task involves extracting place references from narrative texts. To this end, we developed a modular pipeline consisting of two main components: (i) a Named Entity Recognition (NER) module that identifies candidate place names, and (ii) an Entity Linking (EL) module that connects these candidates to Wikidata via SPARQL queries, as shown in Fig. 1.

The NER component supports interchangeable backends, allowing adaptability across different languages and corpus types. Seven configurations are currently implemented, involving various combinations of spaCy 3.8.5, Flair, Geoparser, and REL (see Table 1). These can be used individually or combined (e.g., spaCy + Flair, REL + Flair, Geoparser + spaCy), enabling comparative evaluations and tailored deployments.

In the EL stage, SPARQL queries are generated based on the NER output and filtered to return only geospatial entities (e.g., instances of wd:Q486972, representing human settlements). The module first retrieves the top three candidate entities by textual label matching. These are then ranked by calculating a similarity score between each candidate's description and the surrounding context in which the entity appears—typically the sentence or event, as defined in the gold standard. If no suitable match is identified, a fallback mechanism queries the Wiki API and selects the top-ranked result. The module outputs files in JSON-LD format that contain GeoSPARQL triples, supporting integration with semantic geospatial infrastructures.

To assess the performance of the place-extraction module from unstructured text, we conducted an evaluation using contextual narratives curated within the CRAEFT KB. These narratives, which describe traditional craft practices across Europe, consist of 462 distinct events. To construct a statistically significant gold standard, we selected a representative subset of events using the standard sample size determination formula with finite population correction [8], resulting in a final sample of 237 manually annotated events.

Performance was evaluated using standard information retrieval metrics: precision, recall, and F1-score. These were applied to compare the effectiveness of each configuration in terms of both entity coverage and annotation accuracy. All seven NER configurations of the pipeline were run on the selected gold standard sample. Each configuration was further compared against a state-of-the-art baseline, with an emphasis on maintaining a lightweight and computationally efficient design suitable for standard computing environments. For this purpose, we employed the ReLiK small model [10] as the baseline.

Preliminary results, presented in Table 1, show that the spaCy 3.8.5 transformer model achieves the highest F1 score (0.79) and precision (0.78), indicating the best generalization between the chosen frameworks. Notably, combining spaCy with Flair leads to the highest recall (0.87), which suggests it is especially effective at capturing relevant entities, though with a slight trade-off in precision and a higher risk of overfitting. All configurations outperform the ReLiK small model baseline in F1 score (except for REL and Geoparser) and in precision, highlighting the significant benefits of transformer-based and hybrid approaches

Table 1. Performance of extraction configurations. Bold = improvement over ReLiK-small; underline = highest per column.

Configuration	Precision	Recall	F1 Score
spaCy 3.8.5 transformer model	**0.78**	**0.80**	**0.79**
Flair	**0.73**	**0.81**	**0.77**
Geoparser	**0.75**	0.56	0.64
REL	**0.67**	0.37	0.48
REL+Flair	**0.71**	0.69	**0.70**
spaCy 3.8.5 + Flair	**0.70**	**0.87**	**0.78**
Geoparser + spaCy 3.8.5	**0.70**	0.73	**0.71**
ReLiK small model (baseline)	0.59	0.75	0.66

in geospatial named entity recognition. These findings support the use of more recent, transformer-based NER models for improved geospatial data retrieval. In scenarios where downstream human validation is involved, a higher recall is often preferable (even at the cost of lower precision) because it ensures fewer false negatives (i.e., missed entities), and extra false positives can be quickly discarded by a human reviewer. This is the example case of combining spaCy with Flair. Conversely, configurations with higher precision but lower recall may reduce noise but risk omitting relevant entities, increasing the user's effort to manually recover them, such as the case of spaCy 3.8.5 transformer model. This trade-off should guide model selection based on the intended pipeline context.

In the future, we intend to enhance the evaluation process by conducting detailed execution time tests for each API endpoint. This will allow us to accurately measure and compare the performance, scalability, and responsiveness of the chosen frameworks under different workloads. These tests will also help validate design choices and guide future optimizations, making the API more robust and suitable for production environments.

In addition, we plan to leverage open-source Large Language Models, which can be fine-tuned to the specific domain of Cultural Heritage data. This fine-tuning will improve the accuracy of NER task and disambiguation, ultimately enhancing the quality of the geospatial data retrieved and integrated into the knowledge graph. By experimenting with different model architectures and fine-tuning strategies, we will explore how LLMs can complement and optimize the existing pipeline, improving the overall performance and adaptability of the API.

Finally, we intend to expand our evaluation to include additional datasets, such as the AIDA dataset, as well as other domain-specific resources. Incorporating a wider range of datasets will enable a more comprehensive comparison across different contexts and domains, providing deeper insights into the generalizability and robustness of our approach. This broader evaluation will also highlight potential areas where the pipeline may require adjustments to handle diverse data structures, entity types, or geographic representations, ensuring that the system can be applied effectively in various real-world scenarios.

Acknowledgments. This work was externally supported by Craeft, funded under grant agreement No 101094349 by the European Union's Horizon Europe research and innovation programme, and by ITSERR (Italian Strengthening of the ESFRI RI RESILIENCE), funded under the MUR National Recovery and Resilience Plan funded by the European Union-NextGenerationEU. There was no additional external funding received for this study.

Disclosure of Interests. The authors have no competing interests to declare that are relevant to the content of this article.

References

1. Battle, R., Kolas, D.: Geosparql: enabling a geospatial semantic web. Semant. Web J. **3**(4), 355–370 (2011)
2. Brank, J., Leban, G., Grobelnik, M.: Annotating documents with relevant wikipedia concepts. In: Proceedings of SiKDD, vol. 472 (2017)
3. De Cao, N., Izacard, G., Riedel, S., Petroni, F.: Autoregressive entity retrieval. In: 9th International Conference on Learning Representations, ICLR 2021, Virtual Event, Austria, 3–7 May 2021. OpenReview.net (2021). https://openreview.net/forum?id=5k8F6UU39V
4. De Cao, N., et al.: Multilingual autoregressive entity linking. Trans. Assoc. Comput. Linguist. **10**, 274–290 (2022). https://doi.org/10.1162/tacl_a_00460. https://aclanthology.org/2022.tacl-1.16/
5. Halterman, A.: Mordecai 3: a neural geoparser and event geocoder. arXiv preprint arXiv:2303.13675 (2023)
6. Isaac, A., Haslhofer, B.: Europeana linked open data–data. europeana. eu. Semant. Web **4**(3), 291–297 (2013)
7. Isaksen, L., Simon, R., Barker, E.T., de Soto Cañamares, P.: Pelagios and the emerging graph of ancient world data. In: Proceedings of the 2014 ACM Conference on Web Science, WebSci '14, pp. 197–201. Association for Computing Machinery, New York (2014). https://doi.org/10.1145/2615569.2615693
8. Israel, G.D., et al.: Determining sample size (1992)
9. Luthra, M., Eskevich, M.: Data-envelopes for cultural heritage: going beyond datasheets. In: Proceedings of the Workshop on Legal and Ethical Issues in Human Language Technologies@ LREC-COLING 2024, pp. 52–65 (2024)
10. Orlando, R., Huguet Cabot, P.L., Barba, E., Navigli, R.: Retrieve, read and link: fast and accurate entity linking and relation extraction on an academic budget. In: Findings of the Association for Computational Linguistics: ACL 2024. Association for Computational Linguistics, Bangkok (2024)
11. Van Hulst, J.M., Hasibi, F., Dercksen, K., Balog, K., De Vries, A.P.: Rel: an entity linker standing on the shoulders of giants. In: Proceedings of the 43rd International ACM SIGIR Conference on Research and Development in Information Retrieval, pp. 2197–2200 (2020)

12. Wu, L., Petroni, F., Josifoski, M., Riedel, S., Zettlemoyer, L.: Scalable zero-shot entity linking with dense entity retrieval. In: Webber, B., Cohn, T., He, Y., Liu, Y. (eds.) Proceedings of the 2020 Conference on Empirical Methods in Natural Language Processing (EMNLP), pp. 6397–6407. Association for Computational Linguistics (2020). https://doi.org/10.18653/v1/2020.emnlp-main.519. https://aclanthology.org/2020.emnlp-main.519/
13. Zabulis, X., et al.: A roadmap for craft understanding, education, training, and preservation. Heritage **6**(7), 5305–5328 (2023)

Towards Knowledge Retrieval for Agentic Systems in Cloud: Learning, Analysis and Control

Paul Daniëlse(✉), Shashikant Ilager, and Zhiming Zhao

University of Amsterdam, 1098 XH Amsterdam, The Netherlands
{p.e.g.danielse,s.s.ilager,z.zhao}@uva.nl

Abstract. AI technologies play an important role in diagnosing the performance of a service system and ensuring its continuous operation (AIOps) by detecting anomalies, localizing root causes, and adapting system configurations. Large language models (LLMs) possess broad knowledge of computing systems and offer new opportunities for effective root cause analysis and system adaptation. However, their inference capabilities for complex problems remain challenging. In this abstract, we discuss how a knowledge graph-based representation of performance anomalies, root causes, and adaptation strategies can be used to enhance LLMs in AIOps through an agentic system model.

Keywords: Knowledge modeling · Agentic Systems · AIOps

1 Introduction

Across the world, the use of cloud technologies and infrastructure has become commonplace. The scale at which companies require business solutions varies wildly. On one hand, large teams of DevOps engineers manage hundreds of microservices in a continuous effort to get the most out of the complex and ever-evolving infrastructure. On the other hand, small businesses might not even have a dedicated role for this. Recently, AI has played a crucial role in helping identify problems and pinpoint their root causes, empowering both. This has proven to be effective; however, the continual growth of these systems necessitates the automated control of the computing, network, and storage infrastructure through AIOps. AIOps, a framework for continually improving software deployments, is built upon the industry-standard DevOps *Continuous Integration/Continuous Development* pipeline. It consists of 4 complex main tasks. It aims to automate IT operations through: **monitoring**, **anomaly detection (AD)**, **root cause analysis (RCA)**, and **adaptation**. RCA and AD have been studied extensively [6], and many AI algorithms exist. Recently, there has been an increasing need for adaptation in order to create closed-loop optimization. *How do we find and apply adaptation strategies and monitor the effects in order to mitigate performance bottlenecks?* For this, Generative AI is an enabling technology and

H. Takeda et al. (Eds.): IJCKG 2025, LNAI 16297, pp. 453–460, 2026.
https://doi.org/10.1007/978-981-95-5009-8_32

has accelerated research towards self-healing IT deployments through the use of agentic systems. However, these systems face challenges in storing, retrieving, and updating the required domain knowledge. Currently, prompt engineering manually adds knowledge to the query. Imagine if these agents could store information in a structured and open manner? This would enable the reuse of knowledge from previously encountered situations and facilitate information sharing between systems, thereby fulfilling the ideals set out by the Semantic Web. [1] This would enable continual learning without the need to create fine-tuned models or extensive collections of examples and related documents. But many open questions exist. In this paper, we present a prototype agentic system for AIOps. By giving or removing access to domain knowledge, we demonstrate the need for a structured approach to capturing and searching knowledge when supporting the various stages of AIOps and IT operations in cloud-native ecosystems. Our prototype tackles the adaptation stage of AIOps. By adapting the deployment of a benchmark application within a cloud environment at runtime, we can mitigate or reduce performance-related problems. Furthermore, in the related work, we analyze the state-of-the-art implementations that utilize generative AI in conjunction with domain knowledge to address the RCA and AD stages of AIOps. [2,7]. Finally, we present initial results on applying adaptations with and without domain knowledge provided during the adaptation process.

2 Research Questions

We aim to answer the following main research question: **RQ:** *How can existing domain-specific knowledge be used in agentic systems to support and improve a wide range of use cases in cloud AIOps?* and the following sub-research questions. **RQ1:** *How should knowledge be represented and used in agentic systems for AIOps to improve goal-oriented processes?* To answer this question, various representations of knowledge, for example, graphs, tables, or structured text, can be provided in a prototype system, and their performance in increasing task accuracy can be measured. **RQ2:** *How can knowledge, needed for control and analysis, be retrieved for tasks?* The agentic system will be provided access to a Knowledge Graph, which it can then query with SPARQL. Then, a workflow can be used to first analyze the control problem and identify the related knowledge available in the knowledge base. Then a filtering step will decrease the noise in the selected knowledge. Finally, keeping records of system results enables continual learning and identifying the most effective knowledge to retrieve.

3 Related Work

Generative AI is becoming more popular for finding solutions to multi-objective optimization (MOO) problems in engineering and science. Finding solutions to problems quicker than traditional swarm-based heuristics. Still, techniques like cross-over selection are employed to iteratively improve the solutions. Yao at al. generate efficient solutions to coding problems [11]. In this case, traditional solutions to this MOO problem would not have been possible. Huang et al. [5] apply similar methods to writing code for specific optimization problems and show that they can achieve good results and code that rivals the state-of-the-art implementations. These techniques are promising. For AIOps, this has not yet been studied, but it provides a clear opportunity for generating structured knowledge or, furthermore, using this knowledge to apply adaptation strategies. This problem has been studied from many different angles. He et al. [4] show that using knowledge graphs can significantly improve results when answering complex questions. They provide methods for searching in and using the found knowledge in the generation. Sun et al. [9] shows that even when rigidly structured knowledge is given to generative AI, and then allowed to shift relationships, it inspires better generation results. However, what is lacking from both is a feedback mechanism that allows for updating this knowledge. Recently, knowledge graphs have been able to continually improve tasks like the retrieval of information, accuracy in answering questions, and reduce the hallucinations of large language models (LLMs) significantly. Recent advances have explored ways to enrich knowledge graphs by encoding structural information in an embedding [12]. By doing so, the LLM is able to identify gaps in the knowledge in its context and take action accordingly. Another improvement to these systems stems from the retrieval of knowledge. Paths-over-Graph [10] introduces a pruning step by identifying the paths within the knowledge graph necessary for the reasoning to occur, and by doing so, pruning many irrelevant nodes. In the context of AIOps, these studies show that we are moving towards systems that can identify when their knowledge is missing gaps by looking at structural information. After resolving these gaps, leveraging the reasoning paths to retrieve information on specific platforms and adaptation techniques, and linking them together.

Improving the understanding of subject matter by generative AI pertains to causality and reasoning. Rodling et al. [8] explore various methods and heuristics to give causal explanations of processes given partial information. This research is still missing a way to encode these causal structures in a semi-permanent way for use in knowledge graphs that deal in both semantic and causal information, but it does provide insights on how this could be leveraged in RCA for AIOps.

4 System Model

To support effective knowledge-based control in AIOps, several foundational capabilities must be designed and integrated into the agentic system. These capabilities align with the key stages of the AIOps loop: monitoring, detection, analysis, and adaptation. To support effective knowledge-based control in AIOps, several core requirements must be addressed:

- **Goal Definition:** The system must operate with clearly defined objectives. In our prototype AIOps use case, the goal is to reduce performance anomalies while minimizing resource consumption.

- **Monitoring and Feedback:** A robust monitoring infrastructure is required to observe the system state and detect anomalies. We use Prometheus and Jaeger for real-time metrics and tracing. Feedback from these tools enables the system to evaluate the impact of actions and refine future decisions.
- **Detection and Triggering:** A lightweight anomaly detection mechanism (In our case, an Isolation Forest) must identify significant deviations and initiate the control process.
- **Actuation Capabilities:** The system needs the ability to apply changes at specific control points, such as modifying configuration files or resource limits, to address detected issues.
- **Knowledge Recording and Evaluation:** The system must store knowledge about past actions and their outcomes. This supports continual learning and avoids repeated mistakes, such as misconfiguring APIs or deploying ineffective fixes.

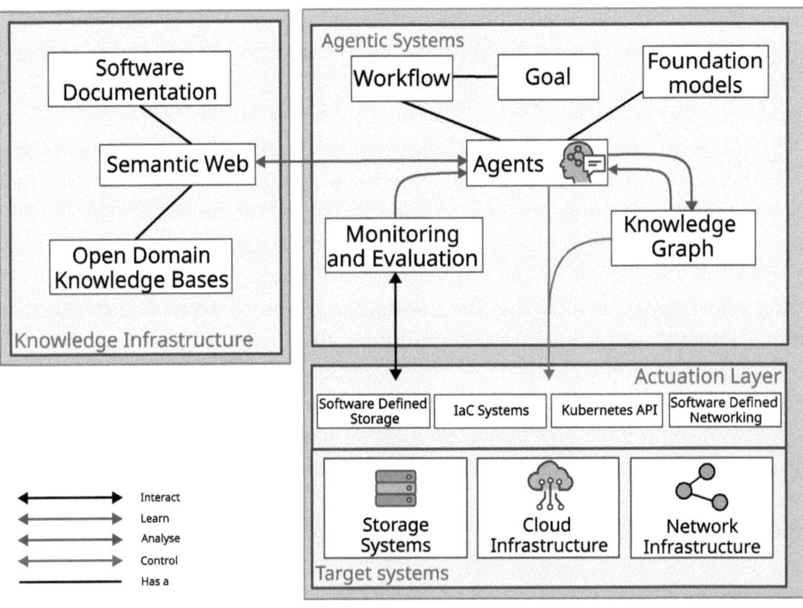

Fig. 1. Overview of Agentic systems for controlling systems in the cloud-edge continuum. Domain knowledge is stored as a knowledge graph in open domain knowledge bases, where agents can learn, analyze gaps, and use domain knowledge to effectively control. The semantic web enables agents to retrieve knowledge from other agents and data sources.

Figure 1 illustrates the full architecture. Agentic components have access to monitoring infrastructure, domain knowledge through a knowledge graph, and actuation endpoints. The Semantic Web enables agents to learn from both external sources, such as documentation and configuration wikis, as well as internal monitoring and evaluation systems. In our prototype, we distill knowledge from the documentation of the runtime systems (Kubernetes, Knative) and application components (Memcached), and store this knowledge in a knowledge graph. We collect specific knobs important for performing adaptation strategies. For example, parameters for changing the Horizontal Scaling behavior, and the resource limits parameters for Vertical Scaling.

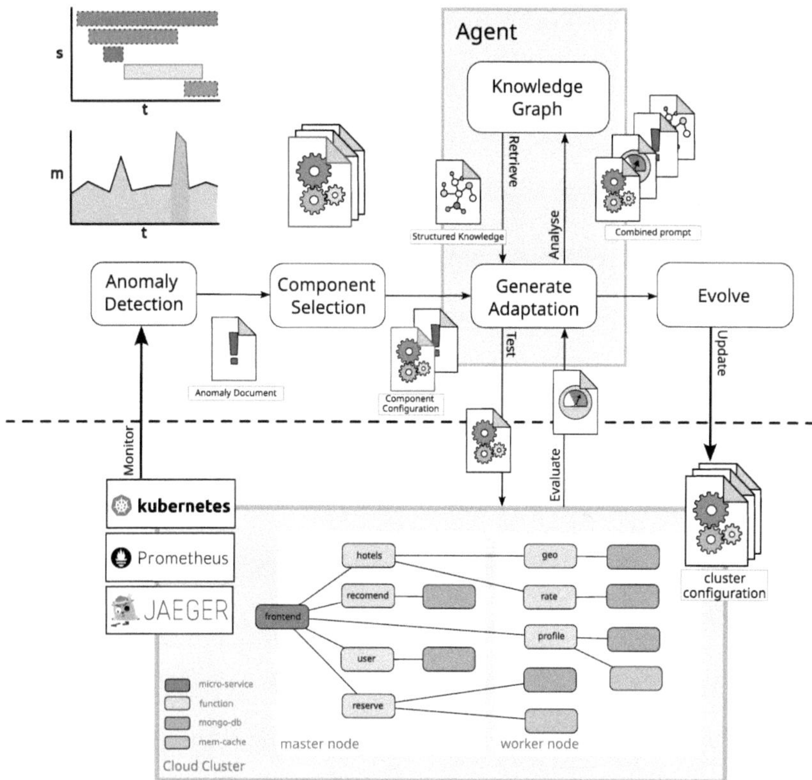

Fig. 2. Prototype agentic system process diagram for optimizing cloud infrastructure environments.

4.1 Prototype System for Cloud AIOps Architecture

The prototype agentic system aims to close the loop between continuous monitoring and updating services, as seen in Fig. 2. A lightweight anomaly detector monitors and classifies request traces[1]. Then, using Shapely values, the anomaly is linked to the relevant services. Together with the service and the timing of the

[1] Jaeger Tracing.

anomaly, the monitoring data is selected to build the query to the Agent. The query consists of three parts: the tracing and monitoring data, the necessary domain knowledge on the frameworks and possible adaptations, and the service configuration files where the adaptation will be applied. Then the agent is asked to resolve the performance anomaly by preparing a prompt with relevant monitoring data, classification of the anomaly, and knowledge on adaptation strategies, platform options, and control knobs.

Within a 10-minute window, typically tens of anomalies occur. For each, a possible fix is applied, and the environment is measured to determine if this resolves the anomaly. Finally, the optimal solutions are selected based on the number of anomalies in the system and CPU resource usage. Then the system will restart the process. To learn which solutions and knobs the agent can use most effectively, a short history of all queries is maintained and provided to the model, which further helps the system learn from recent solutions and optimize them.

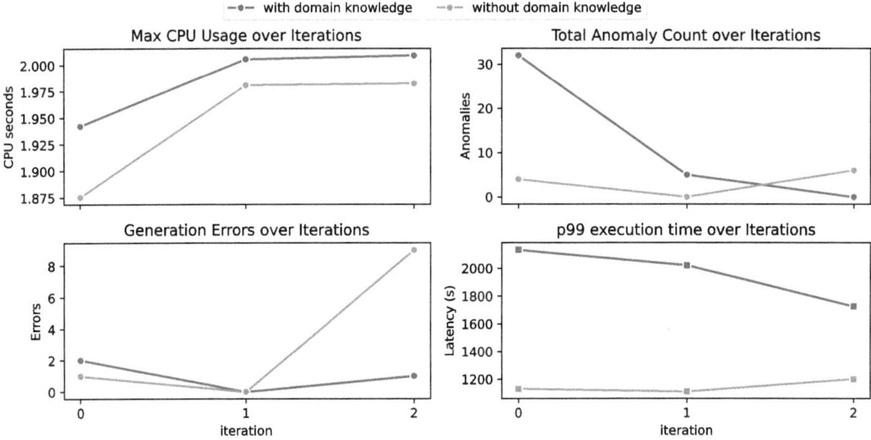

Fig. 3. The performance of the generated configuration over several iterations in closed-loop optimization of performance anomalies.

4.2 Preliminary Results

We evaluate the performance of our system on generating an optimal configuration for the *HotelReservations* benchmark from the DeathStarBench suite [3] on a two-node Kubernetes cluster. We examine the configuration optimization process with and without giving structured knowledge to the agent. Figure 3 shows the anomaly count over many iterations, which slowly goes down and remains stable for the agentic system *with domain knowledge*. The statistics of the agentic system *without knowledge* show a slower reduction. This is attributed to the generation errors that result. Here, the system often creates broken configurations. We can see that the system is slowly improving in terms of p99 latency, albeit at the cost of higher maximum CPU usage.

5 Future Work

This research will continue with the overarching goal of improving the various stages of the AIOps framework. We aim to contribute across three key areas: anomaly visibility, root cause analysis, and adaptive mitigation.
Improving the visibility of Anomalies: Building on recent work in anomaly classification [7], we focus on identifying common patterns and assigning meaningful labels to them based on the origin, diagnosis, or the identified root cause. This enhances interpretability and supports more targeted responses.
Real Causality for Root Cause Analysis: Rather than relying on correlation, we leverage data from system interventions to infer causal structures. These are modeled as causal graphs within the systems knowledge base, enabling accurate and predictive root cause analysis that goes beyond the state-of-the-art systems.
Experimental adaptation to increase knowledge: We employ evolutionary algorithms for both exploration (testing new adaptations) and exploitation (refining known ones). Additionally, reasoning over documentation and historical data may suggest new actuation sites, which the system can validate experimentally.
Taken together, these directions point toward more autonomous, knowledge-driven agentic systems that continually improve through observation, reasoning, and adaptation. Integrating knowledge retrieval not only strengthens each stage of the AIOps cycle but also opens the door to learning systems capable of tackling increasingly complex operational challenges in dynamic environments.

6 Conclusion

Although the results from our prototype seem promising, there are still many areas for improvement. These Agentic systems are the future for automating many fields. They offer numerous opportunities for various tasks and challenges. To meet these, storing and adding knowledge is a fundamental way to engineer these systems. The challenge to ensure that these systems stand the test of time is to enable the learning of new knowledge and new techniques. Our prototype software is publicly available on github[2].

Acknowledgments. This work has been partially funded by EU Horizon-Europe ENVRI-HUB next (101131141), EVERSE (101129744), and BLUECLOUD 2026 (101094227) projects, and by the Dutch Research Council, LTER-LIFE project.

References

1. Berners-Lee, T., Hendler, J., Lassila, O.: Web semantic. Sci. Am. **284**(5), 34–43 (2001)
2. Chen, Y., et al.: Automatic root cause analysis via large language models for cloud incidents. In: Proceedings of the Nineteenth European Conference on Computer Systems, pp. 674–688 (2024)

[2] Our repository: peg-danielse/closing-the-loop.

3. Gan, Y., et al.: An open-source benchmark suite for microservices and their hardware-software implications for cloud & edge systems. In: Proceedings of the Twenty-Fourth International Conference on Architectural Support for Programming Languages and Operating Systems, pp. 3–18. ACM, Providence (2019). https://doi.org/10.1145/3297858.3304013
4. He, J., Ma, M.D., Fan, J., Roth, D., Wang, W., Ribeiro, A.: GIVE: Structured Reasoning with Knowledge Graph Inspired Veracity Extrapolation (2024). http://arxiv.org/abs/2410.08475
5. Huang, Y., Wu, S., Zhang, W., Wu, J., Feng, L., Tan, K.C.: Autonomous multi-objective optimization using large language model. IEEE Trans. Evol. Comput. (2024). https://api.semanticscholar.org/CorpusID:270440958
6. Lee, C., Yang, T., Chen, Z., Su, Y., Lyu, M.R.: Eadro: an end-to-end troubleshooting framework for microservices on multi-source data. In: 2023 IEEE/ACM 45th International Conference on Software Engineering (ICSE), pp. 1750–1762 (2023). https://doi.org/10.1109/ICSE48619.2023.00150
7. Liu, J., et al.: Large Language Models can Deliver Accurate and Interpretable Time Series Anomaly Detection (2024). https://doi.org/10.48550/arXiv.2405.15370. http://arxiv.org/abs/2405.15370
8. Rödling, S., Zečević, M., Dhami, D.S., Kersting, K.: Causal Explanations Over Time: Articulated Reasoning for Interactive Environments (2025). https://doi.org/10.48550/arXiv.2506.03915. http://arxiv.org/abs/2506.03915
9. Sun, J., et al.: Think-on-graph: deep and responsible reasoning of large language model on knowledge graph. In: Kim, B., Yue, Y., Chaudhuri, S., Fragkiadaki, K., Khan, M., Sun, Y. (eds.) International Conference on Representation Learning, vol. 2024, pp. 3868–3898 (2024). https://proceedings.iclr.cc/paper_files/paper/2024/file/10a6bdcabbd5a3d36b760daa295f63c1-Paper-Conference.pdf
10. Tan, X., Wang, X., Liu, Q., Xu, X., Yuan, X., Zhang, W.: Paths-over-graph: knowledge graph empowered large language model reasoning. In: Proceedings of the ACM on Web Conference 2025, pp. 3505–3522 (2025)
11. Yao, S., Liu, F., Lin, X., Lu, Z., Wang, Z., Zhang, Q.: Multi-objective Evolution of Heuristic Using Large Language Model (2025). https://doi.org/10.48550/arXiv.2409.16867. http://arxiv.org/abs/2409.16867
12. Zhang, Y., Chen, Z., Guo, L., Xu, Y., Zhang, W., Chen, H.: Making large language models perform better in knowledge graph completion. In: Proceedings of the 32nd ACM International Conference on Multimedia, MM '24, pp. 233–242. Association for Computing Machinery, New York (2024). https://doi.org/10.1145/3664647.3681327

3T: Triple Stores Triggers

Giannis Vassiliou[1], Georgia Eirini Trouli[1], Nikolaos Papadakis[1], and Haridimos Kondylakis[2](✉)

[1] Hellenic Mediterranean University, Heraklion, Greece
{giannisvas,gtrouli,npapadak}@ics.forth.gr
[2] CSD-UOC & FORTH-ICS, Heraklion, Greece
kondylak@ics.forth.gr

Abstract. Database triggers have been introduced since the 1980s, and their value has been widely recognized. They allow for automating several tasks and enforcing business rules at the database level, ensuring data integrity, auditing, and consistent logic across applications. However, notwithstanding their high value, the use of triggers in mainstream triple stores has been mostly overlooked. The dynamic nature of semantic data, the increasing scale of RDF datasets, and the growing demand for real-time processing in graph-based applications demand efficient methods for real-time data validation and integrity enforcement, change tracking and auditing, and automated inference and reasoning. In this demonstration, we recognize triggers as first-class citizens in the semantic web and introduce a new interpreted language to describe them. We present 3T, an integrated development environment (IDE) and a graphical user interface (GUI) enabling triple-store-independent triggers, allowing their real-time, effective, and efficient execution.

Keywords: Triple Stores · RDF · Knowledge Graph · Triggers

1 Introduction

A database trigger is a procedural code that is automatically executed (or "triggered") in response to certain events on a database table or view. Triggers can be defined to run either before or after the triggering event, depending on the desired behavior. They are commonly used to enforce business rules, maintain data integrity, automate system tasks, handle real-time events, or perform auditing. Triggers have been around since the early days of relational databases, emerging in the 1980s as database management systems (DBMS) like Oracle, IBM DB2, and SQL Server started including procedural extensions.

Despite the tangible benefits, triggers have not been explored in triple stores and related RDF engines [1] primarily due to concerns about performance, as adding triggers can slow down systems that are optimized for fast querying and scalability, especially when dealing with large RDF datasets. While some vendors like Blazegraph ignore triggers altogether, others like Stardog offer rule-based alternatives such as SWRL, which require non-scalable reasoning engines and

H. Takeda et al. (Eds.): IJCKG 2025, LNAI 16297, pp. 461–467, 2026.
https://doi.org/10.1007/978-981-95-5009-8_33

lack real-time responsiveness. Virtuoso advertises triggers, but they are confined to underlying relational layers, inaccessible to RDF developers. Research efforts, particularly in property graphs [2,4], have proposed frameworks like PG-Triggers to define and implement triggers, but practical integration remains challenging. Furthermore, the inherent flexibility and schemaless nature of RDF data also make it challenging to define and manage triggers effectively, since the relationships between triples can be dynamic and complex. Instead, triple stores have focused on optimizing query performance [3], prioritizing data integration [5], interoperability, and inferencing.

However, as RDF datasets grow in size and complexity and are used in a more dynamic landscape than ever, there is an urgent need for methods that help to efficiently ensure data integrity, enforce business rules, and automate workflows in real-time. Exploiting triggers could enable automated reasoning, immediate response to data changes, and effective management of dependencies across interconnected triples, making systems more responsive and efficient. Triggers also support tasks like auditing, data validation, and synchronizing with external systems, which are critical as RDF data is increasingly integrated into large-scale, event-driven architectures and applications requiring real-time insights.

As such, in this paper, we argue that researchers and triple store vendors should capitalize on pre-existing knowledge from the relational world, and enrich the domain ofthe semantic web with the low-hanging fruit of triggers, upgrading them as first-class citizens in the modern world of triple stores. In this demonstration, we present 3T, a system unlocking the trigger functionality for the triple stores, that is triple-store agnostic and enables trigger declaration and execution in an efficient and effective manner. More specifically, our contributions in this paper are the following:

- We demonstrate 3T, the first system enabling the execution of triggers, in a triple-store agnostic method, introducing a new simple interpreted language for describing triggers.
- We also demonstrate an IDE front-end with a GUI allowing the user-friendly management of the stored triggers
- We also provide an API that can be used for interacting with the triple store, hiding the trigger layer and enabling users to build triple-store agnostic programs that are able to exploit the constructed trigger engine.

2 3T System Architecture

An overview of the 3T system is shown in Fig. 1a and consists of three layers. In the data layer, we find the triple store, where the triples of the knowledge graph are stored. Note that our system is triple store agnostic and as such, it can be connected to any triple store available, upgrading it with triggers. In the service layer, we find the 3T trigger engine that responds to events based on the trigger conditions and executes the appropriate commands. All triggers are available in the main memory of the 3T Trigger Engine; however, for persistence, they are also stored in a text file. An API, the 3T API exposes the whole functionality to the GUI layer. As a proof of concept, we have implemented the 3T GUI for recording, updating, and executing triggers from a command line. However, the same API can be used by other external applications for interacting with the triple store, allowing triggers to come into play when our API is used. In the sequel, we present in detail the various components of the system.

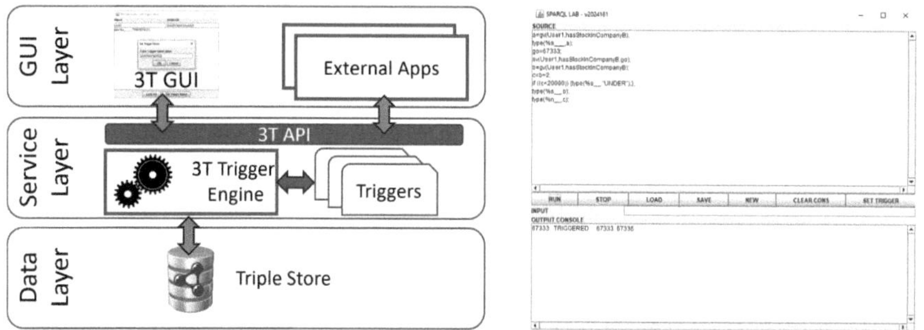

(a) 3T system architecture.

(b) Executing code that fires the Trigger

Fig. 1. Side-by-side view of system architecture and trigger code.

The 3T GUI functions similarly to an Integrated Development Environment (IDE) and is supported by a custom interpreted language. It is created in Java using ANTLR[1] and is shown in Fig. 1b.

The key feature of the GUI is the ability to easily set triggers for a knowledge graph and execute custom code when these triggers are activated. Triggers are monitored based on specific subject-predicate combinations (could also be only one subject or only one predicate) that, when updated, a connected action (a set of code lines) is executed (for example, store a new value in the KG).

After editing their code, users can press the "RUN" button to execute it or use the "STOP" button to halt execution. Other features, such as clearing the console, are also available for ease of use. After pressing the "SET TRIGGER" button, the user is presented with a window where they can add an identifier to the trigger, whereas a text area is provided for writing the code that will be executed when the trigger is activated, with the option to load existing code from a file. Once the trigger is configured, the details are saved to a file, which is automatically loaded the next time the system starts, ensuring the trigger remains active until manually deleted.

When a trigger is set and activated, the associated code (written in our custom language) executes. This code can perform additional operations such as inserting, deleting, or updating triples in the KG. By offering a simple, GUI-driven interface, we provide an effective way to set triggers and automatically maintain the KG in the desired state. To showcase our system, we developed two examples to run by setting triggers in the 3T GUI.

[1] https://www.antlr.org/.

Example 1. The first example is related to brokers and stocks. Assume for example, that a broker, "Broker A," can hold a number of stocks in "Company A." We can set a trigger to execute code whenever "Broker A" updates the number of stocks they hold in "Company A." This code could perform various operations supported by the interpreted language, such as updating the number of stocks "Broker A" holds in another company, like "Company B." To set this trigger, the user would select "Broker A" as the subject and "hasStocksInCompanyA" as the predicate, then write or load the associated code through the application's GUI. An example of relevant actual code is shown in Fig. 1b. The first line reads (using the function "gv" that gets the value) the number of stocks that "User1" has in "CompanyB" in variable "a" whereas the fourth line updates the stocks of "User1" to 67333 (sets the value using the "sv" function). The seventh line includes a trigger that is fired when the condition is met, printing a string; however, updates on the data could be made here.

Example 2. Another illustrative example comes from the chemistry world. Consider the chemical reaction:

```
1 ACID + 1 BASE → 1 SALT + 1 WATER.
```

In the KG, we might assume the following triples:

```
(ACID, hasMolecules, 50);
(BASE, hasMolecules, 40);
(SALT, hasMolecules, 40);
```

We could define a trigger that fires when the number of molecules of ACID exceeds the number of molecules of BASE. When this condition is met, the trigger would execute code to simulate the chemical reaction, producing SALT and WATER, and leaving some remaining ACID (e.g., 10 moles). In this case, the triple *(SALT, hasMolecules, 40)* would be updated accordingly. For this chemistry example, the trigger is set to monitor the subject *ACID* and the predicate *hasMolecules*. Whenever this value changes, the trigger activates, ensuring the reaction adheres to the chemistry rules and updates the KG accordingly.

2.1 The Service Layer

Syntax. In this layer, we find the 3T engine, which, in essence, for each incoming statement checks all triggers loaded in main memory and executes the body of the triggers. The syntax of the language that the engine supports is shown in Listing 1.1.

In our notation, upper-case letters are reserved for terminal symbols; nonterminals are in a lower case, alternatives are separated by the | symbol. The language includes many features like variable declarations, assignments, and arrays. Control flow structures like 'if-then-else' statements and 'for' and 'while' loops are fully supported, enabling conditional logic and iterative processes. Additionally, users can define functions, which can be invoked within the code or even call themselves recursively.

Listing 1.1. 3T Syntax

```
prog: block+
 | (COMMENT)* (function)* stat+;

function: FUNCTION ID '(' ID (',' ID)* ')' block

stat:
 COMMA # ecommand
 | SV '(' ID ',' ID ',' ID ')' COMMA # setvalue
 | COMMENT # comment
 | ID '(' expr (',' expr)* ')' COMMA # procedure
 | ID '=' expr COMMA # assign
 | ID '[' expr ']' '=' expr COMMA # assignarr
 | RETURN expr COMMA # retstat
 | PPRINT '(' printitem+ (',' expr)+ ')' COMMA
 | IF relexpr ifthen (ELSE ifthen)? # ifstat
 | FOR '(' ID expr TO expr ')' ifthen # forstat
 | WHILE relexpr 'DO' block # whilestat
 | ARRAY ID '[' expr ']' COMMA # array;

block: '{' stat+ '}' # blocking;

printitem: (STRPRINT | NUMBERPRINT | NEWL | SPAC);

ifthen: (block COMMA | stat) # blockstat;

expr: expr op=('*' | '/') expr # MulDiv
 | expr op=('+' | '-') expr # AddSub
 | ID '(' expr (',' expr)* ')' # functcall
 | FLOAT # float
 | INT # int
 | ID # id
 | ID '[' expr ']' # idvar
 | STRINGLITERAL # vstring
 | '(' expr ')' # parens
 | GV '(' ID ',' ID ')' # getvalue;

relexpr: '(' ('(' expr op=(EQ | GT | GTE | LT
 | LTE | DIF) expr ')' | relexpr)
 (opp=(LAND | LOR) ('(' expr op=(EQ
 | GT | GTE | LT | LTE | DIF) expr ')'
 | relexpr))* ')';
```

Semantics. In our case, triggers are executed when an insert/update/delete operation is executed, explicitly mentioning the subject/object of the trigger condition. Then the body of the condition is executed. Our language is quite flexible to be able to declare multiple lines of code in the body of the trigger, including additional conditions etc. We next describe the trigger semantics along the classical dimensions.

Granularity. We assume that each trigger execution is linked to a subject and/or predicate that are monitored and that the trigger is fired as soon as relevant data is inserted, updated, or deleted through our API. For each incoming statement, all triggers are checked to identify whether relevant conditions are met in order to execute the trigger body. In the trigger body, all actions are executed as a set.

Action Time. Trigger execution of our engine occurs after the insertion, update, or deletion of information in the triple store, which is executed using our API. In essence, execution takes place after a successful commit and operates within an autonomous transaction.

Chained Execution. Triggers are checked in the order they have been inserted. Multiple triggers might be executed in the order they are found. So far, we have disabled recursive execution of triggers; however, the topic is well explored in relational databases by techniques limiting recursion that will be incorporated in the next version of the system.

API. Furthermore, the service layer includes an API that can be used in order to insert, update or delete triples from the connected triple store, allowing external apps to exploit the trigger mechanisms introduced by 3T.

2.2 The Data Layer

As already stated our system is triple-store independent as long as the triple store is supported by the Apache Jena API. As such it currently supports TDB and TDB2 native stores, Fuseki, Virtuoso, Stardog, and Blazegraph. Although other APIs can also be used for the time being we only support Apache Jena.

3 Demonstration and Conclusions

The purpose of the demonstration is primarily to highlight to technicians, researchers, and triple store providers the usefulness of triggers in RDF knowledge graphs and argue that we should leverage the established knowledge from the relational database domain, enhancing the semantic web sphere by integrating triggers, elevating them to be primary elements in contemporary triple store technology. The demonstration will proceed in four phases:

Phase 1. Triggers Usefulness. The demonstration will start by explaining use-case scenarios from the financial and scientific domains, highlighting the usefulness of triggers.

Phase 2. The Language. Then we will present trigger examples, explaining the constructs of the language and the flexibility it introduces. We will use the 3T GUI for visualizing existing triggers, updating them, and writing new ones, whereas we will show how they are fired when the information in the triple store is inserted, deleted, or updated.

Phase 3. The Semantics. Then we will explain in detail the semantics of 3T triggers, explaining the granularity, the action time, and the chained execution we support.

Phase 4. "Hands-on". Conference participants will then be allowed to write their own triggers, and then perform update operations in the triple store in order to be executed when triggered. The result will be shown both from the 3T GUI and from the triple store.

4 Conclusions

This paper presents 3T, the first system enabling trigger declaration and execution over RDF triple stores. It facilitates real-time automation, data integrity, and business rule enforcement in knowledge graphs (KGs). 3T includes an interpreted trigger language, a user-friendly GUI-based IDE, and an API for integration with external applications, making it accessible to both technical and non-technical users.

Future work includes support for recursive triggers, inference-based triggers, integration of full SPARQL updates in trigger bodies, and extending 3T to real-time RDF streaming scenarios using platforms like RDF4J.

Acknowledgments. The work reported in this paper is implemented in the framework of H.F.R.I call "Basic research Financing (Horizontal support of all Sciences)" under the National Recovery and Resilience Plan "Greece 2.0" funded by the European Union – NextGenerationEU (H.F.R.I. TALE Project, Number: 16819).

References

1. Agathangelos, G., Troullinou, G., Kondylakis, H., Stefanidis, K., Plexousakis, D.: RDF query answering using apache spark: review and assessment. In: 34th IEEE International Conference on Data Engineering Workshops, ICDE Workshops 2018, Paris, France, 16–20 April 2018, pp. 54–59. IEEE Computer Society (2018). https://doi.org/10.1109/ICDEW.2018.00016
2. Bellomarini, L., Bernasconi, A., Ceri, S., Gagliardi, A., Magnanimi, D., Martinenghi, D.: Towards a standard for triggers in property graphs. In: Atzori, M., et al. (eds.) Proceedings of the 32nd Symposium of Advanced Database Systems, Villasimius, Italy, 23–26 June 2024. CEUR Workshop Proceedings, vol. 3741, pp. 70–79. CEUR-WS.org (2024). https://ceur-ws.org/Vol-3741/paper02.pdf
3. Bonifati, A., Dumbrava, S., Kondylakis, H., Troullinou, G., Vassiliou, G.: Progressive querying on knowledge graphs. In: Simitsis, A., Kemme, B., Queralt, A., Romero, O., Jovanovic, P. (eds.) Proceedings 28th International Conference on Extending Database Technology, EDBT 2025, Barcelona, Spain, 25–28 March 2025, pp. 106–118. OpenProceedings.org (2025). https://doi.org/10.48786/EDBT.2025.09
4. Ceri, S., Bernasconi, A., Gagliardi, A., Martinenghi, D., Bellomarini, L., Magnanimi, D.: Pg-triggers: triggers for property graphs. In: Barceló, P., Sánchez-Pi, N., Meliou, A., Sudarshan, S. (eds.) Companion of the 2024 International Conference on Management of Data, SIGMOD/PODS 2024, Santiago, Chile, 9–15 June 2024, pp. 373–385. ACM (2024). https://doi.org/10.1145/3626246.3653386
5. Kondylakis, H., Plexousakis, D.: Exelixis: evolving ontology-based data integration system. In: Sellis, T.K., Miller, R.J., Kementsietsidis, A., Velegrakis, Y. (eds.) Proceedings of the ACM SIGMOD International Conference on Management of Data, SIGMOD 2011, Athens, Greece, 12–16 June 2011, pp. 1283–1286. ACM (2011). https://doi.org/10.1145/1989323.1989477

TCRMQ: Question Answering by Multi-hop SPARQL Queries over Event-Based Knowledge Graphs

Michalis Mountantonakis[1,2](✉) and Yannis Tzitzikas[1,2]

[1] Information Systems Laboratory, FORTH-ICS, Heraklion, Greece
{mountant,tzitzik}@ics.forth.gr
[2] Computer Science Department, University of Crete, Heraklion, Greece

Abstract. There is an increasing trend of combining Knowledge Graphs (KGs) and LLMs for several tasks, including the generation of structured queries from natural questions. Indeed, it is quite challenging to answer natural questions over KGs, especially for a) sophisticated ontologies, since in most cases one has to derive multi-hop queries with property path expressions, and b) questions requiring extra background information that is not included in such KGs. Towards this direction, we present a research prototype, called `TCRMQ` (Text-2-CIDOC-CRM Query), that enables users to express their questions in natural language and to retrieve the desired answer over KGs using the event-based CIDOC-CRM model (an ISO standard used from many Cultural Heritage organizations). `TCRMQ` is based on a novel two-stage method combining Ontology Path Patterns and Knowledge from Large Language Models (LLMs) and can generate multi-hop SPARQL queries from natural language multilingual questions, that can even require external background knowledge to be answered. Its current version supports two CIDOC-CRM KGs containing artworks, and has been evaluated for English and Greek language through a dedicated benchmark. By applying this method over GPT-4 we achieve accuracy 83%, while the baseline has only 32% accuracy.

Keywords: LLMs · CIDOC-CRM · Text to SPARQL · Event Based KGs

1 Introduction

There is an increasing trend of combining Knowledge Graphs (KGs) and Large Language Models for several tasks and domains, e.g., Fact Validation [9], Question Answering [11,13], and many others. A challenging direction is the LLM to generate structured queries from natural questions over KGs, i.e., Text-2-SPARQL approaches [2,3,7], such as for DBpedia [6] and Wikidata [5]. The task is more challenging for KGs using sophisticated ontologies, since in most cases one has to derive *multi-hop* queries with *property path* expressions (Challenge

H. Takeda et al. (Eds.): IJCKG 2025, LNAI 16297, pp. 468–476, 2026.
https://doi.org/10.1007/978-981-95-5009-8_34

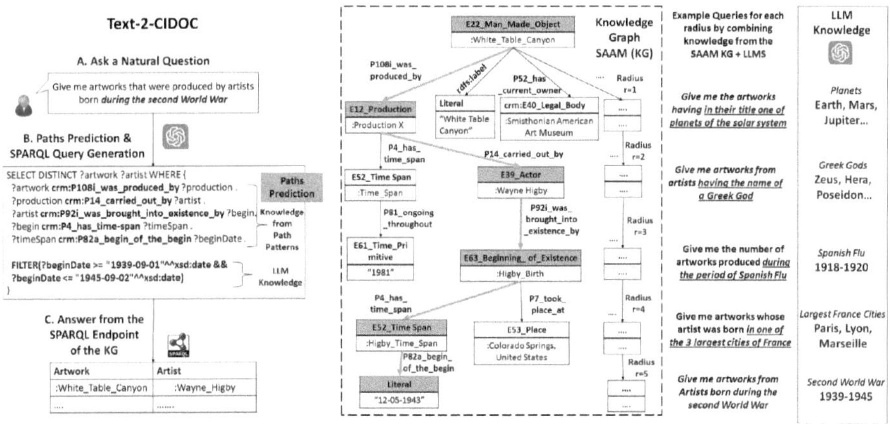

Fig. 1. A natural question requiring knowledge from SAAM KG and from the LLM

A). Also, a user can often express natural questions that require additional background information that is not included in the KG (Challenge B), e.g., missing facts or additional reasoning.

Toward this direction, we present a research prototype, called `TCRMQ` (from Text-to-CIDOC CRM Query), which enables users to express their questions in natural language (even in more languages apart from English) and to retrieve the answer over KGs that use the CIDOC-CRM model [4]. CIDOC-CRM is an event-based model and an ISO standard that is currently used by more than 400 Cultural Heritage (CH) organizations [8]. In Fig. 1, a real part of the SAAM (Smithsonian American Art Museum) KG [14] is shown. We can see long paths that we need to follow due to the event-nature of CIDOC-CRM. In the left side of Fig. 1, the user expressed the question "`Give me artworks that were produced by artists born during the second World War`". This question needs to be converted to SPARQL and it is required to follow a long path with radius (or path length) $r = 5$, for finding the artist of an artwork and his/her birth date (Challenge A). Additionally, we need external knowledge (e.g., see the right side) to answer the question (Challenge B). Specifically, the KG contains information about the artworks and artists of the museum, including their birth date. On the contrary, the KG does not contain information about the dates of Second World War, however, LLMs such as ChatGPT contain the mentioned dates.

To answer such questions and tackling both challenges, `TCRMQ` i) uses a *novel two-stage method* combining Ontology Path Patterns and Knowledge from LLMs [10], to predict the most relevant ontology paths and to formulate a multi-hop SPARQL query for a given natural question (not only in English but also in other languages), ii) evaluates the generated query by sending it to the SPARQL endpoint of the KG and iii) retrieves and presents the results to the user. For the question in Fig. 1 (Step A), `TCRMQ` predicted the most relevant paths (see

```
You are given the following triple patterns of the form  Class->Property->Class or Literal
#Patterns
E12_Production->P14_carried_out_by->E39_Actor
...
E52_Time_Span->P82a_begin_of_the_begin->Literal
#End of Patterns
Give me the classes and properties that will be used to answer the question:
I want artworks from artists born during the second world war
I strictly want the output in the following format: Classes: {} Properties:{}

LLM Output - Predictions
Classes:{E22_Man_Made_Object, E39_Actor, E52_Time_Span}
Properties:{P14_carried_out_by, P82a_begin_of_the_begin}
```

Fig. 2. Predictions Prompt and LLM Output

the orange parts on the right side) and generated a multi-hop SPARQL query from ChatGPT (Step B), that includes both the relevant ontology paths and the desired values for the dates in the FILTER (external knowledge from the LLM). Finally, the query was sent to the endpoint to obtain the response (Step C).

Regarding the contribution, we a) present the research prototype TCRMQ (https://demos.isl.ics.forth.gr/Text2CIDOC), by using the methods in [10], b) provide use cases and scenarios, and c) present evaluation results for 120 questions (by extending the benchmark in [10]) for three ChatGPT models, in English and Greek. Concerning the novelty, this is the first application for query formulation over CIDOC-CRM KGs by using LLMs, that can also yield queries with the right filtering conditions, even if such information is not present in the KG.

The rest of this paper is organized as follows: Sect. 2 describes the methods and the functionality of TCRMQ, Sect. 3 presents the use cases, Sect. 4 shows the evaluation results and Sect. 5 concludes the paper.

2 Methods and Functionality

As a pre-processing step, ontology path patterns up to a given radius r (or path length) are created once (e.g., $1 \leq r \leq 5$), by sending SPARQL queries to the endpoint of each desired KG. For instance, in Fig. 1, a path pattern of $r = 2$ is the following: ⟨E22_Man_Made_Object, P108_was_produced_by, E12_Production, P14_carried_out_by, E39_Actor⟩. Afterwards, the path patterns for each KG are stored to be used in the LLM prompts.

Methods. The demo supports several methods using such ontology path patterns and knowledge from the LLM. Also, a zero shot method is available where no further input is given to the LLM. For each method a single prompt template has been created, apart from the two-stage path patterns method, which sends two different prompts. All the details for the methods are given in [10]. The two-stage method, which is the most effective one, is presented below.

```
You are given the following path patterns of the form
{Class}->Property->{Class}->...->{Class} or (Literal)
### Patterns
{E22_Man_Made_Object} -> P108i_was_produced_by  -> {E12_Production} -> P14_carried_out_by
-> {E39_Actor} -> P92i_was_brought_into_existence_by  -> {E63_Beginning_of_Existence}
-> P4_has_time_span -> {E52_Time_Span} -> P82a_begin_of_the_begin->(Literal)
### End of Patterns

By using properties and classes from the above patterns,  and by never using a class as
a property (or the  opposite), please generate only a SPARQL query (without  explanation)
for answering the question:  I want  artworks from artists born during the second world war
```

Fig. 3. Query Generation Prompt with the predicted path

Two Stage Ontology Path Patterns Method. The key objective is to decrease the number of possible candidate path patterns, because a) it is expensive (in terms of efficiency and monetary cost) to feed a large number of patterns for the LLM, and b) if we feed the LLM with several irrelevant path patterns, it can be confused, which can result in a wrong pattern (and query). We first send an LLM prompt using only the path patterns with $r = 1$ to predict the most relevant classes and properties of the KG to answer a question (for the running example, see Fig. 2 and the predicted paths in orange color in Fig. 1). Afterwards, we select candidate patterns based on the predictions and 3 rules.

- **Rule A**. It keeps only path patterns of any radius r including at least all the predicted classes and all the predicted properties.
- **Rule B**. If Rule A fails, it keeps the path patterns of any r with either all the predicted properties or all the predicted classes.
- **Rule C**. If both rules fail, it keeps all the path patterns of any radius r including at least one predicted class or property.

The final step is to create a prompt (i.e., see Fig. 3) using only the filtered path patterns based on the 3 rules and to ask the LLM to generate the SPARQL query. In our example, Rule A was executed since the path in orange in Fig. 1 includes the 3 predicted classes and the 2 predicted properties.

`TCRMQ` **Functionality**. The user can a) write a question in a natural language (evaluated in English and Greek, but can work with other languages, too), by selecting the desired KG, the path patterns method and the ChatGPT version. Afterwards, B) based on the user selections, the prompt(s) that correspond to the desired method and KG is sent to ChatGPT through its API and a SPARQL query is generated. The SPARQL query can either be sent directly to the SPARQL endpoint, or the user can edit the query, e.g., for asking for something different. Finally, C) the results of the SPARQL query are retrieved and presented, where the user can click on the URLs to browse more information for the entities.

3 Use Cases

The current version of TCRMQ exploits two RDF KGs containing artworks, from the CH domain with millions of triples, SAAM [14] and Kerameikos [1]. Below, we present use cases (UC) and scenarios (see a video in the following link: https://www.youtube.com/watch?v=v2zo_xFOaY8).

UC1. SPARQL Query Generation over a Given KG . We can ask a question (even in several languages using the same methods and prompts) to generate the SPARQL query. This concerns users that are not familiar with RDF and the ontology (e.g., CIDOC-CRM), i.e., they can ask natural questions and browse both the SPARQL query and the results (see the steps of Fig. 1).

UC2. Query Enrichment with Missing Facts. The user can ask questions, such as those on the right side of Fig. 1, requiring external knowledge. In Fig. 1, the LLM was able to generate the SPARQL query by also adding in the FILTER statement the dates of the Second World War. It is worth mentioning that except for adding information in the filter statement of the query (such as dates), there are cases where the LLM needs extra reasoning, e.g., for questions like "Give me artists born in Greece but not in its capital" (see Fig. 4).

UC3. Ontology Learning and Query Editing. The user (e.g., an expert) can learn how an ontology is used in the given KG, e.g., which path patterns are used for expressing an event, and has the ability to edit the generated query, e.g., to restrict the results and to correct possible errors.

4 Experimental Evaluation

We use an extension of the benchmark that is available online in the following page: https://github.com/mountanton/CIDOC-QA-using-LLMs. That page also includes the code and all the results. The benchmark contains 60 natural questions for each KG (in total 120 questions). The questions are divided in 6 categories, each one having 10 questions, i.e., questions requiring to follow a path i) with radius $r = 1$, ii) with $r = 2$, iii) with $r = 3$, iv) with $r = 4$, v) a mixed path (combining paths of different r) and vi) questions that need enriched data from external knowledge (e.g., see questions in the right part of Fig. 1). The last category is the new part of the benchmark. Moreover, we use the same set of questions translated into Greek. We provide experimental results for 3 ChatGPT models (versions v3.5, v4-mini and v4) for both languages by using the zero-shot method, where we give the following prompt "By using the CIDOC-CRM ontology, please produce only a SPARQL query (without explanation) for the question Q" ; the baseline method, where we give to the LLM the list of all the properties and classes of the KG; and the two-stage path patterns method. For these methods, we provide the accuracy; a number in the range

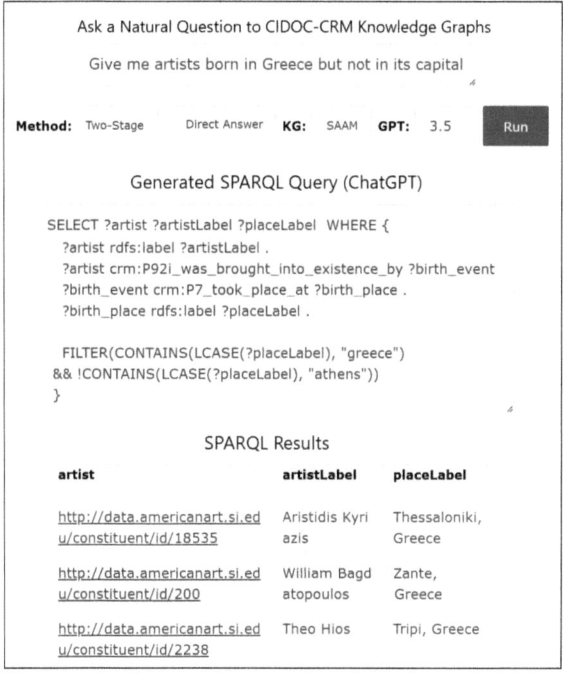

Fig. 4. Screenshots from TCRMQ

[0,1] showing the fraction of questions that we managed to produce the correct SPARQL query. Each generated query was annotated as correct, if it produced the same results as the SPARQL query of the gold standard.

Table 1. Average Accuracy using the two-stage method

Questions Type	GPT-3.5	GPT-4 Mini	GPT-4
$r = 1$ questions	0.75	0.90	0.95
$r = 2$ questions	0.65	0.80	0.90
$r = 3$ questions	0.75	0.55	0.80
$r = 4$ questions	0.50	0.40	0.80
Mixed r questions	0.65	0.50	0.85
Questions with enriched facts	0.45	0.30	0.70
Average	0.62	0.57	0.83

Comparison between the Models. Figure 5 shows that the zero-shot method achieved low accuracy, i.e., less than 0.12 for all models, while the baseline

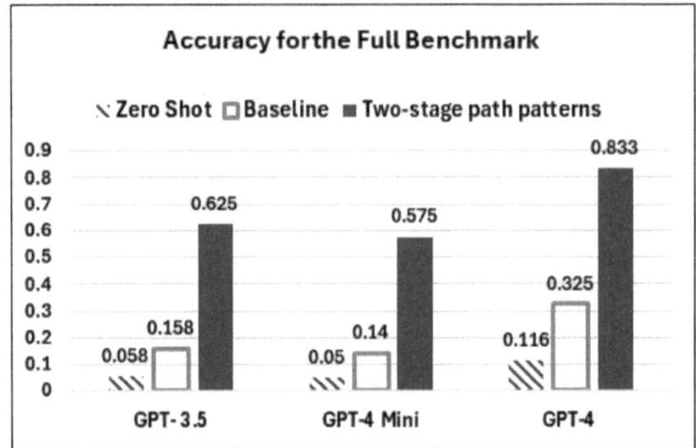

Fig. 5. Accuracy of ChatGPT models

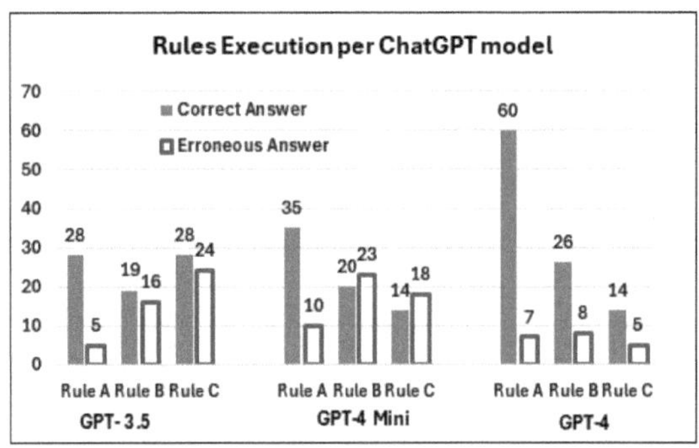

Fig. 6. Rules Execution (two-stage method)

achieved 0.325 for ChatGPT v4 and approximately 0.15 for the other two versions. The highest accuracy (i.e., 0.833) observed by using ChatGPT v4, i.e., it provided the correct SPARQL query for the 83.3% of the 120 queries (100 /120), while the corresponding percentage for the second best, i.e., v3.5, was 62.5%. Regarding questions' categories (for the two-stage method), Table 1 shows that the GPT-4 outperformed the other models in all categories, and especially for the questions requiring either to follow a long path, or enriched facts from the LLM.

Analysis of Results/Errors. In Fig. 6 we can see a kind of ablation study, i.e., the rules executed for each ChatGPT model. Generally, the accuracy when Rule A is executed is much higher compared to those of Rules B and C. Specifically, we can see in Fig. 6, that for GPT-4, the predictions of the first prompt resulted in Rule A 67 times (with accuracy 60/67) whereas for the other models less than 45 times. Therefore, the predictions of the first prompt for GPT-4 were more accurate. Apart from the rules execution, the GPT-3.5 and GPT-4mini models were more vulnerable to i) SPARQL syntax errors or/and hallucination

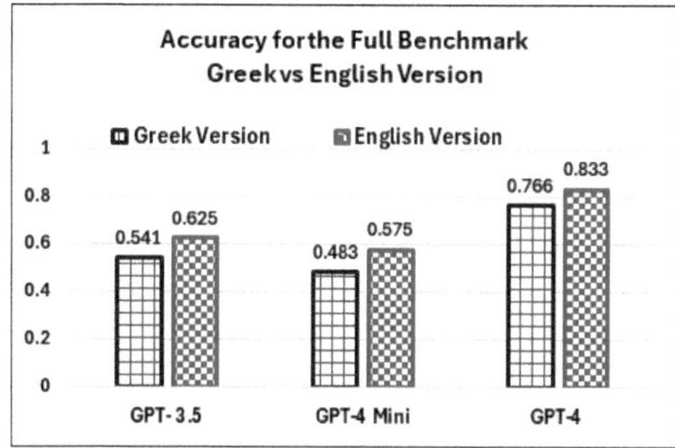

Fig. 7. Results of Greek vs English Benchmark

problems (e.g., for the questions needed missing facts), and ii) the selection of totally wrong paths for each question (see an analysis for GPT-3.5 in [10]). For GPT-4, the main erroneous cases were for questions that the selected path from the LLM was a superset or a subset of the correct one, or filter statements that were syntactically correct but not the desired ones.

English vs Greek Benchmark. The accuracy remains high even by sending the questions in Greek (see Fig. 7), but it is lower compared to English, e.g., 0.766 vs 0.833 for GPT-4. The main reason is that in some questions, greek words in the filter statements were missing and ambiguous greek words resulted in the wrong path.

Cost and Execution Time. We needed on average $\sim 0.01\$$ for each question by using GPT-4 (the cost was much lower when Rule A executed), $\sim 0.001\$$ through GPT-4mini, and $\sim 0.02\$$ for GPT-3.5 (the most expensive in our case). Finally, the execution time is on average 2–3 seconds for the whole process (query generation and answer from the endpoint), regardless of the used GPT model.

5 Concluding Remarks

We shall demonstrate `TCRMQ`, a tool that offers QA over CIDOC-CRM based KGs, by translating the incoming natural question to a multi-hop SPARQL query, and by retrieving the results from the SPARQL endpoint. The tool is based on a *two-stage ontology path patterns method* that can be used in event-based models, and apart from questions requiring data from the KG, it leverages the LLM to generate queries with the right filtering conditions, even for information that is not present in the KG (e.g., dates). For a benchmark with 120 natural questions, we achieved an accuracy of 0.833 through the two-stage method and GPT-4, compared to 0.325 of the baseline. In future work, we plan to i) enrich the benchmark with questions for named entities, ii) use other event-based models, and iii) investigate RAG techniques [12] to exploit external knowledge.

References

1. Kerameikos.org (2021). http://kerameikos.org/. Accessed Aug 2025
2. Allemang, D., Sequeda, J.: Increasing the accuracy of llm question-answering systems with ontologies. In: ISWC, pp. 324–339. Springer, Heidelberg (2024). https://doi.org/10.1007/978-3-031-77847-6_18
3. Diallo, P.A.K.K., Reyd, S., Zouaq, A.: A comprehensive evaluation of neural sparql query generation from natural language questions. IEEE Access (2024)
4. Doerr, M.: The CIDOC CRM, an ontological approach to schema heterogeneity. In: Dagstuhl Seminar Proceedings. Schloss Dagstuhl (2005)
5. Feng, Y., Ding, L., Xiao, G.: GeoQAMap-geographic question answering with maps leveraging LLM and open knowledge base (short paper). In: GIScience 2023. Schloss Dagstuhl-Leibniz-Zentrum für Informatik (2023)
6. Kovriguina, L., Teucher, R., Radyush, D., Mouromtsev, D.: SPARQLGEN: One-shot prompt-based approach for SPARQL query generation (2023)
7. Mecharnia, T., d'Aquin, M.: Performance and limitations of fine-tuned llms in sparql query generation. In: Proceedings of GenAIK Workshop, pp. 69–77 (2025)
8. Mountantonakis, M., Theocharakis, I., Tzitzikas, Y.: Why we need ontology-specific data portals: a case study for cidoc-crm. In: SWODCH (2023)
9. Mountantonakis, M., Tzitzikas, Y.: Real-time validation of ChatGPT facts using RDF knowledge graphs. ISWC Demo Paper (2023)
10. Mountantonakis, M., Tzitzikas, Y.: Generating sparql queries over cidoc-crm using a two-stage ontology path patterns method in llm prompts. ACM J. Comput. Cult. Herit. **18**(1), 1–20 (2025)
11. Pride, D., Cancellieri, M., Knoth, P.: Check for updates CORE-GPT: combining open access research and large language models for credible, trustworthy question answering. In: TPDL 2023, Zadar, Croatia, 26–29 September 2023, Proceedings, vol. 14241, p. 146. Springer, Heidelberg (2023). https://doi.org/10.1007/978-3-031-43849-3_13
12. Sapidis, I., Zervos, V., Mountantonakis, M., Tzitzikas, Y.: Interactive and provenance-aware search and QA over documents using LLMs, RAG and knowledge graph verbalization. In: Proceedings of TPDL, Tampere, Finland (2025)
13. Sun, K., Xu, Y.E., Zha, H., Liu, Y., Dong, X.L.: Head-to-tail: how knowledgeable are large language models (llm)? AKA will llms replace knowledge graphs? arXiv preprint arXiv:2308.10168 (2023)
14. Szekely, P., et al.: Connecting the smithsonian American art museum to the linked data cloud. In: Cimiano, P., Corcho, O., Presutti, V., Hollink, L., Rudolph, S. (eds.) ESWC 2013. LNCS, vol. 7882, pp. 593–607. Springer, Heidelberg (2013). https://doi.org/10.1007/978-3-642-38288-8_40

Workshop Papers

A Knowledge Graph-Driven Interactive 3D Simulation for Craft Training

Alexandros Makris[1(✉)], Ioanna Demeridou[1], Panagiotis Koutlemanis[1], Anastasios Roussos[1], Nikolaos Partarakis[1,2], and Xenophon Zabulis[1]

[1] Institute of Computer Science, Foundation for Research and Technology Hellas (ICS-FORTH), N. Plastira 100, Vassilika Vouton, 70013 Heraklion, Greece
{amakris,idemer,koutle,troussos,zabulis}@ics.forth.gr, partarakis@uom.edu.gr

[2] Departments of Applied Informatics (UoM-DAI), University of Macedonia, 156 Egnatia Street, 546 36 Thessaloniki, Greece

Abstract. The proposed work presents a new approach to craft education that combines semantic knowledge graphs (KGs) with real-time 3D interactive simulations using the Finite Element Method (FEM). Crafting processes are modelled as KGs, outlining each procedural step, possible errors, and their consequences. Learners practise tasks in a virtual environment where their actions are checked against the KG, enabling real-time feedback and correction. After each task, the workpiece is rendered with physically based visualisation to show realistic outcomes. This method enhances learning by providing predictive guidance, error correction, and high-quality visual feedback.

Keywords: knowledge graphs · finite element method · craft training

1 Introduction

Craft education demands hands-on practice for skill mastery. Novices often make errors, hindering learning. We propose a new method: integrating semantic knowledge graphs (KG) with real-time 3D FEM simulations. This approach semantically maps crafting processes in KGs, detailing steps, deviations, and impacts. KG nodes represent distinct process steps, explicitly showing correct actions and predictable, correctable, or irreversible errors.

Our interactive simulation environment utilises this semantic representation, enabling learners to engage in realistic, real-time virtual practice. During interaction, the training environment is instantiated based on the KG. This includes selecting the appropriate tools and materials and initialising the workspace where the action will be exercised. Moreover, recorded enactments of the action from the KG provide action execution exemplars.

Upon completing each task, the virtual workpiece is rendered using physically based techniques, which simulate real-world material appearance through light transport simulation. This high-quality visual feedback reinforces learning outcomes by enabling learners to assess the tangible consequences of their procedural decisions.

H. Takeda et al. (Eds.): IJCKG 2025, LNAI 16297, pp. 479–493, 2026.
https://doi.org/10.1007/978-981-95-5009-8_35

The proposed method demonstrates how semantic KGs can effectively underpin an educationally powerful simulation environment, enhancing craft education through predictive interaction, real-time corrective guidance, and realistic visualisation.

Mastering craft techniques often requires extensive hands-on practice, during which inexperienced learners typically encounter predictable procedural errors. To facilitate learning, we propose integrating a semantically structured representation of crafting processes with interactive simulations employing FEM-based physics modelling.

Our method proceeds through the following logically interconnected steps:

1. Semantic Representation of Craft Processes: A detailed KG encoding the procedural flowchart of the crafting process. Each node in the graph corresponds to specific actions or decision points, enriched with semantic descriptions of process steps, conditions, and potential outcomes.
2. Real-time Interactive FEM Simulation: The KG informs a real-time 3D simulation powered by an efficient FEM solver. Learners interact virtually with the simulated tool and workpiece, enabling practice of specific process steps while receiving real-time feedback based on the anticipated procedural pitfalls.
3. High-Fidelity Rendering for Outcome Assessment: A physically based renderer realistically depicts the resulting workpiece. This visualisation reinforces the learner's understanding by vividly illustrating the impact of correct or incorrect procedural choices.

The proposed approach integrates educational techniques in crafts, combining semantic knowledge representation with interactive physical simulation. KGs are central in orchestrating the dynamic, predictive, and interactive learning experiences.

2 Related Work

2.1 Knowledge Graphs in Education

Domain-specific KGs are important for understanding specialised fields. [1] stresses that KGs are knowledge representations for personalised learning, curriculum design, and content recommendations. In [2], application-specific KGs are reviewed in educational contexts. [3] studies knowledge mapping for KG development.

2.2 Knowledge Graphs in Learning Crafting Skills

Source [4] investigates knowledge management and transfer within arts and crafts organisations. Handmade products rely on tacit knowledge, yet transfer methods lack standardisation. Improved codification, training, and dedicated repositories are crucial for preserving this knowledge. The construction of KGs using textual sources was proposed in [2]; however, this is insufficient for crafting education, where tacit knowledge and materiality require visual and physically informed learning sources.

Source [5] describes a methodology for using KGs to document and preserve crafts using a dedicated ontology [6], extending the CIDOC CRM, and implemented online.

This methodology uses semantic annotation to formalise craft concepts, linking knowledge to multimodal resources. It documents crafting entities: tools, materials, and movements. [7] extends this work by integrating ethnographic observations on crafting actions, products, tools, and materials. The representation links recordings to nominal and verbal descriptions, enabling inference, external links, and enrichment.

Source [8] underscores the importance of well-structured and organised content to craft eLearning, advocating that organising complex crafting processes in a KG ensures logical flow and meaningful connections between learning elements. [9] emphasises the need for systematic, quantitative evaluation of 3D human movement using ground truth data, in real-world scenarios for traditional crafts.

2.3 Physics-Based Simulation and Interactive Training

Physics-based simulation, at the intersection of computer science and physics, uses numerical methods to approximate physical systems governed by fundamental laws like thermodynamics. FEMs are widely used for simulating deformable solids, fluids, and multi-phase interactions. Recent advances use GPU acceleration, allowing the coupling of mechanical and thermal events.

Real-time simulation and training environments exhibit high utility in training environments. Replicating complex systems and scenarios with increasing fidelity and interactivity offers advantages for training across diverse disciplines [10–14].

The importance and applications of real-time simulation and training environments are increasing. A central theme is to replicate complex real-world scenarios for training [15, 16]. Key aspects include the need for realism and the integration of immersive technologies. Simulating realistic physical behaviour involves complex phenomena such as bending, cutting, or tearing of materials, enhances immersion and training effectiveness and is a key challenge for enhancing realism [17].

2.4 Physically Based Rendering in Educational Contexts

Physics-based simulation is crucial in computer graphics for photorealistic animation of cloth, hair, translucent and shiny objects. Computational methods for modelling and simulating traditional crafts safeguard crafting skills. The framework proposed in [18] simulates crafting actions using the FEM. Actions are classified into forming, subtractive, and additive based on their mechanics. In [19], their visualisation for documentation is demonstrated, using light transport simulation to cope with challenging materials such as transparent and metallic.

In [20], an interactive model and simulation system address craft preservation using virtual reality. The system implementation creates a knowledge base of materials, tools, and processes, enabling interactive simulation for training. The digital model supports and enhances traditional craft training compared to purely physical methods.

3 Methodology

3.1 Crafting Process Representation

The crafting process is represented as a Knowledge Graph (KG) based on an ontology. A process schema, derived from ethnographic studies and authored by ethnographers and information scientists, outlines foreseen temporal, causal, or conditional relationships, action sequences, and decision-based divergences. This schema is a graph where nodes are steps and edges define relationships. Errors and corrective actions are modelled as branches. Visualised in UML, the graph shows step inputs and outputs. Enacting this schema generates a process, comprising physical interventions, observations, and judgments, all characterised as actions.

Each node references the (a) tools and workpieces employed, (b) their material properties, (c) requisite conditions, and (d) a set of associations between action execution parameters and action outcomes. The knowledge elements in (a), (b), and (c) are described geometrically and physically, as well as the execution parameters in (d). The action outcomes in (d) are the states of the affected entities during and after the action, encoded as FEM volumetric mesh animations.

These knowledge elements are represented physically and semantically. Physical properties and parameters are numerical, and their values are in the SI system. These are the values of geometrical and material properties over time. Objects and actions are knowledge elements registered to a knowledge structure used as a lexicon. Each elementary representation contains a link to its thesauric entry and its recording; 3D for objects and 4D for actions.

3.2 Craft Process Modelling

In modelling traditional craft processes, it is essential to distinguish between the physical entities involved, the actions performed, the agents responsible, and the underlying procedural knowledge that guides activity. The CIDOC Conceptual Reference Model (CIDOC-CRM) offers a rich semantic framework for expressing these distinctions. We use the CRM as described in Table 1.

Workpieces and Products. Our model represents workpieces, the fundamental entities that transform within the crafting process. Regardless of whether a workpiece originates as raw, unprocessed material or emerges as the refined output of a preceding production stage, it is consistently represented as an instance of the CIDOC-CRM entity ‘E18 Physical Thing’. This classification is deliberately broad, encompassing any tangible item characterised by a persistent physical identity that renders it capable of being subject to the operations in a crafting process.

The designation as E18 ensures that workpieces, irrespective of their origin or stage of completion, are uniformly treated as individual physical entities. This allows for tracking and management of these items throughout their represented processes.

Our model introduces a more specific classification for workpieces that result from intentional human activity in manufacturing. Such workpieces, embodying the tangible outcome of earlier production efforts, are additionally categorised as an instance of 'E24 Physical Man-Made Thing'.

This distinction between E18 and E24 serves a central purpose. It provides a mechanism for tracing the evolving status and provenance of physical objects as they progress through successive phases of transformation. By identifying workpieces that are products of prior manufacturing steps, we maintain a history of their treatment. This is essential for provenance tracking, quality control, and the understanding of manufacturing processes involving multiple stages. The dual classification allows capturing the physical workpiece nature and its status as a product.

Table 1. CIDOC CRM-Based Ontological Mapping of Craft Concepts.

CRM Class	Concept	Description in CRM Context	Role in Craft Model
E18 Physical Thing	Material or workpiece	Any persistent item with a physical identity	Represents all workpieces, whether raw or already transformed
E24 Physical Man-Made Thing	Product of the previous step	A physical object intentionally created or modified by humans; subclass of E18	Used for products of previous steps in the crafting process
E5 Event	General Event	Any phenomenon occurring over time	Superclass of E7, used to model the general event structure of processes
E7 Activity	Actions	A deliberate action performed by an actor to achieve a goal, with a start and end time	Represents individual craft actions, such as shaping or hammering
E52 Time-Span	Temporal context	A defined period with a beginning and an end	Specifies when a crafting action takes place
E53 Place	Spatial context	A location in which something occurred or existed	Specifies where the action is performed
E39 Actor	Practitioner	A person or group that performs activities or bears responsibility	Represents the craft practitioner or team responsible for the action
E29 Design or Procedure	Process schema	A plan, workflow, or design describing how something should be done	Encodes idealised process structures including steps, decisions, and error handling

The selection of CIDOC CRM as the ontological foundation is motivated by its status as a widely adopted standard in the cultural heritage domain, particularly in terms of the semantic interoperability and long-term sustainability of the data. Unlike lightweight schema languages (e.g., JSON-LD-based custom vocabularies) or task-specific ontologies (e.g., OWL ontologies for manufacturing workflows), CIDOC CRM provides a rich and extensible conceptual framework that distinguishes between activities, actors, places, and designed procedures. This makes it suitable for representing craft processes that combine material transformations with cultural and social context. While alternative models such as BPMN or PROV-O focus on process execution or provenance tracking, they lack the deep alignment with cultural heritage semantics required for safeguarding intangible heritage. Basing the craft modelling in CIDOC CRM enables compatibility with existing cultural heritage infrastructures and retains the flexibility to extend the model with domain-specific classes for crafts.

Crafting Actions and Events. In craft processes, individual actions are modelled as instances of 'E7 Activity', which is a subclass of the encompassing 'E5 Event'. E5 Event is the foundational category for any phenomenon with a discernible beginning and end in time. The 'E7 Activity' class refines for deliberate interventions executed by human agents, acting individually or collectively, to instigate a transformation in a material object or achieve a defined outcome.

These activities encompass the operational steps that constitute a craft process. Examples of such steps include shaping raw materials, joining separate components, applying decorative elements, or refining a material.

Each activity within a process does not occur in isolation. It is situated within a spatiotemporal context. To specify the temporal dimension of an activity, the E52 Time-Span class is employed. This class defines the time interval during which the action transpired. Location is captured through the E53 Place class, which enables the identification and recording of the physical environment where the action occurred. Spatiotemporal context provides a comprehensive understanding of the process, allowing for analysis and replication of the actions.

Agents. The individual or collective responsible for executing activities is represented as an instance of the 'E39 Actor' class. This categorisation is deliberately broad to encompass individuals and social groups that act in the crafting process. E39 is central to the agency and the involvement of practitioners. A single craftsman shaping a product, a team collaborating on a complex project, fall under the umbrella of the E39 Actor. Their involvement can range from direct manipulation and creation to oversight, decision-making, and the provision of necessary resources or support.

By modelling practitioners as Actors, we establish a link between the actor and the action. This allows for a comprehensive understanding of the production process, enabling tracking responsibilities, identifying stakeholders, and analysing the impact of human agency on the final result. This way, we provide a consistent framework for representing the roles of practitioners. Figure 1 presents the classes and relations used in our knowledge representation model.

Process Schemas. Execution of crafting actions is underpinned by the concept of a Process Schema, which is modelled as an instance of 'E29 Design or Procedure'. This classification situates Process Schemas under categories of E28 Conceptual Object and E90 Symbolic Object. These classes encompass the realm of non-physical entities, representing artefacts of thought, meaning, and codified knowledge.

An E29 Design or Procedure is defined as a specification detailing a course of action that is formulated to achieve an anticipated result. It does not describe or refer to a past or ongoing event; instead, it exists as a blueprint, a conceptual design, or a structured set of instructions that is interpreted and enacted by an agent. Figure 2 illustrates the KG structure of a craft Process Schema.

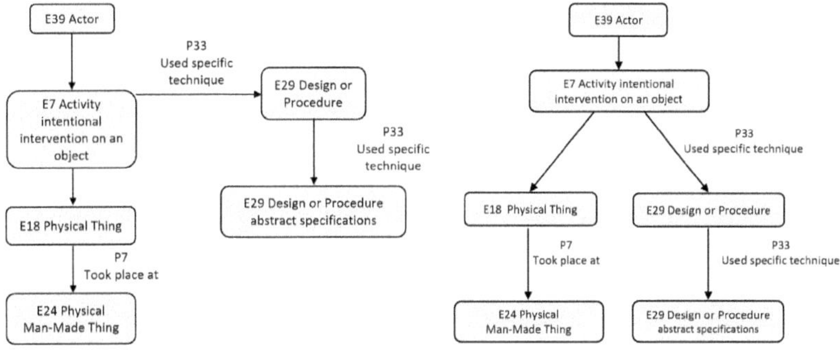

Fig. 1. Knowledge Representation Classes and Relations.

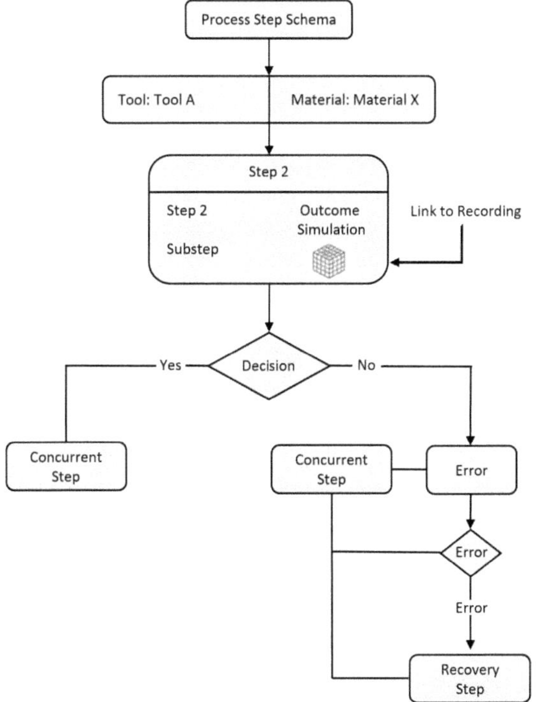

Fig. 2. Knowledge Graph Structure of a Craft Process Schema

We stress the distinction between the Process Schema and the concrete E7 Activity that it informs and directs. An Activity denotes an actual, situated event that unfolds in time and space, invariably involving some physical change. In contrast, an E29 Design or Procedure is a model of such events, existing in the realm of abstract concepts, providing a template for potential actions.

3.3 Knowledge Representation

Each step is associated with at least a set of action execution parameters and its action outcome. The parameters are the numerical values required to represent the action physically and are the same as those simulated in the FEA. They represent physical quantities pertinent to environmental conditions and practitioner actions, such as temperature, gripping tool posture, and induced motion for tools. The state of affected entities during the action is computed in the simulation. In this way, an ostensive representation of actions and their results is achieved.

The presentation of action outcomes is not limited to the exemplary performance of the action. They contain alternative workflows per the outcomes of branching nodes. Common mistakes, e.g. incorrect tool angles or inappropriate material or condition choices, are also represented as diverging paths. These branches terminate in either recovery paths, showing how the mistake can be corrected, or in failure nodes, used to indicate irreversible outcomes. This enables the system to offer predictive guidance and real-time feedback during the simulation.

Craft knowledge encoded in the graph is multimodal. Each action node is linked to exemplars, including video recordings, verbal descriptions, and semantic annotations. These links facilitate a comprehensive and media-rich learning experience, allowing learners to review expert demonstrations in tandem with procedural simulation. A multilingual and hierarchically structured thesaurus ensures semantic consistency and interoperability across linguistic and cultural contexts.

This procedural, material, and contextual organisation of knowledge enables dynamic simulation scenarios that adapt to learner behaviour while preserving the epistemic integrity of traditional craft practices.

3.4 Education

The Craeft eLearning Portal serves as the central access point for all Craeft users, functioning as a comprehensive e-learning platform. It provides a wide array of functionalities designed to support and guide individuals through their craft education journey. The Craeft eLearning platform provided training courses required to master a particular craft.

Courses can be either educational, focusing on theoretical knowledge, or training, providing practical skills through hands-on practice. Lessons can be either online or offline and draw from diverse training materials, craft demonstrations, action visualisations, and action simulations. Course authoring involves linking these sources to lessons, as shown in Fig. 3.

FEM provides simulations based that are initialised using the process schema. The primary objective of these simulations is to visually represent the effects of applied forces and material transformations given a specific set of parameters. The simulator takes two key inputs: the parameters for the simulation and the simulation setup, which is derived directly from the process schema. The simulation results are a collection of object state sequences. These sequences represent the transformation of 3D objects as the virtual action is executed.

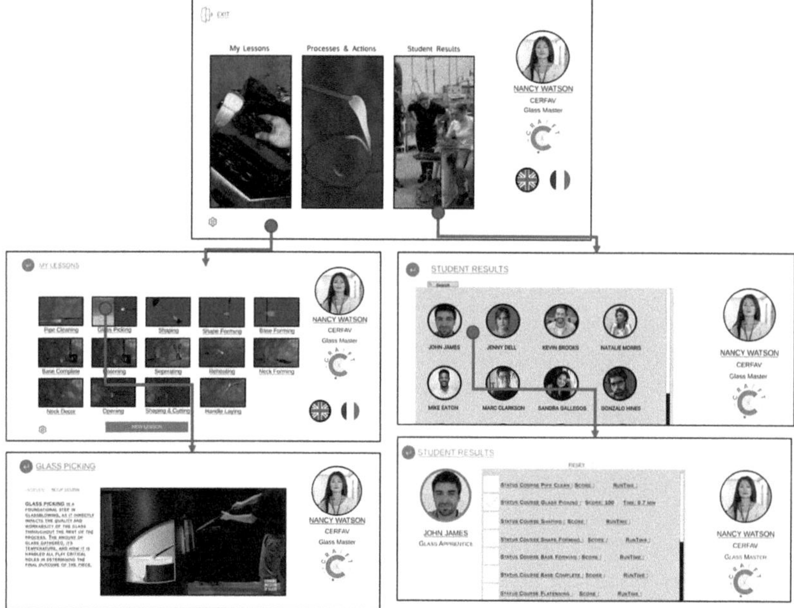

Fig. 3. A course page is generated from the process schema representation.

An application called 'Craft Studio' is the environment for rendering simulator outputs and is a suite of tools for authoring immersive educational and training materials. The Craft Studio is an interactive authoring application implemented in Unity. It provides features for authoring and executing both interactive simulations. The Craft Studio renders the simulation outputs within the Unity game engine.

The process schema is converted to a simulator input. A lesson is compiled into a Lesson Specification File, in JSON format, which is exported and uploaded to the eLearning portal. The theoretical components of the lesson are authored directly within the e-learning portal, containing all necessary material except for the execution aspects. This includes details such as tools, materials, and transformation mechanisms, all designed to appropriately introduce students to the lesson. The execution file for the lesson is the one previously uploaded from the Craft Studio and is linked as a resource within the lesson. Subsequently, the lesson is assigned to apprentices. Apprentices then download the Lesson Specification File and load it into the Apprentice Studio to undertake the lesson.

Workshops, tools, and machines are derived from the KG for craft or processes. The Craft Studio provides a graphical user interface (GUI) for authoring these lessons, making use of these digital assets and renderers. Lesson scenarios are set within a virtual workshop environment. Authoring a Lesson scenario also includes defining its input/output interfaces. Student inputs come from the keyboard, mouse, and haptic devices. System outputs are delivered in 2D, 3D, XR (Extended Reality), and haptic rendering.

3.5 Real-Time FEM Interactive Simulation

The real-time simulation of manipulating deformable objects uses the NVIDIA PhysX simulator [21]. NVIDIA PhysX simulates complex physics interactions, including soft body dynamics, by leveraging the parallel processing capabilities of GPUs. The FEM is used to handle the complexities of soft body dynamics, ensuring that interactions between objects and the environment are faithfully represented. To achieve this, the objects are assigned properties such as elasticity, friction, and mass distribution. Given that the focus is on real-time applications, the set of physical properties is kept small. In the scenarios considered in this paper, the main external force is the one applied by the tool on the object that is manipulated.

We implemented the simulator using the PhysX SDK directly to build a module that gives full control over the performed manipulations. The developed module takes as input the physical properties, geometries, and initial states of the objects and outputs the deformed geometries. Physical properties and initial states of the actors are defined in configuration files. Different sets of parameters are defined for the soft bodies (materials that are manipulated) and rigid bodies (tools). Object geometry is imported using the Wavefront file format. The deformed meshes that result from the manipulations are passed to the subsequent rendering module for interactive visualisation. Figure 4 presents the interaction loop of the FEM simulator, highlighting the data flow between the simulation and rendering modules.

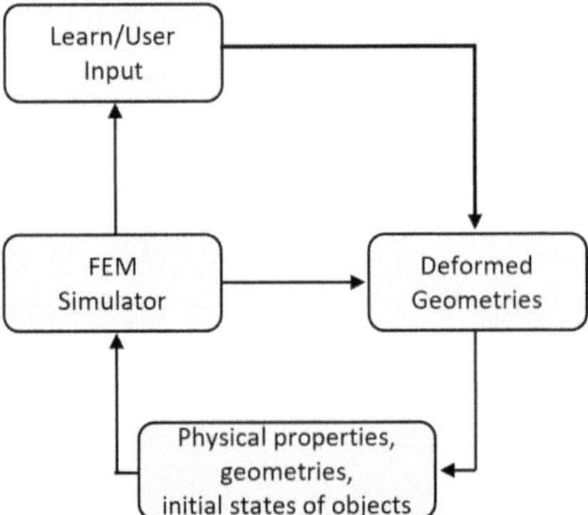

Fig. 4. An illustration showing the interaction loop of the FEM simulator.

The parameters of the simulation are recorded in JSON format and are acquired by the simulation engine during execution. To support interactive real-time simulations of crafting actions, the simulation core has been exported in the form of a DLL library. A Unity3D application has been developed that sets up the simulation environment within a Unity3D scene, and through the DLL, all the components of the scene become available for manipulation via the Unity3D interfaces and through the creation of Unity3D script files. Based on this infrastructure standard, manipulators are integrated for controlling the rotation and translation of the tool. This allows support for repetitive practising of actions and incremental action flows.

Presently, process schemas are translated into simulator input through an intermediate JSON specification. A natural extension is to enable the simulation engine to consume semantic descriptions directly from the knowledge graph at runtime. This would allow the dynamic update according to modifications in the semantic representation, such as for newly added actions or new workflows. This technical integration ensures closer synchronisation between documentation and simulation. We envisage a future pipeline where semantic updates authored in the knowledge graph automatically propagate to the simulation environment, thereby reducing authoring overhead and strengthening the fidelity of training scenarios.

3.6 Physically Based Rendering for Result Visualisation

To reinforce its educational value, the system integrates physically based rendering (PBR) techniques that realistically visualise the workpiece after each interaction. Our rendering module employs light transport simulation informed by the output of the FEM engine, which encodes the material deformation and surface evolution resulting from the learner's actions.

Following the approach detailed in [18], we simulate optical phenomena such as subsurface scattering, refraction, and anisotropic reflection, critical for rendering complex craft materials such as polished wood, metal, glass, and ceramic. These effects reveal surface quality and underlying structure and, thus, carry instructional relevance. The renderer operates on material models parameterised by measurable physical attributes (e.g., roughness, metallicity, index of refraction), enabling meaningful variation in appearance corresponding to process outcomes.

Rendering is performed on mesh outputs derived from FEM-simulated deformation. In our pipeline, the FEM displacement field is used to morph the base mesh, preserving topological fidelity while enabling realistic damage, warping, or compression effects to be visualised. These mesh modifications are shaded using a physically based bidirectional scattering distribution function.

Photorealistic rendering aids immersion and visual appraisal. Learners interpret visual cues as indicators of procedural issues. This feedback links perception and reasoning, fostering understanding of cause-and-effect in craft.

4 Use Case and Evaluation

To demonstrate the proposed method, we consider a use case focused on carving wood and aluminium blocks. This scenario demonstrates how cylindrical and various chisels can be used to manipulate materials. The simulation provides real-time feedback on tool interaction, showing physical effects on blocks via physically based rendering. This use case integrates KGs, interactive FEM, and high-fidelity visualization for craft education.

4.1 Simulation Setup

The PhysX scenes contain objects called actors. Each actor has a physical state and properties that include position, orientation, etc. Actor states evolve due to applied forces and interactions. In our case, two actors are defined at each time step of the simulation: the tool as a rigid body actor and the manipulated material as a soft body actor. Given the actor properties and initial states from the previous section, the simulation is straightforward using the appropriate PhysX SDK API calls.

4.2 Results

For the simulations, we used two types of materials: wood and aluminium. The duration was standardised (5 s) to maintain consistency across experiments. Each scenario involved a tool, either a cylindrical or chisel type, interacting with one of the materials. The initial state of the objects includes their pose in 3D (translation, rotation), and their initial velocities. The tools were driven only by predetermined initial velocities to ensure repeatability and comparability of the results. These initial velocities were chosen to simulate realistic crafting actions. The properties of wood and aluminium were set according to typical physical characteristics of each material within the PhysX simulation environment, as shown in Table 2.

Table 2. Material properties in PhysX Simulations.

Material Property	Timber	Aluminium
Density (mass/volume)	650	2700.0
Dynamic Friction (dimensionless)	0.25	0.25
Young's Modulus (force/area)	5*109	$70*10^9$
Poisson's Ratio (dimensionless)	0.499	0.499
Elasticity Damping (dimensionless)	1.0	1.0
Damping Scale (dimensionless)	1.0	1.0

Simulation results, shown in Figs. 5 and 6, demonstrate high realism for materials like wood and aluminium under cylindrical and chisel tools. Plastic deformation for wood material is not calculated due to the increased processing cost. Different incidence forces and angles caused distinct deformations in each material, highlighting material-specific responses. The simulations accurately model tool-material interactions, providing visual fidelity for realistic craft education.

5 Discussion and Future Work

This framework combines KGs, interactive FEM simulation, and high-fidelity visualization to create an effective learning environment for traditional crafts. It allows learners to practice procedurally, receiving feedback based on authentic domain knowledge. A key benefit is the externalisation of tacit knowledge, structuring craft procedures as ontologically grounded KGs, which supports modular, multilingual, and semantically interoperable representations of craft actions. This enables systematic training workflows and safeguards endangered crafts (Fig. 7).

Fig. 5. Simulation of the effect of cylindrical tools on aluminium and wood surfaces. Each row shows exemplar frames from three simulations. Different contact forces and angles are simulated, showcasing the different impacts on the deformation of the surface.

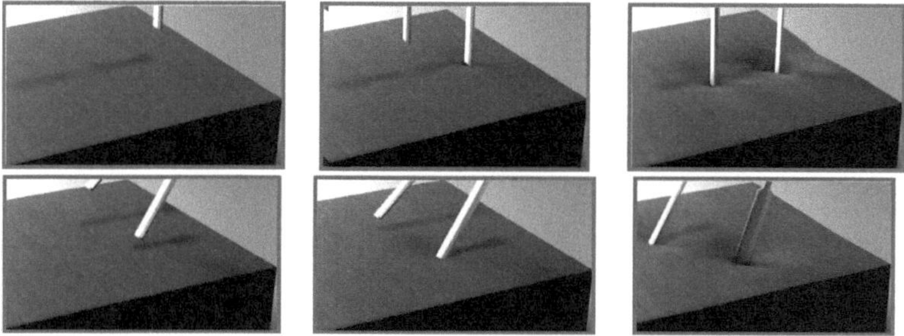

Fig. 6. Simulation of the effect of chisel tools on a wooden surface. Each row shows frames from different simulations. Different contact forces and angles are simulated.

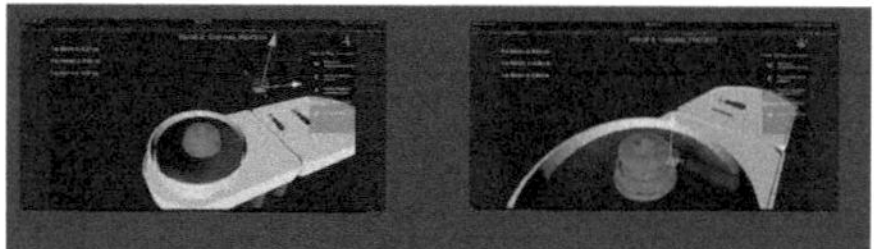

Fig. 7. Manipulation of the tools in 3D.

The simulation environment provides a safe and cost-effective space for experiential learning. Learners may practice complex procedures without the risks associated with material waste or tool misuse. Real-time FEM feedback enhances procedural understanding, while photorealistic rendering of outcomes supports visual assessment and reflection on one's actions. Exemplars and decision modelling within the KG reinforce understanding through situated comparison and error recovery.

Some limitations remain. The current system is optimised for well-defined procedural sequences with observable outcomes. More complex craft processes involving subjective decision-making, fine-motor precision, or ambiguous success criteria may require additional modelling strategies. The accurate simulation of certain materials, particularly fibrous, layered, or porous substances, remains a technical challenge in FEM and rendering. Future work will expand the system's applicability by extending the ontology to new craft domains (e.g., fibre arts, leatherworking). We plan to integrate haptic feedback and motion capture for enhanced realism. User studies will evaluate pedagogical impact, focusing on error awareness, procedural retention, and cross-domain knowledge sharing.

This work is part of the Horizon Europe project Craeft, which brings together computer scientists, heritage experts, and craft practitioners in a coordinated effort to safeguard and transmit traditional craftsmanship in Europe. Within Craeft, the proposed approach documents craft processes in semantically interoperable formats, building knowledge bases enriched with multimodal resources, and enabling innovative learning environments for practitioners and apprentices. By integrating knowledge graphs, FEM-based simulation, and physically based rendering, this work complements project activities such as motion capture benchmarks, multimodal craft dictionaries, and online repositories of craft knowledge.

Acknowledgements. This research was funded by the European Commission in the context of the Horizon Europe research and innovation program in the project Craeft, 101094349.

References

1. Abu-Salih, B., Alotaibi, S.: A systematic literature review of knowledge graph construction and application in education. Heliyon **10**, e25383 (2024). https://doi.org/10.1016/j.heliyon.2024.e25383
2. Weichselbraun, A., Waldvogel, R., Fraefel, A., Van Schie, A., Kuntschik, P.: Building knowledge graphs and recommender systems for suggesting reskilling and upskilling options from the web. Information **13**, 510 (2022). https://doi.org/10.3390/info13110510
3. Fan, X., Sun, Y., Wang, Q., Wang, X.: Research on the application of knowledge graph in higher education of art and design. In: Kaki, S., Majoul, B., Mohd Sharif, M.F., and Syed Mohammed, S.F. (eds.) Proceedings of the 2024 6th International Conference on Literature, Art and Human Development (ICLAHD 2024), pp. 494–501. Atlantis Press SARL, Paris (2024). https://doi.org/10.2991/978-2-38476-319-1_57
4. Osterloh, M., Frey, B.S.: Motivation, knowledge transfer, and organizational forms. Organ. Sci. **11**, 538–550 (2000). https://doi.org/10.1287/orsc.11.5.538.15204
5. Zabulis, X., et al.: A roadmap for craft understanding, education, training, and preservation. Heritage **6**, 5305–5328 (2023). https://doi.org/10.3390/heritage6070280
6. Zabulis, X., et al.: Digitisation of traditional craft processes. J. Comput. Cult. Herit. **15**, 1–24 (2022). https://doi.org/10.1145/3494675
7. Zabulis, X., et al.: Multimodal dictionaries for traditional craft education. Multimodal Technol. Interact. **8**, 63 (2024). https://doi.org/10.3390/mti8070063
8. Partarakis, N., Zabulis, X.: Applying cognitive load theory to Elearning of crafts. Multimodal Technol. Interact. **8**, 2 (2023). https://doi.org/10.3390/mti8010002

9. Olivas-Padilla, B.E., Glushkova, A., Manitsaris, S.: Motion capture benchmark of real industrial tasks and traditional crafts for human movement analysis. IEEE Access **11**, 40075–40092 (2023). https://doi.org/10.1109/ACCESS.2023.3269581
10. Knudson, M.M., et al.: Trauma training in simulation: translating skills from SIM time to real time. J. Trauma Inj. Infect. Crit. Care. **64**, 255–264 (2008). https://doi.org/10.1097/TA.0b013e31816275b0
11. Menghal, P.M., Laxmi, A.J.: Real time simulation: a novel approach in engineering education. In: 2011 3rd International Conference on Electronics Computer Technology, pp. 215–219. IEEE, Kanyakumari (2011). https://doi.org/10.1109/ICECTECH.2011.5941592
12. Chambers, T.L., et al.: Real-time simulation for a virtual reality-based MIG welding training system. Virtual Real. **16**, 45–55 (2012). https://doi.org/10.1007/s10055-010-0170-x
13. Lee, Y., Ko, C., Lee, H., Jeon, K., Shin, S., Han, C.: Interactive plant simulation modeling for developing an operator training system in a natural gas pressure-regulating station. Pet. Sci. **14**, 529–538 (2017). https://doi.org/10.1007/s12182-017-0170-5
14. Vergnano, A., Berselli, G., Pellicciari, M.: Interactive simulation-based-training tools for manufacturing systems operators: an industrial case study. Int. J. Interact. Des. Manuf. IJIDeM. **11**, 785–797 (2017). https://doi.org/10.1007/s12008-016-0367-7
15. Tang, L., et al.: Improving the quality of preclinical simulation training for dental students using a new digital real-time evaluation system. Eur. J. Dent. Educ. **25**, 100–107 (2021). https://doi.org/10.1111/eje.12580
16. Antonovsky, A., Pollock, C., Straker, L.: Identification of the Human Factors Contributing to Maintenance Failures in a Petroleum Operation. Hum. Factors J. Hum. Factors Ergon. Soc. **56**, 306–321 (2014). https://doi.org/10.1177/0018720813491424
17. Wolff, R., Preusche, C., Gerndt, A.: A modular architecture for an interactive real-time simulation and training environment for satellite on-orbit servicing. J. Simul. **8**, 50–63 (2014). https://doi.org/10.1057/jos.2013.10
18. Zabulis, X., et al.: Modelling and simulation of traditional craft actions. Appl. Sci. **14**, 7750 (2024). https://doi.org/10.3390/app14177750
19. Zabulis, X., et al.: Simulation and visualisation of traditional craft actions. Heritage **7**, 7083–7114 (2024). https://doi.org/10.3390/heritage7120328
20. Xu, R., Xu, J.: Research on interactive traditional craft diagram model and simulation system: take Nanjing Rong Hua craft as an example. In: Proceedings of the 2016 2nd International Conference on Advances in Energy, Environment and Chemical Engineering (AEECE 2016). Atlantis Press, Singapore (2016). https://doi.org/10.2991/aeece-16.2016.54
21. NVIDIA: PhysX. (2024). https://github.com/NVIDIA-Omniverse/PhysX

Connecting Cultural Heritage Through Knowledge Graphs: A Case Study on Medieval Manuscripts

Nicolò Pratelli[1], Emanuele Lenzi[1,2](✉), and Valentina Bartalesi[1]

[1] Institute of Information Science and Technologies (ISTI), National Research Council of Italy (CNR), Pisa, Italy
{nicolo.pratelli,emanuele.lenzi,valentina.bartalesi}@isti.cnr.it

[2] Department of Information Engineering (DII), University of Pisa, Pisa, Italy

Abstract. Knowledge Graphs (KGs) are increasingly leveraged in Cultural Heritage (CH) domains to enable structured, interoperable representations of artefacts, people, events, and places. The ISO CIDOC CRM, as a core reference ontology, facilitates semantic alignment and interoperability across diverse CH datasets. However, despite the CIDOC CRM, linking independently developed KGs remains a complex challenge. This paper investigates the process of integrating two KGs collecting knowledge about ancient manuscripts: the Mapping Manuscript Migrations (MMM) and the IMAGO knowledge graphs, both of which adopt CIDOC CRM as their underlying schema. Through this case study, we explore methodological and technical hurdles in aligning entities, reconciling modelling decisions, and preserving provenance and semantics during integration. We discuss strategies for identity resolution, semantic approximation, and metadata enrichment, and propose reusable patterns for future KG linking efforts in the CH domain.

Keywords: Knowledge Graphs · Cultural Heritage · CIDOC CRM · Semantic Interoperability · Linked Data

1 Introduction

Cultural Heritage (CH) institutions have increasingly adopted Knowledge Graphs (KGs) as a means to represent, share, and connect information in a structured and semantically meaningful way [10]. By relying on Semantic Web technologies and formal ontologies such as the CIDOC Conceptual Reference Model (CRM) [17], these institutions aim to achieve data interoperability, enhance discovery, and support complex queries across heterogeneous collections. Through this approach, metadata from libraries, museums, archives, and scholarly projects can be brought together into a unified, machine-readable knowledge space.

CIDOC CRM, as a ISO standard and reference ontology in the CH domain, provides a powerful framework for representing events, agents, objects, and their

H. Takeda et al. (Eds.): IJCKG 2025, LNAI 16297, pp. 494–508, 2026.
https://doi.org/10.1007/978-981-95-5009-8_36

interrelations. Many cultural heritage projects, including Europeana [21], ARIADNE [23], and the Sampo series [20], have adopted CRM-based models to structure their data. However, despite the use of a common reference model, the integration of independently developed KGs remains a complex and non-trivial task. Semantic interoperability is not solely dependent on ontology alignment; it also requires addressing differences in how data are modelled, annotated, and interpreted. Projects often vary in their choices of granularity, in the assumptions about what constitutes a concept (e.g., a "manuscript" or an "event"), the use of identifiers for entities, or the inclusion or lack of provenance and interpretative metadata. As a result, linking KGs that follow the same model in principle can still involve significant technical, conceptual, and curatorial challenges.

This paper investigates these issues through a case study of integrating knowledge stored in the Mapping Manuscript Migrations (MMM) [12] KG into the *Index Medii Aevi Geographiae Operum* (IMAGO) [7] KG. Both KGs are based on CIDOC CRM and aim to describe medieval and early modern manuscripts, yet they differ significantly in modelling scope and detail. IMAGO, developed within a digital humanities framework, emphasises interpretative modelling and layered provenance, whereas MMM focuses on aggregating structured, factual metadata from multiple manuscript repositories. The presented integration process involved retrieving and preparing metadata from both KGs, implementing a Python-based comparison algorithm, and validating potential matches through a combination of automated and manual techniques. This required resolving differences in shelfmark formats, author naming conventions, city and library identifiers, and representations of physical and intellectual manuscript attributes. Through this study, we aim to contribute to the ongoing discourse on semantic integration in CH, highlighting the importance of combining robust technical approaches with domain expertise and curatorial insight.

The paper is structured as follows: first of all, we present a review of related digital libraries and semantic CH projects (Sect. 2), then we describe the specific characteristics of the IMAGO and MMM datasets (Sect. 3), and detail the matching and linking methodology (Sect. 4). We then discuss the challenges encountered (Sect. 5), and conclude with reflections on the implications for semantic interoperability and recommendations for future work (Sect. 6).

2 Semantic Technologies in Cultural Heritage

The evolution of Semantic Web technologies, coupled with adopting the Linked Open Data (LOD) paradigm, has significantly influenced methodologies within the CH domain. These frameworks promote the formal encoding of data and facilitate its interconnection across heterogeneous sources. Central to this shift are the FAIR principles - Findability, Accessibility, Interoperability, and Reusability - which underpin sustainable data management and collaborative research.

2.1 Ontologies

In the domain of CH, the adoption of Semantic Web technologies has been strongly supported by the development of well-established ontologies that provide a common conceptual framework for knowledge representation and interoperability. Among the most influential initiatives, CIDOC CRM stands out as a reference ontology for the formal description of CH information, enabling the integration of heterogeneous datasets and fostering semantic interoperability across institutions. In close relation to CIDOC CRM, FRBRoo (Functional Requirements for Bibliographic Records – object-oriented) [15] and its reformulation, LRMoo (Library Reference Model – object-oriented) [27], provide a harmonized ontology that bridges the domain of cultural heritage (CH) documentation with bibliographic and library data. This integration facilitates cross-domain interoperability among museum, archive, and library resources. Similarly, the Dublin Core Metadata Initiative (DCMI) [29] has played a key role in defining a widely adopted set of metadata terms that facilitate resource discovery, sharing, and long-term preservation. Other initiatives, such as the Europeana Data Model (EDM) [16], build on these foundations to offer a flexible semantic framework that supports aggregation and enrichment of CH resources at a European scale. These ontologies and models represent fundamental milestones in the application of Semantic Web technologies to the CH sector. They not only provide the necessary standards for structuring and linking complex information, but also enable innovative approaches to digital curation, data integration, and knowledge discovery. Their widespread adoption demonstrates the crucial role of ontology-driven approaches in ensuring both semantic consistency and interoperability, which are essential for advancing research, accessibility, and sustainable management of CH data.

2.2 Semantic Web Applications

The CH domain has pioneered numerous ontology-based Digital Libraries (DLs), offering users enriched means of navigating collections. Among these is Storyspace [30], a tool facilitating curatorial storytelling through event-based narrative structures tied to digital artefacts, used by institutions like the Irish Museum of Modern Art (IMMA)[1] and the National Gallery of Ireland[2]. Bletchley Park Text [24], another example, leverages user-input keywords sent via SMS to dynamically construct personalised digital exhibitions, guided by a domain-specific"story" ontology. Labyrinth 3D [14] merges immersive 3D navigation with semantic pathfinding, aligning narrative structures with ontologies such as FRBR and DOLCE. Larger initiatives such as CultureSampo [25] have scaled these concepts significantly. Aggregating data from over 20 Finnish institutions, CultureSampo uses a CIDOC CRM-driven model to semantically interlink 134,000 cultural objects and their contextual relationships. BiographySampo

[1] https://imma.ie/.
[2] https://www.nationalgallery.ie/.

[19] extends this framework to historical biographies, offering structured access and reasoning capabilities. The Mapping Manuscript Migrations (MMM) project further illustrates the potential of ontology harmonisation by integrating diverse manuscript datasets from institutions including Oxford[3] and the Schoenberg Database[4]. These platforms demonstrate how semantic infrastructures foster knowledge discovery, collaborative curation, and data reuse across institutions.

Historical and literary research increasingly benefits from semantic data modelling. Platforms such as Histropedia [13] allow users to construct timelines using linked data from Wikidata[5] and Wikipedia[6]. Chronas[7] provides an interactive cartographic experience, connecting historical events to spatial data enriched with semantic links. The Henri Poincaré Correspondence project [11], comprising over 2,000 letters, utilises an extended Dublin Core[8] schema for annotation and aspires to migrate to CIDOC CRM for enhanced semantic granularity. In ancient geography, Pelagios [28] and Pleiades [5] projects offer linked datasets that associate historical place names using SKOS[9], Dublin Core[10], and FOAF[11], among others. These initiatives enrich geographical narratives of antiquity by integrating various sources under common semantic frameworks. In archaeology, DECHO [2] employs a CIDOC CRM-based ontology to semantically represent artifacts, including 3D models and analytic metadata. ARIADNE and ARIADNEplus [23], as part of the European research infrastructure, introduced the ARIADNE Catalogue Data Model (ACDM) [3], an extension of DCAT designed to describe archaeological datasets in a FAIR-compliant manner. Later, the AO-Cat Ontology [18] was developed and adopted as a formal ontology to model the resources managed by the ARIADNE Research Infrastructure (ARIADNE RI), within the ARIADNEplus project. Both projects have had a significant impact on the archaeological community, not only in Europe but also internationally [26]. The literary domain has likewise embraced semantic frameworks. The DM2E project [4] adapts the EDM to digitised manuscripts with domain-specific properties, while DanteSources [6] applies RDF and SPARQL to model Dante Alighieri's references and sources. It also serves as the backbone for the Hypermedia Dante Network (HDN) [9], which extends the semantic modelling to the Divine Comedy using commentaries from the Dartmouth Dante Project. Additionally, a digital scholarly edition of Paolo Bufalini's notebook [1] uses SPAR Ontologies, Open Annotation, and Dublin Core to establish semantic intertextual relationships within a KG structure. In the broader cultural heritage domain, the In/Tangible European Heritage - Visual Analysis, Curation

[3] https://oerc.ox.ac.uk/.
[4] https://sdbm.library.upenn.edu/.
[5] https://www.wikidata.org/wiki/Wikidata:Main_Page.
[6] https://en.wikipedia.org/wiki/Main_Page.
[7] https://www.chronas.org/.
[8] https://www.dublincore.org/schemas/.
[9] https://www.w3.org/2004/02/skos/.
[10] https://www.dublincore.org/.
[11] http://xmlns.com/foaf/spec/.

and Communication project (INTAVIA)[12] exemplifies the use of Semantic Web technologies to connect heterogeneous datasets [22]. The goal of this project was to integrate tangible objects from museums, archives, and libraries with intangible assets such as biographical information, enabling their mutual contextualization. By employing semantic modelling technologies, INTAVIA created a unified environment for data integration and visual storytelling, allowing cultural heritage professionals and the public to explore objects and life stories together in richer and more meaningful ways. Building on this approach, the ongoing European Cloud for Heritage OpEn Science project (ECHOES)[13] aims to establish the European Collaborative Cloud for Cultural Heritage (ECCCH), a shared platform that supports both tangible and intangible heritage professionals. Semantic Web technologies are central to this goal, as they enable interoperability between heterogeneous datasets, ensure compliance with FAIR principles, and support the creation of semantically enriched Digital Twins.

3 The IMAGO Project

IMAGO is an Italian research project focused on the systematic survey and formal representation of medieval and Renaissance geographic works. It classifies their authors, genres, and contents, while also documenting their manuscript traditions, editions, and bibliographic references. The project also produces critical editions of selected representative works. Beyond its relevance to medieval and digital philology, IMAGO also provides valuable resources for historical and geographical research. For example, within the project, an index of the toponyms cited in the Latin works of Dante Alighieri has been created [8]. The main results of the project is a Web portal[14], allowing users to explore the IMAGO KG using a user-friendly interface. Analysis of the projects in Sect. 2 reveals a trend toward extending and reusing established ontologies (e.g., CIDOC CRM or EDM), rather than developing isolated schemas. Projects such as the Sampo suite, DECHO, ARIADNE, and DanteSources demonstrate the benefits of such alignment, particularly in terms of semantic consistency and FAIR-compliance. Building on these best practices, the IMAGO ontology has been conceived as a CIDOC CRM extension, promoting compatibility with other CRM-based initiatives, including MMM and the Sampo series. IMAGO introduces bespoke classes and properties for representing manuscript metadata and Latin toponyms from the medieval and Renaissance periods. IMAGO adheres to FAIR and LOD principles from inception, ensuring that the KG it supports remains interoperable, extensible, and sustainable for future scholarship.

4 Mapping Manuscripts Between IMAGO and MMM

To enrich the IMAGO KG with information from the MMM, we sought to identify and align manuscripts shared between the two projects. The goal was to

[12] https://intavia.eu/.
[13] https://www.echoes-eccch.eu/.
[14] https://imagoarchive.it/en/index.html.

detect and connect individuals in both datasets that refer to the same physical manuscript. The process shown in Fig. 1 began with the selection of key metadata fields relevant for comparison, which were then extracted from both KGs. A Python-based matching algorithm[15] was then developed to analyse the extracted data, identify potential similarities, and propose candidate correspondences. These candidate matches were subsequently reviewed and validated through a manual inspection.

A major challenge was the structural heterogeneity between the two knowledge bases, arising from differences in data modelling approaches and metadata conventions. Addressing these discrepancies was essential to ensure reliable match detection. After validation, the matched MMM data were transformed into RDF triples and prepared for integration into IMAGO. The enriched dataset was serialised in Turtle format and loaded into the IMAGO KG. To support exploration, a web interface[16] was developed to visualise the newly established links.

4.1 Data Retrieval

The comparison between IMAGO and MMM manuscripts was based on several key attributes, as reported in the following:

- **Shelfmark**: An alphanumeric identifier that uniquely identifies a manuscript within a library. Shelfmarks may include collection-specific labels (e.g., "671", "Fondo Armanni, II F 2", "Canon. Misc. 721").
- **Library**: The institution that holds the manuscript.
- **City**: The location of the holding library.
- **Authors**: Individuals whose works are included in the manuscript.
- **Physical dimensions**: The manuscript's height, width, and number of folios.

To retrieve the relevant data, some SPARQL queries were implemented and performed on both the KGs. The SPARQL queries were written to return all relevant manuscript metadata in a single row per result, using the `GROUP_CONCAT` function. This allowed multiple values (e.g., authors, collections) to be concatenated into a single field using a triple asterisk (`***`) delimiter, enabling easier post-processing.

During query construction, a key modelling difference between the two KBs had to be addressed. In IMAGO, the class `Manuscript` is used to describe specific sections of a manuscript, often corresponding to individual works, rather than the manuscript as a whole. As a result, instances of the class `Manuscript` in IMAGO do not always map directly to a unique physical item. To resolve this, shelfmark and library pairs were used together to reconstruct references to the entire manuscripts.

For MMM, a separate query was required for manuscripts originating from the Schoenberg Database of Manuscripts (SDBM). Unlike other MMM sources,

[15] https://github.com/AIMH-DHgroup/imago-mmm-linking.

[16] https://imagoarchive.it/mmm/mmmSearch.html.

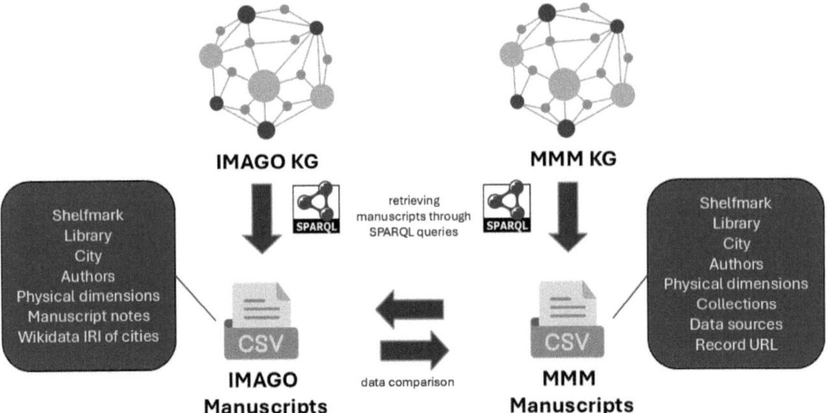

Fig. 1. Workflow of the linking and alignment process between the MMM and IMAGO manuscripts.

which use the `skos:prefLabel` property to store the shelfmark, SDBM employs this property for internal numeric identifiers (e.g., `SDBM_MS_25172`), which do not include shelfmark information. Instead, the actual shelfmark is provided through the `mmm:catalog_or_lot_number` property.

In addition to core comparison fields, auxiliary metadata were retrieved to improve match accuracy and resolve ambiguities as follows:

From IMAGO:

- **Manuscript notes**: Unstructured text often containing information about physical dimensions (height, width, folio count), which are not modelled explicitly in the graph structure.
- **Wikidata IRI for the city**: this metadata is used to reconcile geographic names between IMAGO and MMM by mapping them to a common Wikidata identifier.

From MMM:

- **Collections**: Additional textual context to shelfmarks, which can help infer the holding institution or city when those fields are missing.
- **Data source**: This metadata indicates the original repository (i.e., Bibale, e-Codices, SDBM), useful for interpreting modelling conventions, such as Bibale's embedding of city and library names directly in the manuscript label.
- **Data source entry**: this metadata links to the source entry of the three original repositories, allowing human inspection of authorship or other information not fully captured in the RDF graph.

The SPARQL queries yielded a total of 970 manuscripts from IMAGO and 208,827 from MMM, of which 183,031 were from the SDBM. The results were exported in CSV format and served as the input datasets for the manuscript matching algorithm. Figure 1 illustrates the process of retrieving manuscripts from IMAGO and the MMM KGs. Figure 2 shows the attributes and the metadata of two manuscripts retrieved from IMAGO and the MMM KGs.

5 Data Import and Preprocessing

The manuscript comparison was implemented as part of the matching algorithm. Two CSV datasets, generated from the queries (one for IMAGO and one for MMM) were imported into a Jupyter notebook. Each row was parsed into a Python dictionary representing a manuscript record with its associated metadata. Particular attention was given to the physical dimensions (folios, height, width). In IMAGO, these measurements were embedded in unstructured textual annotations; regular expressions were applied to extract numeric values. In contrast, MMM manuscripts explicitly defined these measurements in structured fields, which were grouped into a sub-dictionary with the keys `"folios"`, `"height"`, and `"width"`. Each processed record was then added to its respective list (IMAGO or MMM), resulting in two structured collections of dictionaries.

An essential preparatory step was the reconciliation of city names used in the two knowledge bases. City data serve as an important filter when identifying potential matches across the large MMM dataset. IMAGO relies on Wikidata IRIs to identify cities, whereas MMM uses custom IRIs. For this reason, comparisons were performed on city labels. However, cities were sometimes represented with different names (e.g., Firenze vs. Florence, Gand vs. Gent). To resolve these variations, Wikidata IRIs from IMAGO were used to retrieve English and alternative labels, which were then stored in a custom dictionary to enable label reconciliation during comparisons.

A similar issue concerned author names. IMAGO primarily employs IRIs from MIRABILE[17], an authoritative digital archive for Medieval Culture, while MMM uses VIAF and Wikidata identifiers. Furthermore, IMAGO favours Latinised forms (e.g., Leo Baptista Albertus), while MMM often uses vernacular forms (e.g., Alberti, Leon Battista). Given the relatively small number of authors (around 50), a manual mapping dictionary was created to align author names across the two datasets.

6 Attribute Matching and Validation

To link manuscripts across IMAGO and MMM, we developed a matching framework that operates in two steps: (1) field-level comparison, where specific attributes of records are evaluated using tailored functions, and (2) full-level validation, where the field scores are combined into acceptance rules. Based on these scores, manuscript pairs are classified as certain or uncertain matches.

[17] https://www.mirabileweb.it/home.

6.1 Matching of Manuscript Attributes

Field-level matching functions were designed to evaluate specific manuscript attributes. The string similarity scores they return (ranging from 0 to 3) are later combined to compute an overall full matching score, which determines whether two records can be considered a match.

Shelfmark. Shelfmarks are the primary identifiers of manuscripts but often vary in format, level of detail, or use of abbreviations. To standardize comparisons, we normalized shelfmarks by (i) converting to lowercase, (ii) removing punctuation and generic terms (e.g., Ms., Cod., olim), (iii) separating alphanumeric elements (e.g., 382D → 382 D), (iv) removing leading zeros, and (v) standardizing format indicators (e.g., 4° → Q). For MMM records containing multiple or alternative shelfmarks, each was compared until a match was found. The function first checked for numeric equivalence, including conversion of Roman to Arabic numerals. If unsuccessful, a more detailed comparison was applied, accounting for abbreviations and acronyms (e.g., NAL 133 vs. Nouv. Acq. Lat. 133). For SDBM entries with encoded shelfmarks (e.g., IMILAD041I/00), only the actual shelfmark portion (e.g., D041I/00, which corresponds to the shelfmark "D 41 inf.") was considered.

Physical Measurements. We also compared three physical attributes: folio count, height, and width. IMAGO values were extracted from notes using regular expressions, while MMM values were explicitly defined. Similarity scores were assigned as follows:

- 3: all values match exactly,
- 2: either the folio count or the dimensions match,
- 1: values differ, but by no more than three units.

Library Name. Library names were compared to account for inconsistencies in naming conventions. While most entries used standardised forms, variants were common (e.g., Corpus Christi College Library vs. Parker Library at Corpus Christi College). We used the Python `difflib` library[18] to compute string similarity and supplemented this with a set of manually curated mappings prepared by a specialist scholar (e.g., Bibliothèque Nationale de France = BNF, Paris).

Author. Author names were matched by checking whether entries shared at least one author. To address name variants, we relied on a reconciliation dictionary prepared by a scholar in Latin Philology. When further disambiguation was needed, alternative names and aliases provided by both IMAGO and MMM were used to improve the matching process.

[18] https://docs.python.org/3/library/difflib.html.

6.2 Full Matching

A full comparison between all the attributes was carried out by combining the string similarity scores obtained from the individual fields described in Sect. 4.1. Each IMAGO record of attributes was tested against potentially relevant MMM records of attributes. The process always began with the shelfmark: only records with compatible shelfmarks were considered for further evaluation. When this condition was satisfied, the scores from the remaining fields were integrated to determine the overall similarity and to classify the match as certain or uncertain. If all three physical measurements (height, width, and folio count) matched exactly, a bonus score was assigned to strengthen the comparison. For cities with large manuscript collections, stricter rules were applied to reduce false positives. The acceptance rules were:

1. **For Paris, Oxford, Cambridge, London, and Vatican City:**
   - A matching library plus either a matching shelfmark or a matching author; or
   - An author match combined with a physical measurement score of 2.
2. **For all other cities:**
   - An author match, or any two other matching fields.

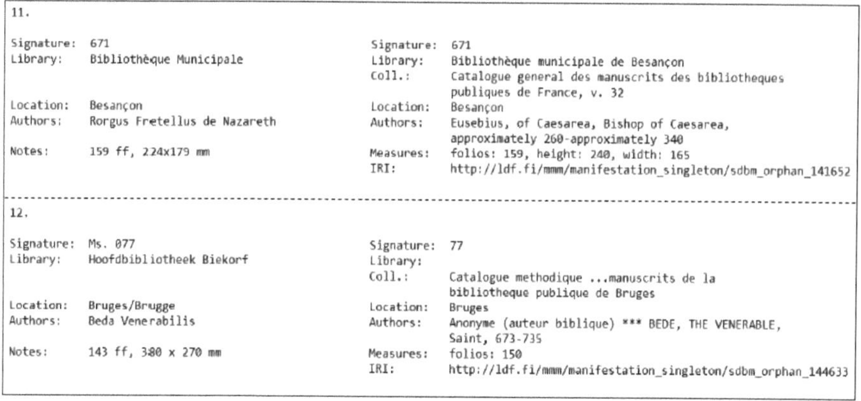

```
11.

Signature:  671                          Signature:  671
Library:    Bibliothèque Municipale      Library:    Bibliothèque municipale de Besançon
                                         Coll.:      Catalogue general des manuscrits des bibliotheques
                                                     publiques de France, v. 32
Location:   Besançon                     Location:   Besançon
Authors:    Rorgus Fretellus de Nazareth Authors:    Eusebius, of Caesarea, Bishop of Caesarea,
                                                     approximately 260-approximately 340
Notes:      159 ff, 224x179 mm           Measures:   folios: 159, height: 240, width: 165
                                         IRI:        http://ldf.fi/mmm/manifestation_singleton/sdbm_orphan_141652
-----------------------------------------------------------------------------------------------
12.

Signature:  Ms. 077                      Signature:  77
Library:    Hoofdbibliotheek Biekorf     Library:
                                         Coll.:      Catalogue methodique ...manuscrits de la
                                                     bibliotheque publique de Bruges
Location:   Bruges/Brugge                Location:   Bruges
Authors:    Beda Venerabilis             Authors:    Anonyme (auteur biblique) *** BEDE, THE VENERABLE,
                                                     Saint, 673-735
Notes:      143 ff, 380 x 270 mm         Measures:   folios: 150
                                         IRI:        http://ldf.fi/mmm/manifestation_singleton/sdbm_orphan_144633
```

Fig. 2. Attributes of two manuscripts retrieved from the IMAGO and MMM KGs, proposed for manual validation.

Each accepted match was categorised as certain (perfect measurement score) or uncertain (all other cases). Results were written to a formatted text file for inspection (see Fig. 2).

6.3 Validation and Results

The algorithm produced 344 candidate matches, which were then manually validated. Metadata consistency was checked, and when necessary, additional information from external resources linked to IMAGO and MMM was consulted. Special attention was given to author names, shelfmarks, and physical measurements. External sources often provided metadata (e.g., titles, dates, materials), which was crucial in resolving uncertain cases.

After manual validation, a total of 319 matches were confirmed. In cases where multiple MMM manuscripts corresponded to the same IMAGO manuscript, only one record was retained to avoid duplication. As a result, 283 IMAGO manuscripts (approximately 30% of the 970 initially compared) were successfully linked to at least one MMM manuscript.

7 Discussion and Challenges

Linking two independently developed knowledge graphs, IMAGO and MMM, posed a variety of challenges, despite both being modelled using the CIDOC CRM ontology. These challenges stemmed from differences in modelling strategies, metadata completeness, naming conventions, and scale. Addressing these discrepancies was essential to achieving accurate and meaningful integration across the two knowledge graphs.

A key difficulty lay in the modelling heterogeneity between the two datasets. IMAGO conceptualises manuscripts at a granular level, often representing individual parts or sections and linking them to specific works. In contrast, MMM models manuscripts as unified physical objects. Furthermore, IMAGO includes interpretative and frequently unstructured metadata, while MMM emphasises structured, factual information. These discrepancies necessitated semantic normalisation to enable effective comparisons. Another major issue was identifier and label misalignment. Authors, cities, and libraries were identified using different schemes: IMAGO often employs MIRABILE IRIs and Latinised author names, while MMM uses VIAF or Wikidata IRIs and vernacular forms. For geographic data, IMAGO frequently reuses Wikidata IRIs, whereas MMM uses custom identifiers. Library names also differ in format and level of specificity. These mismatches required the implementation of reconciliation mechanisms, including manually curated dictionaries. Metadata completeness was also inconsistent. Some records lacked structured fields for key information such as author names, cities, or libraries, or included relevant data only in free-text annotations or external resources. These gaps limited the efficacy of automated matching and necessitated additional preprocessing steps. String variation posed further obstacles, especially with shelfmark notations. These differed in formatting, use of punctuation, acronyms, or historical markers (e.g.,"olim"). Such variations rendered direct string comparisons unreliable, requiring normalisation strategies to remove punctuation, expand acronyms, and standardise shelfmark formats. Scalability was another concern, given the discrepancy in dataset size. IMAGO contains 970 records, whereas MMM includes over 208,000. Even when filtered by city, the number of candidate matches was high, particularly in large hubs such as Paris or Oxford, raising the risk of false positives. This necessitated the use of stricter matching criteria and more precise filtering techniques. A further complication was the absence of shared global identifiers for manuscripts. While shelfmarks are often unique within libraries, they are not standardised across datasets. Consequently, heuristic methods and detailed field-level comparisons were required to infer reliable matches. Granularity differences added another layer of complexity. IMAGO provides detailed descriptions at the level of individual works within manuscripts, while MMM treats manuscripts as single entities. This disparity made it difficult to determine when a partial match referred to the same physical object. Although automated methods initially identified 344 potential matches, human validation proved essential. Through careful manual review, 283 IMAGO records were reliably linked to MMM counterparts. This highlights the critical role of human oversight in semantic data reconciliation. Multilingual and cultural variability also had to be addressed. City and author names appeared in different linguistic forms (e.g., Firenze vs. Florence, Gand vs. Gent), requiring the use of multilingual reconciliation techniques and external resources such as Wikidata.

In summary, the integration of IMAGO and MMM demanded a multifaceted approach involving automated data processing, normalisation, semantic alignment, and manual curation. The primary challenges were rooted in structural and semantic heterogeneity, inconsistent identifiers, and unstandardised metadata. Overcoming these barriers was crucial to ensure the reliability and interoperability of the linked KGs.

8 Conclusions and Future Work

In this paper, we presented a case study on the integration of two knowledge graphs in the Cultural Heritage domain: the IMAGO and the Mapping Manuscript Migrations knowledge graph. Although both are modelled on the CIDOC CRM ontology, the alignment process revealed several challenges related to heterogeneous modelling choices, inconsistent identifiers, and variable metadata quality. Our methodology - combining automated preprocessing, reconciliation of metadata, field-level comparison, and human validation - allowed us to reliably link 283 IMAGO manuscripts with corresponding MMM records.

The study highlights two key lessons. First, the adoption of a common reference ontology such as CIDOC CRM is a necessary but not sufficient condition for interoperability: semantic consistency, shared identifiers, and curatorial practices remain decisive factors. Second, automated approaches can reduce the integration effort, but expert review remains essential, especially in domains where interpretative and historical dimensions play a central role.

Future work will extend in several directions. From a methodological perspective, we aim to refine matching algorithms with probabilistic and machine learning approaches, allowing for more scalable reconciliation of large datasets. From a data perspective, the integration could be expanded beyond manuscripts to include related entities such as works, authors, and geographical places, leveraging external resources like Wikidata and VIAF for stronger identity resolution. Finally, we plan to evaluate how the enriched IMAGO KG supports new forms of scholarly research (e.g., tracing the transmission of geographical knowledge across time and space) and to make the resulting datasets and tools more accessible to the wider Cultural Heritage community.

Acknowledgements. We gratefully acknowledge the contribution of Andrea Fraticelli, whose Master's thesis formed an important part of this work. This work was externally supported by ITSERR (Italian Strengthening of the ESFRI RI RESILIENCE), funded under the MUR National Recovery and Resilience Plan funded by the European Union-NextGenerationEU, and by CRAEFT, funded under grant agreement No 101094349 funded by the European Union's Horizon Europe research and innovation programme. The funders had no involvement in the study's design, data collection and analysis, publication decision, or manuscript preparation.

Disclosure of Interests. The authors have no competing interests to declare that are relevant to the content of this article.

References

1. Paolo bufalini's notebook: – semantic scholarly digital edition. Web site of a digital scholarly edition, Department of Classical Philology and Italian Studies (FICLIT) and Digital Humanities Advanced Research Centre (/DH.arc), University of Bologna (2020). https://projects.dharc.unibo.it/bufalini-notebook/. initial release 2018; updated April 2020, current version January 2021
2. Aliaga, D.G., Bertino, E., Valtolina, S.: Decho—a framework for the digital exploration of cultural heritage objects. J. Comput. Cult. Herit. **3**(3) (2011). https://doi.org/10.1145/1921614.1921619,
3. Aloia, N., Debole, F., Felicetti, A., Galluccio, I., Theodoridou, M.: Mapping the ariadne catalog data model to cidoc crm: bridging resource discovery and item-level access. In: SCIRES-IT (2017). https://doi.org/10.2423/i22394303v7n1p1
4. Baierer, K., et al.: Dm2e: a linked data source of digitised manuscripts for the digital humanities. Semant. Web **8**(5), 733–745 (2017). https://doi.org/10.3233/SW-160234. https://journals.sagepub.com/doi/abs/10.3233/SW-160234
5. Barker, E., Simon, R., Isaksen, L., de Soto Cañamares, P.: The pleiades gazetteer and the pelagios project. In: Berman, M.L., Mostern, R., Southall, H. (eds.) Placing Names: Enriching and Integrating Gazetteers, pp. 97–109. Indiana University Press, Bloomington (2016). http://www.iupress.indiana.edu/product%5finfo.php?products%5fid=808056
6. Bartalesi, V., Meghini, C., Metilli, D., Tavoni, M., Andriani, P.: A web application for exploring primary sources: the DanteSources case study. In: Digital Scholarship in the Humanities, pp. 705–723 (2018)
7. Bartalesi, V., Metilli, D., Pratelli, N., Pontari, P.: Towards a knowledge base of medieval and renaissance geographical Latin works: the imago ontology. Digital Scholarship Human. **37**(1), 34–50 (2022)
8. Bartalesi, V., Pratelli, N., Lenzi, E., Pontari, P.: Using semantic web to create and explore an index of toponyms cited in medieval geographical works. J. Comput. Cult. Herit. **16**(2) (2023). https://doi.org/10.1145/3582263,

9. Bartalesi, V., et al.: A formal representation of the divine comedy's primary sources: the hypermedia dante network ontology. Digital Scholarship Human. **37**(3), 630–643 (2022)
10. de Boer, V.: Knowledge graphs for cultural heritage and digital humanities. In: Proceedings of the 5th Workshop on AnalySis, Understanding and ProMotion of HeritAge Contents, SUMAC '23, p. 3. Association for Computing Machinery, New York (2023). https://doi.org/10.1145/3607542.3617354
11. Bruneau, O., Lasolle, N., Lieber, J., Nauer, E., Pavlova, S., Rollet, L.: Applying and developing semantic web technologies for exploiting a corpus in history of science: the case study of the Henri poincare correspondence. Semant. Web **12**(2), 359–378 (2021). https://doi.org/10.3233/SW-200400. https://journals.sagepub.com/doi/abs/10.3233/SW-200400
12. Burrows, T., Hyvönen, E., Ransom, L., Wijsman, H.: Mapping manuscript migrations: digging into data for the history and provenance of medieval and renaissance manuscripts. Manuscript Stud. J. Schoenberg Inst. Manuscript Stud. **3**(1), 249–252 (2018)
13. Conlen, J., Fu, H., Jehnke, M., Shetty, S., Zhou, Y.: Timeline of everything. VTechWorks, Virginia Tech Institutional Repository (2019). https://vtechworks.lib.vt.edu/items/6c7b84ad-92c6-4bc0-9937-bf9ed665ccf7
14. Damiano, R., Lombardo, V., Lieto, A., Borra, D.: Exploring cultural heritage repositories with creative intelligence: the labyrinth 3d system. Entertain. Comput. **16**, 41–52 (2016). https://doi.org/10.1016/j.entcom.2016.05.002. https://www.sciencedirect.com/science/article/pii/S1875952116300167
15. Doerr, M., Bekiari, C.: Frbroo, a conceptual model for performing arts. In: Proceedings of the 2008 Annual Conference of CIDOC (2008). https://publications.ics.forth.gr/_publications/drfile.2008-06-42.pdf
16. Doerr, M., Gradmann, S., Hennicke, S., Isaac, A., Meghini, C., Sompel, H.: The Europeana data model (EDM). In: World Library and Information Congress: 76th IFLA General Conference and Assembly, pp. 10–15 (2010)
17. Doerr, M., Ore, C.E., Stead, S.: The CIDOC Conceptual Reference Model - A New Standard for Knowledge Sharing (2007). https://doi.org/10.13140/2.1.1420.6400, pages: 56
18. Felicetti, A., Meghini, C., Richards, J., Theodoridou, M.: The ao-cat ontology (2023). https://doi.org/10.5281/zenodo.7818375
19. Hyvönen, E.: BiographySampo – publishing and enriching biographies on the semantic web for digital humanities research. In: Hitzler, P., et al. (eds.) ESWC 2019. LNCS, vol. 11503, pp. 574–589. Springer, Cham (2019). https://doi.org/10.1007/978-3-030-21348-0_37
20. Hyvonen, E.: Digital humanities on the semantic web: Sampo model and portal series. Semant. Web **14**(4), 729–744 (2023). https://doi.org/10.3233/SW-223034. https://journals.sagepub.com/doi/abs/10.3233/SW-223034
21. Isaac, A., Haslhofer, B.: Europeana linked open data data.europeana.eu. Semant. Web **4**(3), 291–297 (2013). https://doi.org/10.3233/SW-120092. https://journals.sagepub.com/doi/abs/10.3233/SW-120092
22. Kusnick, J., Mayr, E., Seirafi, K., Beck, S., Liem, J., Windhager, F.: Every thing can be a hero! narrative visualization of person, object, and other biographies. In: Informatics, vol. 11, p. 26. MDPI (2024)
23. Meghini, C., et al.: Ariadne: a research infrastructure for archaeology. J. Comput. Cult. Herit. **10**(3) (2017). https://doi.org/10.1145/3064527

24. Mulholland, P., Collins, T.D., Zdrahal, Z.: Bletchley park text: Using mobile and semantic web technologies to support the post-visit use of online museum resources. J. Interact. Media Educ. (2005)
25. Makela, E., Hyvonen, E., Ruotsalo, T.: How to deal with massively heterogeneous cultural heritage data lessons learned in culturesampo. Semant. Web **3**(1), 85–109 (2012). https://doi.org/10.3233/SW-2012-0049. https://journals.sagepub.com/doi/abs/10.3233/SW-2012-0049
26. Niccolucci, F., Richards, J.: The ARIADNE Impact. Archaeolingua Foundation (2019). https://doi.org/10.5281/zenodo.4319058
27. Riva, P., Žumer, M., Aalberg, T.: Lrmoo, a high-level model in an object-oriented framework (2022)
28. Vitale, V., Soto, P.d., Simon, R., Barker, E., Isaksen, L., Kahn, R.: Pelagios connecting histories of place. Part i: Methods and tools. Int. J. Human. Arts Comput. **15**(1-2), 5–32 (2021). https://doi.org/10.3366/ijhac.2021.0260
29. Weibel, S.L., Koch, T.: The dublin core initiative. D-Lib Maga. **6**(12) (2000), https://webdoc.sub.gwdg.de/edoc/aw/d-lib/dlib/december00/weibel/12weibel.html
30. Wolff, A., Mulholland, P., Collins, T.: Storyspace: a story-driven approach for creating museum narratives. In: Proceedings of the 23rd ACM Conference on Hypertext and Social Media, HT '12, pp. 89–98. Association for Computing Machinery, New York (2012). https://doi.org/10.1145/2309996.2310012

On Encoding Big Knowledge Graphs as Quantum States and on Running Grover's Algorithm

Giorgos Theodorakis[1,2,3], Michalis Touloupakis[1], and Yannis Tzitzikas[1,2,3](✉)

[1] Computer Science Department, University of Crete, Heraklion, Greece
{csd4881,csd5160}@csd.uoc.gr
[2] Institute of Computer Science (ICS), Foundation for Research and Technology – Hellas (FORTH), Heraklion, Greece
tzitzik@csd.uoc.gr
[3] Center for Quantum Science & Technologies (FORTH-QuTech), Foundation for Research and Technology – Hellas (FORTH), 70013 Heraklion (Crete), Greece

Abstract. To evaluate the potential of Quantum Computing (QC) on real data management tasks in this paper we elaborate on the task of encoding large in size Knowledge Graphs as numerical data for using them for creating the corresponding quantum state. This is a task that cannot be bypassed, and has to be performed using a classical computer. We report experimental results for datasets up to size 800 million triples, in particular for the entire DBpedia. We also report times for creating the corresponding quantum states by superimposing the encoded datasets. Finally, since a significant number of tasks are reduced to Grover's algorithm, we report efficiency results on running this algorithm on datasets with size up to 10^8 bitstrings on a simulator and on IBM Quantum Computer. Even if the output of Grover is not stable on current quantum hardware, this analysis is useful for estimating the efficiency and thus identifying to what tasks the application of QC will be beneficial.

Keywords: quantum computing · knowledge graphs · knowledge management · DBpedia · Grover

1 Introduction

Motivation: Quantum Computing (QC) is actively researched across multiple levels: (a) Physical level: constructing and benchmarking quantum hardware [17]), (b) Algorithmic level: developing suitable algorithms, such as Shor's [20] and Grover's [8], and (c) Application level: mapping real-world problems to quantum algorithms [4]. However, the data and knowledge management perspective is not well investigated. To this end, in this paper we continue the work of [21] on evaluating methods that can be used for storing a big Knowledge Graph (KG) in a quantum computer. The main motivation is that the properties of superposition states, could be exploited for storing a huge amount of triples in a small

H. Takeda et al. (Eds.): IJCKG 2025, LNAI 16297, pp. 509–525, 2026.
https://doi.org/10.1007/978-981-95-5009-8_37

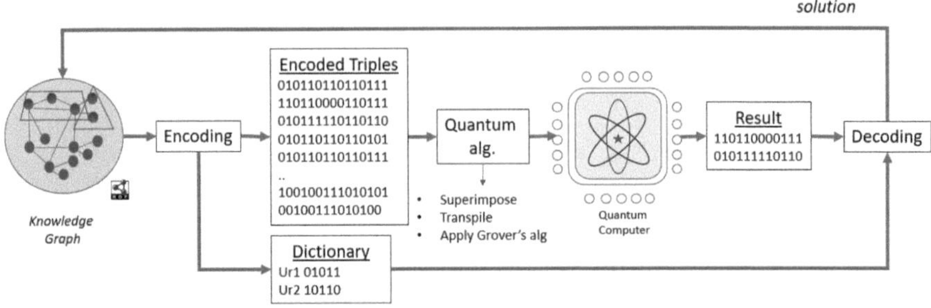

Fig. 1. The workflow of encoding and decoding.

number of qubits. Moreover, a mapping of KG triples to quantum states could enable native quantum algorithms to accelerate various KG tasks.

Novelty: To the best of our knowledge this is the first work that attempts to store a big KG in a quantum computer, and measures the efficiency of Grover's algorithm. The process that we investigate is illustrated in Fig. 1.

Contribution: In brief the main contributions of this paper are: (a) We compare the efficiency of two encoding methods, and report the time required for encoding the entire DBpedia that comprises 800 million triples, (b) we measure the efficiency as regards the creation of the superimposed state and (c) we measure the efficiency of Grover's algorithm on large datasets, upto 100 million bitstrings (of 27 bits each).

The rest of this paper is organized as follows: Sect. 2 discusses in brief the required background, Sect. 3 describes related work, Sect. 4 describes encoding methods, the corresponding requirements and challenges, as well as the proposed method, and Sect. 5 presents experimental results. Subsequently Sect. 6 evaluates superposition, and Sect. 7 the application of Grover's algorithm. Finally, Sect. 8 concludes the paper.

2 Background

Quantum Computing. In QC (Quantum Computing) instead of representing the data with 0/1 bits, we represent data with *qubits.* A qubit is described by a state vector in a two-dimensional complex vector Hilbert space. We express a qubit as an expression $a|0\rangle + b|1\rangle = a \begin{bmatrix}1\\0\end{bmatrix} + b \begin{bmatrix}0\\1\end{bmatrix} = \begin{bmatrix}a\\b\end{bmatrix}$ where $a^2 + b^2 = 1$. Unlike probabilities, where $a, b \in [0, 1]$ and $a+b = 1$, here a and b can be complex numbers, however a^2 and b^2 have a probabilistic interpretation($a^2 + b^2 = 1$). A qubit $a|0\rangle + b|1\rangle$ where $a, b \neq 0$, is *simultaneously* in two states, this is called *superposition*, and only when we measure it, we will get either 0 or 1. As such, a system with n qubits can represent 2^n states at once. The "magic" is that what we will measure (0 or 1) may depend on the states of other qubits, due to the phenomenon of *interference*. We could therefore say that we have a kind of entangled probabilities. In particular, if we have two *entangled* qubits (that in the physical world correspond to two entangled electrons/photons), they might appear to be independent particles in a superposition, but as soon as one of them is measured, the outcome of a measurement of the other is fixed as well. Many computational benefits of QC stem from superposition, i.e. the fact that a quantum system can do processing while the qubits are in a superposition state.

There are not many fundamental quantum algorithms; for an overview see [15] and [11]. The more important ones are: Shor's algorithm [20] that solves the factorization problem in polynomial time, and Grover's algorithm [8] for searching in an unstructured list at the cost of $\sqrt{n}$. We should also mention quantum annealing [18] that relies on specialized hardware (D-Wave) for solving optimization problems.

Knowledge Graphs (KGs). *Knowledge Graphs* (KG) [10] are structured representations of data so that they allow capturing the relationships between different concepts, entities or pieces of information. They rely on a graph-based data model (expressed mainly in RDF/OWL) to capture knowledge in application scenarios that involve integrating, managing and extracting value from diverse data sources.

DBpedia is a prominent project in the Linked Data and Semantic Web community that extracts structured information from Wikipedia and makes this information available on the Web [2]. It serves as a central linking hub in the Linked Open Data (LOD) cloud, facilitating interoperability among datasets by providing common identifiers for entities and concepts [12].

3 Related Work

Works on QC and *data management* have started to appear. There are a few papers that attempt to form the landscape, e.g. [9], however the papers that report results of experiments on real QC hardware are much less. A general remark is that in all of these works, the data management perspective is largely unexplored and quite far from technological and practical exploitation, in the sense that the datasets used are too small. Below we list, and comment in brief, related works:

- [14] proposes a quantum *machine learning* algorithm for knowledge graphs, however the experimental results are performed on classical computers, not quantum computers.
- [13] elaborates on *quantum storage design* for Tables in RDBMS, provides example over *very small* data.
- [3] focuses on *entity matching* with *Quantum Neural Networks*; they evaluate quantum machine learning algorithms for entity matching on a *small handcrafted* dataset and compare them to similar classical algorithms.
- [22] describes a demonstration system for three graph algorithms over D-Wave quantum computers and simulators. Again the datasets were very small:

shortest path algorithm was applied on graphs with 5 vertices, graph isomorphism on graphs with up to 30 vertices, and community detection on graphs with up to 50 vertices.

- Finally, there are a few works on *query optimization*, like [6,16], as well as on topics like cardinality estimation, index selection, transaction scheduling, that are beyond the scope of this paper.

  To the best of our knowledge this is the first work on trying to encode large KGs.

4 The Encoding Method

We investigate the encoding proposed in [21]. In particular, according to the first, denoted by $R1$, each URI is assigned a distinct integer starting from 0. According to $R2$, and since the triple predicates cannot appear as subjects or objects, we assign separately integers to the triple predicates and then to the rest of the URIs. An example is shown in Fig. 2. The comparison of these two encodings is given in [21]. That work, also contributed the framework `QuantumEncoDeco`, that is publicly available[1], enabling anyone to encode/decode RDF graphs, and make QC experiments. In the spectrum of encodings for quantum computing, these encodings fall to what is called *basis encodings* [19].

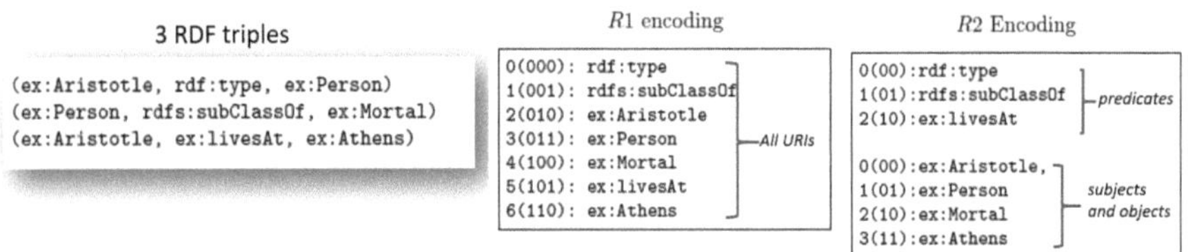

Fig. 2. Example of the two basis encodings for RDF triples.

4.1 Downloading the Input Data

The input data for our experiments was sourced from DBpedia's December 2022 snapshot, which is available on the DBpedia Databus platform[2] The download process was achieved through the Databus API following the steps that are publicly described[3], for self-containedness they are given below: (1) Fetch the query:

[1] https://github.com/YannisTzitzikas/QuantumEncoDeco.

[2] Collection of 155 files with total size 128.43 GB, with 806,600,000 triples, 168,876,007 unique Entities and 54,782 unique Predicates, accessible from https://databus.dbpedia.org/dbpedia/collections/dbpedia-snapshot-2022-12.

[3] https://databus.dbpedia.org/dbpedia/collections/dbpedia-snapshot-2022-12.

```
    query=$(curl -H -C- "Accept:text/sparql" \
https://databus.dbpedia.org/dbpedia/collections/dbpedia-snapshot-2022-12)
```

(2) Execute the query and download all files in the collection:

```
files=$(curl -X -C- POST -H "Accept: text/csv" \
--data-urlencode "query=${query}" \
\url{https://databus.dbpedia.org/sparql} | \
tail -n +2 | sed 's/\r$//' | sed 's/"//g')
while IFS= read -r file ; do wget $file; done <<< "$files"
```

4.2 Encoding Process

The pre-existing version of `QuantumEncoDeco`[4], supports two specialized encoders ($R1$ and $R2$). Both encoders are stateful, meaning that subsequent encoded values depend on the previous ones. The key point is that once a URI is encoded, if we encounter it again, then we should not re-encode it, therefore we need to have stored all the previous encodings in memory with the possibility of running out of memory for huge datasets such as those provided by DBpedia. Additionally, both $R1$ and $R2$ require global knowledge of the total number of unique subjects, predicates, and objects in the dataset. Since the information is available only after scanning the data, the encoding must be performed in two passes:

1. *First Pass - Mapping Formation:* Scan all triples, assign IDs to the components, and collect statistics for the data set.
2. *Second Pass - Final Encoding:* Iterate over all the triples again and write their binary representation using the information collected in the first pass.

In general, the logical skeleton of the two-pass encoders is shown in Algorithm 1. The system was implemented as a configurable pipeline using the Pipe and Filter architectural style, in which discrete processing components (filters) are connected via data channels (pipes) to form a flexible and reusable workflow [7], which in turn enables modular configuration and testing. In this implementation, each encoder proposed in [21], as well as the memory lookup method found in Algorithm 1, can be implemented as a series of independent filters, allowing us to iterate through various configurations without requiring to write much code.

4.3 Challenges and Solutions

The main challenge is how to tackle cases where the graph is too big to fit in main memory. Therefore, we have to find an effective method to implement the storage component and the associated retrieval process, i.e. the implementation of `Memo.lookup(s)`. Below we enumerate various possible solutions to tackle this problem.

---

[4] https://github.com/YannisTzitzikas/QuantumEncoDeco.

Algorithm 1: Encoding RDF Triples from TTL File

```
Input: A file f in TTL format, an encoding method R (either R1 or R2)
Output: A file f_o containing the encoded triples
S ← ∅; P ← ∅; O ← ∅;          ▷ Init. of a set for subjects, predicate, objects
foreach triple t = (s, p, o) in f do
    S → S ∪ {encode(s)}; P → P ∪ {encode(p)}; O → O ∪ {encode(o)};
end
nbits_s ← CalcBits(S, R); nbits_p ← CalcBits(P, R); nbits_o ← CalcBits(O, R)
foreach triple t = (s, p, o) in f do
    s_e ← Mem.lookup(S, s); p_e ← Mem.lookup(P, p); o_e ← Mem.lookup(O, o);
    f_o.append(toBin(s_e, nbits_s), toBin(p_e, nbits_p), toBin(o_e, nbits_o))
end
Function encode(s):
    id ← Mem.lookup(s)
    if id = null then
        nid ← brand new id
        Mem.add((s, nid))
        id ← nid
    end
    return id
```

1. **Linear File Search**: A straightforward approach would be to have a flat file storing encoding mappings in the format `<URI, encoded_value>`. That means the `Mem.lookup(s)` would have to scan the entire file to check for existing entries.
2. **Binary Search File**: Another approach is to store the mappings *sorted lexicographically* again in the format `<URI, encoded_value>`. Each encoding operation involves performing a binary search to check for existing entries. This yields a time complexity of $O(\log n)$ for each lookup and $O(n \log n)$ time complexity overall. While this approach is relatively simple to implement, reconstructing (i.e. resorting) the file after each update is very time consuming, and that will be a major bottleneck as the datasets grow. One way to improve performance is to maintain an in-memory cache and keep as many mappings in memory as possible (to avoid frequent file reconstructions). Although better than the first, this method is not ideal for tackling huge datasets such as those of DBpedia.
3. **Dedicated Key-Value Storage Engine**: We can employ a key-value store like RocksDB[5], an embedded key-value store optimized for write-intensive workloads. This engine provides $O(1)$ average access time through logarithmic-structured merge trees (LSM-trees), with persistent storage on solid-state drives.
4. **Dedicated Triplestore Engine**: Assuming that our application will be used only for RDF Graphs, we can use a triplestore, like Virtuoso[6]. Triplestores

[5] https://github.com/facebook/rocksdb.

[6] https://vos.openlinksw.com/owiki/wiki/VOS/VOSRDFFAQ.

are specifically designed to manage RDF data, representing information as triples. Virtuoso builds multiple indices based on B-Trees.

5. **K-way Merge for First Pass & any other option for Second Pass**: We can make $R1$ and $R2$ configurations not to assign IDs instantly. Instead, we can store each *batch* of URI components in a temporary file in lexicographic order. After all of the batches are stored, we can *merge* them into a single mappings file, while assigning to each position an ID. An illustration of the process is shown in Fig. 3. This approach avoids all the lookups required by other methods and does not waste any number, thus giving us the optimal amount of bits to encode each of our triples. We can implement the merging by storing batches of size N and applying the *K-way merge algorithm* described in Algorithm 2 since each batch is already sorted in lexicographic order. This algorithm runs in $O(N \log k)$, where k is the number of batches and N is the size of each batch. However, even though this method is optimal for producing the mappings, it does not define a way to implement the `Mem.lookup()` function mentioned in Algorithm 1. This is required for the second pass and thus we would need to combine it with another of the methods we described above.

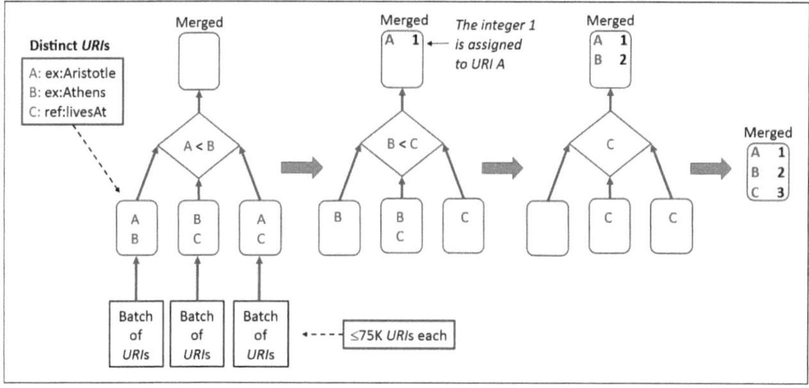

Fig. 3. K-Way Merge.

In the experimental evaluation that follows, we decided to evaluate only the KV-Storage and the K-way Merge methods; the rest were not implemented due to their expected poor run-time characteristics, as justified next:

- Linear File Search, although simple to implement, exhibits $O(n)$ time complexity for each lookup and $O(n^2)$ time complexity overall, becoming prohibitively slow for large-scale knowledge graphs (>1M entities).
- Although the Binary Search method sounds promising, it would still require a lot of time to reconstruct the file after every batch, which will become prohibitively expensive as the batch numbers grow. Also, even if the lookup time exhibits $O(\log n)$ time complexity per lookup, as the file size grows,

Algorithm 2: Merge K Sorted Files into One

```
Input: A Set of sorted files F = {f_1, f_2, ..., f_k}
Output: A file f_m with all entries ∈ F merged and sorted
Let Q be a Min Heap
foreach f_i ∈ F do
    Let e be the first element in f_i
    Insert (e, i) into Q
end
while Q not Empty do
    (e, src) ← Q.pop()
    f_m.append(e)
    if f_src not Empty then
        Let n be the next entry from f_src
        Insert (n, src) into Q
    end
end
```

paging becomes a major concern. Thus, in reality, we would need even more time than in the ideal theoretical condition. The best way to implement this method would be to add in-memory caches, to implement bloom filters, and to highly parallelize it, but that would still be inferior to a Storage Engine that is implemented as a B-Tree or an LSM-Tree.

- As regards using Triplestore engine, it will be harder to implement the Basis encoding per component in comparison to having a dedicated KV Storage Engine to store the component mappings.

5 Encoding: Experimental Results

Hardware Used. All experiments were carried out on a local machine with AMD Ryzen 3 4100 (Renoir), with 16 GB RAM. Both encoders follow the skeleton described in Sect. 4.2, allowing us to measure the execution time of each step independently.

Datasets Used. We decided to start from a 50 GB subset of DBpedia (Sect. 4.1) which we formed by randomly sampling files until we reached the quota. The resulting graph has around 250 million URIs with 45 million unique components (Subjects, Predicates and Objects). Then, by combining the corresponding TTL files into different bundles, we obtained the Datasets described in Table 1, which were used in our experiments. In Table 1, $|E|$ denotes the size of the Entities Set, which comprises all the unique Subjects and Objects within the dataset, $|P|$ denotes the size of the Predicates Set, which is composed by all the Predicates. Data sets A-C are comparable, but D and E show a huge spike in $|P|$ count, because they are built from heterogeneous files spanning many domains.

Even though these sets are composed by randomly selecting files from the Super Set, a trend emerges: the number of unique entities and predicates grow

Table 1. Data Sets Summary

Data Set	Data Set Size ($\|T\|$)	Entities ($\|E\|$)	Predicates ($\|P\|$)
A	25,000,000	14,045,198	15
B	50,000,000	21,109,125	17
C	100,000,000	25,536,231	18
D	160,000,000	30,950,255	16,116
E	200,000,000	36,747,934	39,378
Super Set	250,000,000	38,391,354	41,699
DBpedia	806,600,000	168,876,007	54,782

much slower in respect to the growth of the Data Set Size ($|T|$). This trend can be seen more clearly in Fig. 4 which is a plot of the cumulative count of Unique Components over the number of Triples that we meet, as we traverse the whole Super Set from which we formed the subsets of Table 1. The plot reveals that the

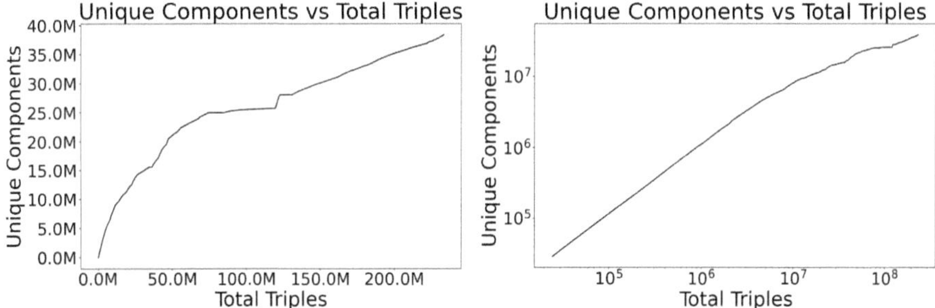

Fig. 4. Unique Components Over Triples of Graph (right: log scale).

Table 2. Time measurements per dataset and encoder (format: HH:mm:ss)

Dataset Set	Encoder	K-way Merge Time (First Pass)	KV-Storage Time (First Pass)	KV-Storage Time (Second Pass)
A	R1	00:02:19	00:03:03	00:01:14
	R2	00:02:21	00:03:01	00:01:16
B	R1	00:04:12	00:03:06	00:03:34
	R2	00:04:16	00:03:05	00:03:42
C	R1	00:08:32	00:09:23	00:31:04
	R2	00:07:57	00:09:44	00:32:39
D	R1	00:13:55	00:14:22	00:42:41
	R2	00:13:52	00:14:28	00:46:24
E	R1	00:17:21	00:18:44	01:01:54
	R2	00:17:10	00:18:41	01:00:21
Super Set	R1	00:20:15	02:00:03	01:40:03
	R2	00:19:52	02:00:45	01:40:45
DBpedia	R1	01:38:10	-	-

set of unique components follows a logarithmic growth with respect to the size of the graph. That means that the ratio of the unique components found when we increase the subset size will always get smaller. Thus, if we had more main memory, we can potentially load all these Datasets without relying on persistent storage.

Comparative Results. Table 2 summarizes the time measurements for each dataset and encoder and all durations are reported in HH:mm:ss format. Both encoders display similar execution times, because their only difference design-wise is that $R2$ applies Basis Encoding separately on the Predicates, whereas $R1$ does not make any distinctions based on the role. Thus, their implementations are almost identical and, more importantly, the storage method used remains the same regardless of the encoding. Storage access is the main bottleneck of the system.

Although both methods in the First Pass seem to have a linear performance up to data set E, when faced with more than 50,000,000 unique components, rocksDBs caching mechanisms struggle to keep up, due to the high memory demands. Also, as it is apparent in Fig. 4, at around the 150,000,000th triple recorded, we see an enormous spike on the amount of Unique Component met. Meaning, the engine constantly tries to re-organize the data inside it and thus making write operations more expensive.

On the other hand, the K-way merge method exhibits an almost linear growth, which can be seen in Table 2. This made it capable of producing the mappings of the whole DBpedia snapshot in 1 h and 38 min, something that would have taken a really long time for the KV-Storage Engine method to produce, which can be clearly seen in Fig. 5. Thus, we marked with a (-) the cells of the other two methods, to denote that we did not produce any results for them.

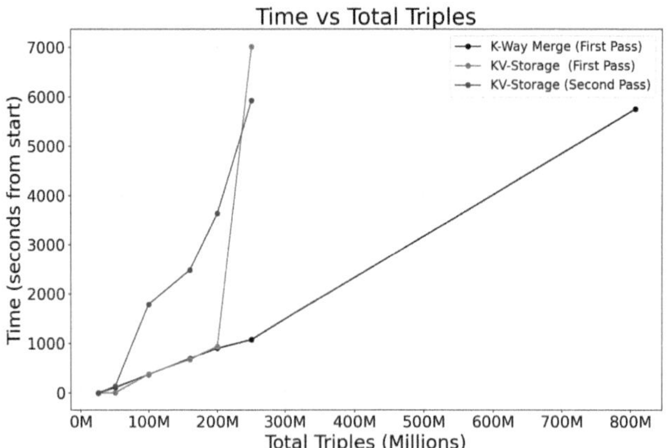

Fig. 5. Plots of Methods Execution Time over $|T|$.

In conclusion, the best combination is to generate the mappings via the K-way merger method and then load them in a key value Storage Engine to produce the final encodings during the Second Pass.

6 On Evaluating Superposition

How. After encoding a KG, the next step is to express it as a quantum state. Let T be the set of triples of our KG. For a $t \in T$, we shall use $\psi(t)$ to denote its encoding (either through $R1$ or $R2$, as described in Sect. 4). We can store these encoded triples by *superimposing* them, i.e. we can store the state $\Psi_T = \frac{1}{\sqrt{|T|}} \sum_{t \in T} \psi(t)$.

Experimental Results. Below we report the times required for creating the superimposed state, assuming that the encoded data are stored in a file. In particular, we measure the time required for reading it, for loading it in main memory, for creating the state vector, for creating the quantum circuit from the state vector, and for the transpilation (the process of transforming a quantum circuit into an equivalent one that is optimized and compatible with a specific quantum hardware).

Hardware Used. All experiments were executed on a local machine (i5-12th gen, 16 GB RAM). We used the Qiskit AerSimulator, for short Aer. Later we shall report measurement using IBM Quantum Computer.

Measurements. We measured file sizes varying from 1,000 bitwords with 10 bits each, to over 10 million bitwords. The bitwords were produced algorithmically (representing the numbers from 0 to 2^{bits}). To ensure statistical accuracy, we iterated the measurements 200 times and took 5 samples, gathering more than 1000 time slots. The results are shown in Table 3. Note that simulator time is the time required by the Aer-simulator function to perform metrics on the quantum circuit we transpiled until the final measurements.

Table 3. Times required for creating the superposition

Data set	Qubits	Bit-strings	Loading time (from file to main mem)	State vector creation	Create circuit from state vector	Transpile	Total time	Simulation time
s1	10	10^3	1,8 ms	9 ms	0.3 ms	541 ms	552 ms	4 ms
s2	14	10^4	18 ms	22 ms	2.6 ms	105 ms	148 ms	10 ms
s3	17	10^5	126 ms	26 ms	17 ms	95 ms	264 ms	30 ms
s4	20	10^6	1000 ms	232 ms	174 ms	235 ms	1650 ms	200 ms
s5	24	10^7	14000 ms	2700 ms	2100 ms	3700 ms	2250 ms	4300 ms
s6	27	10^8	100,000 ms	271,000 ms	324,000 ms	40,000 ms	510,000 ms	15,000 ms

Observations. Notice that the total time for 100 million rows of 27 bits is 510 s, i.e. 8.5 min. A few findings are in order. At first note that as we scale the file size exponentially, the time required for file creation and loading time also grows exponentially, as we would have expected. An interesting observation is that the transpilation seems to have some kind of optimal entry of qubits. Recall that transpilation is the process of transforming an input quantum circuit so that it aligns with the topology of a specific quantum device, while also optimizing its instructions for execution on noisy quantum computers. In Table 3 we can see a strange phenomenon about transpiliation. Although one would expect a linear relationship with "file size" (or the number of bitwords/bitstrings), this is not the case as we get the fastest transpilation time for 17 qubits. The most plausible explanation is that it is related to the optimization process adopted. Briefly, transpilation is the result of a 6-step process in the Qiskit SDK: *Initialization, Layout, Routing, Translation, Optimization, Scheduling*[7]. The time to transpile in Qiskit depends not only on qubit size, number of gates, type of gates, but also on optimization level. So, even if our input file has more bitwords, the structure of the quantum circuit produced may not grow proportionally at first because of hidden caching and low optimization costs. However, at high qubit numbers the dominant cause is the statevector's gate decomposition cost which takes $O(2^n)$ time complexity.

The previous experiments were done using the simulator. Below we report results over the IBM Quantum Computer. Of course the number of qubits that are available is limited, as well as the time that one can use it with no charging. The results are shown in Table 4. In comparison with the results using the simulator (Table 3), here we observe that transpilation is more expensive. On small scales (less than a million bitwords), the execution time explodes because of the transpilation which takes a couple seconds to be performed, whereas when we compute on bigger datasets the hardware takes advantage over the simulator, as we see in s5-s6 data samples. For the total time we can see only in sample 6 that the time differs and that's because the simulator does not take time to measure, rather than it guesses the outcome. On the other hand, real Quantum hardware has to perform measurement operations which are time consuming processes and explains why there is a 6 min difference between Aer and IBM Quantum Computer on the transpilation. A comparative plot of the total time, in log scale, is given in Fig. 6 (left).

As regards measurements over the encoded DBpedia, the entire dataset requires more qubits that those that the Aer simulator supports. Even if the IBM supports 125 qubits, with over 30 qubit Qiskit fails to produce the state vector, due to limitation of gates.

7 On Evaluating Grover's Algorithm

One of the more fundamental QC algorithms is the Grover's algorithm [8] that enables for searching in an unstructured list of n elements at the cost of $\sqrt{n}$. It is

[7] https://quantum.cloud.ibm.com/docs/en/guides/transpile.

Table 4. Times required for creating the superposition and measurement on IBM Quantum hardware (excluding queuing delays).

Data set	Qubits	Bitstrings	Loading time (from file to main mem)	State vector creation	Create circuit from state vector	Transpile	Total time	Execution (1000 shots)
s1	10	10^3	1.8 ms	9 ms	0.3 ms	1500 ms	1600 ms	2000 ms
s2	14	10^4	18 ms	22 ms	2.6 ms	300 ms	350 ms	2000 ms
s3	17	10^5	126 ms	26 ms	17 ms	300 ms	450 ms	3000 ms
s4	20	10^6	1000 ms	232 ms	174 ms	700 ms	2,300 ms	4,000 ms
s5	24	10^7	14,000 ms	2,700 ms	2,100 ms	10,000 ms	30,000 ms	5,000 ms
s6	27	10^8	100,000 ms	**4 min**	**5 min**	**2 min**	**14 min**	6,000 ms

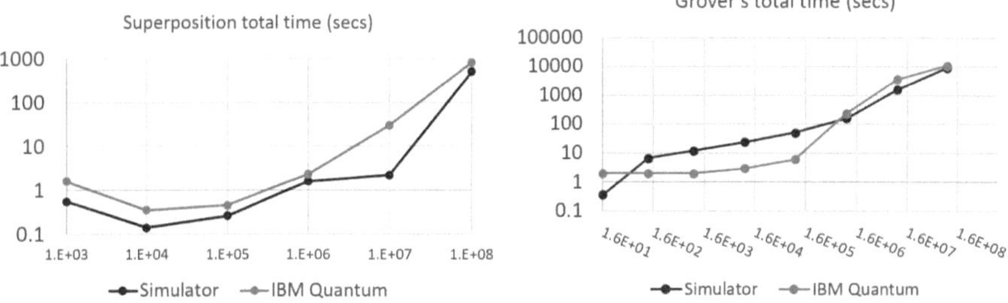

Fig. 6. Aer Simulator vs IBM Quantum Computer: Superposition total time (left), Grover's Alg. total execution time (right).

an iterative algorithm whose number of iterations depends on: (a) the size of the search space N (typically $N = 2^n$ for an n-qubit system), and (b) the number of solutions (or marked items) M. In particular, Grover's algorithm requires approximately: $k \approx \frac{\pi}{4}\sqrt{\frac{N}{M}}$ iterations (i.e. applications of the Grover operator) to maximize the probability of measuring a marked state. If we have one solution, i.e. $M = 1$, then $k \approx \frac{\pi}{4}\sqrt{N}$. If $N = 2^n$ then $k \approx \frac{\pi}{4} \cdot 2^{n/2}$. For example, if $N = 1{,}000{,}000$ and $M = 1$, then we need $k \approx \frac{\pi}{4}\sqrt{1{,}000{,}000} = \frac{\pi}{4} \cdot 1000 \approx 785$ iterations. Essentially, Grover's algorithm amplifies the amplitude of the marked states in each iteration. After about $k \approx \frac{\pi}{4}\sqrt{\frac{N}{M}}$ iterations, the success probability is near 1. If more iterations are applied, the probability starts to decrease due to the sinusoidal nature of the amplitude amplification.

Experimental Results. Here we report experimental results about the application of Grover's algorithm. Note that even if the application of Grover's algorithm on many qubits is not stable on current quantum hardware [1,5], if we have many qubits, such an analysis is useful for estimating the efficiency and thus identify to what tasks the application of QC will be beneficial. The time required by Qiskit's Aer *Simulator* is shown in Table 5, while the times over the IBM quantum computer are shown in Table 6.

In particular, Table 5 reports five distinct timing measures: "Time to run" includes the creation of the superposition, the application of the Grover operator the optimal number of times, and the final measurement, "1 Grover op time" corresponds to a single application of the Grover operator on the given qubits, "Total Ops time" is the cumulative duration of all applications of the Grover operators, measured for a single shot, "100 shots time" extends this measurement to 100 shots. Finally, "Total time" represents the complete runtime of the program, beginning with circuit construction and superposition creation, followed by Grover's iterations and measurement over 100 shots.

Table 5. Grover's algorithm run times in Qiskit's Aer Simulator

Data set	#rows	#qubits	Marked States M	Time to run	1 Grover Op time	Total Ops time	100 shots time	Total time
s1	16	4	1	150 ms	1,8 ms	25 ms	356 ms	370 ms
s2	128	7	1	470 ms	2 ms	38 ms	350 ms	6,600 ms
s3	10^3	10	1	420 ms	2 ms	30 ms	400 ms	12,000 ms
s4	10^4	14	1	482 ms	2 ms	100 ms	500 ms	24,000 ms
s5	10^5	17	1	2,300 ms	2 ms	1000 ms	1300 ms	50,000 ms
s6	10^6	20	1	21,000 ms	2 ms	17,000 ms	18,000 ms	161,000 ms
s7	10^7	24	1	26 **min**	2 ms	25 **min**	24 **min**	26 **min**
s8	10^8	27	1	60 **min**	2 ms	60 **min**	60 **min**	151 **min**

Table 6. Grover's algorithm run times in a IBM QC (execution excludes queuing delays).

Data set	#rows	#qubits	Marked States M	Time to run	1 Grover Op time	Total Ops time	100 shots exec (HW)	Total time (HW)
s1	16	4	1	15 ms	1.8 ms	25 ms	2,000 ms	2,000 ms
s2	128	7	1	470 ms	2 ms	38 ms	2,000 ms	2,000 ms
s3	10^3	10	1	420 ms	2 ms	30 ms	2,000 ms	2,000 ms
s4	10^4	14	1	482 ms	2 ms	250 ms	3,000 ms	3,000 ms
s5	10^5	17	1	2,300 ms	2 ms	2,000 ms	4,000 ms	6,000 ms
s6	10^6	20	1	21,000 ms	2 ms	1 **min**	4,000 ms	4 **min**
s7	10^7	24	1	26 **min**	2 ms	40 **min**	5,000 ms	60 **min**
s8	10^8	27	1	60 **min**	2 ms	120 **min**	6,000 ms	180 **min**

A comparative plot of the total time over Aer simulator and IBM Quantum Computer, in log scale, is shown in Fig. 6(right). We can see that although real hardware is faster in small datasets, it becomes significantly more expensive as we approach datasets of millions. One difficulty in evaluating IBM Quantum hardware performance is the effect of queuing delays. Although the intrinsic execution time of a circuit can be estimated from gate durations and repetition delays, the time a job spends in the queue is highly variable, ranging from seconds to hours depending on system load, and often alters the actual runtime. Because these delays are external and unpredictable, they are excluded from the timing analysis, and no reliable conclusions about end-to-end runtime can be drawn from them.

8 Concluding Remarks

To evaluate the potential of Quantum Computing (QC) for real-world data management tasks, this paper elaborated on encoding large Knowledge Graphs into numerical formats(two basis encodings) for quantum state preparation. We presented experimental (timing) results for encoding KGs up to size 800 million triples. In brief we comparatively evaluated 2 methods for the First Pass, one that requires costly look-ups and writes on disk, using a KV-Storage engine, and one that avoids both by sorting the components lexicographically and then mapping them, using K-way Merge. The K-way Merge method follows linear growth rate and manages to produce all the mappings of the whole DBpedia snapshot we downloaded (Sect. 4.1), which was over 800 million triples, in 1 h and 40 min. In contrast, the KV-Storage approach scales poorly. Even though it managed to keep up with the Sorting method on small subsets up to 200 million triples, its duration grew exponentially while the data size grew. In the end, it managed to produce the mappings of a subset of 250 million triples in 2 h, whereas the K-way Merge approach managed to encode the same data set in just 20 min.

In addition we reported time measurements for superposition, transpilation and for running the Grover's algorithm for datasets up to size 100 million bitstrings. As regards superposition, the total time for creating the quantum state for 100 million rows of 27 bits each, is 8.5 min using the Aer simulator and around 14 min on quantum hardware. The creation of a superimposed state for the entire DBpedia (800 million triples) is not currently feasible: the Aer simulator does not support the required number of qubits, while on IBM Quantum Computer Qiskit fails to produce the state vector, due to limitation of gates. The application of Grover's algorithm over the superposition of 100 million rows of 27 bits each, takes around 2.5 h on Aer simulator and 3 h on current quantum hardware. Finally, we should stress again that we focused on efficiency, not on the accuracy of the current quantum hardware.

The code for the encoding will be released as extension of `QuantumEncoDeco`[8]. The same open repository will be enriched with pointers to the encoded versions of various datasets, to facilitate experimenting with various quantum algorithms.

Acknowledgments. This work was supported by DEDALUS - Data Management using Quantum Computing project (TA 5180519) funded by SUB1.1 Clusters of Research Excellence (CREs), Greece.

[8] https://github.com/YannisTzitzikas/QuantumEncoDeco.

References

1. AbuGhanem, M.: Characterizing Grover search algorithm on large-scale superconducting quantum computers. Sci. Rep. **15**(1), 1281 (2025)
2. Auer, S., Bizer, C., Kobilarov, G., Lehmann, J., Cyganiak, R., Ives, Z.: DBpedia: a nucleus for a web of open data. In: The Semantic Web, pp. 722–735. Springer (2007)
3. Bischof, L., Teodoropol, S., Füchslin, R.M., Stockinger, K.: Entity matching with quantum neural networks (2024)
4. Dalzell, A.M., et al.: Quantum algorithms: a survey of applications and end-to-end complexities. arXiv preprint arXiv:2310.03011 (2023)
5. Daoud, A., Millberg, V.: Performance and noise levels of Grover's algorithm on quantum computers with different qubit connectivities. Technical report, KTH, School of Electrical Engineering and Computer Science (2024)
6. Fankhauser, T., Solèr, M.E., Füchslin, R.M., Stockinger, K.: Multiple query optimization using a gate-based quantum computer. IEEE Access **11**, 114031–114043 (2023)
7. Ford, M.R.N.: Fundamentals of Software Architecture. O'Reilly Media (2020)
8. Grover, L.K.: A fast quantum mechanical algorithm for database search. In: Proceedings of the Twenty-Eighth Annual ACM Symposium on Theory of Computing, pp. 212–219 (1996)
9. Hai, R., Hung, S.H., Coopmans, T., Littau, T., Geerts, F.: Quantum data management in the NISQ era: extended version. arXiv preprint arXiv:2409.14111 (2024)
10. Hogan, A., et al.: Knowledge graphs. ACM Comput. Surv. (CSUR) **54**(4), 1–37 (2021)
11. Jordan, S.P.: Quantum algorithm zoo. https://quantumalgorithmzoo.org
12. Lehmann, J., et al.: DBpedia-a large-scale, multilingual knowledge base extracted from Wikipedia. Semantic Web **6**(2), 167–195 (2015)
13. Li, T., et al.: Quantum storage design for tables in RDBMS. Proc. VLDB Endow (2024). ISSN 2150-8097
14. Ma, Y., Tresp, V.: Quantum machine learning algorithm for knowledge graphs. ACM Trans. Quantum Comput. **2**(3), 1–28 (2021)
15. Montanaro, A.: Quantum algorithms: an overview. NPJ Quantum Inf. **2**(1), 1–8 (2016)
16. Nayak, N., Rehfeld, J., Winker, T., Warnke, B., Çalikyilmaz, U., Groppe, S.: Constructing optimal bushy join trees by solving QUBO problems on quantum hardware and simulators. In: Proceedings of the International Workshop on Big Data in Emergent Distributed Environments, pp. 1–7 (2023)
17. Proctor, T., Young, K., Baczewski, A.D., Blume-Kohout, R.: Benchmarking quantum computers. Nat. Rev. Phys. 1–14 (2025)
18. Rajak, A., Suzuki, S., Dutta, A., Chakrabarti, B.K.: Quantum annealing: an overview. Phil. Trans. R. Soc. A **381**(2241), 20210417 (2023)
19. Rath, M., Date, H.: Quantum data encoding: a comparative analysis of classical-to-quantum mapping techniques and their impact on machine learning accuracy. EPJ Quantum Technol. **11**(1), 72 (2024)
20. Shor, P.W.: Polynomial-time algorithms for prime factorization and discrete logarithms on a quantum computer. SIAM Rev. **41**(2), 303–332 (1999)
21. Tzitzikas, Y., Kondylakis, H.: Knowledge graphs and quantum computing: first blood. In: Proceedings of the 14th International Joint Conference on Knowledge Graphs, IJCKG 2025. Springer (2025)

22. Vuorenkoski, L., Uotila, V.: Graphs on qubits: demonstrating three graph algorithms on quantum computers. In: VLDB 2024 Workshop: The Second International Workshop on Quantum Data Science and Management (QDSM'24), Guangzhou, China (2024)

Evaluation Challenge

HIT-SCIR at IJCKG2025: Multi-step Text-to-SQL System

Dingzirui Wang, Xuanliang Zhang, Keyan Xu, Qingfu Zhu, and Wanxiang Che(✉)

Harbin Institute of Technology, Harbin, China
{dzrwang,xuanliangzhang,kyxu,car,dzrwang}@ir.hit.edu.cn

Abstract. The text-to-SQL task translates natural language questions into SQL queries, simplifying database access. While large language models (LLMs) have shown strong performance, they often struggle with complex reasoning, such as commonsense and numerical reasoning, required for more challenging SQL generation. We propose a new pipeline that enhances SQL generation by incorporating advanced reasoning skills, alongside techniques like entity linking and self-correction. Tested on the Archer dataset, which requires more complex reasoning, our approach improves performance by 32.8% over the baseline, demonstrating its effectiveness in handling challenging queries.

Keywords: Text-to-SQL · Large Language Models

1 Introduction

The text-to-SQL task is crucial for reducing the difficulty of retrieving information from databases [3,8]. Considering the strong performance of LLMs without fine-tuning, methods based on LLMs have become the current mainstream in text-to-SQL tasks [7,10,18]. Specifically, the text-to-SQL task is defined as taking a user question and a related database as input and generating the corresponding SQL query.

Existing work shows that pipeline approaches can effectively enhance text-to-SQL performance [5,16,22]. For example, MAC-SQL [20] performs entity linking, question decomposition, and self-correction, while DART-SQL [14] rewrites the question before entity linking. However, these methods handle relatively simple SQL queries that are semantically consistent with user questions. In real-world applications, generating SQL often requires more complex reasoning, such as commonsense and numerical reasoning, which current methods struggle with [25].

To address these challenges, we propose a new method that enhances the complex reasoning capabilities when generating SQL queries. We also employ techniques like entity linking [13] and self-correction [2] to improve performance on tasks requiring complex reasoning.

H. Takeda et al. (Eds.): IJCKG 2025, LNAI 16297, pp. 529–536, 2026.
https://doi.org/10.1007/978-981-95-5009-8_38

To validate our approach, we conducted experiments on Archer [25], a dataset for text-to-SQL tasks demanding more complex reasoning ability than the previous works [12,23]. Our method improves performance by 32.8% over the baseline, demonstrating its effectiveness.

Our contributions are as follows:

- We propose a new text-to-SQL pipeline that significantly improves performance on Archer;
- Our method achieves a 32.8% improvement over the baseline on the Archer dataset, validating its effectiveness.

2 Methodology

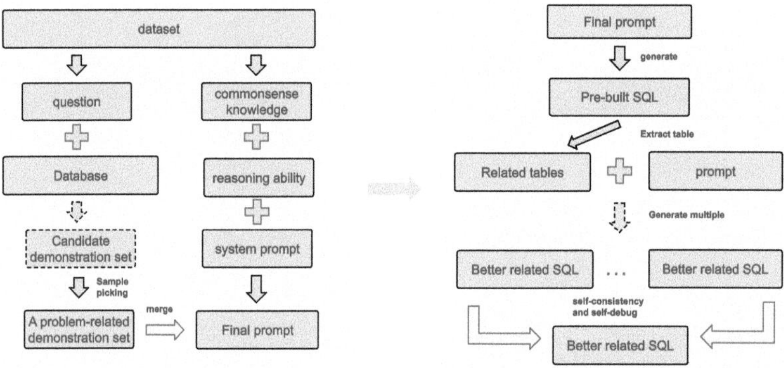

Fig. 1. The illustration of our method.

In this section, we describe the pipeline we designed for enhancing the text-to-SQL task, particularly focusing on addressing the challenges of complex reasoning such as commonsense and numerical reasoning, which is shown in Fig. 1. Our approach involves a combination of advanced prompting techniques and multiple iterative self-improvement methods. The components of our pipeline are outlined below.

To improve the model's ability to generate complex SQL queries, we use a demonstration-based prompting strategy. The prompt format includes not only the user's question but also additional structured information that aids the model in solving complex queries.

We further enhance the effectiveness of the model by introducing a system-level prompt. This explicitly defines the task and constraints that the model should follow when generating SQL queries. The system prompt informs the model about its role as an SQL and DevOps expert and clarifies the conditions under which it should provide its response. Our analysis of error examples revealed a higher frequency of errors in queries involving hypothetical reasoning and year-related calculations. To address these deficiencies, we summarized the recurring issues and supplemented the prompt with specific instructions on hypothetical reasoning and key considerations for performing year-related computations. Additionally, our experiments have shown that using a large number of examples (many-shot) significantly improves the performance of the text-to-SQL task. Based on our experiments, we selected a shot count of 10.

To further refine the generated SQL queries, we incorporate a self-debugging mechanism [2]. This involves multiple rounds of query generation, followed by self-assessment and correction. The model iteratively improves its output by detecting and correcting errors in the SQL, particularly those related to table joins, complex conditions, or logic inconsistencies. This multi-round process helps to enhance the robustness of the generated SQL queries.

Another key component of our method is the self-consistency approach [21]. In this approach, we generate multiple SQL queries for the same question and compare them to ensure consistency across different outputs. By selecting the most consistent query, we reduce the likelihood of generating incorrect or suboptimal SQL statements. Through experiments, we discovered that the model's oracle performance is significantly higher than the results of a self-consistency approach with the same configuration. This insight motivated us to explore a superior method for selecting the final answer. Consequently, we adopted a model-based selection approach, where we input multiple generated SQL queries and their execution results into the large language model, allowing it to select the correct SQL. This method effectively enhances query accuracy.

Finally, we introduce a link-table method to further enhance query accuracy. In this step, we first pre-generate SQL queries and extract relevant tables from them. Using this information, we optimize the prompt by focusing the model's attention on specific tables relevant to the query. This helps in reducing ambiguity and improving the precision of the generated SQL queries.

Our final pipeline integrates all the above components: demonstration-based prompting, system-level prompting, self-debugging, self-consistency, and link-table optimization. These techniques work together to enhance the model's ability to generate SQL queries requiring complex reasoning capabilities. We set hyperparameters based on the complexity of the task and adjust them dynamically during the iterative process.

3 Experiment

3.1 Experiment Settings

In our experiments, we use `Deepseek-Coder-33B` and `GPT-4o` as the primary models. `Deepseek-Coder-33B` is a high-performance language model designed for code generation and understanding tasks, with robust multimodal processing capabilities. This model, trained using advanced deep learning techniques, contains 33 billion parameters, enabling it to efficiently handle tasks such as code generation, debugging, code completion, and cross-language code translation. It excels across various programming languages, particularly in understanding and reasoning through complex syntactic structures, making it highly applicable in fields like automated software development and code optimization. `GPT-4o`, on the other hand, is an optimized version of the `GPT-4o` architecture, focusing on improving accuracy and efficiency in both code generation and natural language processing tasks. Compared to the base version of `GPT-4o`, `GPT-4o` enhances the model's adaptability to language-code multimodal inputs, showing significant advantages in complex language reasoning, context retention, and cross-domain knowledge application. It generates high-quality code and excels in code semantic understanding, error correction, and multi-task code processing, making it a powerful tool for solving problems across various domains. Both models play crucial roles in our experiments, providing solid support and data foundations for our research in code generation, code understanding, and multimodal tasks.

For the demonstration selection, we select 10 databases based on the BM25 similarity with the user question, selecting 5 questions from each database. For the self-debugging process, we set the maximum number of iterations to 3. For self-consistency, we set the temperature to 0.5 and generate 16 outputs for each input.

3.2 Main Experiment

Table 1. The main experimental results of our method.

Method	Dev EX	Test EX
`GPT-3.5` + CT-3 [25]	10.57	15.84
Deepseek-Coder-33B	11.54	–
+ Self-Debug	13.46	–
+ Schem-Linking	14.42	–
+ Self-Consistency	16.35	–
`GPT-4o`	34.62	39.12
+ Refine Prompt	37.86	–
+ Many Shot(10)	39.80	–
+ Model Choose	41.34	48.66

Our main experimental results are shown in Table 1. From the table, we can observe the following insights:

(i) Our proposed approach leads to a considerable performance improvement, yielding a 4.81% increase in performance over the baseline method (`GPT-3.5` + CT-3), which demonstrates the effectiveness of our methodology. Specifically, the model incorporating self-debugging, schema-linking, and self-consistency consistently outperforms the base model across different experimental settings. These

enhancements are particularly evident in the progression from 11.54% for the baseline `Deepseek-Coder-33B` to 16.35% when all improvements are combined.

(ii) The self-debugging mechanism introduces the most notable improvement among the individual techniques, raising performance from 11.54% to 13.46%. This significant increase suggests that the primary challenge lies in the complexity of the original questions, which often leads to the model producing SQL queries that cannot be executed. By iterative refining and debugging the generated SQL, the model can better address these difficulties, thus boosting its performance.

(iii) Schema-linking also plays an essential role in enhancing the model's capability, as it further improves performance to 14.42%. This indicates that the ability to link the input queries to relevant schema components effectively is crucial for solving complex SQL generation tasks. By explicitly modeling these relationships, the model is better equipped to produce accurate and executable SQL queries.

(iv) Self-consistency provides an additional performance boost, raising the performance to 16.35%. This technique, which leverages multiple sampled outputs to ensure consistency, proves highly effective for addressing ambiguity in the input queries. By combining several outputs and selecting the most consistent one, the model achieves more robust performance, reducing the likelihood of incorrect SQL generations.

In conclusion, our approach addresses the key challenges in SQL generation, particularly in handling execution failures, ambiguous schema mapping, and query consistency. These results underscore the potential of combining self-debugging, schema-linking, and self-consistency for improving model performance in complex tasks.

4 Related Work

Research on the text-to-SQL task has garnered significant attention due to its practical applications in enabling users to interact with databases using natural language. This task focuses on the automatic generation of SQL queries based on user questions, while also incorporating the structure of the underlying database schemas. The challenge lies in bridging the gap between the flexibility of human language and the rigid syntax of SQL, making this a complex problem at the intersection of natural language processing (NLP) and database management. As the field evolves, there has been a shift towards utilizing more sophisticated models, driven by advances in language understanding and representation.

In recent years, methods based on LLMs have gained traction as the dominant approach for text-to-SQL tasks, thanks to their remarkable performance even without fine-tuning [7]. One approach is to fine-tune LLMs specifically for text-to-SQL tasks, either by leveraging text-to-SQL datasets directly [11,17,24] or by incorporating data from related tasks such as entity linking and knowledge generation [6,19]. Alternatively, many recent studies explore the use of few-shot inference with LLMs, which delivers superior performance without the need for fine-tuning. Techniques like in-context learning [1,4,16], self-consistency [9,15, 26], and multi-agent systems [20,22] have shown significant promise, allowing LLMs to solve text-to-SQL tasks more efficiently.

5 Conclusion

In this work, we address the challenge of improving the reasoning capabilities required for complex text-to-SQL tasks. While existing approaches prove effective for simple queries that align semantically with user inputs, they often struggle with real-world demands such as commonsense and numerical reasoning. To overcome these limitations, we propose a new text-to-SQL pipeline that incorporates advanced techniques like entity linking and self-correction to enhance performance on complex queries. Our experiments on the Archer dataset, which focuses on advanced reasoning capabilities, show a significant improvement of 32.8% over baseline methods. This demonstrates the effectiveness of our approach in handling more sophisticated SQL generation tasks. In conclusion, our contributions advance the text-to-SQL field, particularly in scenarios requiring complex reasoning. Future work may explore further optimizations to address even broader reasoning challenges across various domains.

References

1. Chang, S., Fosler-Lussier, E.: Selective demonstrations for cross-domain text-to-SQL. In: Bouamor, H., Pino, J., Bali, K. (eds.) Findings of the Association for Computational Linguistics: EMNLP 2023, pp. 14174–14189. Association for Computational Linguistics, Singapore (2023). https://doi.org/10.18653/v1/2023.findings-emnlp.944, https://aclanthology.org/2023.findings-emnlp.944
2. Chen, X., Lin, M., Schärli, N., Zhou, D.: Teaching large language models to self-debug. In: The Twelfth International Conference on Learning Representations (2024). https://openreview.net/forum?id=KuPixIqPiq
3. Deng, N., Chen, Y., Zhang, Y.: Recent advances in text-to-SQL: A survey of what we have and what we expect. In: Calzolari, N., (eds.) Proceedings of the 29th International Conference on Computational Linguistics, pp. 2166–2187. International Committee on Computational Linguistics, Gyeongju, Republic of Korea (2022). https://aclanthology.org/2022.coling-1.190
4. Gao, D., et al.: Text-to-SQL empowered by large language models: a benchmark evaluation. Proc. VLDB Endow. **17**(5), 1132–1145 (2024). https://doi.org/10.14778/3641204.3641221, https://doi.org/10.14778/3641204.3641221
5. Gu, Z., et al.: Interleaving pre-trained language models and large language models for zero-shot nl2sql generation (2023). https://arxiv.org/abs/2306.08891
6. Hong, Z., et al.: Knowledge-to-SQL: Enhancing SQL generation with data expert LLM (2024). https://arxiv.org/abs/2402.11517
7. Hong, Z., et al.: Next-generation database interfaces: a survey of LLM-based text-to-SQL (2024). https://arxiv.org/abs/2406.08426
8. Kanburoğlu, A., Tek, F.: Text-toSQL: a methodical review of challenges and models. Turkish J. Electr. Eng. Comput. Sci. **32**(3), 403–419 (2024). https://doi.org/10.55730/1300-0632.4077, publisher Copyright: TÜBİTAK

9. Lee, D., Park, C., Kim, J., Park, H.: MCS-SQL: leveraging multiple prompts and multiple-choice selection for text-to-SQL generation (2024). https://arxiv.org/abs/2405.07467
10. Li, B., Luo, Y., Chai, C., Li, G., Tang, N.: The dawn of natural language to SQL: are we fully ready? (2024). https://doi.org/10.14778/3681954.3682003, https://arxiv.org/abs/2406.01265
11. Li, H., et al.: Codes: towards building open-source language models for text-to-SQL. Proc. ACM Manag. Data **2**(3) (2024). https://doi.org/10.1145/3654930
12. Li, J., et al.: Can LLM already serve as a database interface? a BIG bench for large-scale database grounded text-to-SQLs. In: Thirty-seventh Conference on Neural Information Processing Systems Datasets and Benchmarks Track (2023). https://openreview.net/forum?id=dI4wzAE6uV
13. Liu, Q., et al.: Awakening latent grounding from pretrained language models for semantic parsing. In: Zong, C., Xia, F., Li, W., Navigli, R. (eds.) Findings of the Association for Computational Linguistics: ACL-IJCNLP 2021. pp. 1174–1189. Association for Computational Linguistics, Online (2021). https://doi.org/10.18653/v1/2021.findings-acl.100, https://aclanthology.org/2021.findings-acl.100
14. Mao, W., et al.: Enhancing text-to-SQL parsing through question rewriting and execution-guided refinement. In: Ku, L.W., Martins, A., Srikumar, V. (eds.) Findings of the Association for Computational Linguistics ACL 2024, pp. 2009–2024. Association for Computational Linguistics, Bangkok, Thailand and virtual meeting (2024). https://aclanthology.org/2024.findings-acl.120
15. Ni, A., et al.: Lever: learning to verify language-to-code generation with execution. In: Proceedings of the 40th International Conference on Machine Learning. ICML'23, JMLR.org (2023)
16. Pourreza, M., Rafiei, D.: DIN-SQL: decomposed in-context learning of text-to-SQL with self-correction. In: Thirty-seventh Conference on Neural Information Processing Systems (2023). https://openreview.net/forum?id=p53QDxSIc5
17. Pourreza, M., Rafiei, D.: DTS-SQL: decomposed text-to-SQL with small large language models (2024). https://arxiv.org/abs/2402.01117
18. Shi, L., Tang, Z., Yang, Z.: A survey on employing large language models for text-to-SQL tasks (2024). https://arxiv.org/abs/2407.15186
19. Sun, Y., et a;.: QDA-SQL: questions enhanced dialogue augmentation for multi-turn text-to-SQL (2024). https://arxiv.org/abs/2406.10593
20. Wang, B., et al.: Mac-SQL: a multi-agent collaborative framework for text-to-SQL (2024). https://arxiv.org/abs/2312.11242
21. Wang, X., et al.: Self-consistency improves chain of thought reasoning in language models. In: The Eleventh International Conference on Learning Representations (2023). https://openreview.net/forum?id=1PL1NIMMrw
22. Xie, Y., et al.: Decomposition for enhancing attention: Improving LLM-based text-to-SQL through workflow paradigm (2024). https://arxiv.org/abs/2402.10671
23. Yu, T., et al.: Spider: a large-scale human-labeled dataset for complex and cross-domain semantic parsing and text-to-SQL task. In: Riloff, E., Chiang, D., Hockenmaier, J., Tsujii, J. (eds.) Proceedings of the 2018 Conference on Empirical Methods in Natural Language Processing, pp. 3911–3921. Association for Computational Linguistics, Brussels, Belgium (2018). https://doi.org/10.18653/v1/D18-1425, https://aclanthology.org/D18-1425

24. Zhang, C., et al.: Finsql: model-agnostic LLMs-based text-to-SQL framework for financial analysis. In: Companion of the 2024 International Conference on Management of Data, pp. 93–105. SIGMOD/PODS '24, Association for Computing Machinery, New York, NY, USA (2024). https://doi.org/10.1145/3626246.3653375, https://doi.org/10.1145/3626246.3653375
25. Zheng, D., Lapata, M., Pan, J.: Archer: a human-labeled text-to-SQL dataset with arithmetic, commonsense and hypothetical reasoning. In: Graham, Y., Purver, M. (eds.) Proceedings of the 18th Conference of the European Chapter of the Association for Computational Linguistics (Volume 1: Long Papers). pp. 94–111. Association for Computational Linguistics, St. Julian's, Malta (2024). https://aclanthology.org/2024.eacl-long.6
26. Zhong, R., Snell, C., Klein, D., Eisner, J.: Non-programmers can label programs indirectly via active examples: a case study with text-to-SQL. In: Bouamor, H., Pino, J., Bali, K. (eds.) Proceedings of the 2023 Conference on Empirical Methods in Natural Language Processing, pp. 5126–5152. Association for Computational Linguistics, Singapore (2023). https://doi.org/10.18653/v1/2023.emnlp-main.312, https://aclanthology.org/2023.emnlp-main.312

OraPlan–SQL: A Planning-Centric Framework for Complex Bilingual NL2SQL Reasoning

Marianne Menglin Liu, Sai Ashish Somayajula, Syed Fahad Allam Shah, Sujith Ravi, and Dan Roth(✉)

Oracle AI, Austin, USA
{marianne.liu,ashish.somayajula,syed.fahad.allam.shah,sujith.ravi, dan.roth}@oracle.com

Abstract. We present ORAPLAN–SQL, our system for the ARCHER NL2SQL Evaluation Challenge 2025, a bilingual benchmark requiring complex reasoning such as arithmetic, commonsense, and hypothetical inference. ORAPLAN–SQL ranked first, exceeding the second-best system by more than 6% in execution accuracy (EX), with 55.0% in English and 56.7% in Chinese, while maintaining over 99% SQL validity (VA). Our system follows an agentic framework with two components: Planner agent that generates stepwise natural language plans, and SQL agent that converts these plans into executable SQL. Since SQL agent reliably adheres to the plan, our refinements focus on the planner. Unlike prior methods that rely on multiple sub-agents for planning and suffer from orchestration overhead, we introduce a *feedback-guided meta-prompting* strategy to refine a single planner. Failure cases from a held-out set are clustered with human input, and an LLM distills them into corrective guidelines that are integrated into the planner's system prompt, improving generalization without added complexity. For the multilingual scenario, to address transliteration and entity mismatch issues, we incorporate entity-linking guidelines that generate alternative surface forms for entities and explicitly include them in the plan. Finally, we enhance reliability through plan diversification: multiple candidate plans are generated for each query, with the SQL agent producing a query for each plan, and final output selected via majority voting over their executions.

Keywords: NL2SQL · LLM Agent · Meta-Prompting · Planner Optimization

1 Introduction

Natural language to SQL (NL2SQL) tasks test the ability of language models to bridge natural language understanding, structured reasoning, and symbolic execution [11,13]. The Archer NL2SQL Evaluation Challenge [12] advances this

M.M. Liu and S.A. Somayajula—Equal contribution; authors listed in no particular order.

H. Takeda et al. (Eds.): IJCKG 2025, LNAI 16297, pp. 537–544, 2026.
https://doi.org/10.1007/978-981-95-5009-8_39

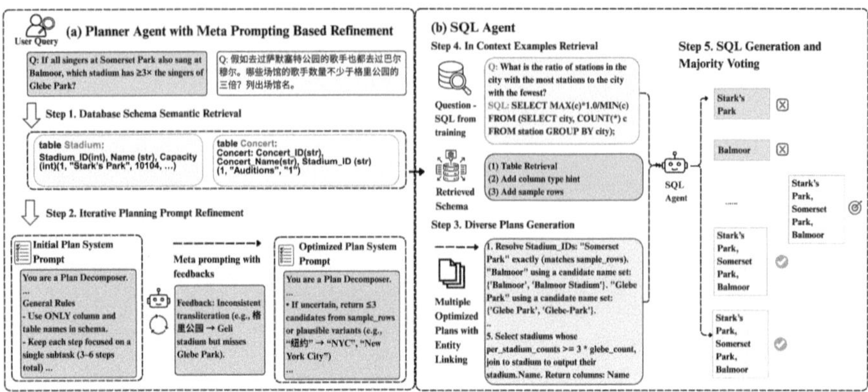

Fig. 1. (a) A Planner Agent with an optimized system prompt, incorporating meta-prompting and entity linking to resolve transliteration and mismatch issues. Diverse plans are generated for robustness. (b) A SQL Agent processes these plans, using retrieved schema and in-context examples to enhance quality, with majority voting selecting the final output from the diverse SQL candidates.

direction by introducing a bilingual benchmark that emphasizes complex reasoning beyond surface-level mapping. Queries involve arithmetic computation, commonsense inference, and hypothetical scenarios, making the task significantly more challenging than standard semantic parsing benchmarks [3,7]. Performance is evaluated along two dimensions: SQL validity (VA) and execution accuracy (EX), which together measure both syntactic correctness and semantic faithfulness: VA is the proportion of syntactically valid SQL queries, while EX is the proportion of generated queries whose execution results exactly match the gold answers [12]. Goal of this work is to advance Text-SQL generation by improving on this benchmark.

Prior work has advanced Text-to-SQL generation [2], yet significant limitations remain in handling diverse reasoning scenarios [5,9]. For example, the query "If the number of students in the CS department increased by 10% last year, how many are currently enrolled?" requires combining hypothetical reasoning with database retrieval. Direct Input → Model → Output pipelines often misapply arithmetic or hallucinate joins, and such errors are difficult to diagnose or correct. ReFoRCE [1] improves robustness through relevant schema retrieval and self-refinement, but still treats Text-to-SQL as a direct mapping, leaving intent interpretation errors opaque. AgenticData [10] introduces a multi-agent planner that decomposes queries into sub-queries, each handled by specialized agents, enabling explicit reasoning but incurring orchestration overhead and brittle handoffs. We argue that both approaches overlook a middle ground: strengthening the system prompt of a single planner so that natural-language plans encode arithmetic, commonsense, and hypothetical reasoning. This approach mitigates the fragility of multi-agent pipelines while preserving explicit reasoning, offering a promising direction for more efficient and robust Text-to-SQL systems [4,8].

To address the limitations of prior works, we present ORAPLAN–SQL, a Text-to-SQL framework that integrates explicit planning with robust SQL generation. As shown in Fig. 1, it consists of two agents, the planner agent that translates a user query into a stepwise plan in natural language and the SQL agent that generates executable SQL from the plan. Our key contributions are threefold: (1) We propose a feedback-guided meta-prompting approach for improving the system prompt of the planner: clustered failure cases from a held-out set, together with human feedback, are provided to an LLM, which then derives generalizable prompt guidelines that both correct observed errors and transfer to unseen cases. (2) We design multilingual entity-linking guidelines in the planner prompt to mitigate transliteration and mismatch issues [6] by generating entity variants before SQL generation. (3) We enhance robustness through plan diversification by generating multiple candidate plans for each query, converting them into SQL, and selecting the final output via majority voting over their executions. These advances highlight the crucial role of planner prompt design in enhancing Text-to-SQL reasoning, especially in complex and multilingual settings. OraPlan–SQL ranked first on the Archer leaderboard, improving execution accuracy by more than 6% over the second-best system (English: 55.0%, Chinese: 56.7%) while maintaining near-perfect validity ($> 99\%$).

2 Methodology

In ORAPLAN–SQL, the Text-to-SQL mapping is decomposed into two stages: a *planner agent* f_{plan} that generates an intermediate natural language plan P, and a *SQL agent* f_{sql} that converts this plan into an executable SQL query $\mathcal{S}$:

$$f_{\text{plan}} : (Q, \mathcal{D}) \mapsto P, \quad f_{\text{sql}} : (P, \mathcal{D}) \mapsto \mathcal{S},$$

where Q is the input query and $\mathcal{D}$ is the database schema. The overall mapping is expressed as

$$f = f_{\text{sql}} \circ f_{\text{plan}}.$$

This decomposition introduces an explicit intermediate plan P to make reasoning steps transparent, while the SQL agent ensures accurate SQL generation. In the following sections, we describe the optimizations applied to each component.

2.1 Enriching Prompts with Schema Retrieval and In-Context Examples

Following ReFoRCE [1], we avoid appending the full database schema, which degrades performance. Instead, we use an embedding-based retriever to select the top-k relevant schema elements, integrated into the planner and SQL agent's system prompts to ensure schema-aware generation and reduce hallucinations. We also apply in-context learning (ICL) by incorporating semantically similar training queries and gold SQL pairs into the SQL agent's system prompt. While ICL is limited to the SQL agent due to the lack of ground-truth plans in the training set, extending it to the planner is a potential future direction.

2.2 Planner Agent and Meta-Prompting Based Refinement

The planner agent is central to OraPlan–SQL, converting a natural language query and retrieved schema into a step-by-step plan. This plan is then passed to the SQL agent, which reliably produces executable SQL. Most errors stem from the planner, making its improvement crucial for system performance.

Instead of relying on multiple specialized sub-agents as in AgenticData [10], we adopt a simpler (yet effective) approach based on analyzing planner outputs on a held-out set to identify common failure modes. These include errors in temporal reasoning, arithmetic, and counterfactual cases. With human input, the errors are clustered, and an LLM derives corrective guidelines that are incorporated into the planner's system prompt to address these errors and generalize to related unseen cases. We refer to this process as *feedback-guided meta-prompting.* It is *meta-prompting* because the LLM does not generate task outputs (plans) directly, but instead produces higher-level guidelines that shape the planner's own system prompt. It is *feedback-guided* because it explicitly leverages both human input and systematic error analysis. By iteratively refining the planner's system prompt with these derived guidelines, the agent improves accuracy of planning across diverse reasoning categories, offering a lightweight yet effective alternative to complex multi-agent planning.

In the bilingual setting, particularly for Chinese, we observed transliteration issues where proper nouns in the query did not align with database entries. To address this, we integrate entity-linking guidelines into the planner's system prompt. These guidelines instruct the planner to generate alternative surface forms for all entities encountered in the query (e.g., "NYC", "New York City"), which are then directly incorporated into SQL generation to improve entity matching and accuracy. Detailed guidelines derived from our meta-prompting approach are presented in Appendix 5.1.

2.3 SQL Agent

The SQL agent converts each natural language plan into an executable query. Its prompt is augmented with the top-5 retrieved schema elements and semantically similar in-context examples, along with minimal guidelines tuning. Together, these enhancements improve schema awareness, ensure syntactic correctness, and keep the generated SQL faithful to the plan.

2.4 Generate SQL for Diverse Plans and Majority Voting

To enhance robustness, we generate diverse plans by setting the planner's temperature to 0.7. Each resulting plan is passed to the SQL agent, executed, and the final answer is determined through majority voting on sql execution results, reducing the impact of errors from any single plan.

3 Experiments

3.1 Main Results

OraPlan–SQL achieves **state-of-the-art bilingual performance** on ARCHER Challenge, substantially outperforming the 2025 runner-up HIT–SCIR. On **test set**, OraPlan–SQL improves execution accuracy (EX) by **6.3 points in English** (54.96% vs. 48.66%) and **12.6 points in Chinese** (56.67% vs. 44.08%). These results highlight the effectiveness of our framework, which leverages guidelines derived through meta-prompting to refine the planner and enhance execution accuracy. Furthermore, OraPlan–SQL achieves nearly identical accuracy across English and Chinese, **closing the cross-lingual gap** observed in prior systems.

Table 1. Execution Accuracy (EX, %) on the Archer NL2SQL Test set: OraPlan–SQL versus 2025 evaluation competitors. EN refers to English and ZH refers to Chinese.

Team	Model	EN	ZH
OraPlan–SQL	GPT-5 (1st)	**54.96**	**56.67**
HIT–SCIR	GPT-4o	48.66	44.08
Nirvana	Claude4+GPT-4o	45.61	41.41
JD5star-2025	DeepSeek-V3.1-Think	43.89	39.12
KAMFU	Qwen3-235b-a22b	43.32	37.40

3.2 Ablation Studies

We conduct ablation studies to assess the impact of individual components in ORAPLAN–SQL by removing or modifying one module and evaluating execution accuracy (EX) on the English Dev set. Results are shown in Table 2.

- **Planner Agent.** To assess the value of explicit planning, we remove the planner agent and directly feed the query, retrieved schema, and ICL examples to the SQL agent. Performance drops from 72.12% to 64.42% (−7.70 points), confirming that intermediate natural language plans are crucial for accurate reasoning.
- **Guidelines from Meta-Prompting.** We evaluate the effect of refining the planner's system prompt with feedback-guided meta-prompting. Without including the generated guidelines in the planner's system prompt, performance is only 44.23%, whereas incorporating guidelines boosts accuracy to 72.12% on English queries. On Chinese queries, accuracy improves from 49.04% to 79.81%. These large gains highlight that the quality of the planner's prompt is as important as the planning step itself.

- **Impact of Choice of Model.** We evaluate the effect of replacing GPT–5 with other model variants for both planning and SQL generation. The general-purpose `GPT-4o` performs the weakest (52.88% EN, 45.19% ZH), while reasoning-oriented models achieve substantially stronger results: `GPT-o3` reaches 67.31% in both languages, and `GPT-5` further improves to 72.12% (EN) and 79.81% (ZH). Importantly, even with `GPT-4o`, OraPlan–SQL outperforms the second-best team HIT–SCIR on the dev set (52.88% vs. 41.34% in EN; 45.19% vs. 31.73% in ZH) and also exceeds JD5star-2025 (47.11% EN, 46.15% ZH), underscoring the robustness of our framework across models.
- **ICL Examples for the SQL Agent.** Adding semantically similar in-context examples to the SQL agent's prompt improves accuracy from 70.19% to 72.12% (+1.93 points), indicating a modest but positive impact on performance.
- **Majority Voting.** Finally, we study the impact of generating diverse plans and applying majority voting over their corresponding SQL executions. With a single plan, accuracy is 71.15%; with multiple plans and majority voting, accuracy rises slightly to 72.12%. While the gain is marginal, this step improves reliability by mitigating errors from individual plans.

Table 2. Ablation and model impact analysis of OraPlan–SQL on the Archer NL2SQL Dev set. Panel (A) shows ablation results on English (EN) with GPT–5, reporting execution accuracy (EX, %) and relative change (ΔEX). Panel (B) shows the effect of different generation models on EX (%) for English (EN) and Chinese (ZH).

(A) Ablation Study (EN)

Variant	EX (%)	ΔEX
OraPlan–SQL	**72.12**	**0.00**
w/o guidelines	44.23	-27.89
w/o planner agent	64.42	-7.70
w/o ICL examples	70.19	-1.93
w/o majority voting	71.15	-0.97

(B) Model Impact

Model	EN	ZH
gpt-4o	52.88	45.19
gpt-o3	67.31	67.31
gpt-5	**72.12**	**79.81**

4 Conclusion

We introduced **OraPlan–SQL**, a planning-driven framework for Text-to-SQL that achieves state-of-the-art results on the Archer benchmark in both English and Chinese with near-perfect SQL validity.

Our approach combines three innovations: (1) feedback-guided meta-prompting that clusters planner errors and distills corrective guidelines, (2) entity-linking guidelines to resolve transliteration and mismatch issues, and (3) plan diversification with consensus execution to improve robustness via majority voting. These contributions make OraPlan–SQL more effective in complex multilingual settings and provide a scalable, reliable solution for Text-to-SQL.

5 Appendix

5.1 Error Case Patterns and Planner Guidelines

Through error analysis of planner outputs, we identified recurring failure modes and distilled targeted mitigation guidelines, illustrated with select cases.

– **Arithmetic and Aggregation.** Planners often make mistakes in handling division, ranking, or projection scope.
Guideline:

```
- When sums, averages, percentages, or totals are needed:
  • Break into steps (e.g., compute per-row value,
    then aggregate with SUM/AVG/COUNT).
  • Percentages must be stated as formula:
    'percentage = 100.0 * part / total'.
  • Guardrail: force float division (use 100.0 not 100).
  • If the question asks for a rate, ratio, or percentage,
    the output may remain float/decimal.
```

– **Hypotheticals and Counterfactuals.** Planners sometimes conflate assumptions with actions, which results in ill-formed queries.
Guideline:

```
- Trigger: if the question has "if... / assume... /
  suppose... / hypothetically...".
- Always output the two lines first:
    • Counterfactual Condition: <the assumption as a filter>
    • Action After Condition: <task with any extra filters>

Decomposition rule:
1. Treat the assumption as a filter on the modified entity
   (entity-side filter).
    • Exclude that entity from the base set, then re-add
      under the assumption (avoid double counting).
2. Apply other filters from the rest of the sentence
   (population-side filters, e.g., occupation, rank, date).
3. Compute the requested metric under this modified dataset.
```

5.2 Cross-Lingual Query Handling with Entity Linking: Translation Vs. Direct Generation

We evaluated two strategies for processing Chinese queries: (i) a *translation-based pipeline* that translates queries into English before planning, and (ii) a *direct generation pipeline* that takes Chinese queries as input and directly generates English plans. The direct generation approach achieved much higher execution accuracy (79.8% vs. 55.8%) because translation decouples the planner from the original linguistic context, where accurate entity resolution depends on native phrasing. Translation artifacts often introduced errors—for instance, 格里公园 (Glebe Park) became *Geli Park*, and 约翰·尼斯尼克 (John Nizinik) was rendered *John Nisinik*—which propagated into planning and execution, causing the performance gap.

References

1. Deng, M., et al.: Reforce: a text-to-SQL agent with self-refinement, format restriction, and column exploration. In: ICLR 2025 Workshop: VerifAI: AI Verification in the Wild (2025)
2. Deng, N., Chen, Y., Zhang, Y.: Recent advances in text-to-SQL: a survey of what we have and what we expect (2022). https://arxiv.org/abs/2208.10099
3. Dou, L., et al.: Towards knowledge-intensive text-to-SQL semantic parsing with formulaic knowledge. In: Proceedings of the 2022 Conference on Empirical Methods in Natural Language Processing, pp. 5240–5253 (2022). https://aclanthology.org/2022.emnlp-main.413.pdf
4. Granado, F., Lotufo, R., Pereira, J.: Raise: Reasoning agent for interactive SQL exploration. arXiv preprint arXiv:2506.01273 (2025). https://arxiv.org/abs/2506.01273
5. Kanburoğlu, A.B.: Text-to-SQL: a methodical review of challenges and models. Turkish J. Electr. Eng. Comput. Sci. **32**(3), 4077 (2024). https://journals.tubitak.gov.tr/elektrik/vol32/iss3/4/
6. Knight, K., Graehl, J.: Machine transliteration. arXiv preprint cmp-lg/9704003 (1997)
7. Liu, T., et al.: Logiccat: a chain-of-thought text-to-SQL benchmark for complex reasoning. arXiv preprint arXiv:2505.18744 (2025). https://arxiv.org/abs/2505.18744
8. Pourreza, M., et al.: Chase-SQL: multi-path reasoning and preference optimized candidate selection in text-to-SQL. arXiv preprint arXiv:2410.01943 (2024). https://arxiv.org/abs/2410.01943
9. Seek.ai: Text2sql: an overview of past, present and future (2024). https://www.seek.ai/blog/a-survey-of-text-to-sqls-past-present-and-future
10. Sun, J., Li, G., Zhou, P., Ma, Y., Xu, J., Li, Y.: Agenticdata: an agentic data analytics system for heterogeneous data. arXiv preprint arXiv:2508.05002 (2025)
11. Yu, A., Yin, Y., Choi, E., et al.: Spider: a large-scale, high-quality benchmark for complex and cross-domain semantic parsing and text-to-sql tasks. In: Proceedings of the 2018 Conference on Empirical Methods in Natural Language Processing (EMNLP), pp. 3581–3591. Association for Computational Linguistics (2018). https://www.aclweb.org/anthology/D18-1371/
12. Zheng, D., Lapata, M., Pan, J.Z.: Archer: a human-labeled text-to-sql dataset with arithmetic, commonsense and hypothetical reasoning. arXiv preprint arXiv:2402.12554 (2024)
13. Zhong, V., Xiong, C., Socher, R.: Seq2sql: Generating structured queries from natural language. In: Proceedings of the 2017 Conference on Empirical Methods in Natural Language Processing (EMNLP), pp. 423–433. Association for Computational Linguistics (2017). https://www.aclweb.org/anthology/D17-1042/

Author Index

H. Takeda et al. (Eds.): IJCKG 2025, LNAI 16297, pp. 545–547, 2026.
https://doi.org/10.1007/978-981-95-5009-8

MIX
Papier aus verantwortungsvollen Quellen
Paper from responsible sources
FSC® C105338

If you have any concerns about our products,
you can contact us on
ProductSafety@springernature.com

In case Publisher is established outside the EU,
the EU authorized representative is:
Springer Nature Customer Service Center GmbH
Europaplatz 3, 69115 Heidelberg, Germany

Printed by Libri Plureos GmbH
in Hamburg, Germany